网站开发案例课堂

Struts 2+Spring+Hibernate+MyBatis 网站开发案例课堂

施 俊 缪 勇 李新锋 编 著

清华大学出版社
北 京

内 容 简 介

本书详细讲解了 Java EE 中 Struts 2、Spring、Hibernate、MyBatis 等经典流行框架的基本知识和使用方法，通过案例课堂的形式深入细致地描述各相关框架的知识点和使用技巧，尤其是框架的相互整合。本书用具体的实例进行演示和展现，使得原本复杂又难以理解的知识，变得通俗易懂、易于学习，可以帮助读者更快地理解和掌握 Java EE 的开发技能和核心技术。为方便读者学习和教学开展，本书还提供了全程真实课程录像和教学 PPT，读者可以边学边看，按计划跟进学习，也可作为教学参考。

本书层次清晰，结构简单，既体现了 Java EE 开发框架的技术特点，又注重灵活运用、举一反三，不仅适合初学者按部就班地学习，也适合网络开发人员作为技术参考，同时，也可作为高等院校计算机相关专业学生的课堂教材。

本书封面贴有清华大学出版社防伪标签，无标签者不得销售。
版权所有，侵权必究。侵权举报电话：010-62782989　13701121933

图书在版编目(CIP)数据

Struts 2+Spring+Hibernate+MyBatis 网站开发案例课堂/施俊，缪勇，李新锋编著. —北京：清华大学出版社，2018（2019.3重印）
（网站开发案例课堂）
ISBN 978-7-302-50060-5

Ⅰ. ①S… Ⅱ. ①施… ②缪… ③李… Ⅲ. ①网页制作工具 Ⅳ. ①TP393.092.2

中国版本图书馆 CIP 数据核字(2018)第 097055 号

责任编辑：杨作梅
装帧设计：李　坤
责任校对：周剑云
责任印制：董　瑾

出版发行：清华大学出版社
　　　　　网　　址：http://www.tup.com.cn, http://www.wqbook.com
　　　　　地　　址：北京清华大学学研大厦 A 座　　邮　　编：100084
　　　　　社 总 机：010-62770175　　　　　　　　邮　　购：010-62786544
　　　　　投稿与读者服务：010-62776969, c-service@tup.tsinghua.edu.cn
　　　　　质量反馈：010-62772015, zhiliang@tup.tsinghua.edu.cn
印 装 者：三河市金元印装有限公司
经　　销：全国新华书店
开　　本：190mm×260mm　　**印　张**：34.25　　**字　数**：825 千字
版　　次：2018 年 7 月第 1 版　　　　　　　　**印　次**：2019 年 3 月第 2 次印刷
定　　价：88.00 元

产品编号：074929-01

前　　言

Struts 2、Hibernate、MyBatis 和 Spring 框架，是目前非常流行的 Java EE 开发框架技术，不仅能用于传统的网络开发，也能用于当今的移动互联开发。为了帮助读者更好、更快速地掌握这些 Java EE 轻量级框架开发技术并能实际运用，本书以课堂授课形式，从环境配置、基础知识、案例讲解、整合开发、综合实例等方面，对 Java EE 的框架技术作了详细讲解，并特别注重教学中的案例引导作用，帮助读者理解和掌握所学的知识。相信读者通过对本书的学习，不仅可以系统地掌握 Java EE 框架整合开发的相关技术，而且可以掌握它们在实际开发中的运用，从而极大地提升 Java EE 开发水平，并能够胜任相关的开发工作。本书配备了 226 个共 60 小时的全过程多媒体教学视频和教学 PPT，以帮助读者按照书中的操作步骤循序渐进地学习，更好地掌握 Java EE 开发技术。

1. 本书主要特色

- 零基础、入门级的讲解

无论您是否从事计算机相关行业，无论您是否接触过 Java EE 网站开发，都能从本书中找到最佳起点。

- 大量实用案例引导

本书在编排上紧密结合深入学习 Java EE 的先后过程，从开发环境搭建开始，逐步带领读者深入地学习各种框架开发技术，通过大量实用案例引导，使读者既能掌握基础知识，又能提高实战技能。

- 服务课堂教学和训练

在章节编排上，充分考虑课堂教学使用，按照学时规划设计讲解内容，并附有专业授课 PPT，教学组织简明轻松，操作有章可循。

- 丰富的配套学习资源

本书赠送大量王牌资源，除了本书所有案例的源代码资源外，还有各种最新的开发包和数据包，下载地址：www.tup.tsinghua.edu.cn。

- 全程同步教学录像

本书提供全过程、无死角同步操作教学录像，涵盖所有章节、所有知识点、所有操作过程，详细讲解每个实例与项目的开发过程及技术关键点，比看书更轻松，而且扩展讲解部分能得到比学习书中内容更多的收获。

- 体贴入微的后续服务

本书由教学一线老师和实践开发人员精心编著，并提供实时技术支持。无论读者在学习过程中遇到任何问题，均可加入 QQ 群 237540430 或通过邮箱 shikham88@163.com 进行提问，专家人员会在线答疑。

2. 本书主要内容

基础知识部分：第 1～5 章，分别从环境搭建、JSP 网页开发技术、Servlet 技术、MVC 开发模式、EasyUI 插件等 5 个方面介绍 Java EE 基础知识。

框架技术部分：第 6～19 章，详细讲解 Struts 2(第 6～10 章)、Hibernate(第 11～15 章)、MyBatis(第 16 章)、Spring(第 17～19 章)框架技术基础知识和应用技巧，是全书的重点内容。

整合和实例部分：第 20～24 章，具体讲解 Struts 2、Hibernate、MyBatis 和 Spring 相互整合操作方法，并通过网上订餐系统的前台、后台和新闻发布系统三个具体实例演示了 Java EE 框架技术的应用。

3. 本书读者对象

- 有一定 Java 基础，但是没有 Java EE 系统开发经验的初学者。
- 有其他 Web 编程语言(如 ASP、ASP.NET)开发经验，欲快速转向 Java EE 开发的程序员。
- 对 JSP 有一定了解，但是缺乏 Java EE 框架开发经验，并希望了解流行开源框架 Struts 2、Hibernate、MyBatis 和 Spring 以及欲对这些框架进行整合的程序员。
- 有一定 Java Web 框架开发基础，需要对 Java EE 主流框架技术核心进一步了解和掌握的程序员。
- 公司管理人员或人力资源管理人员。

4. 作者及致谢

本书由施俊、缪勇和李新锋编写，其中，扬州职业大学的施俊编写了第 1、2、3、4、6、7、8、9、10、11、12、13、14、15 章，扬州职业大学的缪勇编写了第 5、16、17、18、19、20、21、22、23 章，镇江市机关信息技术员李新锋编写了第 24 章。其他参与编写的人员还有王梅、陈亚辉、李艳会、刘娇、王晶晶、游名扬、李云霞、王永庆、蒋梅芳、谢伟等，同时江苏智途科技股份有限公司、扬州国脉通信发展有限责任公司也为本书的编写提供了帮助，在此一一向他们致谢。

由于作者水平有限，书中难免存在疏漏之处，敬请读者批评指正。

编　者

目　　录

第 I 篇　基础知识部分

第 1 章　搭建 Java Web 开发环境 3
1.1　建立 JDK 的环境 4
　　1.1.1　下载与安装 JDK 4
　　1.1.2　配置 JDK 环境变量 5
　　1.1.3　验证 JDK 是否配置 6
1.2　建立 Tomcat 的环境 7
　　1.2.1　下载与安装 Tomcat 7
　　1.2.2　配置 Tomcat 环境变量 8
　　1.2.3　启动与停止 Tomcat 8
　　1.2.4　Tomcat 的目录结构 9
1.3　搭建 Java Web 开发环境 9
　　1.3.1　下载与安装 MyEclipse 9
　　1.3.2　在 MyEclipse 中配置 JDK 11
　　1.3.3　在 MyEclipse 中配置 Tomcat... 12
1.4　创建 MySQL 数据库环境 13
　　1.4.1　下载 MySQL 13
　　1.4.2　安装与配置 MySQL 15
　　1.4.3　使用 MySQL 数据库 17
1.5　创建和发布 Java Web 工程 18
　　1.5.1　创建 Web 项目、设计项目
　　　　　目录结构 18
　　1.5.2　编写页面代码、部署和
　　　　　运行 Web 项目 20
1.6　小结 21

第 2 章　JSP 动态页面开发技术 23
2.1　JSP 技术基础 24
　　2.1.1　JSP 简介 24
　　2.1.2　JSP 页面组成 25
2.2　JSP 内置对象 30
　　2.2.1　什么是 JSP 内置对象 30
　　2.2.2　out 内置对象 31
　　2.2.3　request 内置对象 31
　　2.2.4　response 内置对象 33
　　2.2.5　session 内置对象 35
　　2.2.6　application 内置对象 37
　　2.2.7　其他内置对象 38
2.3　对象的范围 39
　　2.3.1　page 范围 39
　　2.3.2　request 范围 40
　　2.3.3　session 范围 40
　　2.3.4　application 范围 41
2.4　在 JSP 中使用 JavaBean 42
　　2.4.1　为什么需要 JavaBean 42
　　2.4.2　什么是 JavaBean 42
　　2.4.3　封装数据和业务 42
　　2.4.4　JSP 与 JavaBean 44
2.5　EL 表达式 44
　　2.5.1　EL 表达式概述 44
　　2.5.2　EL 表达式的使用 45
　　2.5.3　EL 隐式对象 46
2.6　JSTL 标签 47
　　2.6.1　JSTL 标签概述 47
　　2.6.2　JSTL 标签的使用 48
　　2.6.3　JSTL 核心标签库 48
2.7　小结 53

第 3 章　Servlet 技术 55
3.1　Servlet 简介 56
　　3.1.1　什么是 Servlet 56
　　3.1.2　编写第一个 Servlet 56
　　3.1.3　Servlet 与 JSP 的关系 58
3.2　Servlet 的生命周期 59
3.3　Servlet 的常用类和接口 62

3.4 Servlet 的应用示例 65
3.5 小结 67

第 4 章 使用 MVC 模式实现用户登录 69

4.1 JSP 开发模型 70
  4.1.1 JSP Model I 模式 70
  4.1.2 JSP Model II 模式 70
4.2 MVC 模式概述 71
  4.2.1 为什么需要 MVC 模式 71
  4.2.2 MVC 模式的定义及特点 72
4.3 JDBC 技术 73
  4.3.1 JDBC 简介 73
  4.3.2 通过 JDBC 连接 MySQL
     数据库 73
4.4 使用 MVC 模式实现用户登录模块..... 75
  4.4.1 项目设计简介 75
  4.4.2 模型设计 76
  4.4.3 视图设计 77
  4.4.4 控制器设计 78
  4.4.5 部署和运行程序 79
4.5 小结 79

第 5 章 jQuery EasyUI 插件 81

5.1 EasyUI 概述 82
5.2 Layout 控件 82
5.3 Tabs 控件 83
5.4 Tree 控件 84
5.5 DataGrid 控件 85
5.6 小结 86

第 II 篇 框架技术部分

第 6 章 认识 Struts 2 框架 89

6.1 Struts 2 框架 90
  6.1.1 Struts 2 的由来 90
  6.1.2 Struts 2 的 MVC 模式 90
  6.1.3 Struts 2 控制器 91
  6.1.4 Struts 2 资源的获取 91
6.2 Struts 2 系统架构 92
  6.2.1 Struts 2 框架结构 92
  6.2.2 Struts 2 的核心概念 93
6.3 Struts 2 的基本运行流程 95
  6.3.1 用户登录的处理流程 95
  6.3.2 加载 Struts 2 类库 95
  6.3.3 配置 web.xml 文件加载核心
     控制器 96
  6.3.4 开发视图层页面 96
  6.3.5 开发业务控制器 Action 97
  6.3.6 配置业务控制器 struts.xml 97
  6.3.7 部署运行项目 98
  6.3.8 使用 Struts 2 实现登录功能的
     处理过程 98
6.4 Struts 2 的控制器和组件 99
  6.4.1 核心控制器 99
  6.4.2 业务控制器 99
  6.4.3 模型组件 100
  6.4.4 视图组件 101
6.5 小结 101

第 7 章 Struts 2 的配置 103

7.1 Struts 2 的配置文件 104
  7.1.1 web.xml 文件 104
  7.1.2 struts.xml 文件 104
  7.1.3 struts.properties 文件 106
7.2 Struts 2 的 Action 实现 107
  7.2.1 POJO 的实现 107
  7.2.2 实现 Action 接口 108
  7.2.3 继承 ActionSupport 108
  7.2.4 Struts 2 支持 Java 对象 109
  7.2.5 Struts 2 访问 Servlet API ... 110
7.3 Action 配置 113
  7.3.1 Struts 2 中 Action 的作用 ... 114
  7.3.2 配置 Action 114
  7.3.3 动态方法调用 114

7.3.4 用 method 属性处理调用
方法 115
7.3.5 使用通配符 117
7.4 Result 配置 117
7.4.1 配置 Result 118
7.4.2 Result 的常用结果类型 119
7.4.3 使用通配符动态配置 Result ... 120
7.4.4 通过请求参数动态配置
Result 121
7.5 小结 122

第 8 章 Struts 2 的标签库 123

8.1 Struts 2 标签库概述 124
8.1.1 Struts 2 标签的分类 124
8.1.2 Struts 2 标签库的导入 124
8.2 Struts 2 的 UI 标签 125
8.2.1 UI 标签的模板和主题 125
8.2.2 表单标签的公共属性 125
8.2.3 简单的表单标签 126
8.2.4 其他表单标签 128
8.2.5 非表单标签 128
8.3 Struts 2 的非 UI 标签 129
8.3.1 控制标签 130
8.3.2 数据标签 132
8.4 使用 Struts 2 实现用户注册功能 133
8.4.1 用户注册流程 133
8.4.2 创建用户实体类 133
8.4.3 开发数据访问 DAO 层 133
8.4.4 开发控制层 Action 134
8.4.5 在 struts.xml 中配置 action 135
8.4.6 开发注册页面 136
8.4.7 部署项目 136
8.5 小结 137

第 9 章 OGNL 和类型转换 139

9.1 OGNL 基础 140
9.1.1 数据转移和类型转换 140
9.1.2 OGNL 基础 140
9.1.3 OGNL 常用符号的用法 141

9.2 Struts 2 的类型转换 143
9.2.1 内置类型转换器 143
9.2.2 自定义类型转换器 143
9.2.3 注册自定义类型转换器 145
9.3 小结 150

第 10 章 Struts 2 的拦截器 151

10.1 Struts 2 的拦截器机制 152
10.1.1 为什么需要拦截器 152
10.1.2 拦截器的工作原理 152
10.1.3 拦截器示例 153
10.2 Struts 2 内建拦截器 155
10.2.1 默认拦截器 155
10.2.2 配置拦截器 157
10.2.3 自定义拦截器 158
10.3 自定义权限验证的拦截器 159
10.4 小结 162

第 11 章 Hibernate 初步 163

11.1 Hibernate 概述 164
11.1.1 JDBC 的困扰 164
11.1.2 Hibernate 的优势 164
11.1.3 持久化和 ORM 164
11.1.4 Hibernate 的体系架构 166
11.2 Hibernate 的下载与安装 167
11.3 小结 168

第 12 章 使用 Hibernate 实现数据的
增删改查 169

12.1 基于 XML 映射文件实现数据的
增删改查 170
12.1.1 Hibernate 数据操作流程 170
12.1.2 添加数据 171
12.1.3 加载数据 177
12.1.4 删除数据 178
12.1.5 修改数据 179
12.2 基于 Annotation 注解实现数据的
增删改查 179
12.3 小结 182

第 13 章 使用 Hibernate 实现关联映射和继承映射 183

13.1 基于 XML 映射文件实现关联映射 184
- 13.1.1 单向多对一关联 184
- 13.1.2 单向一对多映射 187
- 13.1.3 双向多对一映射 188
- 13.1.4 双向多对多映射 191
- 13.1.5 双向一对一映射 195

13.2 基于 Annotation 注解实现关联映射 201
- 13.2.1 双向多对一映射 201
- 13.2.2 双向多对多映射 204
- 13.2.3 双向一对一映射 206

13.3 基于 XML 映射文件实现继承映射 209
- 13.3.1 使用 subclass 进行映射 209
- 13.3.2 使用 joined-subclass 进行映射 212
- 13.3.3 使用 union-subclass 进行映射 213

13.4 小结 214

第 14 章 使用 Hibernate 查询数据 215

14.1 使用 HQL 查询数据 216
- 14.1.1 简单查询 216
- 14.1.2 属性查询 217
- 14.1.3 聚集函数 217
- 14.1.4 分组查询 218
- 14.1.5 动态实例查询 219
- 14.1.6 分页查询 219
- 14.1.7 条件查询 220
- 14.1.8 连接查询 221
- 14.1.9 子查询 223

14.2 使用 QBC 查询数据 225
- 14.2.1 简单查询 225
- 14.2.2 分组查询 226
- 14.2.3 聚集函数 227
- 14.2.4 组合查询 228
- 14.2.5 关联查询 229
- 14.2.6 分页查询 230
- 14.2.7 QBE 查询 230
- 14.2.8 离线查询 232

14.3 小结 232

第 15 章 使用 Hibernate 缓存数据 233

15.1 缓存的概念和范围 234
15.2 一级缓存 234
15.3 二级缓存 236
15.4 查询缓存 241
15.5 小结 242

第 16 章 MyBatis 框架 243

16.1 MyBatis 概念与安装 244
16.2 MyBatis 的增删改查 244
16.3 MyBatis 的关联映射 250
- 16.3.1 一对一关联映射 250
- 16.3.2 一对多关联映射 252
- 16.3.3 多对多关联映射 256

16.4 动态 SQL 259
- 16.4.1 if 元素 259
- 16.4.2 if-where 元素 260
- 16.4.3 set-if 元素 261
- 16.4.4 trim 元素 262
- 16.4.5 choose、when、otherwise 元素 264
- 16.4.6 foreach 元素 265

16.5 MyBatis 的注解配置 267
- 16.5.1 基于注解的增删改查 267
- 16.5.2 基于注解的一对一关联映射 269
- 16.5.3 基于注解的一对多关联映射 270
- 16.5.4 基于注解的多对多关联映射 273
- 16.5.5 基于注解的动态 SQL 274

16.6 MyBatis 的缓存 279

- 16.6.1 一级缓存..................................279
- 16.6.2 二级缓存..................................281
- 16.7 小结..282

第 17 章 Spring 的基本应用..................283
- 17.1 认识 Spring 框架................................284
- 17.2 了解 Spring 的核心机制：依赖注入/控制反转................286
- 17.3 小结..289

第 18 章 Spring Bean 的装配模式........291
- 18.1 Bean 工厂 ApplicationContext.........292
- 18.2 Bean 的作用域...................................293
- 18.3 基于 Annotation 的 Bean 装配.........295
- 18.4 小结..297

第 19 章 面向切面编程 (Spring AOP)..................................299
- 19.1 AOP 简介...300
- 19.2 基于 XML 配置文件的 AOP 实现..300
  - 19.2.1 前置通知..................................300
  - 19.2.2 返回通知..................................303
  - 19.2.3 异常通知..................................304
  - 19.2.4 环绕通知..................................305
- 19.3 基于@AspectJ 注解的 AOP 实现....306
- 19.4 小结..308

第 III 篇 整合和实例部分

第 20 章 Spring 整合 Struts 2 与 Hibernate......................................311
- 20.1 基于 XML 配置的 S2SH 整合..........312
  - 20.1.1 环境搭建..................................312
  - 20.1.2 创建实体类及映射文件.....313
  - 20.1.3 Spring 整合 Hibernate...........314
  - 20.1.4 DAO 层开发...........................316
  - 20.1.5 Service 层开发......................317
  - 20.1.6 Action 开发...........................318
  - 20.1.7 Spring 整合 Struts 2..............319
  - 20.1.8 创建页面................................320
- 20.2 基于 Annotation 注解的 S2SH 整合..321
- 20.3 小结..325

第 21 章 Spring MVC..........................327
- 21.1 Spring MVC 概述...............................328
- 21.2 Spring MVC 常用注解......................329
  - 21.2.1 基于注解的处理器..............329
  - 21.2.2 请求映射方式......................331
  - 21.2.3 绑定控制器类处理方法入参..334
  - 21.2.4 控制器类处理方法的返回值类型................................337
  - 21.2.5 保存模型属性到 HttpSession............................338
  - 21.2.6 在控制器类方法之前执行的方法..339
  - 21.2.7 Spring MVC 返回 JSON 数据..339
- 21.3 直接页面转发、自定义视图与页面重定向...................................341
- 21.4 控制器的类型转换、格式化、数据校验...343
- 21.5 Spring MVC 文件上传.....................346
- 21.6 Spring MVC 国际化.........................347
- 21.7 Spring 整合 Spring MVC 与 Hibernate......................................348
  - 21.7.1 环境搭建..................................349
  - 21.7.2 创建实体类............................349
  - 21.7.3 Spring 整合 Hibernate...........349
  - 21.7.4 DAO 层开发...........................351
  - 21.7.5 Service 层开发......................352
  - 21.7.6 控制器开发............................352
  - 21.7.7 Spring 整合 Spring MVC.....353
  - 21.7.8 创建登录页............................355

21.8 Spring 整合 Spring MVC 与 MyBatis 355
 21.8.1 环境搭建 356
 21.8.2 创建实体类 356
 21.8.3 Spring 整合 MyBatis 356
 21.8.4 DAO 层开发 357
 21.8.5 Service 层开发 357
 21.8.6 控制器开发 358
 21.8.7 Spring 整合 Spring MVC 358
 21.8.8 创建页面 360
21.9 小结 360

第 22 章 Spring 整合 Struts 2 与 Hibernate 实现网上订餐系统前台 361

22.1 需求与系统分析 362
22.2 数据库设计 363
22.3 项目环境搭建 365
22.4 Spring 及 Struts 2 配置文件 366
22.5 创建实体类和映射文件 366
22.6 创建 DAO 接口及实现类 371
22.7 创建 Service 接口及实现类 379
22.8 餐品与菜系展示 384
22.9 查询餐品 389
22.10 查看餐品详情 390
22.11 用户登录与注册 392
 22.11.1 用户登录 392
 22.11.2 用户注册 395
22.12 购物车功能 398
22.13 订单功能 405
 22.13.1 生成订单 405
 22.13.2 查看"我的订单" 408
 22.13.3 查看订单明细 410
 22.13.4 删除订单 412
22.14 小结 413

第 23 章 Spring 整合 Spring MVC 与 Hibernate 实现网上订餐系统后台 415

23.1 需求与系统分析 416

23.2 数据库设计 416
23.3 项目环境搭建 417
23.4 Spring 及 Spring MVC 配置文件 418
23.5 创建实体类 418
23.6 创建 DAO 接口及实现类 423
23.7 创建 Service 接口及实现类 432
23.8 后台登录与管理首页面 439
23.9 餐品管理 446
 23.9.1 餐品列表显示 447
 23.9.2 查询餐品 450
 23.9.3 添加餐品 450
 23.9.4 餐品下架 452
 23.9.5 修改餐品 453
23.10 订单管理 455
 23.10.1 创建订单 455
 23.10.2 查询订单 463
 23.10.3 删除订单 467
 23.10.4 修改订单/查看明细 468
 23.10.5 使用 Echarts 显示销售统计 474
23.11 客户管理 476
 23.11.1 客户列表显示 476
 23.11.2 查询客户 479
 23.11.3 启用和禁用客户 480
23.12 管理员及其权限管理 482
 23.12.1 管理员列表显示 482
 23.12.2 新增管理员 483
 23.12.3 设置/修改管理员权限 485
23.13 小结 489

第 24 章 Spring 整合 Spring MVC 与 MyBatis 实现新闻发布系统 491

24.1 系统概述及需求分析 492
24.2 数据库设计 493
24.3 系统环境搭建 495
24.4 系统配置文件 496
24.5 创建实体类 496

24.6 创建 DAO 接口及动态提供类 499
24.7 创建 Service 接口及实现类 506
24.8 新闻浏览 512
    24.8.1 新闻首页 512
    24.8.2 浏览新闻 515
24.9 发表评论 517
    24.9.1 普通用户登录 517
    24.9.2 发表评论 519

24.10 新闻系统后台 520
    24.10.1 管理员登录与后台管理
            首页 520
    24.10.2 新闻管理 522
    24.10.3 评论管理 527
    24.10.4 用户管理 530
24.11 小结 533

第1篇

基础知识部分

- 第1章 搭建 Java Web 开发环境
- 第2章 JSP 动态页面开发技术
- 第3章 Servlet 技术
- 第4章 使用 MVC 模式实现用户登录
- 第5章 jQuery EasyUI 插件

第 1 章 搭建 Java Web 开发环境

搭建软件开发环境是软件开发的第一步,优秀的开发环境能帮助程序员提高开发速度。本章讲述如何搭建 Java Web 开发环境,包括如下组件:Java 开发包(Java Development Kit)、应用服务器 Tomcat、集成开发环境 MyEclipse 和 MySQL 数据库。

1.1 建立 JDK 的环境

JDK(Java Development Kit)，是整个 Java 的核心，包括 Java 运行环境、大量的 Java 工具和 Java 基础类库。主流的集成开发环境(IDE)，比如 Eclipse、NetBeans 等，都基于 JDK，有些 IDE 在安装时内置了 JDK，有些则需要单独安装。JDK 由 Sun 公司开发，现已被 Oracle 公司收购，它为 Java 程序提供了编译和运行环境，不管是做 Java 开发还是安卓开发都需要在计算机上安装 JDK。

1.1.1 下载与安装 JDK

JDK 可以从 Oracle 官网上下载，目前 JDK 的最新版本是 JDK 8 Update 121，下载页面为 http://www.oracle.com/technetwork/java/javase/downloads/index.html，如图 1-1 所示。

单击 Download 按钮后，选中 Accept License Agreement 单选按钮，根据自己的系统类型下载相应的版本，如图 1-2 所示。

图 1-1 JDK 下载页面

图 1-2 选择 JDK 下载版本

> 注意：笔者的系统为 Windows 10 专业版 64 位系统，下载的是 jdk-8u121-windows-x64.exe 安装文件。

下载 JDK 以后即可安装，安装 JDK 8 的步骤如下。

step 01 双击下载的 exe 程序，进入安装向导窗口，单击"下一步"按钮，如图 1-3 所示。

step 02 进入自定义安装窗口，选择相应的功能，这里我们保留默认路径，也可以单击"更改"按钮，修改为其他路径，之后单击"下一步"按钮，如图 1-4 所示。

step 03 JDK 安装完成之后，安装向导还会自动进入外部 JRE 安装窗口。用户可以选择继续安装或取消，若要安装也可以更改外部 JRE 的安装目录，如图 1-5 所示。

step 04 单击"下一步"按钮，安装 JRE，直到最后的完成窗口，如图 1-6 所示。

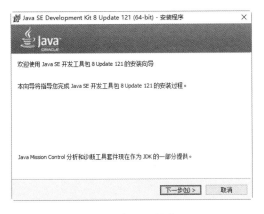

图 1-3　安装向导窗口

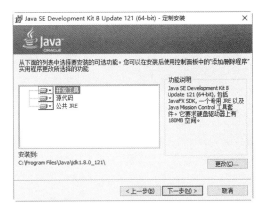

图 1-4　自定义安装窗口

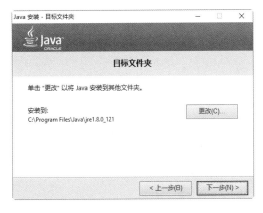

图 1-5　安装 JRE 窗口

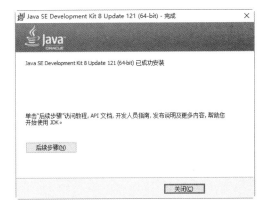

图 1-6　完成窗口

1.1.2　配置 JDK 环境变量

JDK 安装后，如果要在 DOS 控制台窗口下，编译执行 Java 程序，需要对 JDK 进行环境变量配置，配置过程如下。

(1) 右击"我的电脑"，在弹出的快捷菜单中选择"属性"命令(或进入控制面板，选择"系统")，单击左侧的"高级系统设置"，在弹出的"系统属性"对话框中的"高级"选项卡下方单击"环境变量"按钮，弹出"环境变量"对话框，如图 1-7 所示。

(2) 在"系统变量"选项组中，单击"新建"按钮，弹出"新建系统变量"对话框，输入变量名 JAVA_HOME 和变量值 C:\Program Files\Java\jdk1.8.0_121(这是默认的安装路径，可根据自己安装的路径填写)，如图 1-8 所示。

(3) 再次新建系统变量，变量名为 CLASSPATH，变量值为 ".;%JAVA_HOME%\lib\;"（注意，前面的 "." 表示当前路径，此处不可少)，如图 1-9 所示。

(4) 在如图 1-7 所示的"环境变量"对话框中，选择系统变量 Path，单击下方的"编辑"按钮，在弹出的"编辑环境变量"对话框中，单击"编辑文本"按钮，弹出"编辑系统变量"对话框，新增变量值："%JAVA_HOME%;%JAVA_HOME%\bin;"，如图 1-10 所示。

图 1-7 系统属性和环境变量

图 1-8 新建 JAVA_HOME 变量

图 1-9 新建 CLASSPATH 变量

图 1-10 修改 Path

1.1.3 验证 JDK 是否配置

JDK 环境变量配置完成后,在"开始"菜单的"搜索程序和文件"文本框或"运行"对话框中输入 cmd,打开 cmd.exe 程序,在命令提示符后输入 java–version 命令,屏幕上会显示 JDK 的版本信息,表示 JDK 已经配置成功,如图 1-11 所示。

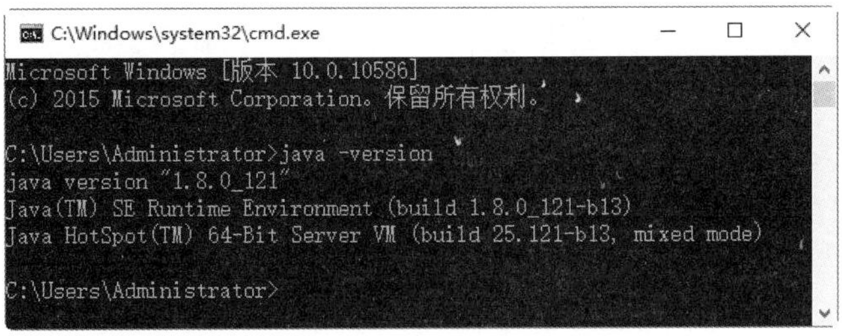

图 1-11 查看 Java 版本测试 JDK 是否配置成功

1.2 建立 Tomcat 的环境

Tomcat 是 Apache 软件基金会(Apache Software Foundation)的 Jakarta 项目中的一个核心项目，是一个免费的开源 Web 容器。随着 Web 应用的发展，Tomcat 被越来越多地应用于商业用途，由 Apache、Sun 和其他一些公司及个人共同开发而成。最新的 Servlet 和 JSP 规范总是能在 Tomcat 中得到体现。目前，官网上的最新版本是 Tomcat 9.0.0.M19，这里我们使用 Tomcat 8.0.43 版本。

1.2.1 下载与安装 Tomcat

从 Apache 官方网站可以获取相应版本，Tomcat 提供了安装版本和解压缩版本的文件，可以根据需要进行下载。

(1) Tomcat 的官网地址为 http://tomcat.apache.org/，如图 1-12 所示。

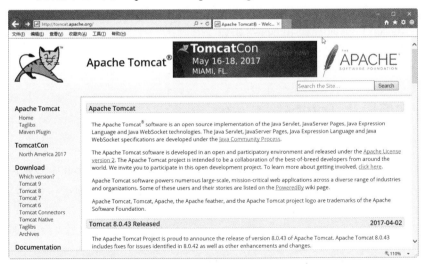

图 1-12 Tomcat 的官网首页

(2) 单击左侧 Download 下方的相应版本 Tomcat 8，进入下载页面，往下拖动滚动条，找到 Tomcat 8.0.43 版本的下载超链接，如图 1-13 所示。

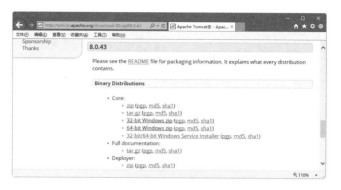

图 1-13　Tomcat 8.0.43 的下载页面

(3) Core 节点下包含 Tomcat 8.0.43 在不同平台的安装文件(根据自己的系统选择)，此处选择 64-bit Windows zip(pgp,md5,sha1)，单击该超链接，即可将其下载到本地计算机。

　　这里下载的是 Tomcat 的免安装版本，在软件开发过程中，结合使用 IDE 开发工具时，建议使用免安装版，安装版一般在实际部署中使用。

1.2.2　配置 Tomcat 环境变量

Tomcat 的免安装版本配置比较简单，解压缩后需要设置 Tomcat 的环境变量，配置的方法与配置 Java 环境变量类似，具体过程如下。

(1) 解压缩 apache-tomcat-8.0.43-windows-x64.zip，将其复制至 C:\Program Files\目录下，也可以放在其他任何地方。

(2) 在"环境变量"对话框的"系统变量"选项组中，新建系统变量 CATALINA_HOME，值设置为 C:\Program Files\apache-tomcat-8.0.43。

(3) 修改系统变量 CLASSPATH，新增值"%CATALINA _HOME%\lib;"，单击"确定"按钮完成配置。

1.2.3　启动与停止 Tomcat

Tomcat 的启动与停止介绍如下。

(1) 解压版 Tomcat 的启动方式为：进入 Tomcat 在本地目录下的 bin 子目录，笔者所用电脑中为 C:\Program Files\apache-tomcat-8.0.43\bin 目录，执行 startup.bat，即可启动服务，效果如图 1-14 所示。shutdown.bat 文件用于关闭 Tomcat 服务。

图 1-14　启动 Tomcat 服务成功

(2) 在浏览器地址栏中输入 http://localhost:8080/(这里 8080 为 Tomcat 的默认端口号，读者可以根据自己的实际配置修改)，进入 Tomcat 的 Web 管理页面，如图 1-15 所示。

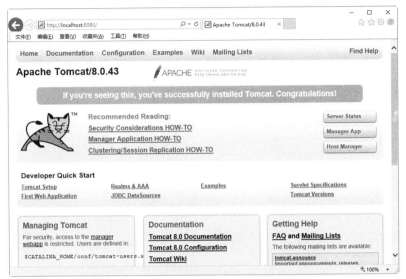

图 1-15 Tomcat 成功配置后出现的管理页面

1.2.4 Tomcat 的目录结构

下面以 Tomcat 8.0.43 版本为例，介绍 Tomcat 的目录结构，如表 1-1 所示。

表 1-1 Tomcat 的目录结构

目 录	说 明
/bin	存放 Tomcat 命令，以.sh 结尾的为 Linux 命令，以.bat 结尾的为 Windows 命令
/conf	存放 Tomcat 服务器的各种配置文件，如 server.xml
/lib	存放 Tomcat 服务器运行过程中需要加载的各种 JAR 文件包
/logs	存放 Tomcat 服务器运行过程中产生的日志文件
/temp	存放 Tomcat 服务器运行过程中产生的临时文件
/work	存放 Tomcat 在运行时的编译后文件，如 JSP 编译后的文件
/webapps	发布 Web 应用，在默认情况下将 Web 应用的文件存放在此目录中

 不同版本的 Tomcat，目录结构略有区别。

1.3 搭建 Java Web 开发环境

1.3.1 下载与安装 MyEclipse

MyEclipse 开发工具的中文官网是 http://www.myeclipsecn.com/，需要注册后才能下载，

根据自己的操作系统选择相应的版本,下载完成后,就可以进行安装了。这里选用 myeclipse-2015-stable-2.0-offline-installer-windows.exe 版本安装,安装步骤如下。

(1) 双击 MyEclipse 安装文件,先进行文件提取解压工作,然后进入许可协议界面,选中 I accept the terms of the license agreement 复选框,单击 Next 按钮,如图 1-16 所示。

(2) 在选择安装路径界面中,默认安装路径为 C:\Users\Administrator\MyEclipse 2015 (这里单击 Change 按钮修改为其他路径),单击 Next 按钮,如图 1-17 所示。

图 1-16　用户许可协议界面

图 1-17　选择安装路径界面

(3) 在选择体系结构界面中,选择 64 Bit(笔者计算机为 64 位,可根据自己计算机系统的实际情况选择),单击 Next 按钮,如图 1-18 所示。

(4) 直至最后的安装完成界面,取消选中 Launch MyEclipse 2015 复选框,单击 Finish 按钮,如图 1-19 所示。

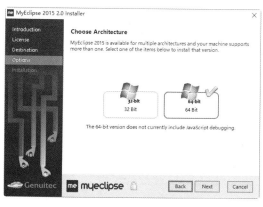

图 1-18　选择体系结构界面

图 1-19　安装完成界面

(5) 第一次启动 MyEclipse 时,会弹出 Workspace Launcher 对话框,要求设置工作空间以存放项目文档,这里可以设置自己的工作空间,将工作空间设置为 C:\Workspaces\MyEclipse 2015,如果同时选中 Use this as the default and do not ask again 复选框,下次启动时就不会再显示设置工作空间对话框了,如图 1-20 所示。

(6) 单击 OK 按钮,进入 MyEclipse 2015 的初始界面,如图 1-21 所示。

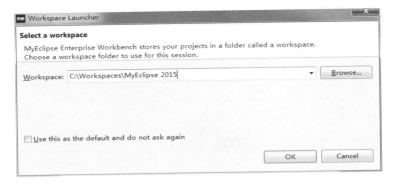

图 1-20　工作空间选择对话框

图 1-21　MyEclipse 2015 的初始界面

1.3.2　在 MyEclipse 中配置 JDK

在 MyEclipse 2015 安装过程中，会默认安装一个 1.7 版本的 JRE，如果想在 MyEclipse 中指定使用后来安装的 1.8 版本的 JRE，可进行如下设置。

(1) 从菜单栏中选择 Window→Preferences(首选项)命令，在弹出的 Preferences 对话框的左侧选择 Java→Installed JREs(已安装的 JRE)，在右侧单击 Add(添加)按钮，在弹出的 Add JRE 对话框中选择 Standard VM 选项，如图 1-22 所示。

(2) 单击 Next 按钮，进入 JRE Definition 设置界面，单击 Directory 按钮，在弹出的界面中指定 JRE 的安装路径，也可在 JRE home 中输入 JRE 安装路径。此处，通过 Directory 按钮找到 jdk_1.8.0_121 版本 JRE 的安装路径。确定后，JRE home、JRE name 和 JRE system libraries 会自动添加进来，单击 Finish 按钮完成添加，如图 1-23 所示。

 提示　　可根据自己计算机的情况，选择已安装的 JRE。

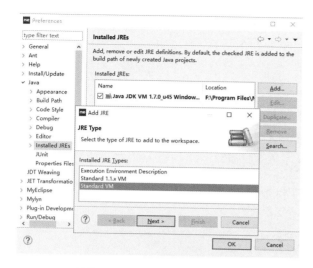

图 1-22　Preferences 对话框

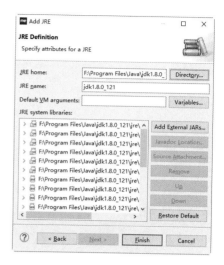

图 1-23　JRE Definition 设置界面

1.3.3　在 MyEclipse 中配置 Tomcat

在 MyEclipse 安装好之后，你会发现系统里配置了相应的 Tomcat 服务器，我们要想使用自己安装的 Tomcat，就需要重新配置。

(1) 在 MyEclipse 的菜单栏中选择 Window→Preferences 命令，弹出 Preferences 对话框，在左侧选择 MyEclipse→Servers→Runtime Environment，单击右侧的 Add(添加)按钮，选择 Tomcat→Apache Tomcat v8.0，并选中 Create a new local server 复选框，如图 1-24 所示。

(2) 单击 Next 按钮，在 Tomcat Server 界面中，单击 Browse 按钮选择 Tomcat 的安装路径，在 JRE 下拉列表框中选择之前添加的 JRE，单击 Finish 按钮完成配置，如图 1-25 所示。

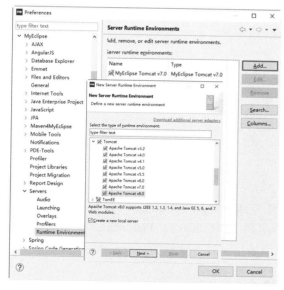

图 1-24　Tomcat 设置界面

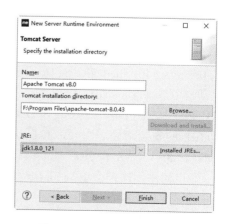

图 1-25　Tomcat Server 界面

（3）配置完成后，在 MyEclipse 主界面下方的 Servers 选项卡中，就可以看见添加的 Tomcat v8.0 Server at localhost 服务器了，如图 1-26 所示。

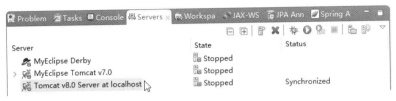

图 1-26　添加完成的 Tomcat 服务器

1.4　创建 MySQL 数据库环境

　　MySQL 是一个小型关系数据库管理系统，也是著名的开放源码的数据库管理系统。MySQL 被广泛地应用在 Internet 上的中小型网站中。由于 MySQL 体积小、速度快、总体运营成本低，许多中小型网站为了降低网站总体运营成本而选择其作为网站数据库。

　　MySQL 由瑞典 MySQL AB 公司开发，后被 Sun 公司收购，现如今 Sun 公司又被 Oracle 公司收购。目前，MySQL 针对不同的用户有不同的版本，分别为社区版和企业版，具体介绍如下。

- MySQL Community Server：社区版完全免费，但是官方不提供技术支持。
- MySQL Enterprise Server：企业版能为企业提供高性能的数据库应用，以及高稳定性的数据库系统，提供完整的数据库提交、回滚、锁机制等功能，但是该版本收费。

> 注意：MySQL Cluster 主要用于建立数据库集群服务器，需要在以上两个版本的基础上使用。

　　MySQL 的版本由 3 个数字标识，如 MySQL-5.7.17。

- 第 1 个数字 5 是主版本号，用于描述文件格式，表示版本 5 的所有发行版都有相同的文件格式。
- 第 2 个数字 7 是发行级别，它与主版本号组合在一起构成了发行序列号。
- 第 3 个数字 17 是此发行系列的版本号，目前 MySQL 5.7.17 是最新版本。

　　MySQL 社区版的性能卓越，搭配 Linux、PHP 和 Apache 可组成良好的 LAMP 开发环境。与大型的关系型数据库(如 Oracle、DB2 和 SQL Server 等)相比，MySQL 的规模小，功能有限，但对于中小型企业和个人学习使用来说，其提供的功能已经足够，本书的后续程序，就是使用 MySQL 数据库作为后台数据库管理系统。

1.4.1　下载 MySQL

　　可以从官网下载 MySQL，其最新版本为 5.7.17。下面介绍如何从官网下载。

　　（1）进入官网主页 http://www.mysql.com/，在官网下载需要注册，单击右上角的 Register 链接，如图 1-27 所示。

图 1-27 MySQL 官网首页

(2) 因现在都属于 Oracle 公司，所以会跳转到 Oracle 的注册页面，填写信息，如图 1-28 所示。

图 1-28 注册页面

(3) 注册成功后，在官网上登录，进入网页 http://dev.mysql.com/downloads/，单击 MySQL Community Server 社区版本(开源免费)，如图 1-29 所示。

图 1-29 版本选择页面

(4) 在"Select Operating System"下拉列表中选择 Microsoft Windows 选项，然后可以选择安装版，也可选择压缩配置版，这里单击"Windows(x86,32-bit),MySQL Installer MSI"右侧的 Download 按钮，如图 1-30 所示。

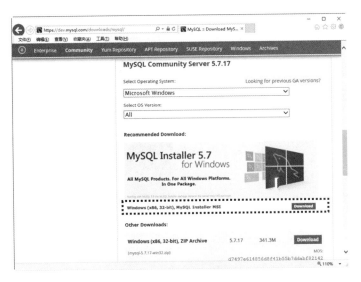

图 1-30　系统版本选择

(5) 进入类型选择页面，这里选择"Windows(x86, 32-bit), MSI Installer(mysql-install-community-5.7.17.0msi)"安装版右侧的 Download 按钮，即可下载，如图 1-31 所示。

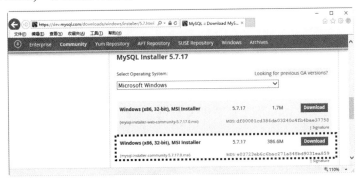

图 1-31　MySQL 下载页面

　　社区版的安装版没有 64bit 的安装程序，32bit 的也可以安装在 64 位系统上。

1.4.2　安装与配置 MySQL

MySQL 的安装与配置过程如下。

(1) 双击 mysql-installer-community-5.7.17.0.msi 安装文件，进入 License Agreement(许可协议)界面，选中 I accept the license terms 复选框，如图 1-32 所示。

(2) 单击 Next 按钮，进入 Choosing a Setup Type(选择安装类型)界面，根据需要选择，这里选择 Custom(自定义)类型，Next 按钮变为可用状态，如图 1-33 所示。

图1-32 用户许可协议界面

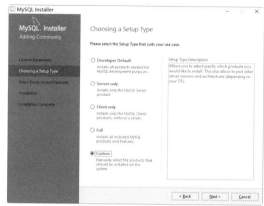

图1-33 选择安装类型界面

(3) 单击 Next 按钮，进入 Select Products and Features(选择产品和功能)界面，在 Available Products 下方组件中，依次展开 MySQL Servers→MySQL Server→MySQL Server 5.7，选中 MySQL Server 5.7.17-X64，单击绿色的右向箭头，就会添加到右侧，选中 MySQL Server 5.7.17-X64，如果单击 Advanced Options 链接，可以在弹出的对话框中修改安装路径，如图 1-34 所示。

(4) 在选择产品和功能界面中单击 Next 按钮后，进入安装界面，单击"Execute"按钮，进行安装。

(5) 安装完成后，进入 Product Configuration(产品配置)界面，单击 Next 按钮，进入 Type and Networking(类型和网络配置)界面，对学习用户来说，在 Config Type 下拉列表框中选择 Development Machine 选项，默认选中 TCP/IP 复选框，Port Number 为 3306，如图 1-35 所示。

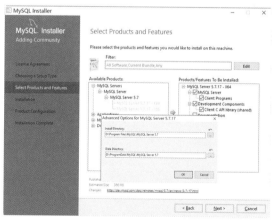

图1-34 选择产品和功能界面

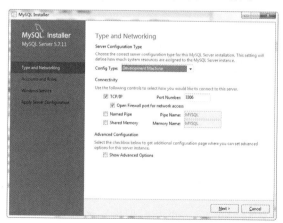

图1-35 类型和网络配置界面

(6) 单击 Next 按钮，进入 Accounts and Roles(账户和角色)界面，设置 MySQL Root 用户的密码，可单击 Add User 按钮，添加用户并设置角色和密码，如图 1-36 所示。

(7) 单击 Next 按钮，进入 Windows Service(Windows 服务)界面，默认选中 Configure MySQL Server as a Windows Service 复选框和 Start the MySQL Server at System Startup 复选框，Windows Service Name 服务名称默认为 MySQL57，可以修改，选中 Standard System Account 单选按钮，如图 1-37 所示。

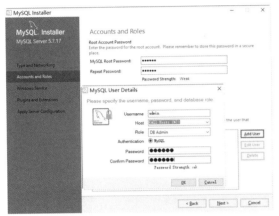

图 1-36　账户和角色界面

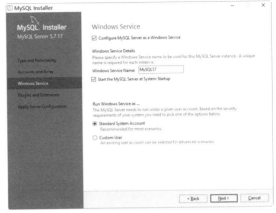

图 1-37　Windows 服务界面

(8) 单击两次 Next 按钮，进入 Apply Server Configuration(应用服务器配置)界面，单击 Execute 按钮，进行安装。安装完成后，单击 Finish 按钮返回到 Product Configuration(产品配置)界面，界面上的状态显示为 Configuration Complete(配置完成)。单击 Next 按钮，进入 Installation Complete(安装完成)界面，单击 Finish 按钮，结束安装。

1.4.3　使用 MySQL 数据库

完成以上任务后，可以进入 MySQL 5.7 Command Line Client 进行测试，以确保正常使用。操作方法：选择"开始"→"所有程序"→MySQL→MySQL Server 5.7→MySQL Command Line Client 命令，出现 DOS 窗口，在其中输入刚刚安装过程中设置的密码，按 Enter 键，出现 mysql> 提示信息，表示已经安装成功，如图 1-38 所示。

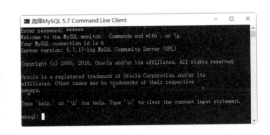

图 1-38　对 MySQL 进行测试

绝大多数的关系数据库都有两个部分：后端作为数据仓库，前端作为用于数据组件通信的用户界面。这种设计非常巧妙，它并行处理两层编程模型，将数据层从用户界面中分离出来，同时运行数据库软件制造商专注于它们的产品强项：数据存储和管理，并为第三方创建大量的应用程序提供了便利，使各种数据库间的交互性更强。MySQL 数据库也不例外，常见的前端工具有 SQLyog、WorkBench、Navicat 等。

SQLyog 是业界著名的 Webyog 公司出品的一款简洁高效、功能强大的图形化 MySQL 数据库管理工具。SQLyog 的官方网址为 https://www.webyog.com/，这里使用 SQLyog 10.2 图形化前端工具操作 MySQL 数据库。

启动 SQLyog 程序，第一次使用，会出现选择语言的界面，如果是汉化版本，这里选择简体中文，显示试用信息，单击"继续"按钮，弹出"连接到我的 SQL 主机"对话框，如图 1-39 所示，这里单击"新建"按钮，设置一个名称(我们输入 My 作为名称)，单击"确定"按钮，在"密码"文本框中输入密码，当然也可先测试链接，如图 1-40 所示。

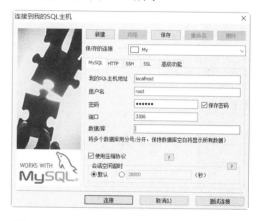

图 1-39　"连接到我的 SQL 主机"对话框(1)　　图 1-40　"连接到我的 SQL 主机"对话框(2)

单击"连接"按钮，进入 SQLyog 主窗口，SQLyog 的界面操作方式与 SQL Server 相似，如图 1-41 所示。

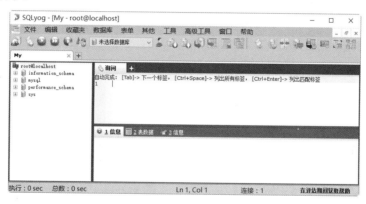

图 1-41　SQLyog 图形界面

1.5　创建和发布 Java Web 工程

安装和配置 MyEclipse 后，就可以通过在 MyEclipse 中创建和发布一个 Web 应用程序来学习 MyEclipse 的大致使用方法了。下面的操作都是基于 MyEclipse 2015 进行的。

1.5.1　创建 Web 项目、设计项目目录结构

(1) 在文件菜单中选择 File→New→Web Project 命令，弹出 New Web Project 对话框。在 Create a JavaEE Web Project 界面的 Project name 文本框中输入 restaurant，在 Java EE version

下拉列表框中选择 Java EE 7-web 3.1 版本，在 Java version 下拉列表框中选择我们自己安装配置的 1.8 版本，在 Target runtime 下拉列表框中选择 Apache Tomcat v8.0，单击 Next 按钮，如图 1-42 所示。

(2) 进入 Java 设置界面，可以在 src 下添加文件夹，这里不用修改，单击 Next 按钮，如图 1-43 所示。

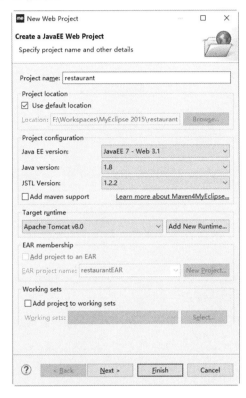
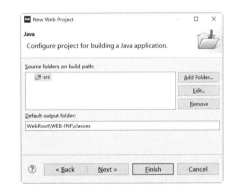

图 1-42　新建 Web 项目对话框　　　　　　图 1-43　Java 设置界面

(3) 进入 Web Module 界面，在其中选中 Generate web.xml deployment descriptor 复选框，单击 Next 按钮，如图 1-44 所示。

 提示　　若在如图 1-42 所示的新建 Web 项目对话框的 Java EE version 下拉列表框中选择 JavaEE 5 – Web 2.5 选项，则 Generate web.xml deployment descriptor 复选框默认是选中的。

(4) 进入 Configure Project Libraries 界面，单击 Finish 按钮，如图 1-45 所示。

 提示　　在这里可以取消选中 Apache Tomcat v8.0 和 JSTL 1.2.2 Library，也可以通过 Add custom JAR 按钮添加自定义的 jar 包。

(5) 完成后，在窗体左侧的包资源管理器视图中，就可以看到 restaurant 项目的目录结构了，如图 1-46 所示。

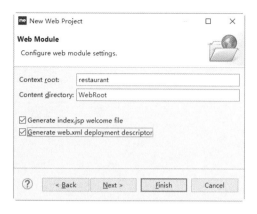
图 1-44　Web 模块设置界面

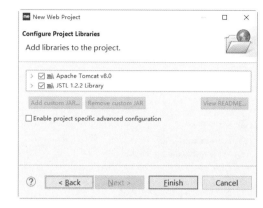
图 1-45　配置项目库文件

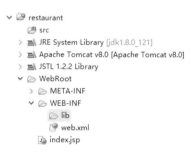
图 1-46　restaurant 项目的目录结构

我们通常把 Java 类文件放在 src 目录下，可在 src 下定义包；把网页文件放在 WebRoot 下，可在根路径下定义文件夹，这样可以方便管理。

1.5.2　编写页面代码、部署和运行 Web 项目

下面使用集成开发工具 MyEclipse 来编写一个 JSP 页面，具体操作步骤如下。

(1) 创建一个 JSP 文件，右击 WebRoot，在弹出的快捷菜单中选择 New→JSP(Advanced Templates)命令，如图 1-47 所示。

(2) 在弹出的对话框中输入文件路径及文件名，这里为了方便，只输入一个 welcome.jsp 页面，直接放在 WebRoot 路径下，如图 1-48 所示。

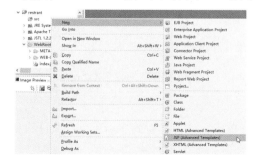
图 1-47　创建 JSP 文件

图 1-48　输入 JSP 文件路径及名称

(3) 单击 Finish 按钮，完成 JSP 页面的创建，当然页面内容需要我们自己编写。在 welcome.jsp 页面的主体部分，编写一个"欢迎来到 Java Web 开发的世界！"提示，并且把字符编码设置为 pageEncoding="utf-8"。

(4) 单击工具栏上的部署图标，弹出 Manage Deployments 对话框，在 Module 下拉列表框中选择需要部署的 restaurant 项目，如图 1-49 所示。

(5) 单击 Add 按钮，在弹出的 Deploy modules 对话框的 Server 下选择系统中安装的 Tomcat v8.0 Server at localhost，然后单击 Finish 按钮，如图 1-50 所示。

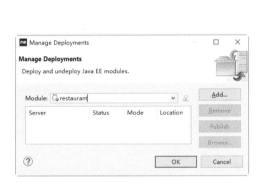

图 1-49　部署 Web 项目

图 1-50　添加 Tomcat 服务器

(6) 启动 Tomcat，在工具栏上启动 Tomcat v8.0 Server at localhost，如图 1-51 所示，此时会在 Console 控制台输出 Tomcat 的启动信息。

(7) 打开浏览器，输入 http://localhost:8080/restaurant/welcome.jsp，按 Enter 键，运行结果如图 1-52 所示。

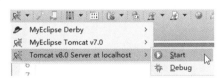

图 1-51　启动 Tomcat

图 1-52　JSP 程序的运行结果

1.6　小　　结

本章详细讲述了搭建 Java Web 环境所需的各种软件的下载及安装方法，包括 JDK、Tomcat、MyEclipse、MySQL，以及在 MyEclipse 中配置 JRE 和 Tomcat 的方法。以上所选择的软件，也是在开发过程中经常用到的组合。最后在 MyEclipse 中创建和发布一个 Web 应用程序，以学习 MyEclipse 的大致使用方法。

第 2 章
JSP 动态页面开发技术

动态网页是指在服务器端运行的程序或者网页,它们会随不同客户、不同时间,返回不同的网页内容。动态网页需要用到服务器端脚本语言,JSP 就是目前流行的一种动态网页技术。动态网页的内容一般存储在数据库中,用户访问动态网页时,JSP 通过读取数据库中的存储数据来动态生成网页内容。本章对 JSP 相关技术进行介绍。

2.1 JSP 技术基础

JSP 是 Java Server Pages 的简称，它是由 Sun 公司倡导，多家公司共同参与建立起来的一种动态网页技术标准。它在动态网页中有着强大而特别的功能，具有跨平台性、易维护性、易管理性等特点。

2.1.1 JSP 简介

1．为什么需要 JSP

静态网页的显示内容是保持不变的，静态网页既不能实现与用户的交互，又不利于系统的扩展。所以，我们需要基于 B/S 技术的动态网页。

使用动态网页，可以动态地输出网页内容、同用户进行交互、对网页内容进行在线更新。这是由 B/S 技术的特点所决定的，如图 2-1 所示。

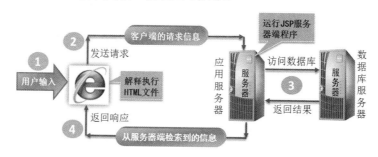

图 2-1　B/S 技术的特点

在 B/S 结构中，浏览器端与服务器端采用请求/响应模式进行交互，这个过程可以分解为如下步骤。

(1) 客户端浏览器接受用户的输入。一个用户在 IE 窗口中输入用户名、密码，单击"登录"按钮，发送对系统的访问请求。

(2) 客户端浏览器向应用服务器端发送请求。客户端把请求消息(包括用户名、密码等信息)发送到应用服务器端，等待服务器端的响应。

(3) 数据处理。应用服务器端通常使用服务器端脚本语言，如 JSP 等，来访问数据库服务器，查询该用户有无访问权限，并获得查询结果。

(4) 发送响应。应用服务器端向客户端浏览器发送响应消息(一般是动态生成的 HTML 页面)，并由访问者的浏览器端解释 HTML 文件，呈现用户界面。

实现动态网页的关键在于运行在应用服务器端的服务器端脚本语言，它可以根据不同用户的请求输出相应的 HTML 页面，然后应用服务器再把这个 HTML 页面返回给客户端。

2．JSP 的执行过程

实际上，JSP 就是指在 HTML 中嵌入 Java 脚本语言，当用户通过浏览器请求访问 Web 应用时，Web 服务器会使用 JSP 引擎对请求的 JSP 进行编译和执行，然后将生成的页面返回给客户端浏览器进行显示。当 JSP 请求提交到服务器时，Web 服务器会通过三个阶段实现处理，执行过程如下。

（1）翻译阶段。当 Web 服务器接收到 JSP 请求时，首先会对 JSP 文件进行翻译，将编写好的 JSP 文件通过 JSP 引擎转换成可识别的 Java 源代码。

（2）编译阶段。经过翻译的 JSP 文件相当于编写好的 Java 源文件，必须将 Java 源文件编译成可执行的字节码文件。

（3）执行阶段。Web 容器接收客户端的请求后，经过翻译和编译两个阶段，生成了可以被执行的字节码文件，此时就进入执行阶段。当执行完成后，会得到请求的处理结果，Web 容器再把生成的结果页面返回到客户端显示。

Web 容器处理 JSP 文件请求的三个阶段，如图 2-2 所示。

Web 容器将 JSP 文件翻译和编译完成后，会将编译好的字节码文件放在内存中，当客户端再一次请求时，就可以重用这个编译好的字节码文件，而无须重新翻译和编译，这样就大大提高了 Web 应用系统的性能。如果对 JSP 文件进行了修改，Web 容器会及时发现，此时 Web 容器就会重新执行翻译和编译过程。因此，JSP 在第一次请求时会比较慢，后续访问速度就会很快。当然，如果 JSP 文件发生了变化，同样需要重新进行编译。

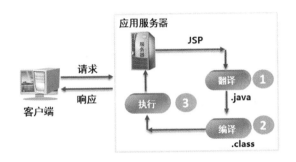

图 2-2　JSP 的执行过程

2.1.2　JSP 页面组成

了解 JSP 的工作原理和执行过程，是学习 JSP 的基础，而使用 JSP 进行动态网页开发，还需要掌握 JSP 页面中包括哪些元素，不同元素具备什么功能。前面谈到，JSP 是通过在 HTML 中嵌入 Java 脚本语言来响应页面动态请求的。下面通过在页面上显示相应日期的示例，展示几个比较常用的 JSP 页面元素。在第 1 章的 restaurant 项目的 WebRoot 下新建 ch02 文件夹，并在该文件夹中新建 showDate.jsp 文件，代码如下：

```
<%@ page language="java" import="java.util.*"
contentType="text/html;charset=utf-8" %>
    <%@ page import="java.text.*" %>
    <html>
        <head>
            <title>输出当前系统日期</title>
        </head>
        <!-- 这是 HTML 注释(客户端可以看到源代码) -->
        <%-- 这是 JSP 注释(客户端不能看到源代码) --%>
        <body>
            你好，今天日期是：
            <%  //使用预定格式将日期转换为字符串
                SimpleDateFormat f = new SimpleDateFormat("yyyy年MM月dd日");
```

```
            String currentTime = f.format(new Date());
        %>
        <%=currentTime %>
        <%! String declare = "这是声明";%>
        <%= declare %>
    </body>
</html>
```

部署项目，通过浏览器访问 http://localhost:8080/restaurant/ch02/showDate.jsp，该示例在浏览器上的运行结果如图 2-3 所示。该示例产生的网页源代码如图 2-4 所示。

图 2-3　在浏览器上查看日期显示　　　　图 2-4　查看示例网页源代码

在该示例中，一共展示了 5 种页面元素，包含静态内容、指令、小脚本、表达式、声明和注释，下面来一一介绍。

1．静态内容

静态内容是指 JSP 页面中的静态文本，它基本上是 HTML 文本，与 Java 和 JSP 无关。

2．JSP 注释

在编程中，添加注释是非常好的习惯，合理详细的注释有利于后期代码的维护以及团队成员的阅读。在 JSP 文件中有三种注释方法。

(1) HTML 注释方法，其格式为<!-- HTML 注释 -->，其中的注释内容在客户端浏览器里是可以看见的。这种注释方法不安全，而且会增加网络负担。

(2) JSP 注释标记，其格式为<%-- JSP 注释 --%>，在客户端查看源代码时看不到注释中的内容，安全性较高。

(3) 在 JSP 脚本中使用注释，和在 Java 类中进行注释的方式是一样的，使用的格式为<% //单行注释%>、<% /* 多行注释 */ %>。

3．JSP 脚本元素

在 JSP 中，将表达式、小脚本、声明统称为 JSP 脚本元素，用于在 JSP 页面中嵌入 Java 代码，实现页面的动态请求。

1) 小脚本

小脚本可以包含任意 Java 片段，形式比较灵活，通过在 JSP 页面中编写小脚本可以执行复杂的操作和业务处理，编写方法就是将 Java 代码片段插入"<%   %>"标记中，在前面的示例中，属于小脚本的代码如下：

```
<%
    //使用预定格式将日期转换为字符串
    SimpleDateFormat f = new SimpleDateFormat("yyyy年MM月dd日");
    String currentTime = f.format(new Date());
%>
```

下面通过小脚本在页面上按行循环输出数组中的值，创建 showArray.jsp 页面，页面部分的小脚本代码如下：

```
<%
    char[] array={'A','B','C','D'};
    for(int i=0;i<array.length;i++){
        out.println(array[i]);
%>
    <br>
<%
    }
%>
```

由于小脚本和 HTML 静态页面混合编写，很容易忘记转行的标签，又不容易被发现。因此大家在编写 JSP 代码时不仅要注意代码的缩进，而且要编写必要的注释。

2）表达式

表达式是对数据的表示，系统将其作为一个值进行计算和显示，当需要在页面中获取一个 Java 变量或者表达式值时，使用表达式是非常方便的。其语法是：<%=Java 表达式 %>。需要注意的是，使用表达式输出数据时，不能在表达式结尾添加分号来代表语句结束。

在 Java 开发过程中，使用 System.out.println()和 System.out.print()向控制台输出信息。在 JSP 网页上，可以使用内置对象 out 把结果输出到页面上，out 对象是 JSP 开发过程中使用最为频繁的对象，使用也是最简单的。

3）JSP 声明

在编写 JSP 页面程序时，有时需要为 Java 脚本定义变量和方法，这时就需要对所使用的变量和方法进行声明。声明一般没有输出，通常与表达式、小脚本一起综合运用。将输出日期的示例进行修改，在同一 JSP 页面中，如果需要在多个地方格式化日期，在 Java 代码中可以增加一个方法来解决，在 JSP 文件中，同样可以声明方法来解决类似问题。在 ch02 的文件夹中新建 show.jsp 文件，具体示例代码如下：

```
<body>
    <%!
        String formatDate(Date date){
            SimpleDateFormat f = new SimpleDateFormat("yyyy年MM月dd日");
            return f.format(date);
        }
    %>
    第一次显示时间：今天是<%=formatDate(new Date()) %> <br>
    第二次显示时间：今天是<% out.print(formatDate(new Date())); %>
</body>
```

部署项目，访问 http://localhost:8080/restaurant/ch02/show.jsp，运行效果如图 2-5 所示。

图 2-5　显示时间的运行结果

4．JSP 指令元素

JSP 指令用来设置与整个 JSP 页面相关的属性，在 JSP 运行时，控制 JSP 页面的某些特性，如网页的编码方式和脚本语言。JSP 指令一般以"<%@"开始，以"%>"结束。

JSP 指令有多种类型，主要有三种：page、include 和 taglib，下面分别进行介绍。

1）page 指令

在 Java 文件中，可以通过两种方式引入其他包中的类。第一种是使用 import 关键字，优点在于：一次引入，处处使用；另外一种方式就是使用完全限定的类名，即类名前必须加上完整的包名。在 JSP 文件中，同样可以采用以上两种方式。通常情况下，我们使用 import 关键字引入 Java 类文件，好处在于：一旦引入，这个 Java 类文件在整个 JSP 文件范围内都可用。

page 指令就是通过设置内部的多个属性来定义 JSP 文件中的全局特性。page 指令只能对当前自身页面进行设置，即每个页面都有自身的 page 指令，如果没有对某些属性进行设置，JSP 容器将使用默认指令属性值。

page 指令一般放在 JSP 页面的第一行，其语法格式如下：

```
<%@ page 属性名1="属性值1" 属性名2="属性值2,属性值3" …… %>
```

示例：

```
<%@ page language="java" import="java.util.*,java.text.*" contentType=
"text/html; charset=utf-8" %>
```

在对同一个属性设置多个属性值时，其间以逗号相互隔开。

page 指令的属性共有 13 个，其中最常用的几个属性的含义如表 2-1 所示。

表 2-1　page 指令常用属性

属　　性	描　　述
language	指定 JSP 页面使用的脚本语言，默认为 Java
import	通过该属性引用脚本语言中使用的类文件
contentType	指定 JSP 页面采用的编码方式，默认为 text/html,charset=ISO-8859-1
pageEncoding	指定页面使用的字符编码，默认为 ISO-8859-1

(1) language 属性。

language 属性用来指定当前 JSP 页面所采用的脚本语言，当前 JSP 版本只能使用 Java 语言。该属性可以不设置，因为 JSP 默认就是采用 Java 作为脚本。language 属性的设置方法如下：

```
<%@ page language="java" %>
```

(2) import 属性。

import 属性在实际开发中使用非常频繁。通过 import 属性可以在 JSP 文件的脚本片段中引用外在的类文件。如果一个 import 属性引入多个类文件，则多个类文件之间要用逗号隔开。import 属性的设置方法如下：

```
<%@ page import="java.util.*,java.text.*" %>
```

(3) contentType 属性。

contentType 属性的设置在开发过程中是非常重要的，且经常被用到，中文乱码一直是困扰开发者的一个问题，而该属性就是用来设置编码格式的，告诉 Web 容器在客户端浏览器上以何种格式显示 JSP 文件以及使用何种编码格式。contentType 属性的设置方法如下：

```
<%@ page contentType="text/html;charset=utf-8" %>。
```

text/html 和 charset=utf-8 之间用分号隔开，它们同属于 contextType 属性值。当设置为 text/html 时，表示该页面以 HTML 页面的格式进行显示。这里设置的编码格式为 utf-8，这样 JSP 页面中的中文就可以正常显示了。

(4) pageEncoding 属性。

该属性用来指定页面所使用的字符编码，如果设置了这个属性，则 JSP 页面的字符编码就是指定的字符集；如果未设置此属性，则使用 contextType 属性的值；如果两个属性均未设置，页面默认使用 ISO-8859-1。

在什么位置插入 page 指令并不重要，因为 page 指令和其他指令一样，只在 JSP 页面编译的时候起作用。page 指令的参数除了以上介绍的 4 个外，还有 extends、session、buffer、autoFlush、isThreadSafe、info、errorPage、isErrorPage、isELIgnored，这些参数的名称是区分大小写的。

2) include 指令

include 指令用于在 JSP 页面中包含一个文件，该文件可以是 JSP 文件、HTML 网页、文本文件或 Java 代码，使用 include 指令可以简化页面代码，提高代码的重用性，语法如下：

```
<%@ include file="被包含文件的url地址" %>
```

例如，在页面中包含 index.jsp 的代码为<%@ include file="../index.jsp"%>。

被包含的文件中，最好不要使用<html>、</html>、<body>、</body>等标签，否则会影响 JSP 网页中的相同标签，从而导致不必要的错误。

例如，许多网站的每个页面都有一个小小的导航条，导航条往往用页面顶端或左边的一个表格制作，同一份 HTML 代码重复出现在整个网站的每个页面上。include 指令是实现该功能非常理想的方法。使用 include 指令，开发者不必再把导航的 HTML 代码复制到每个文件中，从而可以更轻松地完成维护工作。

3) taglib 指令

taglib 指令用于通知 JSP 容器某个页面依赖于自定义标签库，标签库是可用于扩展 JSP 功能的自定义标签的集合，taglib 指令的语法如下：

```
<%@ taglib uri="标签名称空间" prefix="前缀" %>
```

uri 属性指定了 JSP 要在 web.xml 文件中查找的标签库描述符，该描述符是一个标签描述文件(*.tld)的映射。另外，通过 uri 属性直接指定标签描述文件的路径，而无须在 web.xml 文件中进行配置，同样可以使用指定的标记。prefix 属性指定了一个在页面中使用由 uri 属性指定的标签库的前缀，示例如下：

```
<%@ taglib uri="http://java.sun.com/jsp/jstl/core" prefix="c" %>
```

2.2 JSP 内置对象

在当今的 Web 程序页面中，用户交互是必不可少的，JSP 的内置对象就是用于处理浏览器请求的对象，可以直接使用，而不需要自己创建。

2.2.1 什么是 JSP 内置对象

JSP 内置对象，就是当你编写 JSP 页面时，无须做任何声明就可以直接使用的对象。例如，在 2.1.2 小节的示例中出现了如下的代码片段：

```
<%
    char[] array={'A','B','C','D'};
    for(int i=0;i<array.length;i++){
        out.println(array[i]);
    }
%>
```

代码 out.println()可以实现页面的输出显示，但是代码中并没有任何地方声明或者创建这个 out 对象，没有创建就可以直接使用的原因，就是 out 对象是 JSP 的内置对象之一。除了 out 对象之外，在 JSP 中还有其他一些内置对象，如图 2-6 所示。

图 2-6　JSP 常用的内置对象

所谓内置对象就是由 Web 容器加载的一组类的实例,它不像一般的 Java 对象在创建类的实例时,必须要用 new 关键字去构造对象,而是可以直接在 JSP 中使用的对象。

注意

JSP 的内置对象名称均是 JSP 的保留字,不得随便使用。

2.2.2 out 内置对象

out 内置对象是 JSP 在开发过程中使用最为频繁的对象,是 JspWriter 类的实例,用来向客户端输出内容。out 对象常用的方法是 print()和 println(),这个方法用于在页面中打印字符串信息。out 对象除了输出方法外,还有一些其他的方法,这些方法主要用来管理缓冲流或者输出流,如表 2-2 所示。

表 2-2 out 对象的其他方法

方　　法	说　　明
void clear()	清除缓冲区的内容
void clearBuffer()	清除缓冲区的当前内容
void flush()	清除数据流
void close()	关闭输出流
int getBufferSize()	返回缓冲区字节数大小,如不设缓冲区则为 0
int getRemaining()	返回缓冲区还剩余多少可用
boolean isAutoFlush()	返回缓冲区满时,是自动清空还是抛出异常

2.2.3 request 内置对象

request 内置对象是最常用的对象之一,主要用于处理客户端浏览器的请求,其工作原理如图 2-7 所示。

request 对象中包含请求的相关信息,可以在 JSP 页面中通过调用 request 对象的方法获取请求的相关数据。request 对象的常用方法如表 2-3 所示。

图 2-7 request 内置对象的工作原理

表 2-3 request 对象的常用方法

方　　法	说　　明
String getParameter(String name)	根据表单组件名称获取提交数据
String[ ] getParameterValues(String name)	获取表单组件对应多个值时的请求数据
void setCharacterEncoding(String charset)	指定每个请求的编码

续表

方　法	说　明
request.getRequestDispatcher(String path)	返回一个 java.servlet.RequestDispatcher 对象，该对象的 forward( )方法用于转发请求

在 restaurant 项目的 WebRoot/ch02/目录下，创建 reg.jsp 注册页面和 reginfo.jsp 注册提交页面，编程实现用户的注册功能。

注册页面 reg.jsp 的主要代码如下：

```html
<body>
    请输入注册信息<br>
    <form name="form1" method="post" action="ch02/reginfo.jsp">
        用户名：<input type="text" name="username"><br>
        密    码：
            <input type="password" name="password"><br>
        业余爱好：
        <input type="checkbox" name="habit" value="看书">看书
        <input type="checkbox" name="habit" value="玩游戏">玩游戏
        <input type="checkbox" name="habit" value="旅游">旅游
        <input type="checkbox" name="habit" value="看电视">看电视<br>
        <input type="submit" value="提交">  
        <input type="reset" value="取消">
    </form>
</body>
```

注册提交页面 reginfo.jsp 的主要代码如下：

```jsp
<%
    request.setCharacterEncoding("utf-8"); //设置读取字符编码为utf-8
    String username=request.getParameter("username");    //读取用户名
    String password=request.getParameter("password");    //读取密码
    String[] habits=request.getParameterValues("habit");//读取兴趣爱好
%>
<body>
    您输入的注册信息如下：<br>
    用户名为：<%=username %>    <br>
<%
    out.print("密码为: "+password+"<br>");
    out.print("您的兴趣爱好：");
    if(habits!=null){
        for(int i=0;i<habits.length;i++){
            out.print(habits[i]+"  ");
        }
    }
%>
</body>
```

注册页面信息包括用户名、密码、兴趣爱好，如图 2-8 所示；页面提交后，显示用户输入的数据，如图 2-9 所示。

图 2-8　输入注册信息

图 2-9　显示注册信息

2.2.4　response 内置对象

response 对象用于响应客户请求并向客户端输出信息，与内置 request 对象相对应，其工作原理如图 2-10 所示。

图 2-10　response 内置对象的工作原理

response 对象也提供了多个方法用来处理 HTTP 响应，response 对象的常用方法如表 2-4 所示。

表 2-4　response 对象的几个常用方法

方　　法	说　　明
void addCookie()	在客户端添加 Cookie
void setContentType(String type)	设置 HTTP 响应的 contentType 类型
void setCharacterEncoding(String charset)	设置响应所采用的字符编码类型
void sendRedirect(String path)	将请求重新定位到一个不同的 URL 上

在 restaurant 项目的 WebRoot/ch02/目录下，创建登录页面 login.jsp、登录处理页面 control.jsp 和欢迎页面 welcome.jsp。编程实现用户的登录处理，并跳转到欢迎页面。

登录页面 login.jsp 的表单部分代码如下：

```
<form name="form1" method="post" action="ch02/control.jsp">
    用户名：<input type="text" name="username">
    密码：<input type="password" name="password">
    <input type="submit" value="登录">
</form>
```

登录处理页面 control.jsp 接收参数处理请求部分的代码如下：

```
<%
    request.setCharacterEncoding("utf-8");
    String username = request.getParameter("username");
    String password = request.getParameter("password");
    if(username.equals("admin") && password.equals("admin")){
        //此处暂不访问数据库，用户名和密码都为 admin
        response.sendRedirect("welcome.jsp");
    }
%>
```

欢迎页面 welcome.jsp 的主要代码如下：

```
<body>
    欢迎  来到本页面！
</body>
```

在登录页面输入用户名和密码，如图 2-11 所示。单击"登录"按钮，提交至处理页面，并接收用户名和密码参数，进行逻辑判断。符合条件则跳转到欢迎页面，客户端重新建立链接，URL 地址发生了变化，如图 2-12 所示。

图 2-11　登录页面

图 2-12　欢迎页面

如果要在欢迎页面显示登录的用户名，可以使用 request 对象获取用户请求的数据。在 welcome.jsp 页面中显示该用户名，修改 welcome.jsp 的代码，具体如下：

```
<%
    String username=request.getParameter("username");
%>
欢迎<%=username %>，来到本页面！
```

重新部署项目，运行程序，登录后进入欢迎页面，并未显示用户名，如图 2-13 所示。

图 2-13　显示用户信息

重定向是在客户端发挥作用的，客户端重新向服务器请求一个地址链接，由于是新发送的请求，因而上次请求的数据将随之丢失，在地址栏中可以显示转向后的地址。由于服务器重新定向了新的 URL 地址，所以重定向可以理解为是浏览器至少提交了两次请求。

修改登录处理页面 control.jsp，将跳转部分的代码进行修改，代码如下：

```
if(username.equals("admin") && password.equals("admin")){
```

```
request.getRequestDispatcher("welcome.jsp").forward(request, response);
}
```

重新运行程序，再次显示运行结果，如图 2-14 所示。

转发是在服务器端发挥作用的，通过 forward()将提交信息在多个页面间进行传递，整个过程都是在一个 Web 容器内完成的，因而可以共享 request 范围内的数据。而对应到客户端，不管服务器内部如何处理，作为浏览器都只是提交了一个请求，客户端浏览器的 URL 地址栏不会显示转向后的地址。

图 2-14 显示用户的欢迎页面

2.2.5 session 内置对象

我们在上网时，一定有过这样的经历：好不容易找到一个下载地址，可是单击下载时，系统会自动转入登录页面，提示登录。如果是已登录用户，就不会面临这样的问题，系统如何判断用户是否已经登录过该网站呢？JSP 中提供了会话跟踪机制，该机制可以保持每个用户的会话信息，为不同的用户保存自己的数据。

对 Web 开发来说，一个会话就是用户通过浏览器和服务器之间进行的一次通话，它可以包含浏览器与服务器之间的多次请求和响应。当用户向服务器发出第一次请求时，服务器会为该用户创建唯一的会话，会话将一直持续到用户访问结束(即浏览器的关闭)，JSP 提供了一个可以在多个请求之间持续有效的会话对象 session，如图 2-15 所示。

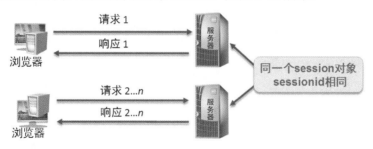

图 2-15 会话过程

session 机制是一种服务器端的机制，在服务器端使用，类似于散列表的结构来保存信息，当程序接收到客户端的请求时，服务器首先检查这个客户端是否已经创建了 session。

在 JSP 中，session 对象用来存储相关用户会话的所有信息。一个用户对应一个 session，并且随着用户的离开，session 中的信息也随之消失。session 对象的常用方法如表 2-5 所示。

表 2-5 session 对象的常用方法

方　　法	说　　明
void setAttribute(String key,Object value)	以 key/value 的形式保存对象值
Object getAttribute(String key)	通过 key 获取对象值

续表

方　法	说　明
void invalidate()	设置 session 对象失效
String getId()	获取 sessionid
void setMaxInactiveInterval(int interval)	设定 session 的非活动时间
int getMaxInactiveInterval()	获取 session 的有效非活动时间(秒为单位)
void removeAttribute(String key)	从 session 中删除指定名称(key)所对应的对象

在 restaurant 项目的 WebRoot/ch02/目录下，创建普通的首页面 index.jsp，访问控制流程如图 2-16 所示。

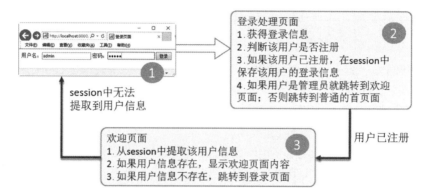

图 2-16　session 对象的访问控制流程

根据访问控制流程，修改 2.2.4 小节中的登录处理页面 control.jsp，在会话中，保存用户信息，如果用户登录成功则跳转到欢迎页面，代码如下：

```
<%
    request.setCharacterEncoding("utf-8");
    String username = request.getParameter("username");
    String password = request.getParameter("password");
    if(username.equals("admin") && password.equals("admin")){
        //此处暂不访问数据库，用户名和密码默认为 admin
        session.setAttribute("LOGINED_NAME", username);//设置登录信息
        session.setMaxInactiveInterval(10*60);//设置 session 的过期时间
        response.sendRedirect("welcome.jsp");
    }else{
        response.sendRedirect("index.jsp");
    }
%>
```

修改 2.2.4 小节中的欢迎页面 welcome.jsp，在欢迎页面中读取会话中的用户信息，并进行校验，校验失败，则返回登录页面，代码如下：

```
<%
    String loginedName=(String)session.getAttribute("LOGINED_NAME");
    if(loginedName==null){
        response.sendRedirect("login.jsp");
    }
```

```
        out.print("欢迎"+loginedName+", 来到本欢迎页面！");
%>
```

重新部署项目，运行登录页面，输入用户名和密码，则跳转到欢迎页面，并显示用户名，地址栏中的地址显示的是 welcome.jsp，如图 2-17 所示。

只要当前浏览器不关闭，在设置的 session 期限内，session 都是有效的，如果已达到期限或者打开一个新的浏览器，则 session 中存储的对象会被释放。

图 2-17 登录后的欢迎页面

2.2.6 application 内置对象

application 对象类似于系统的"全局变量"，可跨越多个浏览器，用于实现用户之间的数据共享。application 对象的常用方法如表 2-6 所示。

表 2-6 application 对象的常用方法

方　法	说　明
void setAttribute(String key,Object value)	以 key/value 的形式保存对象值
Object getAttribute(String key)	通过 key 获取对象值
String getRealPath(String path)	返回相对路径的真实路径

在 restaurant 项目的 WebRoot/ch02/目录下，新建统计显示页面 showCount.jsp，用来实现在网站系统中统计并显示已访问过的人数，具体实现代码如下：

```
<body>
  <%
    Integer count=(Integer)application.getAttribute("count");
    if(count==null){
        count=1;
    }else{
        count=count+1;
    }
    application.setAttribute("count", count);
  %>
  统计访问量：目前有<%=application.getAttribute("count") %>个人访问过本网站！
</body>
```

重新部署项目，运行统计显示页面，后显示有多少个人访问过网站，刷新页面后，或重新打开浏览器访问页面，访问过的人数会增加，运行效果如图 2-18 所示。

图 2-18　访问人数统计页面

2.2.7　其他内置对象

在 JSP 中，除了 out、request、response、session、application 对象外，还有 page、config、exception、pageContext 四个对象。

1．page 对象

page 对象表示当前页面，page 对象就是页面实例的引用，指向当前 JSP 页面本身，类似于 Java 中的 this 关键字。在 JSP 页面中，page 对象使用得较少。

2．config 对象

config 对象用于存放 JSP 页面编译后的初始数据。config 对象代表当前 JSP 配置信息，但 JSP 页面通常无须配置，因此也就不存在配置信息。与 page 对象一样，config 对象在 JSP 页面中也很少使用。

3．exception 对象

exception 对象表示 JSP 页面运行时产生的异常，该对象只有在错误页面(page 指令中设定 isErrorPage 为 true 的页面)中才能使用。

4．pageContext 对象

pageContext 对象代表页面上下文，提供了对 JSP 页面内所有对象及命名空间的访问，并提供访问其他隐含对象的方法，即 request 对象、response 对象、application 对象、config 对象、session 对象、out 对象可以通过访问这个对象的属性来导出，相当于页面中所有功能的集成。pageContext 对象的常用方法如表 2-7 所示。

表 2-7　pageContext 对象的常用方法

方　　法	说　　明
ServletRequest getRequest()	获得 request 对象
ServletResponse getResponse()	获得 response 对象
HttpSession getSession()	获得 session 对象
JspWriter getOut()	获得 out 对象
void setAttribute(String name,Object attribute)	设置 name 属性及属性值

续表

方　法	说　明
void getAttribute(String name)	获取 page 范围内属性的值
void getAttribute(String name,int scope)	获取指定范围内的 name 属性
void include(String relativeUrlPath)	在当前位置包含另一文件
Object getPage()	返回当前页面的 Object 对象

2.3　对象的范围

在 JSP 页面中的对象，无论是用户创建的对象，还是 JSP 的内置对象，都有一个范围，这个范围定义了在什么时间内，在哪些 JSP 页面中可以访问这些对象。在 JSP 中，对象有四种范围：page、request、session 和 application，它们都能借助 setAttribute()方法和 getAttribute()方法来设置和取得其属性，也可通过 removeAttribute()来删除属性。

2.3.1　page 范围

所谓的 page 范围，是指单一 JSP 页面的范围。page 范围内的对象只能在创建对象的页面中访问。在 page 范围内，将数据存入和取出，可以使用 pageContext 对象的 setAttribute()和 getAttribute()方法。page 范围内的对象在客户端每次请求 JSP 页面时创建，在服务器发送响应或请求转发到其他页面或资源后失效。

在 restaurant 项目的 WebRoot/ch02/目录下，创建 rangeOne.jsp 页面和 rangeTwo.jsp 页面。在第一个页面中，可以调用 pageContext 的 setAttribute()方法将一个字符串类型对象保存为 page 范围，然后分别在本页面和另一页面调用 pageContext 的 getAttribute()方法访问具有 page 范围的这个对象。

rangeOne.jsp 页面的代码如下：

```
<%
    String name="pageRange";
    pageContext.setAttribute("name", name);
%>
<h3>rangeOne: <%=pageContext.getAttribute("name") %></h3>
<% pageContext.include("rangeTwo.jsp"); %>
```

rangeTwo.jsp 页面的代码如下：

```
<h3>rangeTwo:
<%=pageContext.getAttribute("name") %>
</h3>
```

部署项目，运行访问 rangeOne.jsp 页面，效果如图 2-19 所示。

注意　　pageContext 对象本身也属于 page 范围，具有 page 范围的对象被绑定到 pageContext 对象中。

图 2-19　page 范围演示

2.3.2　request 范围

相对于在 page 范围内的对象与 pageContext 绑定在一起，request 范围内的对象则与客户端用户的请求绑定在一起，即 request 范围内的对象在页面转发或包含中有效。在该范围内的对象同样可以通过调用 request 对象的 setAttribute()与 getAttribute()方法找到，同时在调用 forward()方法转向的页面或者调用 include()方法包含的页面时，都可以访问 request 范围内的对象。

修改 rangeOne.jsp 页面，代码如下：

```
<%
    String name="requestRange";
    request.setAttribute("name", name);
%>
<h3>rangeOne: <%=request.getAttribute("name") %></h3>
<% pageContext.include("rangeTwo.jsp"); %>
```

修改 rangeTwo.jsp 页面，代码如下：

```
<h3>rangeTwo: <%=request.getAttribute("name") %></h3>
```

部署项目，运行访问 rangeOne.jsp 页面，效果如图 2-20 所示。

注意　　因为请求对象对于客户端的每次用户请求都是不同的，所以对于任何一个新的请求，都要重新创建该范围内的对象。而当请求结束后，创建的对象也就随之消失。

图 2-20　request 范围演示

2.3.3　session 范围

JSP 容器为每一次会话创建一个 session 对象，在会话期间，只要将对象绑定到 session 中，对象的范围就为 session。在会话有效期间，都可以访问 session 范围内的对象。

修改 rangeOne.jsp 页面，代码如下：

```
<%
    String req="requestRange";
    String ses="sessionRange";
    request.setAttribute("req", req);
    session.setAttribute("ses", ses);
    response.sendRedirect("rangeTwo.jsp");
%>
```

修改 rangeTwo.jsp 页面，代码如下：

```
<h3>request: <%=request.getAttribute("req") %></h3>
<h3>session: <%=session.getAttribute("ses") %></h3>
```

运行访问 rangeOne.jsp 页面，效果如图 2-21 所示。

使用 response 对象将页面重定向到 rangeTwo.jsp，在 rangeTwo.jsp 中能够读取 session 对象，由此可见，session 范围内的对象在会话有效期内可以访问，使用 response.sendRedirect()方法重定向到另外一个页面时，相当于重新发起一次请求，而上一次请求中的 request 对象则随之消失。

图 2-21 session 范围演示

2.3.4 application 范围

相对于 session 范围针对一个会话，application 范围则面对整个 Web 应用程序，即当服务器启动后就会创建一个 application 对象，被所有用户所共享。当具有 application 范围的对象被设置值后，在 Web 应用程序的运行期间，所有的页面都可以访问 application 范围内的对象，其范围最大。与前面几个对象类似，application 对象也具有 setAttribute()方法和 getAttribute()方法，用于对该范围内的对象进行存储访问。

修改 rangeOne.jsp 页面，代码如下：

```jsp
<%
    String ses="sessionRange";
    String app="applicationRange";
    session.setAttribute("ses", ses);
    application.setAttribute("app",app);
    response.sendRedirect("rangeTwo.jsp");
%>
```

修改 rangeTwo.jsp 页面，代码如下：

```jsp
<h3>session: <%=session.getAttribute("ses") %></h3>
<h3>application: <%=application.getAttribute("app") %></h3>
```

先运行 rangeOne.jsp 页面，再运行 rangeTwo.jsp 页面，效果如图 2-22 所示。

这时，关闭浏览器再次运行 rangeTwo.jsp，效果如图 2-23 所示。

图 2-22 application 范围演示(1)   图 2-23 application 范围演示(2)

由于 session 范围针对一个会话，当浏览器关闭后会话也随之结束，所以无法读取，而 application 范围针对整个系统的服务，因而数据可以被再次读取。

2.4 在 JSP 中使用 JavaBean

JavaBean 既是一种基于 Java 平台的软件组件，也是一种独立于平台和结构的应用程序编程接口(API)。JSP 搭配 JavaBean 的组合已经成为常见的 JSP 程序标准，广泛应用于各类 JSP 应用程序中。

2.4.1 为什么需要 JavaBean

Java 企业应用是基于组件开发的，就好像我们用积木可以搭建不同的造型。在程序中，Java 是一种面向对象的编程语言，在设计和解决问题时，都是以面向对象的思想进行的。比如，数据库连接类，在这个类中定义了连接方法和关闭方法，对这个类来说，它的使命就是建立连接和关闭连接，是程序的一个组成部分。在 JSP 中调用 JavaBean，有如下两个优点。

(1) 提高代码的可复用性。

对于通常使用的业务逻辑代码，如数据运算和处理、数据库操作等，可以封装到 JavaBean 中。在 JSP 文件中可以多次调用 JavaBean 中的方法来实现快速的程序开发。

(2) 将 HTML 代码和 Java 代码分离，使程序有利于开发维护。

将业务逻辑进行封装，使得业务逻辑代码和显示代码相分离，不会互相干扰，避免了代码又多又复杂的问题，方便了日后的维护。

2.4.2 什么是 JavaBean

JavaBean 是 Java 中开发的可以跨平台的重要组件。JavaBean 在服务器端的应用中表现出强大的生命力，在 JSP 程序中常用来封装业务逻辑、数据库操作等。JavaBean 的本质就是一个 Java 类，只不过这个 Java 类要遵循一些编码的约定。

JavaBean 实际上就是 Java 类，这个类可以重用，从功能上可以分为以下两类：封装数据和封装业务。JavaBean 一般情况下须满足以下要求。

(1) JavaBean 是一个公有类，提供无参的公有的构造方法。

(2) JavaBean 的属性是私有的。

(3) 提供具有公有的访问属性的 getter 和 setter 方法。

符合上述条件的类，我们都可以将其看成 JavaBean 组件。在程序中，开发人员所要处理的无非是业务逻辑和数据，而这两种操作都可以使用 JavaBean 组件。一个应用程序中会使用很多 JavaBean。由此可见，JavaBean 组件是应用程序的重要组成部分。

2.4.3 封装数据和业务

1. 封装数据

来看一个简单的 JavaBean，通过 MyEclipse 集成开发工具，在 restaurant 项目的 src 下创建 com.restaurant.bean 包，在包中创建 Admin 类，类的属性如下：

```
package com.restaurant.bean;
public class Admin {
    private Integer id;              //管理员 id
    private String loginName;        //登录名称
    private String loginPwd;         //登录密码
}
```

在 Admin 类中，可以添加相应的无参数的构造方法。MyEclipse 提供了一个方便快捷的生成 getter 和 setter 的方法，在相应的代码区，选择 source→ Generate Getters and Setters…，在对话框中，选择相应的属性，单击 OK 按钮即可自动添加相应的 getter 和 setter 方法，代码如下：

```
package com.restaurant.bean;
public class Admin {
    private Integer id;              //管理员 id
    private String loginName;        //登录名
    private String loginPwd;         //登录密码
    //无参的构造方法
    public Admin() {
    }
    //添加相应属性的 setter 方法和 getter 方法
    public Integer getId() {
        return id;
    }
    public void setId(Integer id) {
        this.id = id;
    }
    public String getLoginName() {
        return loginName;
    }
    public void setLoginName(String loginName) {
        this.loginName = loginName;
    }
    public String getLoginPwd() {
        return loginPwd;
    }
    public void setLoginPwd(String loginPwd) {
        this.loginPwd = loginPwd;
    }
}
```

这是一个典型的封装数据的 JavaBean。这个 JavaBean 封装了管理员用户的数据，如 id、loginName、LoginPwd 等属性，外部通过 getter/setter 方法可以对这些属性进行操作。

2．封装业务

在编写程序时，一个封装数据的 JavaBean 一般情况下对应着数据库内的一张表(或视图)，JavaBean 的属性与表(或视图)内字段的属性一一对应。同样，相对于一个封装数据的 JavaBean，一般都会有一个封装该类的业务逻辑和业务操作的 JavaBean 相对应。

若与封装数据的 Admin.java 相对应的封装业务 JavaBean 是 AdminControl.java，那么可以在 AdminControl.java 中编写有关管理员表操作的方法。例如，获取 admin 表中的最大编号，结构代码如下。

```java
package com.restaurant.bean;
public class AdminControl {
    // 获取最大的 Id 号，此方法并未实现
    public int getMaxId(){
        int num=0;
        String sql="select max(id) from admin";
        // 省略访问数据库的实现，num 得到最大 ID 号
        return num;
    }
}
```

2.4.4 JSP 与 JavaBean

现在已经掌握了如何创建封装数据的 JavaBean 和封装业务逻辑的 JavaBean，那么在 JSP 页面中如何使用 JavaBean？在 JSP 页面中，可以像使用普通类一样实例化一个 JavaBean 对象，调用它的方法。在项目的 WebRoot/ch02/目录下，新建 showJavaBean.jsp 页面，在 JSP 中引入并使用 JavaBean，代码如下：

```jsp
<%@ page import="com.restaurant.bean.*" %>
<% //使用 JavaBean
AdminControl ac=new AdminControl();
Admin admin=new Admin();
admin.setId(ac.getMaxId()+1);
admin.setLoginName("yzpc");
admin.setLoginPwd("yzpc");
%>
```

在 JSP 中使用 JavaBean 就像在 Java 程序中编写类一样，实例化 JavaBean 后，就可以使用其中的方法了。

2.5 EL 表达式

在 JSP 页面中，为了实现与用户的动态交互，或者控制页面输出，需要在 JSP 页面中嵌入很多 Java 代码，这样不利于对页面的维护和更新，因此 JSP 2.0 引入了 EL 表达式。

2.5.1 EL 表达式概述

EL 的全称是 Expression Language，它是借鉴了 JavaScript 和 XPath 的表达式语言。EL 定义了一系列隐含对象和操作符，使开发人员能够很方便地访问页面的上下文，以及不同作用域内的对象，而无须在 JSP 页面中嵌入 Java 代码，从而使开发人员即使不懂 Java 也能轻松编写 JSP 程序。

EL 表达式提供了在 Java 代码之外，访问和处理应用程序数据的功能，通常用于在某个作用域(page、request、session、application 等)内取得属性值，或者做简单的运算和判断。EL 表达式有如下特点。

(1) 自动类型转换。EL 借鉴了 JavaScript 多类型转换无关性的特点，在使用 EL 得到某个数据时可以自动进行类型转换，因此对于类型的限制更加轻松。

(2) 使用简单。与在 JSP 页面中嵌入 Java 代码相比，EL 表达式使用起来非常简单。

2.5.2 EL 表达式的使用

EL 表达式以"${"开始，以"}"结束，语法如下：

```
${ EL expression }
```

在 showJavaBean.jsp 页面的小脚本中，使用 session 封装管理员对象，代码如下：

```
<% session.setAttribute("admin", admin);
   response.sendRedirect("showEL.jsp");
%>
```

在 restaurant 项目的 WebRoot/ch02/目录下，新建 showEL.jsp 页面。在 showEL.jsp 页面中使用传统方法取得管理员的名称，代码如下：

```
<% Admin admin = (Admin)session.getAttribute("admin");
   String loginName = admin.getLoginName();
%> 用户名：<%=loginName %><br>
```

也可在 showEL.jsp 页面中使用 EL 表达式取得管理员的名称，等价的写法如下：

```
${sessionScope.admin.loginName}
```

两者相比，可以发现，EL 的语法比传统的 JSP 代码更为方便、简洁。EL 提供"."和"[]"两种操作符来存取数据。

(1) 点操作符。

EL 表达式通常由两个部分组成：对象和属性。就像在 Java 代码中一样，在 EL 表达式中也可以用点操作符访问对象的某个属性，例如，通过${sessionScope.admin.loginName}可以访问 admin 对象的 loginName 属性。

(2) "[]"操作符。

与点操作符类似，"[]"操作符也可以访问对象的某个属性，例如，${sessionScope["admin"] ["loginName"] }可以访问管理员对象的登录名称属性。

除此之外，"[]"操作符还提供了更加强大的功能。

- 在属性名中包含特殊字符如"."或"—"等的情况下，就不能使用点操作符来访问，而只能使用"[]"操作符。
- 访问数组，如果有一个对象名为 array 的数组，那么我们可以根据索引值来访问其中的元素，如${array[0]}、${array[1]}等。

在 showEL.jsp 页面中，分别调用 EL 表达式的两种操作符进行国家和城市的输出显示，代码如下：

```
使用EL的"."取得用户名：${sessionScope.ADMIN.loginName } <br>
使用EL的"[]"取得用户名：${sessionScope["ADMIN"]["loginName"] }<br>
<%
   Map countries=new HashMap();        // 定义集合 Map
```

```
            countries.put("CN", "China");
            countries.put("RU", "Russia");
            request.setAttribute("countries", countries);
            List cities=new ArrayList();          // 定义集合 List
            cities.add(0,"BeiJing");
            cities.add("ShangHai");
            request.setAttribute("cities", cities);
        %>
        国家：${countries.CN }<br>
                --城市：${cities[0] }<br>
                --城市：${cities[1] }<br>
        国家：${countries.RU}<br>
```

重新部署项目，在浏览器中访问 http://localhost:8080/restaurant/ch02/showJavaBean.jsp，然后跳转到 showEL.jsp 页面，效果如图 2-24 所示。

图 2-24　EL 表达式操作符的使用

2.5.3　EL 隐式对象

JSP 提供了 page、request、session、application、pageContext 等若干隐式对象。这些隐式对象无须声明，就可以很方便地在 JSP 页面脚本中使用。EL 隐式对象可以分为五类，如表 2-8 所示。

表 2-8　EL 隐式对象的分类

类　别	对象标识符	作　用
JSP 隐式对象	pageContext	提供对页面信息和 JSP 内置对象的访问
作用域访问对象	pageScope	与页面作用域 page 中的属性相关联的 Map 类
	requestScope	与请求作用域 request 中的属性相关联的 Map 类
	sessionScope	与会话作用域 session 中的属性相关联的 Map 类
	applicationScope	与应用程序作用域 application 中的属性相关联的 Map 类
参数访问对象	param	按照参数名称访问单一请求值的 Map 对象
	paramValues	按照参数名称访问数组请求值的 Map 对象
请求头访问对象	header	与请求头名称相对应的字符串的 Map 集合
	headerValues	与请求头名称相对应的字符串数组的 Map 集合
	cookie	所有 cookie 组成的 Map 集合
初始化参数对象	initParam	Web 应用程序上下文初始化参数的 Map 集合

1．JSP 隐式对象

为了能够方便地访问 JSP 隐式对象，EL 表达式引入了 pageContext，它是 JSP 和 EL 的一个公共对象，通过 pageContext 可以访问其他 JSP 内置对象(request、response 等)，这也是 EL 表达式语言把它作为内置对象的一个主要原因。

2．作用域访问对象

在 JSP 页面中定义和设置一个变量，同时指定该变量的作用域，作用域共有四个选项：page、request、session 和 application。在 EL 表达式中，为了访问这四个作用域内的变量和属性，提供了 pageScope、requestScope、sessionScope、applicationScope 这四个作用域访问对象。当使用 EL 表达式访问某个属性时，应该指定查找的范围，如${requestScope.admin}，即在请求(request)范围内查找属性 admin 的值。如果程序中不指定查找范围，则系统会按照 page→request→session→application 的顺序查找。

3．参数访问对象

参数访问对象是与页面输入参数有关的隐式对象，包含 param 和 paramValues 两个对象，通过它可以得到用户的请求参数。两者的不同之处在于，param 对象用于得到请求中单一名称的参数，而 paramValues 对象用于得到请求中的多个值。例如，用户注册时，通常只要填写一个用户名，但可以选择多个兴趣爱好，用户名可以通过 ${param.username}来访问用户名；通过 ${paramValues.habits }可以得到用户所选择的兴趣爱好。

4．请求头访问对象

请求头访问对象用于访问 HTTP 请求头，header 储存用户浏览器和服务器端用来沟通的数据。例如，要取得用户浏览器的版本，可以使用${header["User-Agent"]}。另外在极少情况下，有可能同一标头名称拥有不同的值，此时必须改为使用 headerValues 来取得这些值。要取得 cookie 中设定名称为 userCountry 的值，可以使用${cookie.userCountry}。

5．初始化参数对象

初始化参数对象 initParam，这个映射可用于访问初始化参数的值，初始化参数的值一般都在 web.xml 中设置。例如，一般的方法"String userid = (String)application.getInitParameter("userid");"也可以使用 ${initParam.userid}来取得 userid。

2.6 JSTL 标签

通过 EL 表达式，在一定程度上简化了 JSP 页面开发的复杂度。但 EL 表达式不能实现复杂业务逻辑的处理，业务逻辑的处理还是要通过脚本的方式来实现。JSTL 标签可以不用在 JSP 页面中嵌入 Java 代码，又能在 JSP 中控制程序流程。JSTL 主要提供了五大类标签库：核心标签库、格式标签库、SQL 标签库、XML 标签库和函数标签库。

2.6.1 JSTL 标签概述

JSTL(JavaServer Pages Standard Tag Library，JSP 标准标签库)包含在 JSP 开发中经常用到的一组标签库，这些标签为我们提供了一种不用嵌入 Java 代码，就可以开发复杂 JSP 页面的途径。使用 JSTL 标签库是为了弥补 html 标签的不足和规范自定义标签。使用 JSLT 标签的目的就是不希望在 JSP 页面中出现 Java 逻辑代码。

JSTL 是由 Sun 公司推出、由 Apache Jakarta 组织负责维护的用于编写和开发 JSP 页面的一组标准标签。作为开源的标准技术，它一直在不断地完善。JSTL 标签库包含各种标签，如通用标签、条件判断标签、迭代标签等。

2.6.2 JSTL 标签的使用

1．在工程中引用 JSTL 的 jar 包

在工程中引用 1.1 版本的 JSTL，要添加 jstl.jar 和 standard.jar 两个包，放置到项目的 WebRoot/WEB-INF/lib 目录下即可。

在 MyEclipse 集成开发环境中已经集成了 JSTL，选择 File→New→Web Project 命令，在弹出的 New Web Project 对话框中，在 JSTL Version 下拉列表框中选择 1.2.2 最高版本，如图 2-25 所示。

其余步骤与第 1 章的项目创建一致，最后完成后，MyEclipse 会自动在项目中添加相应版本所需的 jar 包和标签库描述文件。

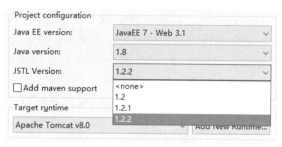

图 2-25　项目中添加 JSTL

2．在需要使用 JSTL 的 JSP 页面中使用 taglib 指令导入标签库描述文件

如果需要使用 JSTL 核心标签库，需要在页面上方增加如下一行指令：

```
<%@ taglib uri="http://java.sun.com/jsp/jstl/core" prefix="c" %>
```

如果需要添加 SQL 标签库、格式标签库、XML 标签库和函数标签库，使用的指令如下：

```
<%@ taglib uri="http://java.sun.com/jsp/jstl/sql" prefix="sql" %>
<%@ taglib uri="http://java.sun.com/jsp/jstl/fmt" prefix="fmt" %>
<%@ taglib uri="http://java.sun.com/jsp/jstl/xml" prefix="x" %>
<%@ taglib uri="http://java.sun.com/jsp/jstl/functions" prefix="fn" %>
```

完成以上步骤，就可以用 JSTL 方便地开发 JSP 页面，而无须嵌入 Java 代码了。

2.6.3 JSTL 核心标签库

核心标签库在 JSTL 中占有十分重要的地位，该标签库的工作是对 JSP 页面一般处理的封装。使用这些标签能够实现 JSP 页面的基本功能，减少编码工作。核心标签库按功能的不同又分为通用标签库、条件标签库、迭代标签库等。

1．通用标签库

通用标签用于在 JSP 页面内设置、删除和显示变量，包含三个标签：<c:set>、<c:out>和<c:remove>。

1) <c:set>标签

<c:set>标签用于定义变量,并将变量存储在 JSP 范围中或者 JavaBean 属性中,其语法格式有如下两种。

(1) 将 value 值存储到范围为 scope 的变量 variable 中,语法格式如下:

```
<c:set var="variable" value="v" scope="scope" />
```

- var 属性的值是设置的变量名。
- value 属性的值是赋予变量的值。
- scope 属性对应的是变量的作用域,可选值有 page、request、session、application。

(2) 将 value 值存储到 target 对象的属性中,语法格式如下:

```
<c:set value="value" target="target" property="property" />
```

- target 属性是操作的对象,可以使用 EL 表达式表示。
- property 属性对应对象的属性名。
- value 属性是赋予对象属性的值。

2) <c:out>标签

<c:out>标签用来显示数据的内容,类似 JSP 中的表达式。但功能更加强大,代码也更加简洁,方便页面维护。语法格式分为指定默认值和不指定默认值两种形式。

(1) 不指定默认值,语法格式如下:

```
<c:out value="value" />
```

value 属性是指需要输出的值,可以用 EL 表达式输出某个变量。

(2) 指定默认值,语法格式如下:

```
<c:out value="value" default="default" />
```

default 属性是 value 属性的值为空时,输出默认的值。

3) <c:remove>标签

<c:remove>标签用于移除指定范围内的变量,作用与<c:set>标签相反,语法格式如下:

```
<c:remove var="value" scope="scope" />
```

- var 属性是指待删除的变量的名称。
- scope 属性是指删除的变量所在的范围,可选值有 page、request、session、application,如果没有指定,则默认为 page。

下面通过示例,从语法的角度看一下如何在 JSP 中应用 JSTL 通用标签。在 restaurant 项目的 WebRoot/ch02/目录下,新建 jstlGeneral.jsp 页面,代码如下:

```
<%@ page language="java" import="java.util.*" pageEncoding="UTF-8"%>
<%@ taglib uri="http://java.sun.com/jsp/jstl/core" prefix="c"%>
<!DOCTYPE HTML PUBLIC "-//W3C//DTD HTML 4.01 Transitional//EN">
<html>
  <head>
    <title>JSTL 通用标签库使用</title>
  </head>
  <body>
```

```
    设置变量之前的 message 值：<c:out value="${message }" default="null" /> <br>
    <c:set var="message" value="Hello World!" scope="page"></c:set>
    设置新值以后的 message 值：<c:out value="${message }"></c:out> <br>
    <c:remove var="message" scope="page" />
    移除变量 message 以后的值：<c:out value="${message }" default="null" />
  </body>
</html>
```

在该示例中，首先使用<c:set>标签在 page 范围内设置一个变量的值，通过<c:out>标签把该变量显示在页面上，然后用<c:remove>标签在 page 范围内删除该变量，并使用<c:out>标签检查该变量是否已经删除。运行页面，显示效果如图 2-26 所示。

图 2-26 使用 JSTL 设置变量

2．条件标签库

对于包含动态内容的 Web 页面，若希望不同类别的用户看到不同形式的内容，就需要用到 JSTL 的另外一个常用标签，即条件标签。

1）<c:if>标签

<c:if>标签用来执行流程的控制，其功能和语言中的 if 完全相同，其语法格式如下：

```
<c:if test="condition" var="varName" scope="scope">
    // 本部分的内容
</c:if>
```

- test 属性是此条件标签的判断条件，当 test 中表达式的结果为 true 时，会执行本部分的内容，如果为 false 则不会执行。
- var 属性定义变量，该变量存放判断以后的结果，该属性可以省略。
- scope 属性是指 var 定义变量的存储范围，可选值有 page、request、session、application，该属性可以省略。

在 restaurant 项目的 WebRoot/ch02/目录下，新建 jstlCondition.jsp 和 doJstlCondition.jsp 页面，实现登录和处理。

jstlCondition.jsp 页面主要是用户登录表单，使用 jstl 的条件标签，其代码如下：

```
<%@ page language="java" import="java.util.*" pageEncoding="UTF-8"%>
<%@ taglib uri="http://java.sun.com/jsp/jstl/core" prefix="c"%>
<!DOCTYPE HTML PUBLIC "-//W3C//DTD HTML 4.01 Transitional//EN">
<html>
  <head>
    <title>JSTL 条件标签库使用</title>
  </head>
  <body>
    <c:set var="isLogin" value="${not empty sessionScope.LOGIN_NAME}"/>
    <c:if test="${not isLogin }">
    <form name="form1" method="post" action="doJstlCondition.jsp">
      用户名：<input type="text" name="username" id="username">
      密码：<input type="password" name="password" id="password">
```

```
        <input type="submit" value="登录">
    </form>
    </c:if>
    <c:if test="${isLogin }">
        <c:out value="${sessionScope.LOGIN_NAME}"/>已经登录!
    </c:if>
 </body>
</html>
```

doJstlCondition.jsp 页面主要使用登录请求处理并使用 session 对象设置封装用户名,其代码如下:

```
<%
    request.setCharacterEncoding("utf-8");
    String username = request.getParameter("username");
    String password = request.getParameter("password");
    if(username.equals("admin") && password.equals("admin")){
        //此处暂不访问数据库
        session.setAttribute("LOGIN_NAME", username);//设置登录信息
        response.sendRedirect("jstlCondition.jsp");
    }else{
        out.print("用户名或密码错误!");
    }
%>
```

重新部署项目,访问 jstlCondition.jsp 页面,如果用户尚未登录,则显示登录页面,如图 2-27 所示。

图 2-27　使用<c:if>判断是否登录(1)

如果用户已登录,则显示登录的用户名,如图 2-28 所示。

图 2-28　使用<c:if>判断是否登录(2)

2)　<c:choose>标签

<c:choose>和<c:when>、<c:otherwise>一起实现互斥条件执行,类似于 Java 中的 if-else。<c:choose>一般作为<c:when>、<c:otherwise>的父标签,示例代码如下:

```
<c:choose>
    <c:when test="${row.v_money<10000 }">
        初级者
    </c:when>
```

```
        <c:when test="${row.v_money>=10000 && row.v_money<20000 }">
            中级者
        </c:when>
        <c:otherwise>
            高级者
        </c:otherwise>
</c:choose>
```

3. 迭代标签库

在 JSP 中，迭代是经常需要用到的操作，主要有两种：<c:forEach>和<c:forTokens>。

1) <c:forEach>标签

通过 JSTL 的<c:forEach>标签，能在很大程度上简化迭代操作，<c:forEach>标签的语法格式如下：

```
<c:forEach var="varName" items="collection" varStatus="statusName" begin="beginIndex" end="endIndex" step="step">
            // …
</c:forEach>
```

- var 属性是对当前成员的引用。即如果当前循环到第一个成员，那么 var 就引用第一个成员；如果当前循环到第二个成员，它就引用第二个成员；以此类推。
- items 指被迭代的集合对象。
- varStatus 属性用于存放 var 引用的成员的相关信息，如索引等。
- begin 属性表示开始位置，默认为 0，该属性可以省略。
- end 属性表示结束位置，该属性可以省略。
- step 属性表示循环的步长，默认为 1，该属性可以省略。

下面以一个管理员信息的显示为例，来体会一下迭代标签给我们带来的便利之处。在 restaurant 项目的 WebRoot/ch02/目录下，创建 jstlIterate.jsp 页面，代码如下：

```
<%@ page language="java" import="java.util.*" pageEncoding="UTF-8"%>
<%@ taglib uri="http://java.sun.com/jsp/jstl/core" prefix="c"%>
<%@page import="com.restaurant.bean.Admin"%>
<%    List adminList=new ArrayList();
    Admin admin=null;        // 2.4.3 小节已创建 Admin 类，此处直接使用
    for (int i = 0; i < 10; i++) {
        admin=new Admin(i+1,"管理员"+(i+1),"密码"+(i+1));
        adminList.add(admin);
    }
    request.setAttribute("ADMINLIST", adminList);
%>
<!DOCTYPE HTML PUBLIC "-//W3C//DTD HTML 4.01 Transitional//EN">
<html>
  <head>    <title>JSTL 迭代标签的使用</title>    </head>
  <body>
    <table border="1" width="80%" align="center">
        <tr>
            <td>ID 号</td>    <td>用户名</td>    <td>密码</td>
        </tr>
        <c:forEach var="admin" items="${requestScope.ADMINLIST }" varStatus="status">    <!-- 循环输出管理员信息 -->
```

```
        <tr <c:if test="${status.index % 2 ==1 }">style=
"background-color:yellow;"</c:if>> <!-- 若为偶数行,更换背景颜色 -->
            <td>${admin.id }</td>
            <td>${admin.loginName }</td>
            <td>${admin.loginPwd }</td>
        </tr>
        </c:forEach>
    </table>
  </body>
</html>
```

该示例的运行效果如图 2-29 所示。

2) <c:forTokens>标签

<c:forTokens>标签专门用于处理字符串的迭代,可以指定一个或多个分隔符,语法格式如下:

图 2-29 <c:forEach>迭代标签示例

```
<c:forTokens items="stringOfTokens" delims="delimiters" var="varName"
begin="begin" end="end" step="step" varStatus="varStatusName">
    相应内容
</c:forTokens>
```

在 jstlIterate.jsp 页面后面添加示例代码,具体如下:

```
<c:forTokens items="China,Russia,France" delims="," var="item">
   <c:out value="${item }"/> <br>
</c:forTokens>
```

2.7 小 结

本章主要讲解了 JSP 技术,包括 JSP 页面组成,JSP 内置对象,对象范围,在 JSP 中使用 JavaBean、EL 和 JSTL 等。通过学习和掌握 JSP 的主要内容,读者能够掌握动态网站的相应开发技术。

第 3 章
Servlet 技术

通过第 2 章的学习,已经了解了 JSP 技术的体系结构和技术内容等知识。在 Internet 上,客户端通过使用 HTTP 协议,向服务器端发送请求信息,服务器对请求数据进行处理,并把处理后的结果反馈给客户端。这些请求和反馈是怎么实现的呢?Servlet 是什么技术,能解决哪些问题?本章将讲解 Servlet 的相关技术。

3.1 Servlet 简介

Servlet 是基于 Java 技术的 Web 组件，它是 JSP 组件的前身，是 Java Web 开发技术的基础和核心组件。

3.1.1 什么是 Servlet

使用 JSP 开发 Web 程序的时候，主要是在 JSP 中写入 Java 代码，当服务器运行 JSP 页面时，执行 Java 代码，动态获取数据，并生成 HTML 代码，最终显示在客户端浏览器上，整个过程如图 3-1 所示。

在 JSP 出现之前，如果想动态生成 HTML 页面，那就只有在服务器端运行 Java 程序，并生成 HTML 格式的内容。运行在服务器端的 Java 程序就是 Servlet，过程如图 3-2 所示。

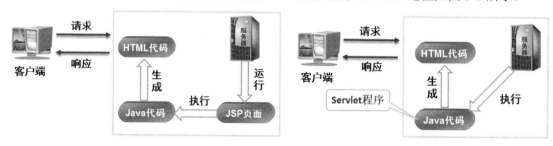

图 3-1　使用 JSP 开发 Web 程序　　　　　图 3-2　使用 Servlet 开发 Web 程序

Servlet 是一个符合特定规范的 Java 程序，运行在服务器端，用于处理客户端请求并做出响应，如图 3-3 所示。

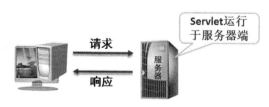

图 3-3　Servlet 运行于服务器端

尽管 Servlet 能够响应任何类型的请求，但在绝大多数的网络应用中，都是客户端通过 HTTP 协议访问服务器端的资源，而编写的 Servlet 也是应用于 HTTP 协议的请求和响应，讲解的重点也就放在和这方面有关的 HttpServlet 类上。

Servlet 具有简单实用的 API 方法、高效率、功能强大、可移植性等特点。

3.1.2 编写第一个 Servlet

了解了 Servlet 的功能和特点，也知道了 Servlet 的定义，那么 Servlet 到底是什么样子呢？符合哪些规范的 Java 程序才是 Servlet 呢？

下面来认识一下 Servlet。首先在 restaurant 项目的 src 中，创建包 com.restaurant.servlet，在 com.restaurant.servlet 包中创建 HelloServletTest.java 的 Servlet 文件，代码如下：

```java
package com.restaurant.servlet;
import java.io.*;
import javax.servlet.*;
import javax.servlet.http.*;
public class HelloServletTest extends HttpServlet {
    public void doGet(HttpServletRequest request, HttpServletResponse response) throws ServletException, IOException {
        response.setContentType("text/html;charset=utf-8");
        PrintWriter out = response.getWriter();
        out.println("<!DOCTYPE HTML PUBLIC \"-//W3C//DTD HTML 4.01 Transitional//EN\">");
        out.println("<HTML>");
        out.println("  <HEAD><TITLE>A Servlet</TITLE></HEAD>");
        out.println("  <BODY>");
        out.print("您好，欢迎来到Servlet的世界！");
        out.println("  </BODY>");
        out.println("</HTML>");
        out.flush();
        out.close();
    }
    public void doPost(HttpServletRequest request, HttpServletResponse response) throws ServletException, IOException {
        doGet(request, response);
    }
}
```

大家可以发现，创建好的 Servlet 已经有了大体的结构，我们只需要简单地修改。需要强调几点：第一，在调用 Servlet 时，首先要在程序中导入 Servlet 所需的包；第二，创建用于 Web 应用的 Servlet 继承自 HttpServlet 类；第三，实现 doGet()或者 doPost()方法。

那么如何访问 Servlet 呢？我们还需要在 restaurant 项目的 WebRoot/web-inf/路径下的 web.xml 文件中进行配置。创建 Servlet 时，MyEclipse 已经为我们在 web.xml 文件中自动添加好配置，配置如下：

```xml
<?xml version="1.0" encoding="UTF-8"?>
<web-app xmlns:xsi="http://www.w3.org/2001/XMLSchema-instance"
xmlns="http://xmlns.jcp.org/xml/ns/javaee"
xsi:schemaLocation="http://xmlns.jcp.org/xml/ns/javaee
http://xmlns.jcp.org/xml/ns/javaee/web-app_3_1.xsd" id="WebApp_ID"
version="3.1">
  <display-name>restaurant</display-name>
  <servlet>
    <servlet-name>HelloServletTest</servlet-name>
    <servlet-class>com.restaurant.servlet.HelloServletTest
    </servlet-class>
  </servlet>
  <servlet-mapping>
    <servlet-name>HelloServletTest</servlet-name>
    <url-pattern>/servlet/HelloServletTest</url-pattern>
  </servlet-mapping>
```

```
</web-app>
```

在 web.xml 配置文件中，<servlet-mapping>节点就是 Servlet 的映射，而<url-pattern>节点则给出了 Web 访问此 Servlet 的 URL 地址。

部署项目，在浏览器地址栏中输入 http://localhost:8080/restaurant/servlet/HelloServletTest，运行效果如图 3-4 所示。

图 3-4　第一个 Servlet 程序运行效果

3.1.3　Servlet 与 JSP 的关系

Servlet 和 JSP 都可以在页面上动态显示数据内容，那么它们之间存在怎样的关系呢？在 restaurant 项目的 WebRoot 路径下创建 MyJsp.jsp 文件，内容如下：

```
<%@ page language="java" import="java.util.*" pageEncoding="UTF-8"%>
<!DOCTYPE HTML PUBLIC "-//W3C//DTD HTML 4.01 Transitional//EN">
<html>
  <head>
    <title>MyJSP </title>
  </head>
  <body>
    This is my JSP page. <br>
  </body>
</html>
```

部署项目，并在浏览器中运行 MyJsp.jsp 后，在 Tomcat 的安装目录下的\work\Catalina\localhost\restaurant\org\apache\jsp 下会生成一个 MyJsp_jsp.java 文件，主要内容如下：

```
public final class MyJsp_jsp extends
org.apache.jasper.runtime.HttpJspBase ...{
    // 省略中间代码
  public void _jspService(final javax.servlet.http.HttpServletRequest request, final javax.servlet.http.HttpServletResponse response)
        throws java.io.IOException, javax.servlet.ServletException {
    //省略定义的其他变量
    javax.servlet.jsp.JspWriter _jspx_out = null;
    try {
      response.setContentType("text/html;charset=UTF-8");
      out = pageContext.getOut();
      out.write("\r\n");
      out.write("<!DOCTYPE HTML PUBLIC \"-//W3C//DTD HTML 4.01 Transitional//EN\">\r\n");
      out.write("<html>\r\n");
      out.write("  <head>\r\n");
      out.write("    <title>My JSP</title>\r\n");
      out.write("  </head>  \r\n");
      out.write("  <body>\r\n");
      out.write("    This is my JSP page. <br>\r\n");
      out.write("  </body>\r\n");
      out.write("</html>\r\n");
    } catch (java.lang.Throwable t) {
```

```
            // 捕获异常
        }   // 进行其他处理
    }
}
```

从示例中可以看出，MyJsp 在运行时首先解析成一个 Java 类 MyJsp_jsp.java，该类继承自 org.apache.jasper.runtime.HttpJspBase 类，而 HttpJspBase 类又继承自 HttpServlet 的类，因此我们可以得出一个结论，就是 JSP 在运行时会被 Web 容器翻译为一个 Servlet。

3.2　Servlet 的生命周期

为了在应用程序中更好地使用 Servlet，接下来了解 Servlet 的生命周期。所谓的生命周期，就是 Servlet 从创建到销毁的过程，包括如何加载和实例化、初始化、处理请求和如何被销毁。

1．加载和实例化

Servlet 容器负责加载和实例化 Servlet。当客户端发送一个请求时，Servlet 容器会查找内存中是否存在该 Servlet 的实例，如果不存在，就创建一个 Servlet 实例。如果存在该 Servlet 的实例，就直接从内存中读取出该实例来响应请求。

Servlet 类的加载是在 Servlet 被第一次请求时执行，主要就是将 Servlet 对应的 class 字节码文件载入内存，该阶段仅执行一次。

实例化 Servlet 是 Servlet 容器创建 ServletConfig 对象，ServletConfig 对象包含 Servlet 的初始化配置信息，此外 Servlet 容器还会使得 ServletConfig 对象与当前 Web 应用的 ServletContext 对象关联。Servlet 容器根据 Servlet 类的位置加载 Servlet 类，成功加载后，由容器创建 Servlet 的实例。

2．初始化

在 Servlet 容器完成 Servlet 实例化后，Servlet 容器将调用 Servlet 的 init()方法进行初始化，初始化的目的是让 Servlet 对象在处理客户端请求前完成一些初始化工作。例如：设置数据库连接参数，建立 JDBC 连接，或者是建立对其他资源的引用。init() 方法在 javax.servlet.Servlet 接口中定义。对于每一个 Servlet 实例，init()方法只被调用一次。

3．服务

Servlet 被初始化以后，就处于能响应请求的就绪状态。当 Servlet 容器接收到客户端请求时，调用 Servlet 的 service()方法处理客户端请求。Servlet 实例通过 ServletRequest 对象获得客户端的请求。通过调用 ServletResponse 对象的方法设置响应信息。该阶段客户端请求一次执行一次，具体执行几次，取决于客户端的请求次数。

4．销毁

Servlet 的实例是由 Servlet 容器创建的，所以实例的销毁也是由容器来完成的。Servlet 容器判断一个 Servlet 实例是否应当被释放时(容器关闭或需要回收资源)，容器就会调用 Servlet 的 destroy()方法，destroy()方法用于指明哪些资源可以被系统回收，而不是由 destroy()方法直接进行回收。

Servlet 的生命周期过程和相应的方法如图 3-5 所示。

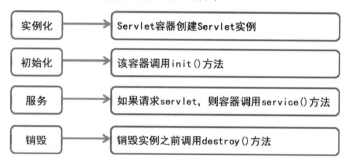

图 3-5　Servlet 的生命周期

在 Servlet 的生命周期中，Servlet 的加载、实例化和销毁只会发生一次，因此 init()和 destroy()方法只能被 Servlet 容器调用一次，而 Service()方法的执行次数取决于 Servlet 被客户端访问的次数。

为了使读者对 Servlet 的生命周期有一个深入的理解，下面来看一个有关 Servlet 的生命周期的实例。在 restaurant 项目中的 src 目录下的 com.restaurant.servlet 包中，创建 LifeServlet.java 的 Servlet 文件，程序代码如下：

```java
package com.restaurant.servlet;
import java.io.IOException;
import javax.servlet.ServletException;
import javax.servlet.http.*;
public class LifeServlet extends HttpServlet {
    // 构造方法
    public LifeServlet() {
        super();
        System.out.println("实例化时，LifeServlet()构造方法被调用");
    }
    // 初始化方法
    public void init() throws ServletException {
        System.out.println("初始化时，init()方法被调用");
    }
    // doGet()方法
    public void doGet(HttpServletRequest request, HttpServletResponse response) throws ServletException, IOException {
        System.out.println("处理请求时，doGet()方法被调用");
    }
    // doPost()方法
    public void doPost(HttpServletRequest request, HttpServletResponse response) throws ServletException, IOException {
        System.out.println("处理请求时，doPost()方法被调用");
    }
    // service 服务方法
    protected void service(HttpServletRequest arg0, HttpServletResponse arg1) throws ServletException, IOException {
        System.out.println("请求 Servlet 时，service()方法被调用");
    }
    // 销毁方法
```

```
public void destroy() {
    super.destroy();
    System.out.println("释放系统资源时，destroy()方法被调用");
}
}
```

运行上面示例的代码，根据 web.xml 中设置的访问 LifeServlet 的 URL，在浏览器地址栏中输入 http://localhost:8080/restaurant/servlet/LifeServlet。因为 LifeServlet 只在控制台进行输出，未对请求给出响应，运行效果如图 3-6 所示。

图 3-6　第一次访问 LifeServlet 结果界面

不管是 post 还是 get 方法提交，都会在 service 中处理，然后，由 service 交由相应的 doPost 或 doGet 方法处理，如果重写了 service 方法，就不会再处理 doPost 或 doGet 了。在重写的 sevice()方法中，可以自己转向 doPost()或 doGet()方法。

紧接着再重新提交一次请求，看一下控制台的运行变化，如图 3-7 所示。

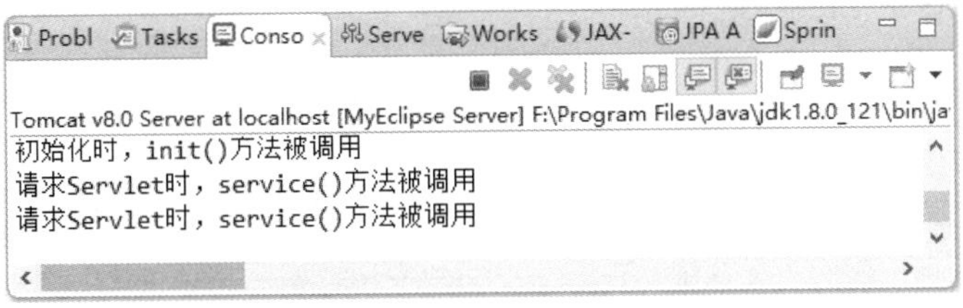

图 3-7　第二次访问 LifeServlet 结果界面

当第二次提交请求时，只是再次执行了 service()方法，Servlet 的 init()方法并没有执行，这说明 init()方法只有在加载当前的 Servlet 时被执行，并且只被执行一次，以后不再执行。

那么，destroy()方法什么时候被执行呢？现在停止 Tomcat 服务，再来观察控制台输出的信息，如图 3-8 所示。

图 3-8　Web 服务器停止服务时的结果页面

在服务器停止的时候，或者是系统回收资源时，Servlet 容器会先调用 Web 应用中所有的

Servlet 对象的 destroy()方法，然后再销毁 Servlet 对象。此外容器还会销毁与 Servlet 对象关联的 ServletConfig 对象。

3.3 Servlet 的常用类和接口

使用 Servlet API 可以开发 HTTP Servlet 或其他 Servlet，Servlet API 包含在两个包内。javax.servlet 包中的类和接口支持通用的不依赖协议的 Servlet，包括 Servlet、ServletRequest、ServletResponse、ServletConfig、ServletContext 接口及抽象类 GenericServlet。javax.servlet.http 包中的类和接口是用于支持 HTTP 协议的 Servlet API。

1．Servlet 接口

Servlet 接口定义了 Servlet 需要实现的所有方法，包括 init()、service()、destroy()方法，以及 getServletInfo()和 getServletConfig()方法。Servlet 接口的常用方法如表 3-1 所示。

表 3-1　Servlet 接口的常用方法

方　　法	说　　明
public void init(ServletConfig config)	由 Servlet 容器调用，用于完成 Servlet 对象在处理客户请求前的初始化工作
public void service(ServletRequest req, ServletResponse res)	由 Servlet 容器调用，用来处理客户端请求
public void destroy()	由 Servlet 容器调用，释放 Servlet 对象所使用的资源
public ServletConfig getServletConfig()	返回 ServletConfig 对象，该对象包含此 Servlet 初始化和启动参数
public String getServletInfo()	返回有关 Servlet 的信息，如作者、版本和版权。返回的字符串是纯文本

2．ServletConfig 接口

在 Servlet 初始化时，Servlet 容器使用 ServletConfig 对象向该 Servlet 传递信息。ServletConfig 接口的常用方法如表 3-2 所示。

表 3-2　ServletConfig 接口的常用方法

方　　法	说　　明
public String getServletName()	返回一个 Servlet 实例的名称，该名称由服务器管理员提供
public String getInitParameter(String name)	获取 web.xml 中设置的以 name 命名的初始化参数值
public ServletContext getServletContext()	返回 Servlet 的上下文对象引用

一个 Servlet 只有一个 ServletConfig 对象。

3．GenericServlet 抽象类

抽象类 GenericServlet 实现了 Servlet 接口和 ServletConfig 接口，给出了除 service()方法之外的其他方法的简单实现，它定义了通用的、不依赖于协议的 Servlet。GenericServlet 抽象类的常用方法如表 3-3 所示。

表 3-3　GenericServlet 抽象类的常用方法

方　　法	说　　明
public void init(ServletConfig config)	调用 Servlet 接口中的 init()方法。此方法还有一无参的重载方法，其功能与此相同
public String getInitParameter(String name)	返回名称为 name 的初始化参数的值
public ServletContext getServletContext()	返回 ServletContext 对象的引用

通常只需要重写不带参数的 init()方法，如果重写 init(ServletConfig config)方法，那么应该包含 super.init(config)这句代码。如果要编写一个通用的 Servlet，只要继承自 GenericServlet 类，实现 service()方法即可。

4．HttpServlet 抽象类

抽象类 HttpServlet 继承自 GenericServlet 类，具有与 GenericServlet 类似的方法和对象，支持 HTTP 的 post 和 get 方法，并提供与 HTTP 相关的实现。HttpServlet 能够根据客户发出的 HTTP 请求，进行相应处理，并得到相应的结果，然后这个相应的结果会被自动封装到 HttpServletRequest 对象中。根据 HTTP 协议中规定的请求方法，HttpServlet 抽象类分别提供了处理请求的相应方法，如表 3-4 所示。

表 3-4　HttpServlet 的抽象类常用方法

方　　法	说　　明
public void service(ServletRequest req, ServletResponse res)	调用 GenericServlet 类中 service()方法的实现
public void service(HttpServletRequest req, HttpServletResponse res)	接收 HTTP 请求，并将它们分发给此类中定义的 doXXX 方法
public void doXXX(HttpServletRequest req, HttpServletResponse res)	根据请求方式的不同，分别调用相应的处理方法，如 doGet()、doPost()等

HttpServlet 类是一个抽象类，如果需要编写 Servlet 就一定要继承 HttpServlet 类，从中将需要响应到客户端的数据封装到 HttpServletResponse 对象中。

5．ServletRequest 接口和 HttpServletRequest 接口

当客户请求时，由 Servlet 容器创建 ServletRequest 对象(用于封装客户的请求信息)，这个对象将被容器作为 service()方法的参数之一传递给 Servlet，Servlet 能够利用 ServletRequest 对

象获取客户端的请求。ServletRequest 接口的常用方法如表 3-5 所示。

表 3-5 ServletRequest 接口的常用方法

方 法	说 明
public Object getAttribute(String name)	获取名称为 name 的属性值
public void setAttribute(String name, Object object)	在请求中保存名称为 name 的属性
public void removeAttribute(String name)	清除请求中名字为 name 的属性

HttpServletRequest 接口位于 javax.servlet.http 包中，继承自 ServletRequest 接口，通过该接口同样可以获取请求中的参数。HttpServletRequest 接口除了继承 ServletRequest 接口中的方法外，还增加了一些用于读取请求信息的方法，增加的方法如表 3-6 所示。

表 3-6 HttpServletRequest 接口的常用方法

方 法	说 明
public String getContextPath()	返回请求 URL 中表示请求上下文的路径，上下文路径是请求 URL 的开始部分
public Cookie[ ] getCookies()	返回客户端在此次请求中发送的所有 cookie 对象
public HttpSession getSession()	返回和此次请求相关联的 session，如果没有给客户端分配 session，则创建新的 session
public String getMethod()	返回此次请求所使用的 HTTP 方法的名字，如 GET、POST

6．ServletResponse 接口和 HttpServletResponse 接口

Servlet 容器在接收客户端请求时，除了创建 ServletRequest 对象用于封装客户端请求信息外，还创建了一个 ServletResponse 对象，用来封装响应数据，并且将这两个对象一并作为参数传递给 Servlet。Servlet 利用 ServletRequest 对象获取客户端的请求数据，经过处理后由 ServletResponse 对象发送响应数据。ServletResponse 接口的常用方法如表 3-7 所示。

表 3-7 ServletResponse 接口的常用方法

方 法	说 明
public PrintWriter getWriter()	返回 PrintWriter 对象，用于向客户端发送文本
public String getCharacterEncoding()	返回在响应中发送的正文所使用的字符编码
public void setCharacterEncoding()	设置发送到客户端的响应的字符编码
public void setContentType(String type)	设置发送到客户端的响应的内容类型，此时响应的状态属于尚未提交

HttpServletResponse 接口与 HttpServletResquest 接口类似，HttpServletResponse 接口也继承自 ServletResponse 接口，用于对客户端的请求执行响应，它除了具有 ServletResponse 接口的常用方法外，还增加了新的方法，如表 3-8 所示。

表 3-8　HttpServletResponse 接口的常用方法

方　法	说　明
public void addCookie(Cookie cookie)	增加一个 cookie 到响应中,这个方法可多次调用,设置多个 cookie
public void addHeader(String name, String value)	将一个名称为 name、值为 value 的响应报头添加到响应中
public void sendRedirect(String location)	发送一个临时的重定向响应到客户端,以便客户端访问新的 URL
public void encodeURL(String url)	使用 sessionID 对用于重定向的 URL 进行编码

7. ServletContext 接口

一个 Servlet 对象表示一个 Web 应用的上下文,Servlet 使用 ServletContext 接口定义的方法与它的 Servlet 容器进行通信。Servlet 容器厂商负责提供 Servlet 接口的实现,容器在应用程序加载时创建 ServletContext 对象,ServletContext 对象被 Servlet 容器中的所有 Servlet 共享。ServletContext 接口的常用方法如表 3-9 所示。

表 3-9　ServletContext 接口的常用方法

方　法	说　明
public getInitParameter(String name)	获取名称为 name 的系统范围内的初始化参数值,可在部署描述中使用<context-param>定义
public void setAttribute(String name, Object object)	设置名称为 name 的属性
public void getAttribute(String name)	获取名称为 name 的属性
public getRealPath(String path)	获取相对路径的真实路径
public void log(String message)	记录一般日志信息

3.4　Servlet 的应用示例

前面的章节已经介绍了使用 JSP 来接收 HTML 表单信息。同样,Servlet 可以接收从浏览器传递的信息,从而实现客户端与服务器端的交互。下面通过示例来演示 Servlet 如何获取表单信息。该示例由一个 HTML 网页和一个 Servlet 程序组成。用户在 HTML 网页的表单中输入用户信息,包括用户名、密码和兴趣爱好,提交表单到 Servlet,Servlet 程序会接收这些信息,并输出信息到浏览器中。

在 restaurant 项目的 WebRoot 目录下新建 ch03 文件夹,并在该文件夹中新建 register.jsp 页面,在 src 目录下的 com.restaurant.servlet 包中,新建 RegisterServlet.java 的 Servlet 文件。

注册页面 register.jsp 的代码如下:

```
<body>
    请输入注册信息<br>
```

```html
<form name="form1" method="post" action="servlet/RegisterServlet">
    用户名：<input type="text" name="username"><br>
    密  码: <input type="password" name="password"><br>
    兴趣爱好：
        <input type="checkbox" name="habit" value="看书">看书
        <input type="checkbox" name="habit" value="玩游戏">玩游戏
        <input type="checkbox" name="habit" value="旅游">旅游
        <input type="checkbox" name="habit" value="看电视">看电视
          <br>
        <input type="submit" value="提交">  
        <input type="reset" value="取消">
</form>
</body>
```

RegisterServlet.java 的 Servlet 文件代码如下：

```java
package com.restaurant.servlet;
import java.io.IOException;
import java.io.PrintWriter;
import javax.servlet.ServletException;
import javax.servlet.http.*;
public class RegisterServlet extends HttpServlet {
   public void doGet(HttpServletRequest request, HttpServletResponse response)    throws ServletException, IOException {
       response.setContentType("text/html;charset=utf-8");
       PrintWriter out = response.getWriter();
       request.setCharacterEncoding("utf-8");//设置字符编码为utf-8
       String username=request.getParameter("username");//读取用户名
       String password=request.getParameter("password");//读取密码
       String[] habits=request.getParameterValues("habit");
       out.println("<!DOCTYPE HTML PUBLIC \"-//W3C//DTD HTML 4.01 Transitional//EN\">");
       out.println("<HTML>");
       out.println("  <HEAD><TITLE>注册信息显示</TITLE></HEAD>");
       out.println("  <BODY>");
       out.print("您提交的注册信息如下：<br/>");
       out.print("用户名为："+username+"<br/>");
       out.print("密码为："+password+"<br/>");
       out.print("您的兴趣爱好：");
       if(habits!=null){
           for(int i=0;i<habits.length;i++){
               out.print(habits[i]+"  ");
           }
       }
       out.println("  </BODY>");
       out.println("</HTML>");
       out.flush();
       out.close();
   }
   public void doPost(HttpServletRequest request, HttpServletResponse response) throws ServletException, IOException {
       doGet(request, response);
   }
}
```

RegisterServlet 在 web.xml 文件中的配置信息如下：

```
<servlet>
  <servlet-name>RegisterServlet</servlet-name>
<servlet-class>com.restaurant.servlet.RegisterServlet</servlet-class>
</servlet>
<servlet-mapping>
  <servlet-name>RegisterServlet</servlet-name>
  <url-pattern>/servlet/RegisterServlet</url-pattern>
</servlet-mapping>
```

部署项目，在浏览器的地址栏中输入 http://localhost:8080/restaurant/ch03/register.jsp，运行结果如图 3-9 所示；输入信息后单击"提交"按钮，运行效果如图 3-10 所示。

图 3-9　输入注册信息

图 3-10　Servlet 显示注册信息

3.5　小　　结

本章主要讲解了 Servlet 的相关知识。Servlet 是一个 Java 程序，它在服务器端运行，接收和处理用户请求，并做出响应。Servlet 的生命周期包括加载和实例化、初始化、服务和销毁几个阶段。本章还介绍了 Servlet 的常用类和接口，并通过注册程序示例讲解了 Servlet 的执行过程。

第 4 章
使用 MVC 模式实现用户登录

前面两章我们已经学习了 JSP 和 Servlet 技术，使用它们可以进行动态网站的开发。大家已经了解了 JSP 技术是在 Servlet 技术的基础上形成的，它的主要任务是简化页面的开发。在编写程序的时候，我们把大量的 Java 代码写在了 JSP 页面中，以方便程序控制和业务逻辑的操作，而这违背了 JSP 技术的初衷，为程序员和前端开发人员带来很大困扰。为了解决这个问题，在进行项目设计时可以采用 MVC 设计模式。

4.1 JSP 开发模型

利用 JSP 进行 Java Web 应用开发有两种开发模型：JSP Model I 和 JSP Model II。

4.1.1 JSP Model I 模式

1．传统的 JSP Model I

在早期的 Java Web 应用开发中，JSP 文件既要负责处理业务逻辑和控制程序的运行流程，还要负责数据的显示，即用 JSP 文件独立自主地完成系统功能的所有任务。传统的 JSP Model I 模式如图 4-1 所示。

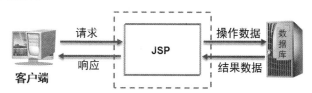

图 4-1 传统的 JSP Model I 模式

2．改进的 JSP Model I

改进的 JSP Model I 利用 JSP 页面与 JavaBean 组件共同协作来完成系统功能的所有任务，JSP 文件负责程序的流程控制逻辑和数据显示逻辑任务，JavaBean 负责处理业务逻辑任务。改进的 JSP Model I 模式如图 4-2 所示。

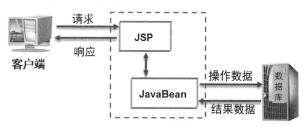

图 4-2 改进的 JSP Model I 模式

4.1.2 JSP Model II 模式

JSP Model II 利用 JSP 页面、Servlet 和 JavaBean 组件分工协作共同完成系统功能的所有任务。其中，JSP 负责数据显示逻辑任务，Servlet 负责程序流程控制逻辑任务，JavaBean 负责处理业务逻辑任务。实际上，JSP Model II 就是采用 MVC 设计模式思想设计的。JSP Model II 模式如图 4-3 所示。

Model II 模式体现了基于 MVC 的设计思想，简单地说，就是将数据显示、流程控制和业务逻辑处理分离，使之相互独立。

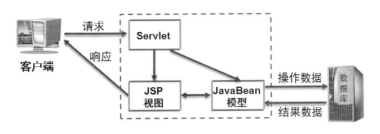

图 4-3　JSP Model II 模式

4.2　MVC 模式概述

4.2.1　为什么需要 MVC 模式

在 JSP 的开发过程中，需要经常访问数据库进行数据验证或读取数据，我们通常是把访问数据库的代码单独放在一个 Java 类中，所有有关数据访问的逻辑和业务逻辑交给它来完成，这样就加重了代码重用度页面的维护困难。

在创建项目时，必须考虑到美工美化界面的问题。如果在 JSP 中实现所有操作，HTML 与 Java 代码混和交织在一起，美工就会一头雾水。如果美工要对这个页面进行美化，而他又不懂 JSP，他所想的就是在页面上尽可能少地出现 Java 代码，将流程控制和数据显示分离，这样他就可以很好地完成美化页面的工作了。也就是在 JSP 页面中只是显示数据，有关程序控制的功能，由 Servlet 来完成。

每一个组件和技术都有自身的功能和特点。在编写程序时，我们应该根据它们的功能来设计它们的作用，就好像我们在餐厅吃饭，服务员把菜谱提供给顾客，顾客根据菜谱点菜，然后把菜单交给服务员，服务员则将菜单交给后厨，厨师做好菜后，把菜交给服务员，由服务员把菜端给顾客，如图 4-4 所示。

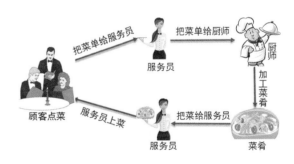

图 4-4　顾客点菜过程

服务员是这个过程的组织者和控制器 (Controller)。服务员负责接待顾客，并把菜谱显示给顾客，把顾客的点菜内容(类似于用户的请求)，交给厨师加工(类似于进行数据访问和处理业务的 Java 类)，最后服务员把菜肴端给顾客(类似于一个响应的 JSP)。

在这个过程中，顾客先看到的是菜谱，之后才会是相应的菜肴。在程序中，用户能够看到的就是 HTML、JSP 页面，这部分称作视图(View)。当服务员把顾客点菜内容交给厨师后，厨师根据不同的菜，采用不同的原料和配料来加工菜肴。这类似于在程序中，根据用户提交的不同请求数据，访问数据库或者进行业务逻辑处理，这部分称为模型(Model)。

在程序设计中，把采用模型、视图、控制器的设计方式称为 MVC 设计模式。

4.2.2 MVC 模式的定义及特点

1．什么是设计模式

设计模式是一套被反复使用、成功的代码设计经验的总结。模式必须是典型问题的解决方案。设计模式为某一类问题提供了解决方案，并优化了代码，使代码更容易让别人理解，提高重用性，从而保证代码的可靠性。

2．MVC 设计模式

MVC 是一种流行的软件设计模式，它把系统分为以下三个模块。

(1) 模型。对应的组件是 JavaBean(Java 类)，可以分为业务模型和数据模型，它们代表应用程序的业务逻辑和状态。

(2) 视图。对应的组件是 JSP 或 HTML 文件，提供可交互的客户界面，向用户显示数据模型。

(3) 控制器。对应的组件是 Servlet，响应客户的请求，根据客户的请求来操作模型，并把模型的响应结果经由视图展现给客户。

MVC 设计模式中模型、视图和控制器三者之间的关系如图 4-5 所示。

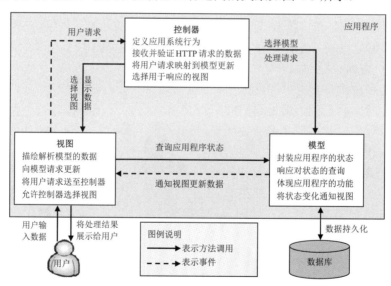

图 4-5　MVC 模式各层关系图

3．MVC 模式的特点

MVC 模式的特点如下。

(1) 各司其职、互不干涉。在 MVC 模式中，三个层各司其职，所以如果哪一层的需求发生了变化，就只需要更改相应层中的代码，而不会影响其他层。

(2) 有利于开发中的分工。在 MVC 模式中，由于按层把系统分开，因此能更好地实现开发中的分工。网页设计人员可以开发 JSP 页面，对业务熟悉的开发人员可以开发模型中相关业务处理的方法,而其他开发人员可以开发控制器,以进行程序控制。

(3) 有利于组件的重用。分层后更有利于组件的重用,如控制层可独立成一个通用的组件,视图层也可做成通用的操作界面。MVC 最重要的特点就是把显示和数据分离,这样就增加了各个模块的可重用性。

4.3 JDBC 技术

在 Java 中如何实现把各种数据存入数据库,从而长久保存呢?Java 是通过 JDBC 技术实现对各种数据库的访问的,换句话说,JDBC 充当了 Java 应用程序与各种不同数据库之间进行对话的媒介。

4.3.1 JDBC 简介

JDBC(Java DataBase Connectivity,Java 数据库连接),由一组使用 Java 语言编写的类和接口组成,可以为多种关系数据库提供统一访问。Sun 公司提供了 JDBC 接口的规范——JDBC API,而数据库厂商或第三方中间件厂商根据该接口规范提供针对不同数据的具体实现——JDBC 驱动。

JDBC 的工作原理如图 4-6 所示。

从图 4-6 中可以看到 JDBC 的几个重要组成要素。最顶层是我们自己编写的 Java 应用程序,Java 应用程序可以使用集成在 JDK

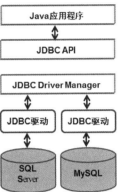

图 4-6 JDBC 的工作原理

中的 java.sql 和 javax.sql 包中的 JDBC API 来连接和操作数据库。JDBC API 由 Sun 公司开发,其提供了 Java 应用程序与各种不同数据库交互的标准接口,如 Connection 接口、Statement 接口、ResultSet 接口等,开发者使用这些 JDBC 接口进行各类数据库操作。JDBC Driver Manager 由 Sun 公司提供,它负责管理各种不同的 JDBC 驱动,位于 JDK 的 java.sql 包中。

JDBC 驱动由各个数据库厂商或第三方中间件厂商提供,负责连接各种不同的数据库。比如访问 SQL Server 和 Oracle 时需要不同的 JDBC 驱动,这些 JDBC 驱动都实现了 JDBC API 定义的各种接口。在开发 Java 应用程序时,我们只需要正确加载 JDBC 驱动,正确调用 JDBC API,就可以进行数据库访问了。

开发一个 JDBC 应用程序,基本需要以下几个步骤。

(1) 加载 JDBC 驱动。
(2) 与数据库建立连接。
(3) 发送 SQL 语句,并得到返回结果。
(4) 处理返回结果。
(5) 关闭数据库连接。

4.3.2 通过 JDBC 连接 MySQL 数据库

在实际编程过程中,使用 JDBC 访问 MySQL 数据库有两种较为常用的驱动方式。一种是

JDBC-ODBC 桥方式，适合于个人开发与测试，通过 ODBC 与数据库连接，JDK 中已经包括了 JDBC-ODBC 桥连接的驱动接口，所以不需要额外下载 JDBC 驱动程序，而只需要配置 ODBC 数据源即可；另一种是纯 Java 驱动方式，它直接同数据库进行连接，在生产型开发中，推荐使用这种方式，纯 Java 驱动方式由 JDBC 驱动直接访问数据库，驱动程序完全用 Java 语言编写，运行速度快，而且具有了跨平台特点。

使用纯 Java 驱动方式访问 MySQL 数据库，首先要从 MySQL 官方网站(https://dev.mysql.com/downloads/connector/j/)下载 MySQL 的 JDBC 驱动。将下载的压缩文件解压缩，可以得到一个文件名为 mysql-connector-java-5.1.42-bin.jar 的文件，这就是我们所需的 JDBC 驱动，后面会用到它。注意：5.1.42 是版本编号，读者下载的驱动版本可能不同，所以这个编号也会不同，但并不影响使用。

在项目中，添加 JDBC 驱动有两种方法：一种是直接将 mysql-connector-java-5.1.42-bin.jar 文件放置到项目的 WebRoot/WEB-INF/lib/目录中即可(对于 Web 项目，采用此方法)；另一种是构建路径，选择项目并右击，从弹出的快捷菜单中选择 Build Path→Add External Archives 命令，如图 4-7 所示。在弹出的文件选择对话框中，浏览找到下载解压的 mysql-connector-java-5.1.42-bin.jar 文件，然后单击打开，在项目中就会添加相应的引用。

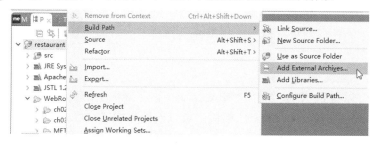

图 4-7　选择 Add External Archives 命令

使用 SQLyog 10.2 图形化前端工具创建一个 restrant 数据库和一个 admin 数据表，并为该表设置三个字段：Id、LoginName、LoginPwd，如图 4-8～图 4-10 所示。

图 4-8　创建数据库 restrant

图 4-9　创建数据表 admin

图 4-10　添加字段项

4.4　使用 MVC 模式实现用户登录模块

MVC 设计模式是一种很好的程序设计模式，有助于理解业务逻辑、划分程序模块、提高代码利用率、提高程序设计速度和效率。本节通过用户登录的示例来帮助读者进一步理解和掌握这种方法。

4.4.1　项目设计简介

在使用 MVC 模式进行编程时，要注意各个组件的分工和协作。当客户端发送请求时，服务器端 Servlet 接收请求数据，并根据数据调用模型中相应的方法访问数据库，然后把执行结果返回给 Servlet，Servlet 根据结果转向不同的 JSP 或 HTML 页面，以响应客户端请求。应注意在视图(JSP)中，不要进行业务逻辑和程序控制的操作，视图只是显示动态内容，不做其他操作。模型和控制器也是一样的，它们有各自的"工作内容"，应该让它们各尽其责。

登录模块是大家比较熟悉的，但是以前是以 JSP 形式实现，现在我们通过 JSP、Servlet、JavaBean 的 MVC 模式实现用户登录程序。基于 MVC 模式的 Web 应用的基本工作流程可以分为如下四个步骤。

(1) 用户通过页面视图发出请求。
(2) 控制器接收请求后，调用相应的模型来处理具体的业务。
(3) 控制器根据返回的结果选择相应的视图组件来反馈结果。
(4) 视图根据接收到的结果将信息显示给用户。

4.4.2 模型设计

模型设计就是 JavaBean 的设计,在模型开发结构中分为三块:com.restaurant.bean 包中存放实体类,com.restaurant.dao 包中存放数据访问类,com.restaurant.service 包中存放业务逻辑类。在 com.restaurant.bean 包中,新建 Admin 的实体类;在 com.restaurant.dao 包中,新建 BaseDAO 类和 AdminDAO 类(继承 BaseDAO 类),分别用来实现数据库连接和数据访问;在 com.restaurant.service 包中,新建 AdminService 类,用来实现业务逻辑处理。本案例的业务逻辑简单,在 dao 和 service 层没有设计接口,当开发的 Web 应用较复杂时,可以设计数据访问和业务逻辑的接口层。

1. Admin.java 实体类

代码如下:

```java
package com.restaurant.bean;
public class Admin {
    private Integer id;           //管理员id
    private String loginName;     //登录名
    private String loginPwd;      //登录密码
    //省略相应属性的getter、setter 方法以及构造方法
}
```

2. BaseDao 数据库连接类

代码如下:

```java
package com.restaurant.dao;
import java.sql.*;
public class BaseDAO {
    //数据库连接
    public Connection getConnection(){
        Connection conn=null;
        try {
            Class.forName("com.mysql.jdbc.Driver");//加载驱动
            conn=DriverManager.getConnection("jdbc:mysql://localhost:3306/restrant","root","123456");//访问数据库,得到数据库连接对象
        } catch (Exception e) {
            e.printStackTrace();
        }
        return conn;
    }
    //对象关闭
    public void closeAll(Connection conn,PreparedStatement pstmt,ResultSet rs){
        if(rs!=null){
            try {
                rs.close();
            } catch (Exception e) {
                e.printStackTrace();
            }
        }
```

        //省略关闭pstmt对象和conn对象
    }
}

3．AdminDao 数据访问类

代码如下：

```
package com.restaurant.dao;
import java.sql.*;
public class AdminDAO extends BaseDAO{
    Connection conn = null;
    PreparedStatement pstmt=null;
    ResultSet rs=null;
    public boolean login(String loginName,String loginPwd){
        boolean isLogin=false;
        String sql="select * from admin where loginName=? and loginPwd=?";
        try {
            conn=this.getConnection();
            pstmt=conn.prepareStatement(sql);
            pstmt.setString(1, loginName);
            pstmt.setString(2, loginPwd);
            rs=pstmt.executeQuery();
            if (rs.next()) {
                isLogin=true;
            }
        } catch (Exception e) {
            e.printStackTrace();
        }finally{
            this.closeAll(conn, pstmt, rs);
        }
        return isLogin;
    }
}
```

4．AdminService 业务逻辑

代码如下：

```
package com.restaurant.service;
import com.restaurant.dao.AdminDao;
public class AdminService {
    AdminDao adminDAO=new AdminDAO();
    public boolean login(String loginName,String loginPwd){
        return adminDAO.login(loginName, loginPwd);
    }
}
```

4.4.3 视图设计

视图设计即页面设计，在 restaurant 项目中的 WebRoot 目录下新建 ch04 文件夹，并在该文件夹中新建 login.jsp、info.jsp 页面。

（1）login.jsp 页面文件供用户提交用户名和密码。代码如下：

```jsp
<%@ page language="java" import="java.util.*" pageEncoding="UTF-8"%>
<!DOCTYPE HTML PUBLIC "-//W3C//DTD HTML 4.01 Transitional//EN">
<html>
  <head><title>MVC 示例——登录页面</title></head>
  <body>
    <form name="form1" method="post" action="servlet/LoginServlet">
        用户名:<input type="text" name="loginName"><br><br>
        密码:<input type="password" name="loginPwd"><br><br>
        <input type="submit" value="登录">
        <input type="reset" value="取消">
    </form>
  </body>
</html>
```

(2) info.jsp 页面文件用来显示登录成功后的信息。代码如下：

```jsp
<body>
    登录成功，欢迎 ${requestScope.LOGIN_NAME}！
</body>
```

4.4.4 控制器设计

使用 Servlet 作为控制器，作用是接收用户的请求数据，选择合适的 Controller 处理具体业务，处理完成后，根据 Model 返回的结果选择一个 View 显示数据。在 com.restaurant.servlet 包中，新建一个 Servlet 文件 LoginServlet.java，代码如下：

```java
package com.restaurant.servlet;
import java.io.IOException;
import java.io.PrintWriter;
import javax.servlet.ServletException;
import javax.servlet.http.*;
import com.restaurant.service.AdminService;
public class LoginServlet extends HttpServlet {
    public void doGet(HttpServletRequest request, HttpServletResponse response) throws ServletException, IOException {
        request.setCharacterEncoding("utf-8");
        response.setContentType("text/html;charset=utf-8");
        PrintWriter out = response.getWriter();
        String loginName=request.getParameter("loginName");
        String loginPwd=request.getParameter("loginPwd");
        AdminService adminService=new AdminService();
        boolean isLogin=adminService.login(loginName,loginPwd);
        if (isLogin) {
            request.setAttribute("LOGIN_NAME", loginName);
       request.getRequestDispatcher("../ch04/info.jsp").forward(request, response);
        }else{
            out.print("登录失败！");
        }
    }
    public void doPost(HttpServletRequest request, HttpServletResponse response) throws ServletException, IOException {
        doGet(request, response);
    }
}
```

4.4.5　部署和运行程序

项目完成后的文件结构如图 4-11 所示。

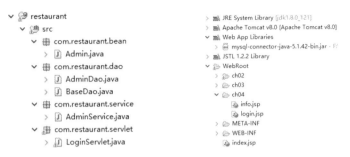

图 4-11　Web 项目的文件结构

在 MyEclipse 中部署项目后，在浏览器的地址栏中输入 http://localhost:8080/restaurant/ch04/login.jsp 网址，可以看到输入界面，输入用户名和密码：admin 和 123456，如图 4-12 所示。单击"登录"按钮，可以看到登录成功并显示用户名的页面，如图 4-13 所示。

图 4-12　登录输入页面

图 4-13　登录成功

如果用户名和密码输入错误，则会提示登录失败，如图 4-14 所示。

图 4-14　登录失败

4.5　小　　结

本章主要讲解了 MVC 框架的基本概念和 JDBC 的基础知识，并结合前面的 JSP 和 Servlet 知识，讲解了如何使用 JSP + Servlet + JavaBean 完成 MVC 框架，并用一个具体的登录示例演示了它们之间的关系和使用方法。这为后续的 Struts 2、Spring 等框架的学习奠定了基础，希望读者多多练习，认真领会。

第 5 章
jQuery EasyUI 插件

　　jQuery 是 JavaScript 的一个基础框架,考虑到框架的通用性和代码文件的大小,jQuery 仅仅集成了 JavaScript 中最为核心和常用的功能。目前,在 jQuery 的基础上已开发出众多插件,这些插件均以 jQuery 为核心编写而成。本章介绍的 jQuery EasyUI 框架便是其中之一。因此,读者学习和使用 jQuery EasyUI 框架后,也有利于学习和使用其他 UI 框架。

5.1 EasyUI 概述

EasyUI 是在 jQuery 的基础上开发的一个 UI 插件,目的在于让 Web 开发者快捷地构建出功能丰富且美观的用户界面。开发者无须编写复杂的 JavaScript,也无须对 CSS 样式有深入的了解。开发者只需要有一些 HTML 和 jQuery 基础,就可以轻松地开发出较好的软件界面。EasyUI 控件的种类很多,由于篇幅,这里仅介绍本书项目案例篇中的项目用到的几种常用控件。可以从官方网站下载 jQuery EasyUI 插件,本书以 jquery-easyui-1.5.1 版本来介绍。下载 jquery-easyui-1.5.1.zip 文件,解压后的目录中主要包含 jquery.min.js、jquery.easyui.min.js 两个文件和 demo、locale、plugins 和 themes 四个目录。demo 目录下包含 jQuery EasyUI 官方提供的例子;locale 目录下包含语言本地化 JavaScript 文件;plugins 目录下包含 EasyUI 提供的各个功能的文件。themes 目录下包含样式和图片文件目录。

5.2 Layout 控件

使用 EasyUI 的 Layout 控件可以实现页面布局,布局是有五个区域(北区 north、南区 south、东区 east、西区 west 和中区 center)的容器。中间的区域面板是必需的,边缘的区域面板是可选的。每个边缘区域面板可通过拖曳边框调整尺寸,也可以通过点击折叠触发器来折叠面板。布局可以嵌套,因此用户可以建立复杂的布局。

使用 Layout 控件实现一个简单布局的过程如下。

(1) 创建 Web 项目 easyui_demo,将 EasyUI 所需的文件事先存放到文件夹 EasyUI 中,再将该文件夹拷贝到项目的 WebRoot 目录下。EasyUI 文件夹的内容如图 5-1 所示。

(2) 新建页面 layout.jsp,在页面的<head></head>元素中引用相关的 css 和 js 文件,代码如下:

```
<head>
<link href="EasyUI/themes/default/easyui.css" rel="stylesheet"
    type="text/css" />
<link href="EasyUI/themes/icon.css" rel="stylesheet" type="text/css" />
<link href="EasyUI/demo.css" rel="stylesheet" type="text/css" />
<script src="EasyUI/jquery.min.js" type="text/javascript"></script>
<script src="EasyUI/jquery.easyui.min.js" type="text/javascript"></script>
<script src="EasyUI/easyui-lang-zh_CN.js" type="text/javascript"></script>
</head>
```

(3) 在页面 layout.jsp 的<body></body>元素中添加如下代码:

```
<body>
    <div class="easyui-layout" style="width:700px;height:350px;">
        <div data-options="region:'north'" style="height:50px">这是北区
north</div>
        <div data-options="region:'south',split:true" style="height:50px;">
这是南区 south</div>
        <div data-options="region:'east',split:true" title="East"
```

```
            style="width:100px;">这是东区 east</div>
        <div data-options="region:'west',split:true" title="West"
            style="width:100px;">这是西区 west</div>
        <div
            data-options="region:'center',title:'Main Title',iconCls:'icon-
ok'">
            这是中区 center</div>
    </div>
</body>
```

(4) 部署项目并启动 Tomcat，在浏览器中浏览页面 layout.jsp，效果如图 5-2 所示。

图 5-1　EasyUI 文件夹的内容　　　　图 5-2　Layout 控件效果

5.3　Tabs 控件

使用 Tabs 控件可以实现选项卡布局，一般用于中部选项卡。在项目 easyui_demo 中创建页面 tabs.jsp，在页面的<head></head>元素中引用相关的 css 和 js 文件。

在页面 tabs.jsp 的<body></body>元素中编写如下代码：

```
<body>
    <div class="easyui-tabs" style="width:700px;height:250px">
        <div title="选项卡 1" style="padding:10px">
            页面 1
        </div>
        <div title="选项卡 2" style="padding:10px">
            页面 1
        </div>
        <div title="选项卡 3" data-options="iconCls:'icon-help',closable:true"
            style="padding:10px">页面 3</div>
    </div>
</body>
```

在浏览器中浏览页面 tabs.jsp，效果如图 5-3 所示。

图 5-3　Tabs 控件效果

5.4　Tree 控件

Tree 控件是 Web 页面中将数据分层以树形结构显示的。Tree 控件在页面上以<ul></ul>标签标识。在项目 easyui_demo 中创建页面 tree.jsp，在页面的<head></head>元素中引用相关的 css 和 js 文件。

在页面 tree.jsp 的<body></body>元素中编写如下代码：

```
<body>
    <!-- 定义 ul -->
    <ul id="tt"></ul>
    <script type="text/javascript">
        // 为 Tree 控件指定数据源
        $('#tt').tree({
            url : 'tree_data.json'
        });
    </script>
</body>
```

在项目的 WebRoot 目录下创建一个 JSON 格式的文件 tree_data.json，作为 Tree 控件的数据源，代码如下：

```
[
    {
        "id": 1,
        "text": "订餐系统管理后台",
        "fid": 0,
        "children": [
            {
                "id": 2,
                "text": "餐品管理",
                "fid": 0,
                "children": [
                    {
                        "id": 3,
                        "text": "餐品列表",
```

```
                    "fid": 0
                },
                {
                    "id": 4,
                    "text": "餐品类型列表",
                    "fid": 0
                }
            ]
        }, {
            "id": 12,
            "text": "退出系统",
            "fid": 0
        }
    ]
}
]
```

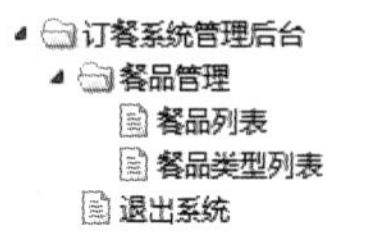

在浏览器中浏览页面 tree.jsp，效果如图 5-4 所示。

图 5-4　Tree 控件效果

5.5　DataGrid 控件

DataGrid 控件以表格的形式显示数据，并为选择、排序、分组和编辑数据提供了丰富的支持。数据网格(DataGrid)的设计目的是减少开发时间，且不要求开发人员具备过多的 JavaScript 和 CSS 等方面的知识。它是轻量级的，但是功能丰富。它的特性包括单元格合并，多列页眉，冻结列和页脚，等等。

在项目 easyui_demo 中创建页面 datagrid.jsp，在页面的<head></head>元素中引用相关的 css 和 js 文件。

在页面 datagrid.jsp 的<body></body>元素中编写如下代码：

```
<body>
    <table id="newsinfoDg" class="easyui-datagrid"></table>
    <script type="text/javascript">
        $(function() {
            $('#newsinfoDg').datagrid({
                fit : true,
                fitColumn : true,
                rownumbers : true,
                singleSelect : false,
                url : 'datagrid_data.txt',
                columns : [ [ {
                    title : '',
                    field : 'productid',
                    align : 'center',
                    checkbox : true
                }, {
                    field : 'unitcost',
                    title : 'unitcost',
                    width : 50
                }, {
                    field : 'status',
```

```
                    title : 'status',
                    width : 60
                }, {
                    field : 'listprice',
                    title : 'listprice',
                    width : 50
                }, {
                    field : 'attr1',
                    title : 'attr1',
                    width : 200
                }, {
                    field : 'itemid',
                    title : 'itemid',
                    width : 100
                } ] ]
            });
        });
    </script>
</body>
```

在项目的 WebRoot 目录下创建文件 datagrid_data.txt，作为 DataGrid 控件的数据源，代码如下：

```
[{"productid":"FI-SW-01","unitcost":10.00,"status":"P",
"listprice":36.50,"attr1":"Large","itemid":"EST-1"},
{"productid":"K9-DL-01","unitcost":12.00,"status":"P",
"listprice":18.50,"attr1":"Spotted Adult Female","itemid":"EST-10"},
// 由于篇幅，此处省略了其他数据
```

在浏览器中浏览页面 datagrid.jsp，效果如图 5-5 所示。

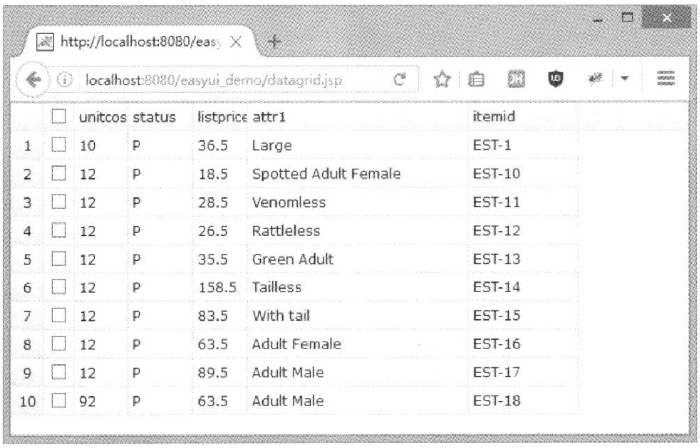

图 5-5　DataGrid 控件效果

5.6　小　　结

本章介绍了 jQuery EasyUI 插件中的 Layout、Tabs、Tree 和 DataGrid 这四个控件的基本用法。在本书项目案例篇中还将结合具体项目，更深入地学习有关 EasyUI 控件的用法。

第II篇

框架技术部分

- 第 6 章　认识 Struts 2 框架
- 第 7 章　Struts 2 的配置
- 第 8 章　Struts 2 的标签库
- 第 9 章　OGNL 和类型转换
- 第 10 章　Struts 2 的拦截器
- 第 11 章　Hibernate 初步
- 第 12 章　使用 Hibernate 实现数据的增删改查
- 第 13 章　使用 Hibernate 实现关联映射和继承映射
- 第 14 章　使用 Hibernate 查询数据
- 第 15 章　使用 Hibernate 缓存数据
- 第 16 章　MyBatis 框架
- 第 17 章　Spring 的基本应用
- 第 18 章　Spring Bean 的装配模式
- 第 19 章　面向切面编程(Spring AOP)

第 6 章
认识 Struts 2 框架

　　Struts 2 由传统的 Struts 1 和 WebWork 两个经典的 MVC 框架发展而来。全新的 Struts 2 与 Struts 1 差别较大，但是相对于 WebWork，Struts 2 的变化很小。实际上，WebWork 和 Struts 社区已经合二为一，即现在的 Struts 2 社区。

6.1 Struts 2 框架

Struts 2 框架是广泛流行的一个 MVC 开源框架。Struts 2 框架充分发挥了 Struts 1 和 WebWork 两种技术的优势，抛弃了原来 Struts 1 的缺点，使得 Web 开发更加容易。

6.1.1 Struts 2 的由来

2001 年 7 月，Struts 1 正式发布，成为 Apache Jakarta 的子项目之一，它只有一个中心控制器，采用 XML 定制转向的 URL，采用 Action 来处理逻辑。在 2005 年的 JavaOne 大会上，Struts 开发者和用户经过讨论，决定基于 XWork 开发一个新框架，这就是后来的 Struts 2。

Struts 2 虽然是在 Struts 1 的基础上发展起来的，但它并没有继承 Struts 1 的设计理念。Struts 2 使用了 WebWork 的设计理念，并且吸收了 Struts 1 的部分优点，对 Struts 1 和 WebWork 两大框架进行了整合，建立了一个兼容 WebWork 和 Struts 1 的 MVC 框架。使原来使用 Struts 1 和 WebWork 的开发人员都能够很快过渡到使用 Struts 2 框架进行开发。

在使用上，Struts 2 更接近 WebWork 的使用习惯，因为 Struts 2 使用了 WebWork 的设计核心而不是 Struts 1 的设计核心。两个框架的优势得到了互补，让 Struts 2 拥有了更广阔的前景。不仅 Struts 2 自身更加强大，还对其他框架下开发的程序提供了很好的兼容性。

6.1.2 Struts 2 的 MVC 模式

由于 Struts 2 的架构本身就是来自 MVC 思想，所以在 Struts 2 的架构中能够找到 MVC 的影子。在 Struts 2 中，视图层对应视图组件，通常是指 JSP 页面，也适用于 velocity、freemarker 等其他视图显示技术。模型层对应业务逻辑组件，它通常用于实现业务逻辑及与底层数据库的交互等。控制层对应系统核心控制器和业务逻辑控制器。系统核心控制器为 Struts 2 框架提供的 StrutsPrepareAndExecuteFilter，是一个起过滤作用的类，能根据请求自动调用相应的 Action。而业务逻辑控制器是开发者自定义的一系列 Action，在 Action 中负责调用相应的业务逻辑组件来完成调用处理。Struts 2 的 MVC 实现如图 6-1 所示。

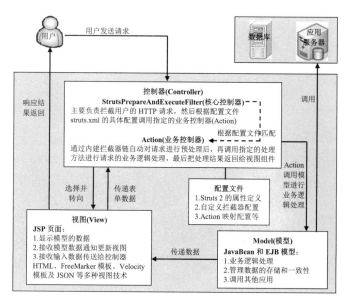

图 6-1 Struts 2 的 MVC 实现

6.1.3 Struts 2 控制器

　　Struts 2 的控制器是整个 Struts 2 框架的核心，由 StrutsPrepareAndExecuteFilter 核心控制器和 Action 业务控制器两个部分组成。在 Struts 2 中通过拦截器来处理用户的请求，从而允许用户的业务逻辑控制器和 Servlet 分离，在处理请求的过程中以用户的业务逻辑控制器为目标，创建一个控制器代理，控制代理回调业务控制器中的 execute()方法来处理用户的请求，该方法的返回值决定了 Struts 2 以怎样的视图资源呈现给用户。Struts 2 的控制器体系概略图如图 6-2 所示。

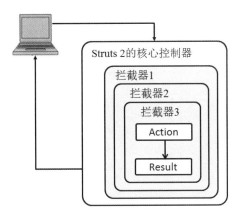

图 6-2　Struts 2 的控制器体系概略图

6.1.4 Struts 2 资源的获取

　　登录 Struts 官方网站 http://struts.apache.org/，单击蓝色的 Download 按钮；或者直接进入 http://struts.apache.org/download.cgi 的最新版下载页面，如图 6-3 所示。

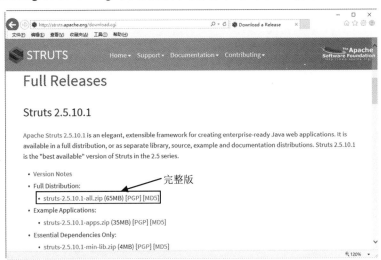

图 6-3　Struts 2 下载页面

在 Full Releases Struts 2.5.10.1 的下方找到 Full Distribution: struts-2.5.10.1-all.zip(65MB)的超级链接，单击下载。在这里以 Struts 2.5.8 的版本进行介绍，将 Struts 2.5.8 压缩包进行解压缩操作后，文件夹内容如下：

（1）apps 文件夹：存放官方的 Struts 2 示例程序，这些程序可以作为学习者的资料，为开发者提供了很好的参照。各示例均为 war 文件，可通过解压缩软件进行解压缩操作。

（2）docs 文件夹：存放官方提供的 Struts 2 文档，包括 Struts 2 API、Struts 2 Tag 等。

（3）lib 文件夹：存放 Struts 2 框架的核心类库，以及 Struts 2 的第三方插件。

（4）src 文件夹：存放 Struts 2 项目对应的源代码。

安装 Struts 2 框架相对比较容易，Struts 2.5.8 框架目录中的 lib 文件夹下有 93 个 jar 包文件。Struts 2 项目所依赖的主要 jar 包文件说明如表 6-1 所示。

表 6-1　Struts 2 项目所依赖的主要 jar 包文件说明

文 件 名	说　明
struts2-core-2.5.8.jar	Struts 2 框架的核心类库，Struts 2 的构建基础
ognl-3.1.12.jar	Struts 2 使用的一种表达式语言类库
freemarker-2.3.23.jar	Struts 2 标签模板使用类库
commons- logging-1.1.3.jar	Struts 2 的日志管理组件依赖包
commons-io-2.4.jar	Struts 2 的输入输出，可看是 java.io 扩展
commons-lang3-3.4.jar	包含一些数据类型工具，可看成 java.lang.*扩展
javassist-3.20.0-GA.jar	JavaScript 字节码解释器
commons-fileupload-1.3.2.jar	Struts 2 的文件上传组件依赖包
log4j-api-2.7.jar	显示程序运行日志

注：Struts 2.5 版本不再提供 xwork-core-*.jar，其功能整合到了 struts-core-*.jar 包中。

6.2　Struts 2 系统架构

Struts 2 在不断地发展和演变，其版本也在不断地更新，截至编写本书时，Struts 2 正式发布的版本为 Struts 2.5.10.1。

6.2.1　Struts 2 框架结构

Struts 2 的官方文档里附带了 Struts 2 的框架结构图，展示了 Struts 2 的框架结构中的内部模块以及运行流程，其大量使用拦截器来处理用户的请求，这些拦截器组成了一个拦截器链，会自动对请求进行一些通用性的功能处理，如图 6-4 所示。

接下来介绍 Struts 2 的体系结构。

当 Web 容器收到一个请求时，它将请求传递给一个标准的过滤器链，其中包括 ActionContextCleanUp 过滤器及其他过滤器(如集成 SiteMesh 的插件)，这是非常有用的技术。接下来，需要调用 FilterDispatcher，用它调用 ActionMapper 确定请求调用哪个 Action，

ActionMapper 返回一个收集了 Action 详细信息的 ActionMapping 对象。

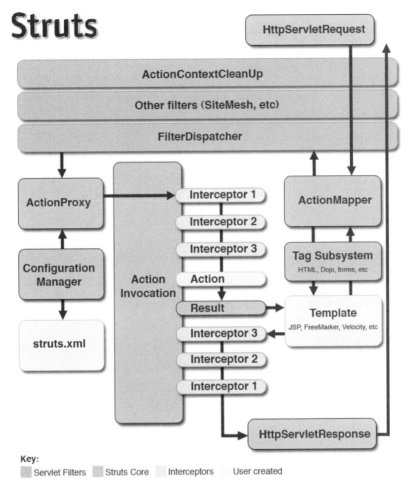

图 6-4 Struts 2 框架结构

接下来 FilterDispatcher 将控制权委派给 ActionProxy，ActionProxy 调用配置管理器 (ConfigurationManager)从配置文件中读取配置信息，然后创建 ActionInvocation 对象。实际上，ActionInvocation 的处理过程就是 Struts 2 处理请求的过程。创建 ActionInvocation 的同时，填充了需要的所有对象和信息，它在调用 Action 之前会依次调用所有配置的拦截器。

一旦 Action 执行返回结果字符串，ActionInvocation 负责查找结果字符串对应的 Result，然后执行这个 Result。通常情况下 Result 会调用一些模板(JSP 等)来呈现页面。

之后拦截器会被再次执行(顺序和 Action 执行之前相反)，最后响应，被返回给在 web.xml 中配置的那些过滤器(FilterDispatcher 等)。

6.2.2 Struts 2 的核心概念

在上面的 Struts 2 框架结构图中，可以看到很多 Struts 2 的模块，有些核心组件是必须要掌握的，这些组件组成了应用程序的功能，也构成了框架本身。

(1) FilterDispatcher，Struts 2 的前端控制器，作为 MVC 模式中的控制器部分，在开发时，只要在项目中的 web.xml 配置文件中配置一次即可。如有其他过滤器，该配置部分通常放在最后。在 Struts 2.1.3 以后的版本中，控制器名称为 StrutsPrepareAndExecuteFilter。

(2) Action 业务类，作为 MVC 中的模型部分，既封装业务数据，也负责处理用户的请求。Action 类中的 execute 方法是默认的动作处理方法。

(3) Result 结果，表示 Action 业务类执行后，要跳转的页面。Struts 2 本身支持多种结果类型，如 jsp、velocity、freemarker 等，在同一个 Web 应用中，各种结果类型可以混用。

(4) Interceptor 拦截器，是 Struts 2 框架中的重要概念。Struts 2 的许多功能都是由拦截器完成的，每一个 Struts 2 工程都使用了拦截器，包括 Struts 2 自带的内建拦截器与默认拦截器。

(5) ActionContext、值栈与 OGNL。虽然 ActionContext 没在框架图中出现，在每个 Action 刚开始运行的时候，Struts 2 都会单独为它建立一个 ActionContext，把所有能访问的数据，包括请求参数(request 的 parameter)、请求的属性(request 的 Attribute)、会话(session)信息等，都放到 ActionContext 中。在以后赋值、取值的时候，就只需要访问 ActionContext 就可以了，所以说 ActionContext 可以被认为是每个 Action 拥有的一个独立的内存数据中心。

OGNL(Object-Graph Navigation Language)对象图导航语言，是一种功能强大的表达式语言(Expression Language，EL)。通过简单一致的表达式语法，可以存取对象的任意属性，调用对象的方法，遍历整个对象的结构图，实现字段类型转化等功能。

值栈可用来容纳多个对象，用来存放一些临时对象。使用 OGNL 访问值栈中的对象属性时，指定属性的引用会引用更靠近值栈栈顶方向的对象，后进栈的对象会覆盖早进栈的对象。简单来说，Struts 2 用值栈为 Struts 2 做了很多引用上的简化，主要是缩短了 OGNL 表达式的长度。值栈也可以作为一个内存数据中心，来存放一些 Struts 2 标签临时定义的数据。

(6) Struts 2 标签。Struts 2 的标签库使用简单，功能强大，简化了页面开发的工作。并且与 Struts 2 框架的其他部分也非常自然地结合，如验证、国际化等。

(7) 自动类型转换。在 Action 类中，可以有多种方式来对应页面的数据，从而自动获取页面的值。但从 request 参数里接收的值都是 String 字符串类型，而 Action 类中的属性可以是各种类型。这就需要 Struts 2 的类型转换机制来支持，可以节省开发者的时间。Struts 2 内置了大量的类型转换方式，还可以自己实现特殊的类型转换器。

(8) 国际化。通常简称 i18n，i 和 n 是英文单词 internationalization 的首尾字符，18 为中间的字符数。Struts 2 非常自然地实现了国际化，只要按照 Struts 2 的要求，把不同语言信息，放到对应的位置即可。

(9) 验证框架。一个稳定、成熟的 Web 系统，服务器端验证是必不可少的。Struts 2 提供了验证框架，在真正调用业务逻辑 Action 之前，对从客户端传递过来的数据进行校验。如果用户提交的数据不符合要求，就不会调用业务逻辑。

6.3 Struts 2 的基本运行流程

Struts 2 框架由三个部分组成：核心控制器 StrutsPrepareAndExecuteFilter、业务控制器和用户实现的业务逻辑组件。在这三个部分里，核心控制器 StrutsPrepareAndExecuteFilter 由 Struts 2 框架提供，而业务控制器和业务逻辑组件需要程序员去实现。下面通过用户登录的示例来讲解 Struts 2 的基本运行流程。

6.3.1 用户登录的处理流程

采用 Struts 2 框架以后，业务请求不再提交给服务器端的 JSP 或 Servlet。下面通过使用 JSP + Struts 2 实现用户的登录验证，来讲解 Struts 2 的运行流程。实现登录功能的 Struts 2 框架的运行流程如图 6-5 所示。

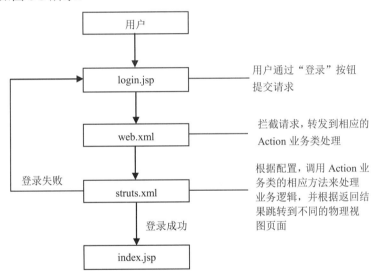

图 6-5 实现登录功能的 Struts 2 框架的运行流程

用户在登录页 login.jsp 中输入用户名和密码后，单击"登录"按钮提交表单信息；读取 web.xml 文件，加载 Struts 2 的核心控制器；根据提交的 Action，在 Struts 2 的 struts.xml 中查找匹配相应的 Action 配置，若没有找到指定 Action 元素的 method 属性值，系统会调用默认方法 execute() 来完成对客户端的登录请求处理。如果登录成功，则返回 success 字符串，否则返回 input 字符串。根据返回结果，在 struts.xml 配置文件中，查找相应的映射，跳转到 index.jsp 首页面。

6.3.2 加载 Struts 2 类库

Struts 2.5.8 中涉及四个基本类库，即 struts2-core-2.5.8.jar、ognl-3.1.12.jar、freemarker-2.3.23.jar、commons-logging-1.1.3.jar；五个附加类库，即 commons-io-2.4.jar、commons-lang3-

3.4.jar、javassist-3.20.0-GA.jar、commons-fileupload-1.3.2.jar、log4j-api-2.7.jar；一个 MySQL 数据库的驱动包，即 mysql-connector-java-5.1.42-bin.jar，将这 10 个 jar 包一起复制到 restaurant 项目下的\WebRoot\WEB-INF\lib 路径下即可。选择 restaurant 项目并右击，在弹出的快捷菜单中选择 Refresh(刷新)命令，在项目下的 Web App Libraries 中可以看见所添加的 jar 包，这样 Struts 2 包就加载成功了，如图 6-6 所示。

6.3.3 配置 web.xml 文件加载核心控制器

Struts 2 将核心控制器 StrutsPrepareAndExecuteFilter 设计成过滤器，是 Struts 2 框架的核心组件，作用于整个 Web 应用程序，因此需要在 web.xml 中进行配置。修改项目\WebRoot\WEB-INF\路径下的 web.xml 文件，添加 Struts 2 核心控制器，配置代码如下：

图 6-6 查看添加的 jar 包

```xml
<?xml version="1.0" encoding="UTF-8"?>
<web-app xmlns:xsi="http://www.w3.org/2001/XMLSchema-instance"
xmlns="http://xmlns.jcp.org/xml/ns/javaee"
xsi:schemaLocation="http://xmlns.jcp.org/xml/ns/javaee
http://xmlns.jcp.org/xml/ns/javaee/web-app_3_1.xsd" id="WebApp_ID"
version="3.1">
  <display-name>restaurant</display-name>
  <!--此处省略其他已有的配置 -->
  <filter>   <!-- 添加配置 Struts 2 框架的核心控制器 -->
    <filter-name>struts2</filter-name>    <!-- 过滤器名 -->
    <!-- 配置 Struts 2 的核心控制器的实现类 -->
    <filter-class>
org.apache.struts2.dispatcher.filter.StrutsPrepareAndExecuteFilter
    </filter-class>
  </filter>
  <!-- 让 Struts 2 的核心控制器拦截所有请求 -->
  <filter-mapping>
    <filter-name>struts2</filter-name>    <!-- 过滤器名 -->
    <url-pattern>/*</url-pattern>         <!-- 匹配所有请求 -->
  </filter-mapping>
</web-app>
```

Struts 2.1.3 以下版本的核心控制器为 org.apache.struts2.dispatcher.FilterDispatcher，Struts 2.1.3 到 2.5 之间版本的核心控制器为 org.apache.struts2.dispatcher.ng.filter.StrutsPrepareAndExecuteFilter。

6.3.4 开发视图层页面

在 restaurant 项目的 WebRoot 路径下新建 ch06 文件夹，并在其中新建 login.jsp 和

index.jsp 页面，设计登录页面的登录表单和主页面的提示信息。

登录页面的表单部分的代码如下：

```
<h3>用户登录</h3>
<form name="form1" method="post" action="login.action">
    用户名：<input type="text" name="loginName">  <br><br>
    密  码：<input type="password" name="loginPwd"><br><br>
    <input type="submit" value="登录">
    <input type="reset" value="取消">
</form>
```

主页面主要就是提示信息"欢迎 XX，来到 Struts 2 的世界！"，代码如下：

```
欢迎 ${param.loginName }，来到 Struts 2 的世界！
```

6.3.5 开发业务控制器 Action

对编程人员来说，使用 Struts 2 框架，主要工作就是编写 Action 类。Action 是由用户定义的业务控制器，在 src 文件夹下新建包 com.restaurant.action，在该包中新建 LoginAction 类，并继承 ActionSupport，代码如下：

```
package com.restaurant.action;
import com.opensymphony.xwork2.ActionSupport;
public class LoginAction extends ActionSupport {
    private String loginName;
    private String loginPwd;
    // 省略属性 loginName、loginPwd 的 getter、setter 方法
    @Override           // 默认方法
    public String execute() throws Exception {
        // 登录的用户名和密码判断，此时暂不访问数据库
        if ("admin".equals(loginName)&&"123".equals(loginPwd)) {
            return "success";        //返回 success 字符串
        }else{
            return "input";          //返回 input 字符串
        }
    }
}
```

Action 可以是一个普通的 JavaBean，它有两个属性：loginName 和 loginPwd。Action 类变量的命名必须与 login.jsp 中的文本输入框的 name 属性匹配。在实际开发中，Action 类一般都继承自 Struts 2 提供的 com.opensymphony.xwork2.ActionSupport 类，以便简化开发。

6.3.6 配置业务控制器 struts.xml

编写好 Action 的代码后，还要进行配置才能让 Struts 2 识别这个 LoginAction，在 src 路径下，新建 struts.xml 文件(注意位置和大小写)，设置包名、action 请求名称以及对应的 Action 类，根据返回结果进行逻辑视图和物理视图之间的映射。最终的 struts.xml 配置文件内容、所对应的类和视图结构如图 6-7 所示。

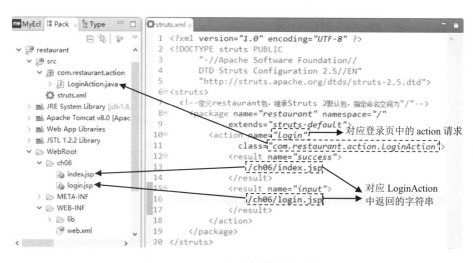

图 6-7　Action 类、视图结构的对应关系

6.3.7　部署运行项目

单击工具栏上的 Manage Deployments…按钮，在对话框中的 Module 右侧下拉列表中选择 restaurant 项目，单击 Add 按钮，在对话框中选择相应的 Tomcat 服务器，单击 Finish 按钮。部署成功后，在浏览器的地址栏中输入 http://localhost:8080/restaurant/ch06/login.jsp，进入登录页面，如图 6-8 所示。输入用户名和密码，如果正确的话，进入登录成功页面，如图 6-9 所示。

图 6-8　用户登录表单　　　　　　　　　　　图 6-9　登录成功页面

6.3.8　使用 Struts 2 实现登录功能的处理过程

Struts 2 进行登录处理的整个运行流程如下。

(1) 通过浏览器，运行登录页面，输入用户名和密码，单击"登录"按钮，向服务器提交用户输入的用户名和密码信息。

(2) 读取 web.xml 配置文件，加载 Struts 2 的核心控制器 StrutsPrepareAndExecuteFilter，对用户请求进行拦截。

（3）根据用户提交表单中的 Action，在 struts.xml 配置文件中查找匹配相应的 Action 配置，这里会查找 name 属性值为 login 的 Action 配置，并且把已经拦截的请求发给相对应的 LoginAction 业务类来处理。

（4）在 struts.xml 配置文件中没有指定 Action 元素的 method 属性值，此时，系统会调用默认方法 execute()来访问数据库完成对客户端的登录请求处理。若登录成功，则返回 success 字符串，否则返回 input 字符串。

（5）根据返回结果，在 struts.xml 配置文件中查找相应的映射，配置 LoginAction 时，指定了<result name="success">/ch06/index.jsp</result>。因此，当 LoginAction 类的 execute()方法返回 success 字符串时，则转向/ch06/index.jsp 页面，否则转向/ch06/login.jsp 页面。

6.4 Struts 2 的控制器和组件

Struts 2 框架是基于 MVC 模式的，基于 MVC 模式框架的核心就是控制器对所有请求进行统一处理。控制器包括核心控制器和业务控制器；组件包括模型组件和视图组件。

6.4.1 核心控制器

StrutsPrepareAndExecuteFilter 控制器是 Struts 2 框架的核心控制器，该控制器负责拦截所有的用户请求。当用户请求到达时，该控制器会过滤用户的请求，所有请求将被交给 Struts 2 框架处理。当 Struts 2 框架获得用户请求后，根据请求的名字决定调用哪部分业务逻辑组件。例如，对于 login 请求，Struts 2 调用 login 所对应的 LoginAction 业务类来处理该请求。

在 Struts 2 中，业务 Action 在 struts.xml 文件中定义，在该配置文件中定义 Action 时，定义了该 Action 的 name 属性和 class 属性。其中 name 属性决定了 Action 处理哪个用户请求；class 属性决定了该 Action 所对应的实现类。

由前面的代码可知，action 的 name 为 login，用户请求页面 login.jsp 的 Action 应该为 login。代码如下：

```
<form name="form1" action="login.action" method="post">
    <!-- 省略其他代码 -->
</form>
```

Struts 2 用于处理用户请求的 Action 实例，并不是用户实现的业务控制器，而是 Action 代理——因为用户实现的业务控制器并没有与 Servlet API 耦合，显然无法处理用户请求。而 Struts 2 框架提供了系列拦截器，该系列拦截器负责解析 HttpServletRequest 请求中的请求参数，传给 Action，并调用 Action 中的 execute()方法处理用户请求。

6.4.2 业务控制器

Action 就是 Struts 2 的业务逻辑控制器，负责处理请求并将结果输出给客户端。对开发人员来说，使用 Struts 2 框架，主要的编码工作就是编写 Action 类。Struts 2 并不要求编写的 Action 类必须继承 ActionSupport，也无须实现任何 Action 接口。可以编写一个普通的 Java 类作为 Action 类，只要该类含有一个返回字符串的无参 execute()方法即可。在处理客户端请求之前，Action 需要获取请求参数。Action 类中通常包含 execute()方法，当业务控制器处理完请求后，根据处理结果，该方法返回一个字符串——每个字符串对应一个视图名。

Struts 2 采用了 JavaBean 的风格，要访问数据，就要给每个属性提供 getter 和 setter 方法。每一个请求参数和表单提交的数据都可以作为 Action 的属性，因此可以通过 setter 方法来获得请求参数或通过表单提交的数据。

在 restaurant 项目的 login.jsp 页面的登录部分，定义用户名和密码的输入框，并指定它们的 name 属性分别为 loginName 和 loginPwd；而在 LoginAction 业务类中定义两个属性：loginName 和 loginPwd，分别对应登录页面表单中两个元素的 name 属性值，并为属性设置 setter 和 getter 方法。当客户端发送的表单请求被 StrutsPrepareAndExecuteFilter 转发给该 Action 时，该 Action 就自动通过 setter 方法获得从表单提交过来的数据信息。

编程人员开发出系统所需的业务控制器后，还需要配置 Struts 2 的 Action，即需要在 struts.xml 中配置 Action 的如下三部分。

(1) Action 中所处理的 URL。
(2) Action 组件所对应的实现类。
(3) Action 返回的逻辑视图和物理资源之间的对应关系。

每个 Action 都要处理一个包含指定 URL 的用户请求，当 StrutsPrepareAndExecuteFilter 拦截到用户请求后，根据请求的 URL 和 Action 处理 URL 之间的对应关系来处理转发。

6.4.3 模型组件

对 Struts 2 框架而言，通常没有为模型组件的实现提供太多的帮助。Java EE 应用中的模型组件，通常是指系统的业务逻辑组件。而隐藏在系统的业务逻辑组件下面的，可能还包含 DAO、领域对象等组件。

通常，MVC 框架里的业务控制器会调用模型组件的方法来处理用户请求。也就是说，业务逻辑控制器不会对用户请求进行任何实际处理，用户请求最终由模型组件负责处理。业务控制器只是中间负责调度的调度器，这也是称 Action 为业务控制器的原因。图 6-10 显示了这种核心控制器调用业务逻辑组件的处理流程。

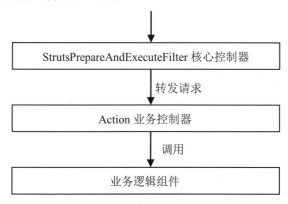

图 6-10 核心控制器调用业务逻辑组件的处理流程

在图 6-10 中,可以看到 Action 调用业务逻辑组件的方法。当核心控制器需要获得业务逻辑组件的实例时,通常并不会直接获取业务逻辑组件的实例,而是通过工厂模式来获得业务逻辑组件的实例;或者利用其他 IoC(控制反转)容器(如 Spring 容器)来管理业务逻辑组件的实例。

6.4.4 视图组件

视图是 MVC 中一个非常重要的因素,Struts 2 可以使用 HTML、JSP、FreeMarker 等多种视图技术。Action 业务类在处理完客户端请求后,会返回一个字符串,作为逻辑视图名。逻辑视图并未与任何视图技术关联,仅仅是返回一个字符串。在 struts.xml 配置文件中,要为 Action 元素指定系列<result.../>子元素,每个<result.../>子元素定义一个逻辑视图和物理视图之间的映射,根据返回的字符串,指向对应的视图组件,显示处理结果,情况如下。

(1) Action 向视图组件输出数据信息,然后由视图组件将这些数据信息显示出来。例如,在 LoginAction 类中获得用户输入的用户名和密码信息,登录成功后跳转到首页面/ch06/index.jsp。

(2) Action 并没有向视图组件输出数据信息,只是根据处理结果进行简单的页面跳转。例如,在登录示例中,当登录失败后跳转到登录页面/ch06/login.jsp。

Struts 2 默认使用 JSP 作为视图资源,在登录示例中,使用 JSP 技术作为视图,故配置<result.../>子元素时没有指定 type 属性。若需要使用其他视图技术,可在配置<result.../>子元素时,指定相应的 type 属性,如<result name="success" type="freemarker">。

6.5 小　　结

本章主要介绍了 Struts 2 框架的基础知识,关于 Struts 2 的 MVC 设计模式,Struts 2 的工作原理,获取 Struts 2 资源以及 Struts 2 系统架构。通过登录示例讲解了 Struts 2 的基本运行流程,以及涉及的相关控制器和组件。Struts 2 框架极大地简化了程序员的工作,只需要简单配置即可开发 Java Web 应用程序。

第 7 章
Struts 2 的配置

通过上一章的学习,我们已经了解了 Struts 2 的基本使用情况。本章带领大家深入了解 Struts 2 的配置,重点介绍 struts.xml 文件中各元素的含义,只有掌握配置文件的用法,才能更好地使用和扩展 Struts 2 框架的功能。

7.1 Struts 2 的配置文件

配置文件也是 Struts 2 应用程序的核心部分。Struts 2 框架的配置文件分为内部使用和供开发人员使用两类。内部配置文件由 Struts 2 框架自动加载，对其自身进行配置；其他的配置文件由开发人员使用，用于对 Web 应用进行配置，配置文件包括 web.xml、struts.xml、struts.properties 等。

7.1.1 web.xml 文件

准确地说，web.xml 并不是 Struts 2 框架特有的文件，作为部署描述符，web.xml 是所有 Java Web 应用程序都需要的核心文件。

Struts 2 框架需要在 web.xml 中配置其核心控制器——StrutsPrepareAndExecuteFilter，用于对框架进行初始化，以及处理所有的请求，对于核心控制器 StrutsPrepareAndExecuteFilter 的配置请参见 6.3.3 小节。

不同版本的 Struts 2 的核心控制器是不同的，如 Struts 2.1.3 以下版本的核心控制器为 org.apache.struts2.dispatcher.FilterDispatcher，Struts 2.1.3 到 Struts 2.5 之间版本的核心控制器为 org.apache.struts2.dispatcher.ng.filter.StrutsPrepareAndExecuteFilter，而 Struts 2.5 以上版本的核心控制器为 org.apache.struts2.dispatcher.filter.StrutsPrepareAndExecuteFilter。

7.1.2 struts.xml 文件

Struts 2 的核心配置文件就是 struts.xml 配置文件，由程序开发人员编写，包含 action、result 等配置，主要负责管理 Struts 2 框架的业务控制器 Action。

struts.xml 文件通常放在/WEB-INF/classes/目录下，在该目录下的 struts.xml 文件可以被 Struts 2 框架自动加载。如果是在 MyEclipse IDE 环境下，进行 Struts 2 的配置，一定要将 struts.xml 文件放到项目的 src 文件夹的根目录下。在使用 MyEclipse 部署到 Tomcat 等 Web 容器的时候，才会自动将 struts.xml 刷新到/WEB-INF/classes/文件夹的下面。一个典型的 struts.xml 文件代码如下：

```xml
<?xml version="1.0" encoding="UTF-8" ?>
<!DOCTYPE struts PUBLIC "-//Apache Software Foundation//DTD Struts
Configuration 2.5//EN" "http://struts.apache.org/dtds/struts-2.5.dtd">
<struts>
    <constant name="常量名" value="常量的值" />
    <include file="包含的文件名"></include>
    <package name="包名" namespace="命名空间名" extends="继承包名" >
        <action name="action请求名" class="包名.Action类名" method="方法名">
            <result name="返回的字符串值1">/视图资源1</result>
            <result name="返回的字符串值2">/视图资源2</result>
        </action>
    </package>
    …
</struts>
```

(1) constant 元素。该元素用于常量的配置，可以改变 Struts 2 的一些行为，从而满足不同应用的需求，constant 元素包括 name(表示常量的名称)和 value(表示常量的值)属性。

例如，处理中文乱码问题时，可以通过在 struts.xml 文件中设置常量的方法解决。代码如下：

```
<constant name="struts.i18n.encoding" value="utf-8" />
```

(2) include 元素。在大部分 Web 应用里，随着应用规模的增加，系统中的 Action 数量也大量增加，导致 struts.xml 配置文件变得非常臃肿。为了避免 struts.xml 文件过于庞大，提高 struts.xml 文件的可读性，可以将一个 struts.xml 配置文件分解成多个配置文件，然后在 struts.xml 文件中包含其他配置文件。

例如，将 struts-part1.xml 文件通过手动的方式导入 struts.xml 文件中。代码如下：

```
<!--通过include包含其他xml的配置文件-->
<include file="struts-part1.xml"></include>
…
```

<include>元素引用的 xml 文件必须是完整的 Struts 2 配置文件，实际上在<include>元素引用文件时，是单独地解析每个 xml 文件。

(3) package 元素。Struts 2 框架会把 action、result 等组织在一个名为 package(包)的逻辑单元中，从而简化维护工作，提高重用性，每一个包都包含 Action、Result 等定义。

Struts 2 的包很像 Java 中的包，但不同的是，Struts 2 中的包可以"继承"定义好的包，从而继承原有包的所有定义(包括 Action、Result 等的配置)，并且可以添加自己包的配置。

- name：该属性为必需的，并且是唯一的，用来指定包的名称(可以被其他包引用)。
- extends：该属性类似 Java 的 extends 关键字，指定要扩展的包。
- namespace：该属性是一个可选属性，该属性定义该包中 Action 的命名空间，默认命名空间用""表示，以"/"表示根命名空间。

(4) action 元素。用于配置 Struts 2 框架的"工作单元"Action 类，action 元素将一个请求的 URL(Action 的名字)对应到一个 Action 类。name 属性是必需的，用来表示 Action 的名字；class 属性是可选的，用于设定 Action 类。

(5) result 元素。用来设定 Action 类处理结束后，系统下一步要做什么，name 属性表示 result 的逻辑名，用于与 Action 类返回的字符串进行匹配，result 元素的值用来指定物理视图即对应的实际资源的位置。代码如下：

```
<action name="login.action" class="com.restaurant.action.LoginAction">
    <result name="success">/index.jsp</result>
</action>
```

在开发过程中，一般情况下，我们所定义的包应该总是扩展 struts-default 包。struts-default 包由 Struts 2 框架定义，其中配置了大量常用的 Struts 2 的特性。没有这些特性，就连简单的在 action 中获取请求数据都无法完成。

struts-default.xml 文件是 Struts 2 框架的默认配置文件，为框架提供默认设置，该配置文件会自动加载。struts-default 包在 struts-default.xml 文件中定义，该文件的结构如下：

```xml
<?xml version="1.0" encoding="UTF-8" ?>
<!DOCTYPE struts PUBLIC
    "-//Apache Software Foundation//DTD Struts Configuration 2.5//EN"
    "http://struts.apache.org/dtds/struts-2.5.dtd">
<struts>
    <constant name="struts.excludedClasses" value="" />
    <bean class="com.opensymphony.xwork2.ObjectFactory" name="struts"/>
    …
    <bean type="ognl.PropertyAccessor" name="java.util.HashMap" class="com.opensymphony.xwork2.ognl.accessor.XWorkMapPropertyAccessor"/>
    <package name="struts-default" abstract="true">
        <result-types>
            <result-type name="chain" class="com.opensymphony.xwork2.ActionChainResult"/>
            …
        </result-types>
        <interceptors>
            <interceptor name="alias" class="com.opensymphony.xwork2.interceptor.AliasInterceptor"/>
            …
            <!-- Basic stack -->
            <interceptor-stack name="basicStack">
                <interceptor-ref name="exception"/>
                …
            </interceptor-stack>
            …
        </interceptors>
        <default-interceptor-ref name="defaultStack"/>
        <default-class-ref class="com.opensymphony.xwork2.ActionSupport" />
    </package>
</struts>
```

上述代码只列出了 struts-default.xml 文件的基本结构，我们自己写的 struts.xml 是不是和它很相似。我们可以在项目的 Web App Libraries 下的 struts-core-2.5.8.jar 中，找到 struts-default.xml 文件。Struts 2 框架每次都会自动加载 struts-default.xml 文件，此文件定义了 Struts 2 的默认包，里面包含许多需要的拦截器和结果，通过这些配置，Struts 2 自动帮助完成属性注入等工作。

7.1.3　struts.properties 文件

Struts 2 框架除了 struts.xml 文件外，还包含 struts.properties 文件，该文件定义了 Struts 2 框架的常量(也称为 Struts 2 属性)，开发者可以通过该文件管理 Struts 2 的常量，以满足需求。

struts.properties 文件是一个标准的 Properties 文件，该文件放在和 struts.xml 同样的目录中。在 MyEclipse 中，编译时会自动将 src 下的 struts.properties 文件编译后加载到 WEB-INF/classes 路径下。

该文件包含系列的 key-value 对象，每个 key 就是一个 Struts 2 常量，该 key 对应的 value 就是一个 Struts 2 常量值。例如，前面通过在 struts.xml 文件中使用<constant…/>元素设置常量，下面介绍在 struts.properties 文件中实现常量的赋值，代码如下：

```
struts.i18n.encoding=utf-8
struts.devMode=true
```

还有一个 struts-plugin.xml 文件是 Struts 2 插件使用的配置文件,如果不是用插件开发,不需要编写这个配置文件。

7.2 Struts 2 的 Action 实现

Action 是 Struts 2 应用的核心,用于处理用户的请求,因此 Action 也被称为业务控制器。每个 Action 类就是一个工作单元,Struts 2 框架负责将用户的请求与相应的 Action 匹配,如果匹配成功,则调用该 Action 类对用户请求进行处理,而匹配规则需要在 Struts 2 的配置文件中声明。在 Struts 2 框架下实现 Action 类有如下三种方式。

- 普通的 POJO 类,该类包括一个无参数的 execute()方法,返回值为字符串。
- 实现 Action 接口。
- 继承 ActionSupport 类。

7.2.1 POJO 的实现

在 Action 中,如果需要传递的参数有多个(如登录示例中的用户名和密码字段等),就需要在 Action 中定义相应的属性来记录这些信息,这样就会变得很麻烦。如果使用 POJO,将不用在 Action 类中定义这些属性,而采用类似 JavaBean 的方式,从而使代码变简洁。

在 Struts 2 中,Action 可以不继承特殊的类或不实现任何特殊的接口,仅仅是一个 POJO。POJO 全称是 Plain Ordinary Java Object(普通的 Java 对象),只要具有一部分 getter/setter 方法的类就可以称作 POJO。POJO 是一个简单、正规的 Java 对象,包含业务逻辑处理或持久化逻辑等,不具有任何特殊角色和不继承或不实现任何其他 Java 框架的类或接口。

在 POJO 中,要有一个无参数的 execute()方法,还要有一个公共的无参数的构造方法,默认的构造方法就可以,定义格式如下:

```
public String execute() throws Exception {
    …
}
```

execute()方法的要求如下。

- 方法的作用范围为 public。
- 返回一个字符串,就是指示的下一个页面的 Result。
- 不需要传入参数。
- 可以抛出 Exception 异常,当然也可以不抛出异常。

也就是说,任意一个满足上述要求的 POJO 都可算作 Struts 2 的 Action 实现,但在实际的开发中,通常会自己编写 Action 类实现 Action 接口或继承 ActionSupport 类。

7.2.2 实现 Action 接口

虽然 Struts 2 框架并没有强加很多要求,但为了让 Action 类更规范,使不同的开发人员编写的 execute()方法返回的字符串风格一致,Struts 2 提供了一个 Action 接口,用于定义 Action 类应该实现的通用规范。

可以在\struts-2.5.8\src\core\src\main\java\com\opensymphony\xwork2 的路径下找到 Action 接口的定义规范,标准的 Action 接口的代码如下:

```java
package com.opensymphony.xwork2;
public interface Action {
    // 以下定义处理完请求后返回的字符串常量
    public static final String SUCCESS = "success";
    public static final String NONE = "none";
    public static final String ERROR = "error";
    public static final String INPUT = "input";
    public static final String LOGIN = "login";
    // 以下定义用户请求处理的抽象方法 execute()
    public String execute() throws Exception;
}
```

该接口规范定义 Action 需要包含一个抽象方法 execute(),该方法返回一个字符串,除此之外该方法还预定义了五个字符串常量,可用于返回一些预定的 result。

自己编写的 Action 类通常都要实现 com.opensymphony.xwork2.Action 接口,并实现 Action 接口中的 execute()方法,代码示例如下:

```java
import com.opensymphony.xwork2.Action;
public class HelloAction implements Action{
//省略
public String execute() throws Exception {
    …
}
}
```

开发者在自己编写的 Action 类中,用其他字符串作为逻辑视图名也是可以的。

7.2.3 继承 ActionSupport

由于 Action 接口简单,为开发者提供的帮助较小,Struts 2 框架为 Action 接口提供了一个实现类 ActionSupport。该类提供了许多默认方法,如默认处理用户请求的方法、数据校验的方法、获取国际化信息的方法等。ActionSupport 类是 Struts 2 默认的 Action 处理类,可以在\struts-2.5.8\src\core\src\main\java\com\opensymphony\xwork2 的路径下找到 ActionSupport 类文件,部分代码如下:

```java
package com.opensymphony.xwork2;
import com.opensymphony.xwork2.*;
import org.apache.logging.log4j.LogManager;
…
public class ActionSupport implements Action, Validateable, ValidationAware,
```

```java
TextProvider, LocaleProvider, Serializable{
    protected static Logger LOG =
LoggerFactory.getLogger(ActionSupport.class);
    private final ValidationAwareSupport validationAware = new
ValidationAwareSupport();
    private transient TextProvider textProvider;
    private Container container;
    //收集校验错误的方法
    public void setActionErrors(Collection<String> errorMessages){
        validationAware.setActionErrors(errorMessages);
    }
    //返回校验错误的方法
    public Collection<String> getActionErrors() {
        return validationAware.getActionErrors();
    }
    //默认 input()方法，返回 input 字符串
    public String input() throws Exception {
        return INPUT;
    }
    //默认处理用户请求的方法，默认返回 success 字符串
    public String execute() throws Exception {
        return SUCCESS;
    }
    //输入校验方法，这是一个空方法，需要用户自己实现这个方法
    public void validate() {
    }
    //省略其他方法，读者可查看相应的 Struts 2 帮助文档
}
```

ActionSupport 实现了 Action 接口和很多的实用接口，选择从 ActionSupport 继承，可以大大地简化 Action 的开发。在 struts.xml 中，如果<action>元素中没有填写 class 属性，那么默认 ActionSupport 类作为 action 的处理类。

7.2.4 Struts 2 支持 Java 对象

Struts 2 框架支持使用 Java 对象来接收用户输入的数据，还是以登录为例，在开发过程中通常以实体对象(JavaBean)来保存信息，在 LoginAction 中同样可以使用 JavaBean 来接收用户的输入，示例代码如下：

```java
public class LoginAction extends ActionSupport {
    private Users user;          // Users 实体类对象
    // 省略属性的 getter、setter 方法
    @Override
    public String execute() throws Exception {
        // 省略代码
        // 返回字符串
    }
}
```

在使用 Java 对象时，首先新建 com.restaurant.entity 包，并创建一个 Users 的 Java 对象，包括 loginName 和 loginPwd 属性，并设置相应的 getter、setter 方法和构造方法。

现在只需要修改登录页面的表单内容，LoginAction 就可以使用 JavaBean 来接收用户输入的数据了，修改部分如下：

```
用户名：<input type="text" name="user.loginName"><br><br>
密  码：<input type="password" name="user.loginPwd"><br>
```

根据 Struts 2 框架的数据转移机制，传递 user.*请求参数等同于调用 LoginAction.getUser().setLoginName(……)。我们都注意到 LoginAction 类中，并没有创建任何 Users 类的实例，按一般常识，程序应该抛出异常才对，但是在 Struts 2 框架中是不会有问题的，Struts 2 框架会自动实例化任何用于填充数据的对象。

7.2.5 Struts 2 访问 Servlet API

在 Web 开发中，经常会用到 Servlet API 中的对象，Struts 2 框架可以让我们直接访问和设置 Action 类及传递的数据，这就大大降低了与 Servlet API 的耦合。但在某些情况下，可能需要在 action 中访问 Servlet API 中的对象，例如用户登录成功，就应该将用户信息保存到 HttpSession 对象中。Struts 2 访问 Servlet API 中的对象有如下几种方式。

1．与 Servlet API 的解耦访问方式

为了避免与 Servlet API 耦合在一起，Struts 2 框架对 Servlet API 中的 HttpServletRequest、HttpSession 和 ServletContext 进行了封装，构造了三个 Map 对象来替代。在 Action 类中，可以直接访问 HttpServletRequest、HttpSession 和 ServletContext 对应的 Map 对象。Struts 2 提供了 com.opensymphony.xwork2.ActionContext 类获取 Servlet API 中对应的 Map 对象。ActionContext 是 Action 执行的上下文，在 ActionContext 中保存了 Action 类执行所需要的一组对象，可以通过 HttpServletRequest、HttpSession 和 ServletContext 获取对应的 Map 对象。

（1）public Object get(Object key)：ActionContext 类没有提供 getRequest()这样的方法来获取封装了 HttpServletRequest 对象的 Map 对象，需要为 get()方法传递 request 参数，示例如下：

```
ActionContext ac=ActionContext.getContext();
Map request=(Map)ac.get("request");
```

（2）public Map getSession()：获取封装了 HttpSession 对象的 Map 对象，示例如下：

```
ActionContext ac=ActionContext.getContext();
Map session = ac.getSession();
```

（3）public Map getApplication()：获取封装了 ServletContext 对象的 Map 对象，示例如下：

```
ActionContext ac=ActionContext.getContext();
Map application = ac.getApplication();
```

除了利用 ActionContext 来获取 HttpServletRequest、HttpSession 和 ServletContext 对应的 Map 对象这种方式外，Action 类还可以实现某些特定的接口，让 Struts 2 在运行时向 Action 实例注入 HttpServletRequest、HttpSession 和 ServletContext 对应的 Map 对象，这些接口

如下。

(1) org.apache.struts2.interceptor.RequestAware：向 Action 实例中注入 HttpServletRequest 对象对应的 Map 对象，该接口只有一个方法：public void setRequest(Map request)。

(2) org.apache.struts2.interceptor.sessionAware：向 Action 实例中注入 HttpSession 对象对应的 Map 对象，该接口只有一个方法：public void setSession(Map session)。

(3) org.apache.struts2.interceptor.ApplicationAware：向 Action 实例中注入 ServletContext 对象对应的 Map 对象，该接口只有一个方法：public void setApplication(Map application)。

2．与 Servlet API 的耦合访问方式

直接访问 Servlet API 将使 Action 类与 Servlet API 耦合在一起，众所周知，Servlet API 对象是由 Servlet 容器构造的，与这些对象绑定在一起，测试过程中必须有 Servlet 容器，这就不便于 Action 类的测试，但有时候，确实需要直接访问这些对象。Struts 2 提供了直接访问 Servlet API 对象的方式，即直接获取 org.apache.struts2.ServletActionContext 类，该类是 ActionContext 类的子类，该类的几个方法如下。

(1) static PageContext getPageContext()：得到 PageContext 对象，对应内置对象 page。

(2) public static HttpServletRequest getRequest()：得到 HttpServletRequest 对象。

(3) public static ServletContext getServletContext()：得到 ServletContext 对象。

(4) public static HttpServletResponse getResponse()：得到 HttpServletResponse 对象。

ServletActionContext 类并没有定义获得 HttpSession 对象的方法，HttpSession 对象可以通过 HttpServletRequest 对象来得到。

除了利用 ServletActionContext 来直接获取 Servlet API 对象外，Action 类还可以实现特定的 xxxAware 接口，由 Struts 2 框架向 Action 实例注入 Servlet API 对象。

(1) org.apache.struts2.util.ServletContextAware：向 Action 实例中注入 ServletContextAware 对象，该接口只有一个方法：public void setServletContext(ServletContext context)。

(2) org.apache.struts2.interceptor.ServletRequestAware：向 Action 实例中注入 HttpServletRequest 对象，该接口只有一个方法：void SetServletRequestAware(HttpServletRequest request)。

(3) org.apache.struts2.interceptor.ServletResponseAware：向 Action 实例中注入 HttpServletResponse 对象，该接口只有一个方法。

3．Struts 2 访问 Servlet API 示例

下面通过示例来演示 Struts 2 访问 Servlet API 的过程。

(1) 在 restaurant 项目的 src 中新建 com.restaurant.entity 包，并在该包中新建一个 Users 的实体类，暂时设定登录名和登录密码两个字段，代码如下：

```
package com.restaurant.entity;
public class Users {
    private String loginName;
    private String loginPwd;
    // 省略属性 loginName、loginPwd 的 getter、setter 方法
}
```

(2) 在 com.restaurant.action 包中新建一个 MessageAction 的业务类，该类继承

ActionSupport 并实现 ServletRequestAware 接口，代码如下：

```java
package com.restaurant.action;
import java.util.Map;
import javax.servlet.http.HttpServletRequest;
import org.apache.struts2.ServletActionContext;
import org.apache.struts2.interceptor.ServletRequestAware;
import com.opensymphony.xwork2.*;
import com.restaurant.entity.Users;
public class MessageAction extends ActionSupport implements ServletRequestAware {
    private Users user;
    // 省略属性 user 的 getter、setter 方法
    private HttpServletRequest request;
        @Override     //注入 HttpServletRequest 对象
    public void setServletRequest(HttpServletRequest request) {
        this.request=request;
    }
    @Override       // 默认方法
    public String execute() throws Exception {
        ActionContext ac=ActionContext.getContext();
        Map session=ac.getSession();
        // 登录的用户名和密码判断
        if ("admin".equals(user.getLoginName()) && "123".equals(user.getLoginPwd())) {
            session.put("LOGIN_USER", user);
            ac.put("success", "登录成功，通过 ActionContext 类访问 Servlet API！");
            request.setAttribute("messageAware", "您好，通过 xxxAware 接口访问 Servlet API！");
            return "success";         //返回 success 字符串
        }else{
            ac.put("error", "用户名或密码错误，登录失败！");
            ServletActionContext.getRequest().setAttribute("messageSAC", "您好，通过 ServletActionContext 类直接访问 Servlet API！");
            return "error";          //返回 error 字符串
        }
    }
}
```

（3）在 WebRoot 目录下新建一个 ch07 文件夹，新建 login.jsp、success.jsp、error.jsp 文件。

登录页面 login.jsp 的表单部分代码如下：

```
<form name="form1" method="post" action="messageAction">
    用户名：<input type="text" name="user.loginName"><br><br>
    密  码：<input type="password" name="user.loginPwd">
        <br><br>
    <input type="submit" value="登录">
    <input type="reset" value="取消">
</form>
```

登录成功页面 success.jsp 的代码如下：

```
<body>
```

```
    欢迎 ${sessionScope.LOGIN_USER.loginName},
    <p align="center">${requestScope.success }</p>
    <p align="center">${requestScope.messageAware }</p>
</body>
```

登录失败页面 error.jsp 的代码如下：

```
<p align="center">${requestScope.error }</p>
<p align="center">${requestScope.messageSAC }</p>
```

(4) 在 struts.xml 文件中添加相应的配置如下：

```
<package name="restaurant" namespace="/" extends="struts-default">
    <!-- Struts 2 访问 Servlet API 示例的配置 -->
    <action name="messageAction"
            class="com.restaurant.action.MessageAction">
        <result name="success">/ch07/success.jsp    </result>
        <result name="error">/ch07/error.jsp</result>
    </action>
</package>
```

(5) 重新部署项目，在浏览器中输入 http://localhost:8080/restaurant/ch07/login.jsp，运行的登录页的效果如图 7-1 所示，登录成功页面如图 7-2 所示，登录失败页面如图 7-3 所示。

图 7-1 登录页面

图 7-2 登录成功页面　　　　　　　　图 7-3 登录失败页面

7.3　Action 配置

在 struts.xml 文件中，需要对 Struts 2 的 Action 类进行相应的配置，struts.xml 文件可以比喻成视图和 Action 之间联系的纽带。每个 Action 都是一个业务逻辑处理单元，Action 负责接收客户端请求、处理客户端请求，最后将处理结果返回给客户端，这一系列过程都是在 struts.xml 文件中进行配置才得以实现的。

7.3.1 Struts 2 中 Action 的作用

对 Struts 2 程序应用的开发者而言，Action 才是应用的核心，开发者需要提供大量的 Action 类，并在 struts.xml 文件中配置 Action。Action 主要有如下三个作用。

(1) 为给定的请求封装需要做的实际工作(调用特定的业务处理类)。

可以把 Action 看作控制器的一部分，它的主要职责就是控制业务逻辑，通常 Action 使用 execute()方法实现这一功能。

(2) 为数据的转移提供场所。

Action 作为数据转移的场所，也许你认为这会使 Action 变得复杂，但实际上这使得 Action 更简洁，由于数据保存在 Action 中，在控制业务逻辑的过程中可以非常方便地访问到它们。

(3) 帮助框架决定由哪个结果呈现请求响应。

Action 的最后一个职责是返回结果字符串。Action 根据业务逻辑执行的返回结果判断返回何种结果字符串，根据框架 Action 返回的结果字符串选择相应的视图组件呈现给用户。

7.3.2 配置 Action

Action 映射是框架中的基本"工作单元"。Action 映射就是将一个请求的 URL 映射到一个 Action 类，当一个请求匹配某个 Action 名称时，框架就使用这个映射来确定如何处理请求。在 struts.xml 文件中，通过<action>元素对请求的 Action 地址映射和 Action 类进行配置。<action>元素的属性介绍如下。

(1) name：必选属性，指定客户端发送请求的地址映射名称。
(2) class：可选属性，指定 Action 实现类所在的包名+类名。
(3) method：可选属性，指定 Action 类中的处理方法名称。
(4) converter：可选属性，应用于 Action 的类型转换器的完整类名。

在实际开发中，通常都是将每个<package>放在一个单独的文件中，例如叫作 struts-xxx.xml，最后由 struts.xml 通过<include>元素引用这些 struts-xxx.xml 文件。

7.3.3 动态方法调用

在实际应用中，随着应用程序不断地扩大，我们不得不管理数量庞大的 Action。例如，一个系统中，用户的操作可分为登录和注册两部分，若一个请求对应一个 Action，我们就需要编写两个 Action 处理用户的请求。在具体开发过程中，为了减少 Action 的数量，通常在一个 Action 中编写不同的方法(必须遵守与 execute()方法相同的格式)处理不同的请求，如编写 LoginAction，其中 login()方法处理登录请求，register()方法处理注册请求。此时可以采用动态方法调用 DMI(Dynamic Method Invocation)来处理。动态方法调用是指表单元素的 action 并不是直接等于某个 Action 的名称。

采用动态方法调用时，在 Action 的名字中使用"！"来标识要调用的方法名称，格式如下：

```
<form action="Action 名字!方法名字">
```

使用动态方法调用的方式将请求提交给 Action 时，表单中的每个按钮提交事件都可交给同一个 Action，只是对应 Action 中的不同方法。这时，在 struts.xml 文件中只需要配置该 Action，而不需要配置每个方法，格式如下：

```
<action name="Action 名字" class="包名.Action 类名" >
    <result> 物理视图 URL </result>
</action>
```

官网不推荐使用这种方式，建议大家尽量不要使用，因为动态方法的调用可能会带来安全隐患(通过 URL 可以执行 Action 中的任意方法)，所以在确定使用动态方法调用时，应该确保 Action 类中的所有方法都是普通的、开放的方法。基于此原因，Struts 2 框架提供了一个属性的配置，用于禁止调用动态方法。在 struts.xml 中，通过<constant>元素将 struts.enable.DynamicMethodInvocation 设置为 false，来禁止动态方法调用，代码如下：

```
<constant name="struts.enable.DynamicMethodInvocation" value="false"/>
```

要在 Struts 2.5 版本中使用动态方法调用，除了将上面的属性设置为 true 外，还要添加语句<global-allowed-methods>regex:.*</global-allowed-methods>，不然就会出现错误。

7.3.4　用 method 属性处理调用方法

在 struts.xml 文件中配置<action>元素时，若省略 method 属性，则调用的是 execute()方法；若指定 method 属性，则可以让 Action 调用指定的方法来处理用户的请求，而不是使用 execute()方法来处理。下面通过示例来演示指定 method 属性处理用户的登录和注册。新建 login1.jsp、success1.jsp、error1.jsp 和 register1.jsp 文件，新建 LRAction，修改 struts.xml 配置文件，步骤如下：

(1) 在/WebRoot/ch07/目录下，新建登录页面 login1.jsp，代码如下：

```
<form name="form1" method="post" action="loginAction">
    用户名：<input type="text" name="user.loginName"><br><br>
    密    码：
    <input type="password" name="user.loginPwd"><br><br>
    <input type="submit" value="登录">
    <input type="button" value="注册" onclick="javascript:window.location.href='registerAction'">
</form>
```

在 WebRoot/ch07 目录下新建 success1.jsp 和 error1.jsp 页面，给出"登录成功！"和"登录失败！"的提示；新建 register1.jsp 页面，这里我们只要给出简单提示即可。

(2) 在 com.restaurant.action 包中新建一个 LRAction 业务类，代码如下：

```
package com.restaurant.action;
```

```java
import com.opensymphony.xwork2.ActionSupport;
import com.restaurant.entity.Users;
public class LRAction extends ActionSupport {
    private Users user;
    // 省略属性 user 的 getter、setter 方法
    public String login(){
        if ("admin".equals(user.getLoginName())&&
"123".equals(user.getLoginPwd())) {
            return "success";           //返回 success 字符串
        }else{
            return "error";             //返回 input error 字符串
        }
    }
    public String register(){
        return "register";
    }
}
```

(3) 修改 struts.xml 文件的配置代码如下：

```xml
<!-- method 方法动态调用 -->
<action name="loginAction" class="com.restaurant.action.LRAction" method="login">
    <result name="success">/ch07/success1.jsp </result>
    <result name="error">/ch07/error1.jsp </result>
</action>
<action name="registerAction" class="com.restaurant.action.LRAction" method="register">
    <result name="register">/ch07/register1.jsp </result>
</action>
```

重新部署 restaurant 项目，运行 http://localhost:8080/restaurant/ch07/login1.jsp，登录页面效果如图 7-4 所示。成功页面、失败页面和注册页面主要就是提示语句。

图 7-4　登录页面

上面定义的两个请求分别为 loginAction 和 registerAction，它们所对应的业务处理类都是 com.restaurant.action.LRAction。但 method 属性指定 login 处理逻辑的方法是 login()方法，而 method 属性指定 register 处理逻辑的方法是 register()方法。

使用 method 属性可以指定任意方法请求(只要该方法和 execute 方法具有相同的格式)。从安全角度出发，建议采用 method 属性来实现用同一个 Action 的不同方法处理不同的请求，这样的处理方式会减少 Action 的实现类。但随着 Action 的增多，会导致大量的 Action 配置，因此这样做容易导致重复，而使用通配符是一种解决 Action 配置过多的很好的方法。

7.3.5 使用通配符

在使用 method 属性时，由于在 Action 类中有多个业务逻辑处理方法，在配置 Action 时，就需要使用多个 action 元素。在实现同样功能的情况下，为了减轻 struts.xml 配置文件的负担，就需要借助通配符映射。

Struts 2 提供了通配符 "*"，利用通配符可以在定义 Action 的 name 属性时使用模式字符串(即用 "*" 代表一个或多个任意字符串)，接下来就可以在 class、method 属性以及<result>子元素中使用{N}形式的表达式，代表前面第 N 个星号 "*" 所匹配的字符串。使用通配符的原则是约定高于配置，它实际上是另一种形式的动态调用方法。在项目中，有很多命名规则是约定的，如果使用通配符就必须有一个统一的约定，否则通配符将无法成立。示例代码如下：

```
<action name="*Action" class="com.restaurant.action.LRAction" method="{1}">
    <result name="success">/ch07/success1.jsp </result>
    <result name="error">/ch07/{1}1.jsp </result>
    <result name="register">/ch07/{1}1.jsp </result>
</action>
```

在 action 元素的 name 属性中使用了星号(*)，允许这个 Action 匹配所有以 Action 结束的 URL，如/loginAction.action。配置该 action 元素时，还指定了 method 属性，该属性使用了一个表达式{1}，该表达式的值就是 name 属性值中第一个"*"的值。例如，当请求为/loginAction.action 时，通配符匹配的是 login，那么这个 login 值将替换{1}，最终请求/loginAction.action，将由 LRAction 类中的 login()方法执行。

这里的 name 属性值只有一个"*"，还可以有两个、三个、四个，比如可以写成 name ="*_*"，这样就有两个"*"，此时我们就可以使用{1}、{2}分别表示每个"*"的内容，示例代码如下：

```
<action name="*_*" class=" com.restaurant.action.{1}Action" method="{2}">
    <result name="success">/success.jsp</result>
    <result name="input">/{2}.jsp</result>
</action>
```

上面配置了一个模式为 "*_*" 的 Action，即只要匹配该模式的请求，都可以被 Action 处理。其中，class 属性中的 "{1}"，匹配模式 "*_*" 中的第 1 个 "*"；method 属性中的 "{2}" 匹配模式 "*_*" 中的第 2 个 "*"。例如，Login_login.action 会调用 LoginAction 处理类的 login()方法来处理请求。

7.4 Result 配置

Struts 2 的 Action 处理用户请求结束后，返回一个普通字符串——逻辑视图名，必须在 struts.xml 文件中完成逻辑视图和物理视图的映射，才可以让系统转到实际的视图资源。

7.4.1 配置 Result

逻辑视图和物理视图之间的映射是通过在 struts.xml 中配置<result …/>元素来实现的。Result 的配置由两部分组成：一部分是 result 所代表的实际资源的位置及 result 名称；另一部分就是 result 的类型，由 result 元素的 type 属性进行设定。

根据<result>元素在 struts.xml 文件中所在位置的不同，可以将 result 分为以下两种。

1. 局部 result

前面我们配置的<result>是作为<action>的子元素出现的，此时 result 元素称为局部 Result。<result>元素可以有 name 和 type 属性，但这两种属性都不是必需的，具体介绍如下。

(1) name 属性：指定逻辑视图的名称，默认值是 success。

(2) type 属性：指定返回的视图资源的类型，不同的类型代表不同的结果输出，默认值是 dispatcher，表示支持 JSP 视图技术。

示例代码如下：

```xml
<action name="loginAction" class="com.restaurant.action.LoginAction">
    <!-- 配置名称为 success 的结果映射，结果类型为 dispatcher -->
    <result name="success" type="dispatcher">
        <param name="location">/ch07/success.jsp</param>
    </result>
</action>
```

配置<result>元素时如果没有指定 name 和 type 属性值，则系统将使用默认的 name 属性值(success)和默认的 type 属性值(dispatcher)。param 子元素的 name 属性有如下两个值。

(1) location：指定该逻辑视图所对应的实际视图资源。

(2) parse：指定在视图资源名称中是否可以使用 OGNL 表达式。默认值为 true，表示可以使用；如果为 false，表示不支持 OGNL 表达式。

在配置局部 result 时，代码可以简化如下：

```xml
<action name="loginAction" class="com.restaurant.action.LoginAction">
    <result>/ch07/success.jsp </result>
</action>
```

2. 全局 result

因为局部 result 只能由本<action>元素访问，不能被其他 Action 使用，在有些情况下，多个 Action 可能需要访问同一个结果，这时我们需要配置全局 Result 来满足多个 Action 共享一个结果的要求。全局 result 在 package 元素的<global-results>子元素中指定，全局 result 的作用范围是对所有的 Action 都有效。配置全局 result 的示例代码如下：

```xml
<package name="restaurant " namespace="/" extends="struts-default">
    <!-- 配置全局 result -->
    <global-results>
        <result name="success">/ch07/success.jsp</result>
    </global-results>
    <action name="loginAction" class="com.restaurant.action.LoginAction"/>
</package>
```

如果一个 Action 中包含与全局 result 同名的局部 result，则局部 result 会覆盖全局 result。即当 Action 处理完用户请求后，首先搜索当前 Action 中的局部 result，当没有匹配的局部 result 时，才会搜索全局 result。在<action>元素中配置的<result>子元素与在<global-results>中配置的<result>子元素属性都是相同的，只是两者的作用范围不同。

7.4.2 Result 的常用结果类型

在 Struts 2 框架中调用 Action 请求处理之后，就要向用户呈现结果视图，Struts 2 支持多种类型的视图，这些视图是由不同的结果类型来管理的。

1．常用结果类型

1) dispatcher 类型

dispatcher 是最常用的结果类型，也是默认的结果类型。Struts 2 在后台使用 Servlet API 的 RequestDispatcher 转发请求。如果不设置 result 元素的 type 属性，默认的 type 类型为 dispatcher，使用 dispatcher 类型其实是将请求转发(forward)到指定的 JSP 资源。对于 dispatcher 的使用范围，除了可以配置常用的 JSP 外，还可以配置其他 Web 资源，比如 Servlet 等。

2) redirect 类型

redirect 类型将请求用来重定向(redirect)到指定的视图资源，该资源可以是 JSP 文件，也可以是相应的 action 请求。使用 redirect 结果类型时，系统实际上会调用 HttpServletResponse 对象的 sendRedirect()方法重定向指定视图资源。

在使用 redirect 时，用户要完成一次与服务器之间的交互，浏览器需要发送两次请求。下面修改 7.3.4 小节中的配置文件的登录部分，演示如何使用 redirect 类型，修改前面的 struts.xml 配置文件如下：

```xml
<!-- 配置映射，使用redirect类型的type -->
<action name="loginAction" class="com.restaurant.action.LRAction"
method="login">
    <result name="success" type="redirect">/ch07/success1.jsp</result>
    <result name="error" type="dispatcher">/ch07/error1.jsp </result>
</action>
```

重新部署程序，运行 http://localhost:8080/restaurant/ch07/login1.jsp，登录成功页面的地址栏显示 success1.jsp，登录失败页面的地址栏显示 loginAction，如图 7-5 和图 7-6 所示。

图 7-5　登录成功页面　　　　　　　　图 7-6　登录失败页面

在上述配置中，result 元素使用 redirect 类型时，在 Action 处理请求后，将重新生成一个新的请求。

3) redirectAction 类型

redirectAction 类型和 redirect 类型的后台工作原理一样，即都是利用 HttpServletResponse 的 sendRedirect()方法将请求重新定向到指定的 URL。但 redirectAction 类型主要用于重定向到 Action。即请求处理完成后，如果需要重定向到一个 Action，那么就使用 redirectAction 类型。

4) chain 类型

chain 类型是一种特殊类型的视图结果，用于在一个 Action 执行完之后链接到另一个 Action 中继续执行，新的 Action 使用上一个 Action 的上下文(ActionContext)，数据也会被传递。

在 Struts 2 开发中，chain 类型也是经常用到的一种结果类型。比如在 Servlet 开发中，一个请求被一个 Servlet 处理后，不是直接产生响应，而是把这个请求传递到下一个 Servlet 继续处理，直到需要的多个 Servlet 处理完成后，才生成响应返回。

在 Struts 2 开发中，也会产生这样的需要，一个请求被一个 Action 处理过后，不是直接产生响应，而是传递到下一个 Action 中继续处理，此时就需要使用 chain 这种结果类型了。

2．其他 Result 类型

除了上面提到的这些 Result，Struts 2 还提供了其他 Result 类型，比如同 velocity、xslt 等的结合，下面做简单的介绍。

(1) freemarker 类型。用来整合 freemarker 模板结果类型。FreeMarker 是一个纯 Java 模板引擎，是一种基于模板来生成文本的工具。

(2) velocity 类型。用来处理 velocity 模板。Velocity 是一个模板引擎，可以将 Velocity 模板转化成数据流的形式，直接通过 Java Servlet 输出。

(3) xslt 类型。用来处理 XML/XSLT 模板，将结果转换成 XML 输出。

(4) httpheader 类型。用来控制特殊 HTTP 行为。

(5) stream 类型。用来向浏览器进行流式输出。

7.4.3 使用通配符动态配置 Result

所谓动态结果，就是在配置时，你不知道执行后的结果是哪一个，在运行时才能知道哪个结果作为视图显示给用户。前面介绍 Action 配置的时候，可以通过在<action/>元素的 name 属性中使用通配符，在 class 或 method 中使用表达式，以便我们动态地决定 Action 的处理类以及处理方法。除此之外，我们还可以在配置<result/>的时候使用表达式动态地调用视图资源，在本章使用通配符配置 Action 的示例中，已经使用通配符动态配置 Result 了，看看下面的配置片段：

```
<action name="*Action" class="com.restaurant.action.LRAction"
    method="{1}">
    <result name="success">/ch07/success1.jsp</result>
    <result name="error">/ch07/{1}1.jsp</result>
    <result name="register">/ch07/{1}1.jsp</result>
</action>
```

上面的代码片段有一个名称为*Action 的 Action，这个 Action 可以处理任何*Action 模式的 Action 请求。例如，有一个用户请求 loginAction，对应的处理类是 LoginAction，处理这个请求的方法就是 login。当系统处理完请求后，返回一个 success 字符串找到与之对应的物理视图，我们采用了动态的视图资源，则访问的就是 login_success.jsp 视图资源文件；如果系统处理完请求后，返回一个 error 字符串，则访问 login_error.jsp 视图资源文件。

7.4.4 通过请求参数动态配置 Result

除了通配符以外，在配置时还可以使用表达式，在运行时，框架根据表达式的值来确定要使用哪个结果。配置<result/>元素不仅可以使用${1}表达式来指定视图资源，还可以使用${属性名}的方式来指定视图资源。在后面的这种配置下，${属性名}中的属性名对应 Action 中的属性名称，而且不仅可以使用这种简单的表达式形式，还可以使用完全的 OGNL 表达式，如${属性名.属性名.属性名}，看如下的配置代码：

```xml
<package name="restaurant " extends="struts-default" namespace="/">
    <default-action-ref name="login"></default-action-ref>
    <action name="*Action" class="com.restaurant.action.LoginAction" method="login">
        <result name="success">/{1}.jsp?userName=${name}</result>
    </action>
</package>
```

当返回转发 JSP 视图资源的时候会附带一个参数，${name}其中的 name 就是 LoginAction 的成员变量。

在 restaurant 项目中，演示通过请求参数动态配置 Result，在/WebRoot/ch07/目录下新建页面 input.jsp，如图 7-7 所示。

图 7-7 动态 Result 配置运行效果

用户输入一个 JSP 页面的文件名称，随后系统转向该响应的资源，input.jsp 页面的代码如下：

```html
<form name="form1" action="pageAction" method="post">
    输入目标页面文件的名称：<input type="text" name="pageName">
    <input type="submit" value="转入"><br><br>
    注意：输入 ch07 路径下存在的文件名称，跳转到相应页面。<br>
        否则，则出现相应的文件不存在的错误。
</form>
```

处理该请求的 PageAction 相对简单，提供一个属性封装相应请求参数，代码如下：

```java
package com.restaurant.action;
```

```
import com.opensymphony.xwork2.ActionContext;
import com.opensymphony.xwork2.ActionSupport;
public class PageAction extends ActionSupport {
    private String pageName;
    //省略属性的setter、getter赋值和取值方法
    @Override
    public String execute() throws Exception {
        ActionContext.getContext().put("info","您已经成功转向到"+pageName+".jsp
页面！");
        return super.execute();
    }
}
```

上面的 execute()方法返回一个 success 字符串，然后在 struts.xml 中配置该 Action，配置文件如下：

```
<action name="pageAction" class="com.restaurant.action.PageAction">
    <result name="success">/ch07/${pageName}.jsp</result>
</action>
```

上面在配置<result>元素的实际资源时，使用表达式来指定实际的资源，要求对应的 Action 类中也要包含 pageName 属性。新建 welcome.jsp 页面，通过${requestScope.info}的 EL 表达式在页面上显示信息。

重新部署项目，当我们在 input.jsp 页面文本框中输入 welcome，然后单击转入，系统将提示转到 welcome.jsp 页面，如图 7-8 所示。

如果在 input.jsp 的页面中，输入任意字符串，然后执行跳转。例如，输入 abc，系统将转入 abc.jsp 页面，但是我们在 ch07 的路径下并没有提供 abc.jsp 页面资源，因此将看到 404 错误，表示无法找到资源，如图 7-9 所示。

图 7-8　跳转成功

图 7-9　未找到指定页面

7.5　小　　结

本章深入介绍了 Struts 2 的配置文件、Action 实现、Action 配置和 Result 配置的相关知识，重点介绍了 Struts 2 的 Action 实现、Action 配置和 Result 配置。在 Struts 2 的 Action 实现中重点介绍了 Struts 2 访问 Servlet API；在 Action 配置中讲解了 Action 的动态调用、指定 method 属性、使用通配符等配置方法；对于 Result 配置部分，介绍了 Result 的常用结果类型等知识。

第 8 章
Struts 2 的标签库

　　JSTL 标签库能够简化 JSP 的编写,避免 JSP 中嵌入大量的 Java 脚本,可以将显示和控制逻辑分离。常见的 Web 层框架,包括 Struts 2,都提供了自己特有的标签库。Struts 2 的标签库非常丰富,大大简化了数据输出和页面效果生成,同时还能完成一些基本的流程控制功能。

8.1 Struts 2 标签库概述

Struts 2 提供了功能强大的标签库，而且远远超过传统标签库的基本功能：数据显示和数据输出。使用 Struts 2 标签库不仅简化了数据的输出，还可以提供大量的标签来生成页面效果。与 Struts 1 的标签库相比，Struts 2 的标签库更加易用和强大。

8.1.1 Struts 2 标签的分类

Struts 2 的标签非常多，Struts 2 把所有的标签都定义在 URI 为"/struts-tags"的命名空间下，在使用上并没有分类，为了介绍方便，将其按功能大致分为以下三类。
(1) UI 标签：主要用来生成 HTML 元素的标签。
(2) 非 UI 标签：主要用于数据访问、流程控制的标签。
(3) Ajax 标签：用于支持 Ajax(Asynchronous JavaScript And XML)的标签。
对于 UI 标签还可以进一步细分，具体如下。
(1) 表单标签：主要用于生成 html 页面的 form 元素，以及普通表单元素的标签。
(2) 非表单标签：主要用于生成页面上的树、Tab 等标签。
对于非 UI 标签，按其功能可分为如下两类。
(1) 数据访问标签：主要用来提供与数据访问相关的功能。
(2) 流程控制标签：主要用来完成条件逻辑、循环逻辑的控制，也可用于对集合的操作。

8.1.2 Struts 2 标签库的导入

Struts 2 提供的 Struts2-core-2.5.8.jar 文件中包含标签的处理类和描述文件。解压此文件，在 META-INF 路径下可以找到 struts-tags.tld 文件，该文件就是 Struts 2 的标签库描述文件。Struts 2 标签的使用随 Web 容器版本的差异而不同。

在使用 Struts 2 标签时，首先需要导入标签库，导入语法与使用 JSTL 标签相同。在 JSP 页面中使用标签库时，必须使用 taglib 指令引入标签库，代码如下：

```
<%@ taglib prefix="s" uri="/struts-tags" %>
```

在上述代码中，prefix="s"指定了使用此标签库时的前缀，uri="/struts-tags"指定了标签库描述文件的路径。如果项目采用的 Servlet 版本是 2.3 或以下，需要在 web.xml 中增加对标签库的定义，代码如下：

```
<taglib>
  <taglib-uri>/struts-tags</taglib-uri>
  <taglib-location>/WEB-INF/lib/struts2-core-2.5.8.jar</taglib-location>
</taglib>
```

如果项目使用的 servlet 版本是 2.4 及其以上，则无须在 web.xml 中增加标签库定义，因为 Web 应用会自动读取该 struts-tags.tld 文件信息。

8.2　Struts 2 的 UI 标签

Struts 2 的 UI 标签主要用来生成 HTML 元素，表单标签用来向服务器提交用户输入信息，绝大部分表单标签都有相应的 HTML 标签与其对应，通过表单标签可以简化表单开发，还可以实现 HTML 中难以实现的功能。

8.2.1　UI 标签的模板和主题

所谓模板，就是一些代码，在 Struts 2 中通常是用 FreeMarker 来编写的，标签使用这些代码能渲染生成相应的 HTML 代码。

一个标签在使用时需要确定显示的数据，以及最终生成何种风格的 HTML 代码，这些是由 FreeMarker 的模板来定义的，每个标签都有自己对应的 FreeMarker 模板。这组模板在 Struts 2 核心 jar 包(Struts2-core-2.5.8.jar)的 template 包中，如图 8-1 所示。

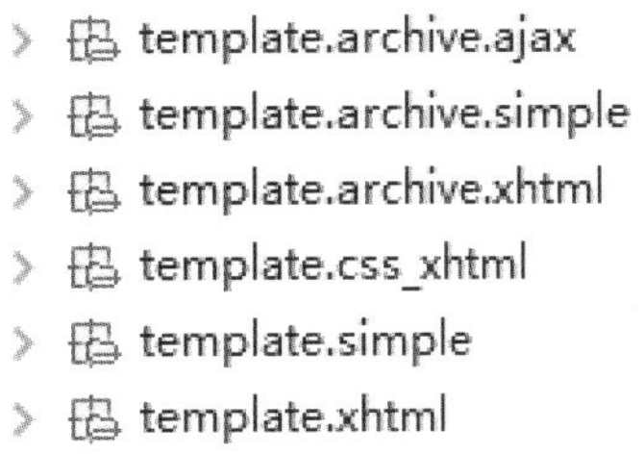

所谓主题，是指一系列模板的集合。通常情况下，一个系列的模板有相同或类似的风格，这样才能保证功能或视觉效果的一致性。Struts 2 标签使用一个模板来生成最终的 HTML 代码。如果使用不同的模板，那么同一

- template.archive.ajax
- template.archive.simple
- template.archive.xhtml
- template.css_xhtml
- template.simple
- template.xhtml

图 8-1　Struts2-core-2.5.8.jar 包的结构

个标签所生成的 HTML 代码也不一样，也意味着不同的标签所生成的 HTML 代码的风格也可能不一样。在 Struts 2 中，可以通过设置主题切换标签所生成的 HTML 代码的风格，主题来自模板文件，而模板在 Struts 2 核心 jar 包的 template 包中已经定义。

Struts 2 提供了 4 种内建的主题，可以满足大多数的应用。这 4 种内建主题分别介绍如下。

(1) simple。simple 主题的功能较弱，只提供简单的 HTML 输出。

(2) xhtml。xhtml 主题是在 simple 主题基础上的扩展，是 Struts 2 的默认主题，这个主题的模板通过使用一个布局表格提供了一种自动化的排版机制。

(3) css_xhtml。css_xhtml 主题是在 xhtml 主题基础上的扩展，它在功能上强化了 xhtml 主题在 CSS 样式上的控制。

(4) ajax。ajax 主题是在 css_xhtml 主题基础上的扩展，它在功能上主要强化了 css_xhtml 主题在 Ajax 方面的应用。

8.2.2　表单标签的公共属性

Struts 2 的表单元素标签包含非常多的属性，但有很多属性完全是通用的。Struts 2 的表单标签用来向服务器提交用户输入信息，在 org.apache.struts2.components 包中都有一个对应的类，所有表单标签对应的类都继承自 UIBean 类。Struts 2 表单标签的通用属性如表 8-1 所示。

表 8-1 Struts 2 表单标签的通用属性

属性名	主 题	数据类型	说 明
title	simple	String	设置表单元素的 title 属性
disabled	simple	String	设置表单元素是否可用
label	xhtml	String	设置表单元素的 label 属性
labelPosition	xhtml	String	设置 label 元素显示位置，可选值：top 和 left(默认)
name	simple	String	设置表单元素的 name 属性，与 Action 中的属性名对应
value	simple	String	设置表单元素的值
cssClass	simple	String	设置表单元素的 class 属性
cssStyle	simple	String	设置表单元素的 style 属性
required	xhtml	Boolean	设置表单元素为必填
requiredposition	xhtml	String	设置必填标记(默认标记为*)相对于 label 元素的位置，可选值：left 和 right(默认)
tabindex	simple	String	设置表单元素的 tabindex 属性

8.2.3 简单的表单标签

熟悉 HTML 的读者对于<form>、<select>等 HTML 标签应该是耳熟能详的。我们来看一组简单的 Struts 2 表单标签，这些标签都可以在 HTML 表单元素中找到一一对应的标签，可以通过和 HTML 标签的对比来学习 Struts 2 标签，标签的对应列表如表 8-2 所示。

表 8-2 简单的表单标签的对应列表

标 签	HTML 对应标签	说 明
<s:form>…</s:form>	<form>	表单标签
<s:textfield />	<input type="text" >	单行文本框
<s:password />	<input type="password">	密码输入框
<s:textarea />	<textarea>	文本框
<s:submit />	<input type="submit">	提交按钮
<s:reset />	<input type="reset">	重置按钮
<s:select />	<select>	下拉列表框
<s:radio />	<input type="radio">	单选按钮
<s:checkbox />	<input type="checkbox">	复选框

我们在 restaurant 项目下的 WebRoot 中新建一个 ch08 文件夹，在该文件夹中新建一个 simpleFormTag.jsp 文件，通过简单表单标签开发一个员工登记表页面，来介绍这些标签的使用方法及相应的作用，代码如下：

```
<%@ page language="java" import="java.util.*" pageEncoding="utf-8"%>
<%@taglib prefix="s" uri="/struts-tags"%><!--加载Struts2 标签库 -->
```

```
<!-- 省略部分代码 -->
<body>
  <center>
  <h3>注册登记表</h3>
  <s:form action="register" method="post">  <!--表单标签-->
     <s:textfield name="loginName" label="姓名"></s:textfield>
     <s:password name="loginPwd" label="口令"/>
     <s:select name="degree" label="学历"
          list="{'高中及以下','大学专科','大学本科','研究生及以上'}"/>
     <s:radio name="sex" label="性别" list="{'男','女'}"></s:radio>
     <s:textarea name="protocol" label="注册协议" value="这里省略协议"/>
     <s:checkbox name="love" label="同意员工登记协议"/>
     <s:submit value="提交"></s:submit>
     <s:reset value="重置"/>
  </s:form>
  </center>
</body>
```

部署项目，在浏览器中输入 http://localhost:8080/restaurant/ch08/simpleFormTag.jsp，运行效果如图 8-2 所示。

- IE8.0 及以下版本的浏览器对 Struts 2 的简单表单标签的显示支持不好。
- 查看静态页面源代码，Struts 2 标签不仅仅被解析为"<input />"，同时在标签外面多了<table /><tr /><td />标签，代码如下：

图 8-2 简单表单标签

```
<form id="register" name="register"
action="register" method="post">
   <table class="wwFormTable">  <!--表单标签-->
      <tr>
         <td class="tdLabel">
            <label for="register_loginName"
class="label">姓名:
            </label>
         </td>
         <td class="tdInput">
            <input type="text" name="loginName" value=""
               id="register_loginName"/>
         </td>
      </tr>
      <!-- 省略其他代码 -->
   </table>
</form>
```

这是由于使用 Struts 2 时，使用了 Struts 2 的默认主题，我们可以通过配置文件 struts.xml，对其风格进行更改，添加如下代码：

```
<struts>
   <!-- 设置用户界面主题，默认值为 xhtml 风格 -->
   <constant name="struts.ui.theme" value="simple"/>
   <!-- 省略其他代码 -->
</struts>
```

8.2.4 其他表单标签

除了上述的简单表单标签以外，还有一些表单标签，这里简单介绍如下。

(1) <s:checkboxlist>标签：用来生成复选框组，该复选框组包含多个复选框。也就是说，通过 checkboxlist 标签可以生成一系列 HTML 中的<input type="checkbox"> 标签。

(2) <s:combobox>标签：用来生成一个单行文本框和下拉列表框的组合，而且这两个元素对应同一个参数。其中，以单行文本框中的值作为请求参数的值，而下拉列表框只起到一个辅助输入的作用，并没有 name 属性，也不会产生请求参数。

(3) <s:optgroup>标签：用于生成一个下拉列表框的选项组，因此，该标签必须嵌套在<s:select>标签中使用。因此可以在一个<s:select>标签中添加多个<s:optgroup>标签。

(4) <s:doubleselect>标签：用于生成一个级联列表框(两个下拉列表框)，当选择第一个下拉列表框中的内容时，第二个下拉列表框中的内容会随之改变，这两个下拉列表框是相互关联的。

(5) <s:file>标签：用于创建一个文件选择框，生成 HTML 中的<input type="file" />标签，除了公共属性外，该标签还有一个名称为 accept 的属性，用于指定接收文件的 MIME 类型。

(6) <s:token>标签：主要用来防止多次提交表单，可以避免刷新页面时多次提交。该标签不会在页面上进行任何输出，也没有属性，生成一个 HTML 隐藏域，每次加载页面时，隐藏域的值都不同。

(7) <s:updownselect>标签：与 select 标签非常相似，不同的是，updownselect 标签在生成列表框的同时生成 3 个按钮，分别代表上移、下移和全选。

(8) <s:optiontransferselect>标签：与 updownselect 标签很相似，只不过会生成两个列表框，每个列表框都可以对选项进行上移、下移、全选等操作，而且在这两个列表框之间可以进行左移、右移等操作。

8.2.5 非表单标签

Struts 2 的非表单标签主要用来在页面中生成非表单的可视化元素，输出在 Action 中封装的信息。例如，输出一些错误的提示信息，这些标签可以给程序开发带来便捷。

1. <s:actionerror>、<s:actionmessage>和<s:fielderror>标签

<s:actionerror>、<s:actionmessage>和<s:fielderror>标签分别用来显示动作错误信息、动作信息和字段错误信息。如果信息为空，则不显示，具体功能如下。

(1) actionerror：如果 Action 实例的 getActionErrors()方法的返回不为 null，则该标签负责输出该方法返回的系列错误。

(2) actionmessage：如果 Action 实例的 getActionMessages()方法的返回不为 null，则该标签负责输出该方法返回的系列消息。

(3) fielderror：如果 Action 实例存在表单域的类型转换错误、校验错误，则该标签负责输出这些错误提示。

在 restaurant 项目中的 com.restaurant.action 包中创建 ErrorAction 类，代码如下：

```java
package com.restaurant.action;
import com.opensymphony.xwork2.ActionSupport;
public class ErrorAction extends ActionSupport {
    @Override
    public String execute() throws Exception {
        this.addActionError("ActionError 错误信息 1");
        this.addActionError("ActionError 错误信息 2");
        this.addActionMessage("ActionMessage 普通信息 1");
        this.addActionMessage("ActionMessage 普通信息 2");
        this.addFieldError("fielderror1", "字段错误信息 1");
        this.addFieldError("fielderror2", "字段错误信息 2");
        return SUCCESS;           //返回 success 字符串
    }
}
```

在 struts.xml 中配置 ErrorAction 类，配置如下：

```xml
<action name="errorAction" class="com.restaurant.action.ErrorAction">
    <result name="success">/ch08/showErrorAction.jsp</result>
</action>
```

在 ch08 的文件夹中新建 showErrorAction.jsp，使用标签输出相关信息，代码如下：

```xml
<s:actionerror></s:actionerror>
<s:actionmessage/>
<s:fielderror value="fielderror1"/><!-- 有无 value 属性效果一样 -->
```

重新部署项目，在浏览器地址栏中输入 http://localhost:8080/restaurant/errorAction，运行效果如图 8-3 所示。

2．<s:component>标签

使用<s:component>标签可以自定义组件。例如，当需要多次使用某段代码时，就可以考虑将这段代码定义成一个自定义组件，而后在页面中使用<s:component>标签来调用自定义组件。自定义组件是基于主题和模板管理的，因此在使用<s:component>标签时，常常需要指定如下 3 个属性。

图 8-3　信息提示标签

(1) theme 属性：用来指定自定义组件所使用的主题，若未指定，则默认使用 xhtml。
(2) templateDir 属性：用来指定自定义组件所使用的主题目录，默认使用 template。
(3) template 属性：用来指定自定义组件所使用的模板文件。

另外还可以在 component 标签内嵌套 param 标签，向模板传入参数信息。

8.3　Struts 2 的非 UI 标签

Struts 2 的非 UI 标签包括控制标签和数据标签。控制标签主要用于完成流程控制，以及

对 ValueStack 的控制。数据标签主要用于访问 ValueStack 中的数据。

8.3.1 控制标签

控制标签可以完成输出流程控制，如分支、循环等操作，也可以完成对集合的合并、排序等操作。

1. <s:if>、<s:elseif>、<s:else>标签

这三个标签主要用来进行分支语句控制，它们都将根据一个 boolean 表达式的值决定是否计算、输出标签体的内容。这三个标签可以组合使用，只有<s:if…/>标签可以单独使用，具体用法格式如下：

```
<s:if test="表达式 1">
        标签内容
</s:if>
<s:elseif test="表达式 2">
        标签内容
</s:elseif>
<s:else>
        标签内容
</s:else>
```

可以看出，上面的<s:if><s:elseif><s:else>对应 Java 结构中的 if/elseif/else，对于<s:if><s:elseif>必须指定一个 test 属性，该 test 属性用来设置标签的判断条件，其值为 boolean 型的条件表达式。

2. <s:iterator>标签

该标签主要用来对集合数据进行迭代，根据条件遍历集合类中的数据，这里的集合包含 List、Set 和数组，也可对 Map 类型的对象进行迭代输出。<s:iterator>标签的属性如下。

- value：可选属性，指定被迭代的集合，被迭代的集合通常都使用 OGNL 表达式指定；如果没有指定 value 属性，则使用 ValueStack 栈顶的集合。
- id：可选属性，该属性指定了集合里元素的 ID(现已用 var 替代)。
- status：可选属性，指定迭代时的 IteratorStatus 实例，通过该实例可以判断当前迭代元素的属性，如是否是最后一个，以及当前迭代元素的索引等。
- Begin：开始迭代的索引位置，开始索引从 0 开始。
- End：结束迭代的索引时位置，集合元素的个数要小于或等于此结束索引。
- Step：迭代的步长，每次迭代时索引的递增值，默认为 1。

下面通过示例来演示，在 restaurant 项目的 WebRoot 目录下的 ch08 文件夹中新建 showIterator.jsp 页面，并在该页面中使用<s:iterator>标签，代码如下：

```
<s:iterator value="{'故人西辞黄鹤楼,','烟花三月下扬州。'}" var="poem">
    <s:property value="poem"/><br>
</s:iterator>    <hr>
<s:iterator value="#{'1001':'Java 程序设计','1002':'JSP 程序设计',
'1003':'SSH 框架技术'}" var="bookName">
```

```
            <s:property value="key"/>
            <s:property value="value"/><br>
      </s:iterator>       <hr>
      <s:iterator value="{'清华大学','复旦大学','北京大学','南京大学'}" var="university" status="stat">
            <s:if test="#stat.odd">    <!-- 判断当前索引是否为奇数 -->
                <s:property value="#stat.count"/> <s:property /><br>
            </s:if>
      </s:iterator>         <hr>
      <table border="1">
         <tr><td>序号</td><td>出版社</td></tr>
         <s:iterator value="{'清华大学出版社','人民邮电出版社','北京大学出版社','电子工业出版社'}" var="publisher" status="stat">
           <tr>
             <s:if test="#stat.index%2==0">
                 <td><s:property value="#stat.count"/></td>
               <td style="background-color:red;">
                  <s:property value="publisher"/></td>
             </s:if>
             <s:else>
                <td><s:property value="#stat.count"/></td>
                <td style="background-color:gray;">
                   <s:property value="publisher"/></td>
             </s:else>
           </tr>
         </s:iterator>
      </table>
```

部署项目,在浏览器的地址栏中输入 http://localhost:8080/restaurant/ch08/showIterator.jsp,运行效果如图 8-4 所示。

3. 其他控制标签

除了上述的控制标签外,还有一些控制标签,相对使用得比较少,介绍如下。

图 8-4 使用 Iterator 遍历标签的效果

(1) <s:append>标签:用于拼接多个集合对象,组成一个新的集合。通过这种拼接,从而允许通过一个<s:iterator…/>标签完成对多个集合的迭代。使用<s:append…/>标签时,需要指定一个 id 属性,该属性确定拼接生成的新集合的名字。除此之外,<s:append…/>标签可以接收多个<s:param…/>子标签,每个子标签指定一个集合,<s:append…/>标签负责将<s:param…/>标签指定的多个集合拼接成一个集合。

(2) <s:merge>标签:该标签的用法看起来非常像 append,都是用来将多个结合组合成一个新集合。都有一个 id 属性(var 属性),用来设置新集合的名称,这两个标签的不同点在于组合集合的方式。

(3) <s:sort>标签:sort 标签用来对指定集合中的元素进行排序。sort 标签并不提供自己的排序规则,而是由开发者提供排序规则。排序规则是一个实现 java.util.Comparator 接口的类。

(4) <s:generator>标签：用来将指定的字符串按照指定分隔符分隔成多个子字符串，并将这些子字符串放置到一个集合对象中。转换后的集合也可以使用 iterator 标签来迭代输出。

(5) <s:subset>标签：用来截取集合中的部分元素，从而形成一个新的集合。新的集合是源集合的子集。

8.3.2 数据标签

数据标签主要用于各种与数据访问相关的功能以及 Action 的调用等。数据标签包含的标签有 action、bean、date、debug、i18n、include、param、push、set、text、url、property 等。

(1) <s:action>标签：允许在 JSP 页面中直接访问并调用 Action，要调用 Action 就需要指定 Action 的 name 及 namespace 等属性。还可以通过 executeResult 属性选择是否将处理结果包含在当前页面中。

(2) <s:property>标签：作用就是输出指定值，在上面的程序中已经使用。property 标签输出 value 属性指定的值，如果没有指定 value 属性，则默认输出 ValueStack 栈顶的值。

(3) <s:param>标签：主要用来为其他标签提供参数，如 append、bean、merge 等标签。

(4) <s:bean>标签：允许直接在 JSP 页面创建 JavaBean 实例。通常，该标签和 param 标签结合起来使用。bean 标签创建实例，param 标签则为实例传入指定参数值。

(5) <s:date>标签：用来格式化输出指定的日期或者时间，也可用于输出当前日期值与指定日期值之间的时间差。

(6) <s:set>标签：用来定义一个新的变量，并将一个已知的值赋给这个新变量，同时将这个新变量放到指定范围内，如 session 范围等，等同于 JSP 中的 setAttribute()方法。

(7) <s:url>标签：url 标签用来生成一个 URL 地址，可以通过在其标签体中添加 param 标签传递请求参数，从而指定 URL 发送请求参数。

(8) <s:include>标签：用来在当前页面中包含另一个页面(或者 Servlet)，类似于 JSP 程序中的<%@ include file="" %>、<jsp:include file="" >。

(9) <s:debug>标签：主要用于辅助测试，可以用来输出服务器对象中的信息，如 request 范围的属性、session 范围的属性等。使用 debug 标签只有一个 id 属性，这个属性仅仅是该元素的一个引用 id。

(10) <s:push>标签：用来将指定值放到 ValueStack 的栈顶，设置完成后，可以很方便地访问该值。

(11) <s:i18n>标签：主要用来进行国际化资源文件绑定，然后将其放入 ValueStack 值栈。该标签可以用来加载国际化资源文件，然后在<s:text>标签或者表单标签中使用 key 属性来访问国际化资源文件。

(12) <s:text>标签：主要用来输出国际化资源文件信息，当 JSP 页面中用<s:i18n>标签指定国际化资源文件后，就可以使用<s:text>标签来输出 key 值对应的 value 值。

8.4 使用 Struts 2 实现用户注册功能

用户注册是网站的常见功能，本部分采用 Struts 2+JDBC+JavaBean 开发模式来实现注册功能，接收注册信息。如果正常接收则显示用户注册成功的相关信息，并在信息界面中显示，否则显示用户注册失败。

8.4.1 用户注册流程

用户注册功能可以分解为以下关键环节。

(1) 在注册页面，输入注册信息，提交表单，被 action 所匹配的 Action 类所接收，执行 Action 中的相应方法。

(2) 根据方法执行返回的结果字符串，在 struts.xml 配置文件中查找相应的映射，跳转到相应的 URL。

8.4.2 创建用户实体类

根据数据库中的用户表，修改上一章在 com.restaurant.entity 包中创建的 Users.java，Users 类中的字段名称应与数据库中的 users 表的字段一一对应，包括 Id(id 号)、LoginName(用户名)、LoginPwd(密码)、TrueName(真实姓名)、Email(电子邮件)、Phone(电话)、Address(地址)、Status(状态)。Users 类就是一个典型的 JavaBean，包含私有属性的字段、公有属性的 getter 和 setter 方法。此处对于注册功能，只需要部分字段，Users 实体类的代码如下：

```java
package com.restaurant.entity;
public class Users {
    private Integer id;
    private String loginName;
    private String loginPwd;
    private String trueName;
    private String email;
    private String phone;
    private String address;
    private Integer status;
    // 省略属性的getter、setter方法
}
```

8.4.3 开发数据访问 DAO 层

在前面第 4 章中讲解 MVC 模式时，已经实现用户登录功能，在 com.restaurant.dao 包中已经创建了实现连接数据库和关闭对象功能基本类 BaseDAO，下面在 com.restaurant.dao 包中定义 UserDAO 接口，声明一个 addUsers(Users user)方法，代码如下：

```java
package com.restaurant.dao;
import com.restaurant.entity.Users;
public interface UserDAO {
```

```java
    public int addUsers(Users user);
}
```

新建 com.restaurant.dao.impl 包,并在其中定义 UserDAOImpl 类,以继承 BaseDAO 并实现 UserDAO 接口,代码如下:

```java
package com.restaurant.dao.impl;
import java.sql.*;
import com.restaurant.dao.BaseDAO;
import com.restaurant.dao.UserDAO;
import com.restaurant.entity.Users;
public class UserDAOImpl extends BaseDAO implements UserDAO {
    private Connection conn = null;
    private PreparedStatement pstmt = null;
    private ResultSet rs = null;
    @Override
    public int addUsers(Users user) {
        int result=0;
        String sql="insert into users(loginName,loginPwd,trueName,email,phone,address,status) values(?,?,?,?,?,?,?)";
        try {
            conn=this.getConnection();              //数据库连接
            pstmt=conn.prepareStatement(sql);       //预编译处理
            pstmt.setString(1, user.getLoginName());
            pstmt.setString(2, user.getLoginPwd());
            pstmt.setString(3, user.getTrueName());
            pstmt.setString(4, user.getEmail());
            pstmt.setString(5, user.getPhone());
            pstmt.setString(6, user.getAddress());
            pstmt.setInt(7, 1);
            result = pstmt.executeUpdate();
        } catch (Exception e) {
            e.printStackTrace();
        } finally {
            this.closeAll(conn, pstmt, rs);
        }
        return result;
    }
}
```

8.4.4 开发控制层 Action

在 src 下的 com.restaurant.action 包中新建 RegisterAction 类,并继承 ActionSupport,代码如下:

```java
package com.restaurant.action;
import com.opensymphony.xwork2.ActionSupport;
import com.restaurant.dao.UserDAO;
import com.restaurant.dao.impl.UserDAOImpl;
import com.restaurant.entity.Users;
```

```java
public class RegisterAction extends ActionSupport {
    private Users user;
    private String repassword;
    //省略字段的setter、getter方法
    public String execute() throws Exception{
        UserDAO userDAO=new UserDAOImpl();
        int result=0;
        if (user.getLoginName()!=null && user.getLoginPwd()!=null && user.getLoginPwd().equals(repassword)) {
            result=userDAO.addUsers(user);   //直接调用数据访问层的方法
        }
        String back;
        if (result!=0){
            back="success";
        }else{
            back="input";
        }
        return back;
    }
    @Override      //服务器端校验
    public void validate() {
        if (user.getLoginName()==null||
"".equals(user.getLoginName().trim())) {
            this.addFieldError("loginName", "用户名不能为空！");
        }
        if (user.getLoginPwd().length()==0) {
            this.addFieldError("loginPwd", "密码不能为空！");
        }
        if (!user.getLoginPwd().equals(repassword)) {
            this.addFieldError("repassword", "确认密码和密码不一致！");
        }
        //省略其他字段的校验
    }
}
```

在该类中，提供了 user、repassword 属性，并提供了相应的 setter 和 getter 方法。

提示

在注册页面中需要提供密码、确认密码两个输入框，防止用户错误输入密码，但是确认密码属性仅是视图层提供给用户的，并不应该属于 Users 类，所以在 Action 中直接添加 repassword 属性，来接收输入的确认密码。

8.4.5　在 struts.xml 中配置 action

在前面 struts.xml 的基础上，添加注册功能 Action 请求配置，代码如下：

```xml
<action name="register" class="com.restaurant.action.RegisterAction">
    <result name="success">/ch08/register_success.jsp</result>
    <result name="input">/ch08/register.jsp</result>
</action>
```

在 Struts 2 中，通过 input 字符串来指定当用户输入出现错误时需要返回的页面。

8.4.6 开发注册页面

在/WebRoot/ch08/路径下，新建网站注册页面 register.jsp，代码如下：

```jsp
<%@ page language="java" import="java.util.*" pageEncoding="UTF-8"%>
<%@ taglib prefix="s" uri="/struts-tags" %>
…
  <body>
    <h3><font color="blue">填写注册信息</font></h3>
    <font color="red" size="3px"><s:fielderror/></font>
    <s:form name="form1" action="register" method="post">
      <s:textfield name="user.loginName" label="登录名称"/>
      <s:password name="user.loginPwd" label="登录密码"/>
      <s:password name="repassword" label="确认密码"/>
      <s:textfield name="user.trueName" label="真实姓名"/>
      <s:textfield name="user.email" label="电子邮件"/>
      <s:textfield name="user.phone" label="联系电话"/>
      <s:textfield name="user.address" label="联系地址"/>
      <s:submit value="注册"/>
    </s:form>
  </body>
</html>
```

需要注意 repassword 和 user 属性的不同形式。例如，loginName 属性必须使用 user.loginName 的形式，其中 user 代表 RegisterAction 中的 user 属性。而 repassword 因为是 RegisterAction 的属性，在注册页面中直接使用其名字即可。

在/WebRoot/ch08/路径下，新建注册成功的页面 register_success.jsp，代码如下：

```jsp
<%@ page language="java" import="java.util.*" pageEncoding="UTF-8"%>
<%@ taglib prefix="s" uri="/struts-tags" %>
…
  <body>
    <h3>用户注册详细信息</h3>
    登录名称：<s:property value="user.loginName" /><br>
    登录密码：<s:property value="user.loginPwd" /><br>
    真实姓名：<s:property value="user.trueName" /><br>
    电子邮件：<s:property value="user.email" /><br>
    联系电话：<s:property value="user.phone" /><br>
    联系地址：<s:property value="user.address" /><br>
  </body>
```

8.4.7 部署项目

重新部署项目，在浏览器地址栏中输入 http://localhost:8080/restaurant/ch08/register.jsp，运行注册页面，未填信息时，给出提示，如图 8-5 所示；密码和确认密码不同时，也给出提示，如图 8-6 所示；填写正确，单击"注册"按钮后，显示相应注册信息，如图 8-7 所示。

图 8-5　未填信息时提示　　图 8-6　密码不一致提示　　图 8-7　注册成功提示

8.5　小　　结

在 Struts 2 框架中，视图层主要是由丰富的标签组成的。本章主要讲解了 Struts 2 标签的用法，包括如何通过标签库来改进 JSP 页面的数据显示。本章重点介绍了 Struts 2 标签库的用法，详细地讲解了 Struts 2 的表单标签、非表单标签、控制标签、数据标签各个参数的实际用途及意义，并且用详细的示例代码演示了标签的使用，让读者能够有一个直观的认识。

第 9 章
OGNL 和类型转换

　　Struts 2 提供了内置类型转换器，可以自动对客户端传来的字符串进行类型转换。开发者也可以开发自己的类型转换器，实现更复杂的类型转换。OGNL 在视图层工作，可以简化数据的访问操作。Struts 2 框架使用 OGNL 作为默认的表达式语言。本章从传入数据的转移和类型转换的视角来全面了解OGNL。

9.1 OGNL 基础

9.1.1 数据转移和类型转换

在开发 Web 应用程序中最常见的是，从基于字符串的 HTTP 请求向 Java 语言的不同数据类型移动和转换数据。我们都知道从表单数据向不同数据类型转换数据是件很乏味的工作，这个乏味的工作随着从字符串向各种 Java 类型的对象转换而变得复杂，将字符串解析为浮点型或者将字符串"组装"成各种 Java 对象，这些工作没有一点意思，但这些任务又都是"基础设施"，所有这些转换都是为真正的工作做准备。

数据转移和类型转换实际上发生在请求处理周期的两端，我们已经看到了框架将数据从基于文本的 HTTP 请求转移到 Action 类的 JavaBean 属性，相同的事情同样发生在另一边，当结果呈现给用户时，这些转移到 JavaBean 属性中的数据又"回到"页面。虽然我们没有过多思考实现过程，但数据真的又从 Java 类型转换回了字符串。

数据转移和类型转换是 Web 应用程序与生俱来的部分，几乎每一个 Web 应用程序的每个请求都会发生。相信不会有人反对将这些乏味的工作交给框架自动完成。Struts 2 的数据转移和类型转换机制功能强大，并且秉承了 Struts 2 框架的优点，非常容易扩展。那么是谁帮助 Struts 2 提供的这个强大功能呢？这就是 OGNL。

9.1.2 OGNL 基础

OGNL(Object-Graph Navigation Language，对象图导航语言)，是一种功能强大的表达式语言，提供了存取对象属性、调用对象方法、遍历对象结构图等功能，可以简化数据的访问操作。OGNL 是 Struts 2 内建的表达式语言，大大加强了 Struts 2 数据的访问功能。在前面部分，我们已经使用过 OGNL 进行显示。OGNL 在视图层工作，用来取代页面中的 Java 脚本，可以简化数据的访问操作。

与 JSP 2.0 中内置的 EL 相比，它们都属于表达式语言，用于进行数据访问，但是 OGNL 的功能更加强大，提供了许多 EL 所不具备的功能，比如强大的类型转换功能、访问方法、操作集合对象、跨集合投影等。

OGNL 是一种强大的技术，它被集成在 Struts 2 框架中用来帮助实现数据转移和类型转换。OGNL 在 Struts 2 中，就是基于字符串的 HTTP 输入/输出与 Java 对象内部处理之间的"黏合剂"，它的功能非常强大。尽管看起来可以在没有真正了解 OGNL 的情况下使用框架，但是学习了 OGNL 工具，将有助于我们提高效率。OGNL 在框架中主要有两种功能：表达式语言和类型转换器。

使用 OGNL 表达式能够将表单字段名绑定到对象(Action 对象)中的具体属性，Action 对象被放在叫作值栈(ValueStack)的对象上，通常出现在表单输入的 name 属性或者 Struts 2 标签的各种属性中。OGNL 提供了一个简单的语法，将表单或 Struts 2 标签与特定的 Java 数据绑定起来，用来将数据移入、移出框架，如我们学习过的，页面中 <input type="text"

name="user.loginName">的输入对应 Action 类中 Users 对象的 loginName 属性。登录页面输入框的 name 用到的名字就是 OGNL 表达式，在欢迎页面中使用<s:property value="user.loginName"/>。两个 user.loginName 表达式都是相同的，但前一个保存对象属性的值，后一个是取得对象属性的值。

除了表达式语言处，我们一直使用 OGNL 作为类型转换器，每次数据进入和流出框架，页面中数据的字符串版本和 Java 数据类型之间都发生转换，到目前为止，我们一直都是用 Struts 2 框架提供的内置的类型转换器。OGNL 融入 Struts 2 框架，如图 9-1 所示。

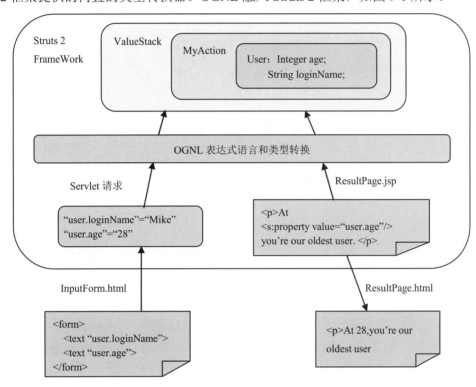

图 9-1 Struts 2 数据的流入与流出

图 9-1 展示了数据流入和流出 Struts 2 框架的路径。数据从 InputForm.html 页面中的 HTML 表单开始，用户提交一个请求，Struts 2 框架处理请求并返回用户的响应(ResultPage.jsp)。为了突出感兴趣的内容，图 9-1 中采用伪标记和伪代码的形式表示。

9.1.3 OGNL 常用符号的用法

OGNL 要结合 Struts 标签来使用。由于比较灵活，也容易把人给弄晕，尤其是%、#、$这三个符号的使用。由于$广泛应用于 EL 中，这里重点介绍%和#符号的用法。

1．#符号的用途

(1) 访问非根对象属性，如 OGNL 上下文和 Action 上下文，由于 Struts 2 中的值栈被视为根对象，所以访问其他非根对象时，需要加#前缀。

例如，#session.msg 表达式，实际上，#相当于 ActionContext.getContext()；#session.msg 表达式相当于 ActionContext.getContext().getSession().getAttribute("msg")。

(2) 用于过滤和投影(projecting)集合，例如，persons.{?#this.age>20}、books.{?#this.price>35}。

(3) 用来构造 Map，在前面<s:iterator>标签示例中，我们就使用过#符号构造 Map，例如 "#{'key1':'value1', 'key2':'value2', 'key3':'value3'}"，这种方式常用在给 radio 或 select、checkbox 等标签赋值上。如果要在页面中取一个 Map 的值可以这样写：

```
<s:property value="#myMap['1001']"/>
<s:property value="#myMap['1002']"/>
```

2．%符号的用途

该符号是在标签的属性值被理解为字符串类型时，告诉执行环境%{}里的是 OGNL 表达式。%符号的用途是在标志的属性为字符串类型时，计算 OGNL 表达式的值，代码如下：

```
<h3>构造 Map</h3>
<s:set name="foobar" value="#{'foo1':'bar1', 'foo2':'bar2'}" />
<p>The value of key "foo1" is
<s:property value="#foobar['foo1']" /></p>
<p>不使用%: <s:url value="#foobar['foo1']" /></p>
<p>使用%: <s:url value="%{#foobar['foo1']}" /></p>
```

运行结果如下：

```
The value of key "foo1" is bar1
不使用%: #foobar['foo1']
使用%: bar1
```

这说明 Struts 2 里不同的标签对 OGNL 表达式的理解是不一样的。当有的标签"看不懂"类似"#foobar['foo1']"这样的语句时，就要用%{}把其括进去，"翻译"一下。

在 JSP 页面中，"%{"表示 OGNL 表达式开始，"}"表示 OGNL 表达式结束。例如，根对象中的对象和属性可通过如下方式访问：

```
%{ Object.field }
```

此外，利用%还可以取出值栈中 Action 对象的方法，用法如下：

```
%{ getText('key') }
```

3．$符号的用途

(1) 在国际化资源文件中，引用 OGNL 表达式。例如，国际化资源文件中的代码，reg.agerange=年龄必须在${min}与${max}之间。

(2) 在 Struts 2 配置文件中，引用 OGNL 表达式，例如下面的配置：

```
<action name="saveUser" class="userAction" method="save">
   <result type="redirect">listUser.action?msg=${msg}</result>
</action>
```

9.2　Struts 2 的类型转换

在 Servlet 或 JSP 页面中，类型转换工作是由程序员自己完成的，比如可以通过下面的语句完成字符串类型和整型、字符串类型和日期类型之间的类型转换。代码如下：

```
String sage=request.getParameter("age");
int age=Integer.parseInt(sage);
String sbirth=request.getParameter("birthday");
DateFormat sdf=new SimpleDateFormat("yyyy-MM-dd");
Date birthday=sdf.parse(sbirth);
```

可以看出，类型转换的工作是必不可少的、非常乏味的，而且也是重复性的，如果有一个好的类型转换机制，将大大节省开发时间，提高开发效率。

作为一个成熟的 MVC 框架，Struts 2 提供了非常强大的类型转换功能，提供了多种内置类型转换器，可以自动对客户端传来的数据进行类型转换，这一过程对开发者来说是完全透明的。另外，Struts 2 还提供了很好的扩展性，如果内置类型转换器不能满足应用需求，开发者可以简单地开发出自己的类型转换器。

9.2.1　内置类型转换器

在客户端页面中输入的数据都被视为字符串类型，如输入的年龄，在 Struts 2 中会自动转为整型，这就是 Struts 2 强大的类型转换功能，非常方便。Struts 2 提供了一些内置的类型转换器，可以处理大多数常用的类型转换，主要包括如下几种。

(1) String：将 int、long、double、boolean、String 类型的数组或 java.util.Date 类型转换为字符串。

(2) boolean/Boolean：在字符串和布尔值之间进行转换。

(3) char/Character：在字符串和字符之间进行转换。

(4) int/Integer、float/Float、long/Long、double/Double：在字符串和数值型的数据之间进行转换。

(5) Date：在字符串和日期类型之间进行转换。具体的输入输出格式与当前的 Locale 相关。

(6) 数组(array)和集合(List、Map)：在字符串数组和数组对象、集合对象之间进行转换。

但强大的 Struts 2 内置类型转换也有不完善的情况，如输入/输出日期的格式必须与当前的 Locale 有关，如果输入的格式不符合要求，那么 Struts 2 框架也无能为力。当然，类型转换是可以扩展的。接下来，我们就扩展 Struts 2 的类型转换。

9.2.2　自定义类型转换器

随着互联网的不断普及，用户体验已经成为网站吸引用户的主要手段。在程序中，要填入坐标和时间，用户不希望分别填写 X 坐标和 Y 坐标，而是希望以某种格式(使用工具将经纬度转换为坐标格式直接输入，如(134.56, 156.79)。用户希望以任何正确的时间格式输入的时间都能够成功发布,如登记日期输入框中输入 2014/04/13 或 "2014 年 4 月 13 日" 都可以,而不是某种特定的时间格式。对 Java 的基本数据类型以及一些系统类(如 Date 类、集合类),Struts 2 提供了内置类型转换功能,但也有一定的限制。对于坐标这样的用户自定义类,Struts 2 还没有智能到可以进行自动类型转换,内置的日期类型转换对输入输出格式是有要求的。如果希望 Struts 2 更智能一些,能够对多种格式的日期进行转换,该怎么办呢?可以通过自定义类型转换器完成,由开发者指定输入格式及转换逻辑。

1.创建自定义类型转换器

Struts 2 提供了一个开发人员编写自定义类型转换器时可以使用的基类:org.apache.struts2.util.StrutsTypeConverter。StrutsTypeConverter 类是抽象类,继承 DefaultTypeConverter 类。在 Struts 2 API 文档中,StrutsTypeConverter 类的继承结构如图 9-2 所示。

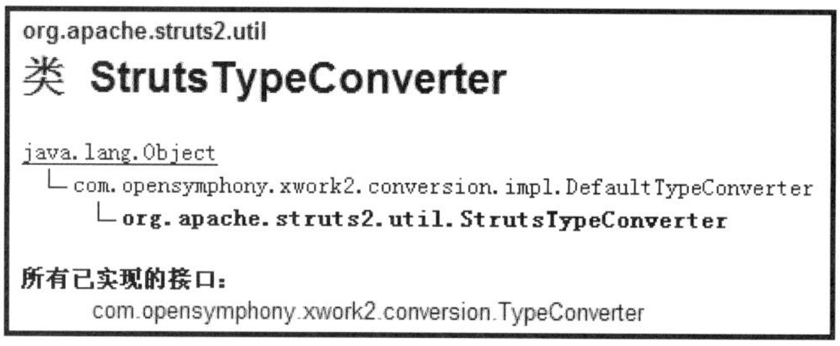

```
org.apache.struts2.util
类 StrutsTypeConverter

java.lang.Object
  └com.opensymphony.xwork2.conversion.impl.DefaultTypeConverter
      └org.apache.struts2.util.StrutsTypeConverter

所有已实现的接口:
  com.opensymphony.xwork2.conversion.TypeConverter
```

图 9-2 StrutsTypeConverter 类的继承结构

StrutsTypeConverter 类定义了两个抽象方法,用于不同的转换方向,分别介绍如下。

(1) public Object convertFromString(Map context, String[] values, Class toType)。

将一个或多个字符串转换为指定的类型,参数 context 是表示 Action 上下文的 Map 对象,values 是要转换的字符串值,toType 是要转换的目标类型。

(2) public String convertToString(Map context, Object object)。

将指定对象转化为字符串,参数 context 是表示 Action 上下文的 Map 对象,参数 object 是要转换的对象。

如果继承 StrutsTypeConverter 类编写自定义类型转换器,需要覆盖这两个抽象方法。

2.配置自定义类型转换器

自定义了类型转换器后,还必须进行配置,将类型转换器和某个类或属性通过 properties 文件建立关联。Struts 2 提供了两种方式来配置转换器:一是应用于全局范围的类型转换器;二是应用于特定类的类型转换器。

(1) 应用于全局范围的类型转换器。

要指定应用于全局范围的类型转换器,需要在 classpath 的根路径下(通常是 WEB-INF/classes 目录,在开发时对应 src 目录)创建一个名为 xwork-conversion.properties 的属性文件,其内容如下:

转换类全名 = 类型转换器类全名

(2) 应用于特定类的类型转换器。

要指定应用于特定类的类型转换器，需要在特定类的相同目录下创建一个名为 ClassName-conversion.properties 的属性文件(className 代表实际的类名)，其内容如下：

特定类的属性名 = 类型转换器类全名

9.2.3 注册自定义类型转换器

下面按照创建和配置类型转换器的方法创建三个自定义类型转换器，分别是不同时间格式的类型转换器、逗号分隔的 x,y 两个数值坐标格式的类型转换器、复选框选择形式的集合类型格式的类型转换器，如图 9-3 所示。

图 9-3 类型转换要求

(1) 在 restaurant 项目中的 com.restaurant.entity 包中，新建 Point 类和 Hobby 类；在 src 下新建 com.restaurant.converter 包，在该包中新建 PointConverter 类、DateConverter 类和 HobbyConverter 类，并继承 StrutsTypeConverter 类；在 com.restaurant.action 包中，新建 RegAction 类和 RegAction-conversion.properties 文件；在 src 目录下，新建 xwork-conversion.properties 文件；在 WebRoot 目录下新建 ch09 文件夹，并在其中新建 reg.jsp 和 success.jsp 文件。其目录结构如图 9-4 所示。

图 9-4 自定义类型转换器目录结构

(2) 新建坐标格式的类型转换器。

首先，创建坐标类 Point，只有 x 和 y 两个属性，代码如下：

```
package com.restaurant.entity;
public class Point {      //坐标类
```

```
private double x;  //X坐标
private double y;  //Y坐标
//省略x，y属性的setter、getter方法
}
```

其次，针对坐标类 Point 的类型转换，创建类型转换器 PointConverter，继承自 StrutsTypeConverter，要求用户以 x,y 的格式输入，分别输出 x 坐标和 y 坐标。该类型转换器只应用于 RegAction 类，实现代码如下：

```
package com.restaurant.converter;
import java.util.Map;
import org.apache.struts2.util.StrutsTypeConverter;
import com.restaurant.entity.Point;
//坐标类型转换类
public class PointConverter extends StrutsTypeConverter {
    @Override       // 将字符串转换为坐标类型
    public Object convertFromString(Map context, String[] values, Class toType) {
        // 获取X、Y坐标
        String str = values[0];
        String xy[] = str.split(",");
        double x = Double.parseDouble(xy[0]);
        double y = Double.parseDouble(xy[1]);
        Point point = new Point();    // 构建坐标对象
        point.setX(x);
        point.setY(y);
        return point;                 // 返回坐标对象
    }
    @Override    //将坐标对象转换为字符串
    public String convertToString(Map context, Object object) {
        Point point = (Point) object;
        double x = point.getX();
        double y = point.getY();
        String str = "(" + x + "," + y + ")";
        return str;              //返回字符串
    }
}
```

> **注意**：values 的类型是 String 数组，而不是 String，即使客户端只输入了一个字符串，也被当作字符串数组处理(当然此时只有一个元素)。因为用户请求参数可能是字符串形式，如姓名、年龄，也可能是字符串数组形式，如爱好、课程等复选框，因此考虑到最通用的情况，将所有的请求参数都视为字符串数组。

然后，在 RegAction 类的同一个目录下创建 RegAction-conversion.properties 属性文件，并添加如下内容：

```
point=com.restaurant.converter.PointConverter
```

> **注意**：其中属性文件中的 key 为 RegAction 类中的 point 属性名，而不是类型 Point 或其他。

(3) 新建日期类型的类型转换器。

首先，针对日期类 java.util.Date 进行类型转换，创建日期类型转换类 DateConverter，该类继承自 StrutsTypeConverter。要求客户端可以使用 "yyyy-MM-dd"、"yyyy/MM/dd" 或者 "yyyy 年 MM 月 dd 日" 中的任意形式输入，并且以 "yyyy-MM-dd" 的格式输出，该类型转换器应用于全局范围，实现代码如下：

```java
package com.restaurant.converter;
import java.text.DateFormat;
import java.text.SimpleDateFormat;
import java.util.Date;
import java.util.Map;
import org.apache.struts2.util.StrutsTypeConverter;
import com.opensymphony.xwork2.conversion.TypeConversionException;
public class DateConverter extends StrutsTypeConverter {
    private final DateFormat[] dfs = {  // 支持转换的多种日期格式
        new SimpleDateFormat("yyyy年MM月dd日"),
        new SimpleDateFormat("yyyy-MM-dd"),
        new SimpleDateFormat("yyyy/MM/dd"),
        new SimpleDateFormat("yyyy.MM.dd"),
        new SimpleDateFormat("yy/MM/dd"),
        new SimpleDateFormat("MM/dd/yy")
        //还可以加更多类型
    };
    @Override      //将指定格式字符串转换为日期类型
    public Object convertFromString(Map context, String[] values, Class toType) {
        String dateStr = values[0];          //获取日期的字符串
        for(int i=0;i<dfs.length;i++){       //遍历日期支持格式，进行转换
            try {
                return dfs[i].parse(dateStr);
            } catch (Exception e) {
                continue;
            }
        }
        //如果遍历完毕后仍没有转换成功，抛出转换异常
        throw new TypeConversionException();
    }
    @Override      //将日期转换为指定格式的字符串
    public String convertToString(Map context, Object object) {
        Date date = (Date) object;
        // 输出的格式是 yyyy-MM-dd
        return new SimpleDateFormat("yyyy-MM-dd").format(date);
    }
}
```

然后，在 src 目录下新建文件 xwork-conversion.properties，并添加如下内容：

`java.util.Date=com.restaurant.converter.DateConverter`

> **注意** 其中，属性文件中的 key 为 Date 类的完整类名，而不是属性名 birthday 或其他。

(4) 新建复选框选择的集合类型格式的类型转换器。

首先，创建兴趣类 Hobby，只有一个属性 hobby，代码如下：

```java
package com.restaurant.entity;
public class Hobby {
    private String hobby;
    // 必须提供默认构造器
    // 否则出现实例化异常，即java.lang.InstantiationException:
    public Hobby() {
    }
    //省略hobby属性的setter、getter方法
}
```

其次，针对兴趣爱好的复选框进行转换格式输出，创建 HobbyConverter 类，该类继承自 StrutsTypeConverter。在页面中选择爱好后，把这些爱好存储到 List 容器里负责类型转换，实现代码如下：

```java
package com.restaurant.converter;
import java.util.ArrayList;
import java.util.List;
import java.util.Map;
import org.apache.struts2.util.StrutsTypeConverter;
import com.restaurant.entity.Hobby;
public class HobbyConverter extends StrutsTypeConverter {
    @Override
    public Object convertFromString(Map context, String[] values, Class toType) {
        List list = new ArrayList();
        for(int i=0;i<values.length;i++){
            Hobby hobby = new Hobby();
            String str=values[i];
            hobby.setHobby(str);
            list.add(hobby);
        }
        return list;
    }
    @Override
    public String convertToString(Map context, Object object) {
        List list =(List)object;
        StringBuffer result= new StringBuffer();
        for(int i=0;i<list.size();i++){
            Hobby h = (Hobby)list.get(i);
            result.append(h.getHobby()+"  ");
        }
        return result.toString();
    }
}
```

然后，在 src 目录下已经创建的 xwork-conversion.properties 文件中，添加以下内容：

```
com.restaurant.entity.Hobby=com.restaurant.converter.HobbyConverter
```

其中属性文件中的 key 为 Hobby 类的完整类名，而不是属性名 hobby。

(5) 在 WebRoot 路径下新建 ch09 文件夹，并在其中创建表单提交页面 reg.jsp，使用 Struts 2 表单标签，页面代码如下：

```jsp
<%@ page language="java" import="java.util.*" pageEncoding="UTF-8"%>
<%@ taglib prefix="s" uri="/struts-tags" %>
……
    <h3><font color="blue">信息录入</font></h3><hr>
    <s:form action="reg">
        <s:textfield name="name" label="名称"/>
        <s:textfield name="age" label="年龄"/>
        <s:textfield name="birthday" label="生日"/>
        <s:textfield name="point" label="坐标"/>
        <s:checkboxlist label="爱好" name="hobby"
        list="{'读书','跳舞','游泳','唱歌'}" value="{'读书','唱歌'}" />
        <s:submit value="提交" ></s:submit>
        <s:reset value="重置"></s:reset>
    </s:form>
```

新建 success.jsp 页面，用于显示 reg.jsp 页面提交后的信息，代码如下：

```jsp
<font color="blue">注册信息如下</font>
<hr>
名称：<s:property value="name"/><br>
年龄：<s:property value="age"/><br>
生日：<s:property value="birthday"/><br>
X 坐标：<s:property value="point.x"/><br>
Y 坐标：<s:property value="point.y"/><br>
兴趣爱好：
<s:iterator value="#request.hobby" var="v">
    <s:property/>  
</s:iterator>
```

(6) 创建 RegAction 类，主要用于应对请求处理，代码如下：

```java
package com.restaurant.action;
import java.util.Date;
import java.util.List;
import com.opensymphony.xwork2.ActionSupport;
import com.restaurant.entity.Point;
public class RegAction extends ActionSupport {
    private String name;
    private int age;
    private Date birthday;
    private Point point;
    private List hobby;
    // 省略属性的 setter、getter 方法
    @Override
    public String execute() throws Exception {
        return super.execute();    //返回 success 字符串
    }
}
```

(7) 配置 struts.xml 文件，作用是配置提交的请求对应哪个 Action 处理，处理完后，转发到哪个页面，并定义一个中文常量，代码如下：

```
<action name="reg" class="com.restaurant.action.RegAction">
    <result name="success">/ch09/success.jsp</result>
    <result name="input">/ch09/reg.jsp</result>
</action>
```

(8) 重新部署程序后，在浏览器地址栏中输入 http://localhost:8080/restaurant/ch09/reg.jsp，运行结果如图 9-5 所示。在页面中填入相关信息后，单击"提交"按钮，则会进入 success.jsp 页面，这时已经调用了自定义的类型转换器，如图 9-6 所示。

图 9-5　信息输入界面　　　　　　　图 9-6　信息显示界面

9.3　小　　结

本章主要介绍了 OGNL 基础知识和类型转换。我们清楚地知道基于 B/S 模式的应用程序要完成数据之间的交互，必须要进行数据类型的转换，否则将出现 B/S 两端类型不兼容问题，从而无法完成数据之间的交互，其转换的基础则是 OGNL。

OGNL 将页面中的元素和对象的属性绑定在一起，把页面提交的字符串自动转换成对应的 Java 基本类型数据并放入"值栈"中，而用户可以通过 OGNL 表达式或者 Struts 2 标签从"值栈"中获得这些属性的值。

总而言之，Struts 2 是很好的 MVC 框架的实现者，它对视图层和非视图层提供了强有力的类型转换机制，从而让开发者运用自如。

第 10 章
Struts 2 的拦截器

拦截器(Interceptor)是一种可以在请求之前或者之后执行的 Struts 2 组件,是 Struts 2 的核心组成部分。Struts 2 框架的绝大多数功能都是通过拦截器来实现的,如数据校验、转换器、国际化、上传、下载等。拦截器是动态拦截 Action 调用的对象。

10.1 Struts 2 的拦截器机制

拦截器(Interceptor)是 Struts 2 的一个重要特性，在访问某个 Action 或 Action 的某个方法、字段之前或之后实施拦截，并且 Struts 2 拦截器是可插拔的。Struts 2 实际上是 WebWork 的升级版本，拦截器处理机制也来源于 WebWork，并按照 AOP(Aspeet Oriented Programming，面向切面编程)思想设计。AOP 是 OOP(Object Oriented Programming，面向对象程序设计)的一种完善和补充，是软件技术和设计思想发展到一定阶段的自然产物。

10.1.1 为什么需要拦截器

任何优秀的 MVC 框架都会提供一些通用的操作，如请求数据的封装、类型转换、数据校验、解析上传的文件、防止表单的多次提交等。早期的 MVC 框架将这些操作都写死在核心控制器中，而这些常用的操作又不是所有的请求都需要实现的，这就导致了框架的灵活性不足，可扩展性较差。

Struts 2 将它的核心功能放到拦截器中实现而不是集中放在核心控制器中实现，把大部分控制器需要完成的工作按功能分开定义，每个拦截器完成一个功能，而完成这些功能的拦截器可以自由选择、灵活组合，需要哪些拦截器，只需要在 struts.xml 配置文件中指定，从而增强了框架的灵活性。

拦截器的方法在 Action 执行之前或者执行之后自动地执行，从而将通用的操作动态地插入 Action 执行的前后，这样有利于系统的解耦，这种功能的实现类似于我们自己组装的电脑，变成了可插拔式。需要某一功能就"插入"一个这个功能的拦截器，不需要这个功能就"拔出"这一拦截器。可以任意地组合 Action 提供的附加功能，而不需要修改 Action 的代码。

如果有一批拦截器经常固定在一起使用，可以将这些小规模功能的拦截器定义成为大规模功能的拦截器栈(拦截器栈是根据不同的应用需求定义的拦截器组合)。从结构上看，拦截器栈相当于多个拦截器的组合，而从功能上看，拦截器栈也是拦截器，同样可以和其他拦截器(或拦截器栈)一起组成更大规模功能的拦截器栈。

通过组合不同的拦截器，我们能够以自己需要的方式来组合 Struts 2 框架的各种功能；通过扩展自己的拦截器，我们可以"无限"扩展 Struts 2 框架。

10.1.2 拦截器的工作原理

拦截器能够在 Action 执行前后拦截它，类似于 Servlet 中的过滤器。拦截器围绕着 Action 和 Result 的执行而执行，拦截器的工作方式如图 10-1 所示。

Struts 2 拦截器的实现原理和 Servlet 过滤器的实现原理类似，以链式执行，对真正要执行的方法(execute())进行拦截。首先执行 Action 配置的拦截器，在 Action 和 Result 执行之后，拦截器再一次执行(与先前调用的顺序相反)，在此链式的执行过程中，每一个拦截器都可以直接返回，从而终止余下的拦截器、Action 及 Result 的执行。

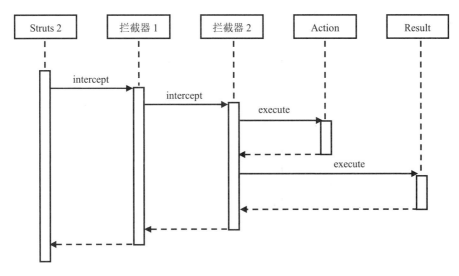

图 10-1　Struts 2 拦截器的工作方式

当 ActionInvocation 的 invoke()方法被调用时，开始执行 Action 配置的第一个拦截器，invoke()方法总是映射到第一个拦截器，ActionInvocation 负责跟踪执行过程的状态，并且把控制权交给合适的拦截器。ActionInvocation 通过拦截器的 intercept()方法将控制权转交给拦截器。

拦截器的执行过程可以看作一个递归的过程，后续拦截器继续执行，最终执行 Action，这些都是通过递归调用 ActionInvocation 的 invoke()方法实现的。每个 invoke()方法被调用时，ActionInvocation 都查询执行状态，调用下一个拦截器，直到最后一个拦截器，invoke()方法会执行 Action。

拦截器有一个三阶段的、有条件的执行周期，具体过程如下。

(1) 做一些 Action 执行前的预处理。拦截器可以准备、过滤、改变或者操作任何可以访问的数据，包括 Action。

(2) 调用 ActionInvocation 的 invoke()方法将控制权转交给后续的拦截器或者返回结果字符串终止执行。如果拦截器决定请求的处理不应该继续，可以不调用 invoke()方法，而是直接返回一个控制字符串，这种方式可以停止后续的执行，并且决定哪个结果呈现给客户端。

(3) 做一些 Action 执行后的处理。此时拦截器依然可以改变可以访问的对象和数据，只是此时框架已经选择了一个结果呈现给客户端了。

10.1.3　拦截器示例

下面通过一个具体示例来讲解拦截器的三个阶段，过程如下。

1. 编写 MyInterceptor 拦截器类

在 restaurant 项目的 src 目录下，新建 com.restaurant.interceptor 包，并在其中新建 MyInterceptor.java 拦截器类，继承 AbstractInterceptor 类，代码如下：

```java
package com.restaurant.interceptor;
import com.opensymphony.xwork2.ActionInvocation;
import com.opensymphony.xwork2.interceptor.AbstractInterceptor;
public class MyInterceptor extends AbstractInterceptor {
    @Override
    public String intercept(ActionInvocation invocation) throws Exception {
        System.out.println("自定义拦截器开始运行！");
        long startTime=System.currentTimeMillis();
        System.out.println("开始时间为："+startTime);
        String result=invocation.invoke();
        System.out.println("自定义拦截器已经结束！");
        long endTime=System.currentTimeMillis();
        System.out.println("结束时间为："+endTime);
        System.out.println("程序执行花费了："+(endTime-startTime));
        return result;
    }
}
```

2. 定义 MyAction 的业务类

在项目的 src 路径下的 com.restaurant.action 包中，新建 MyAction 的业务处理类，代码如下：

```java
package com.restaurant.action;
import com.opensymphony.xwork2.ActionSupport;
public class MyAction extends ActionSupport {
    @Override
    public String execute() throws Exception {
        System.out.println("程序正在执行 Action 中的 execute()方法！");
        return super.execute();    //返回 success 字符串
    }
}
```

3. 配置 struts.xml

在 struts.xml 中添加配置，定义一个 myInterceptor 拦截器，并在 my 的 action 配置中使用该拦截器，代码如下：

```xml
  <interceptors>
    <interceptor name="myInterceptor" class="com.restaurant.interceptor.MyInteceptor" />
  </interceptors>
    <!-- 引用自定义拦截器配置 -->
<action name="my" class="com.restaurant.action.MyAction">
    <result name="success">/ch10/my.jsp</result>
    <interceptor-ref name="myInterceptor"/>
</action>
```

4．新建页面

在项目的 Webroot 路径下，新建 ch10 文件夹，并在 ch10 文件夹中新建 my.jsp 页面，在该页面中给出简单提示"自定义拦截器已经执行！"。

5．部署项目运行程序

部署项目，在浏览器中输入 http://localhost:8080/restaurant/my，地址栏显示 my 的 Action 请求，浏览器显示 my.jsp 的页面内容，并在控制台输出内容如下：

```
自定义拦截器开始运行！
开始时间为：1502995135117
程序正在执行Action中的execute()方法！
自定义拦截器已经结束！
结束时间为：1502995135189
程序执行花费了：72 毫秒
```

该拦截器记录动作执行所花费的时间，代码很简单，intercept()方法是拦截器执行的入口方法，它接收 ActionInvocation 的实例。

当 intercept()方法被调用时，该拦截器开始记录开始时间(也就是进行预处理的工作)，接着该拦截器调用 ActionInvocation 实例的 invoke()方法，将控制权交给剩余的拦截器和动作，因为记录执行时间没有理由终止执行，所有该拦截器总是调用 invoke()方法。

在调用 invoke()方法后，该拦截器等待这个方法的返回值。虽然结果字符串会告诉该拦截器哪个结果会被呈现，但并未指出 Action 是否执行(可能剩余的拦截器终止了执行操作)。不管 Action 是否执行，invoke()方法返回时，就表明某个结果已经被呈现了(响应页面已经发给客户端了)。

获取结果字符串之后，该拦截器记录了执行的用时，并在控制台输出。此时拦截器可以使用结果字符串做一些操作，但是在这里不能停止或者改变响应。对该拦截器而言，它不关心结果，所以不查看返回的结果字符串。该拦截器执行到最后，返回从 invoke()方法获取的结果字符串，从而使递归又回到拦截器链，使前面的拦截器继续执行它们的后续处理工作。

10.2 Struts 2 内建拦截器

在运行 Action 的 execute()方法时，会发现 Action 的属性已经有值了，而且这些值和请求的参数值是一样的。这说明，在 execute()方法之前，已经把用户请求中的参数值和 Action 的属性做了一个对应，并且把请求中的参数值赋值到 Action 的属性上，这个功能由默认配置的拦截器来实现。这些默认配置的拦截器，称为内建的拦截器，也可称为预定义的拦截器。

10.2.1 默认拦截器

在 Struts 2 中，内建了大量的拦截器，这些拦截器以 name-class 对的形式配置在 struts-default.xml 文件中。name 是拦截器的名称，就是我们所引用的名字；class 则指定了该拦截器所对应的实现，只要我们自己定义的包继承了 Struts 2 的默认 struts-default 包，就可以使用默认包中定义的内建拦截器，否则必须自己定义这些拦截器。默认拦截包括以下几个。

1) params 拦截器

params 拦截器提供了框架必不可少的功能，将请求中的数据设置到 Action 中的属性上。

2) staticParams 拦截器

staticParams 拦截器是将在配置文件中通过 action 元素的子元素 param 设置的参数设置到对应的 Action 的属性中，如下面示例所示的配置文件代码：

```xml
<action name="example" class="com.restaurant.action.ExampleAction" >
    <param name="exampleField">Example</param>
    <result>/success.jsp</result>
</action>
```

staticParams 拦截器被调用时，会将通过 param 元素设置的参数值赋给对应的 Action 属性。

3) fileUpload 拦截器

fileUpload 拦截器将文件和元数据从多重请求(multipart、form-data)转换为常规的请求数据，以便能将它们设置在对应的 Action 的属性上。

4) servletConfig 拦截器

servletConfig 拦截器提供了一种将源于 Servlet API 的各种对象注入 Action 中的简洁方法。Action 必须实现相应的接口，servletConfig 拦截器才能将对应的 Servlet 对象注入 Action 中。表 10-1 列出的接口可以由 Action 实现，用来取得 Servlet API 的不同对象。

表 10-1 获取 Servlet API 对象的接口

接口	作用
ServletContextAware	设置 ServletContext
ServletRequestAware	设置 HttpServletRequest
ServletResponseAware	设置 HttpServletResponse
ParameterAware	设置 Map 类型的请求参数
RequestAware	设置 Map 类型的请求(HttpServletRequest)属性
SessionAware	设置 Map 类型的会话(HttpSession)
ApplicationAware	设置 Map 类型的应用程序作用域对象(ServletContext)

5) validation 拦截器

validation 拦截器执行数据验证。

6) workflow 拦截器

workflow 拦截器提供当数据验证错误时终止执行流程的功能。

7) exception 拦截器

exception 拦截器捕获异常，并且能够根据类型将捕获的异常映射到用户自定义的错误页面。该拦截器应该位于定义的所有拦截器的第一位。

Struts 2 框架定义了许多有用的拦截器，我们只介绍了其中比较常用的一部分，在实际开发中如果有需要，可以查阅 struts-default.xml 文件，了解更多的自带拦截器。Struts 2 框架除了提供这些有用的拦截器之外,还为我们定义了一些拦截器栈,在开发 Web 应用程序时,可以直接引用这些拦截器栈,而无须自己定义拦截器。

在 struts-default.xml 中定义的一个非常重要的拦截器栈是 defaultStack。在 struts2-core-2.5.8.jar 包中的根目录下找到 struts-default.xml 文件,在<interceptors>元素下可以找到内建拦截器和拦截器栈,其中 defaultStack 拦截器栈组合了多个拦截器,这些拦截器的顺序经过精心的设计,能够满足大多数 Web 应用程序的需求。只要在定义包的过程中继承 struts-default 包,那么 defaultStack 拦截器栈将是默认的拦截器的引用,代码如下:

```xml
<interceptors>
<!-- 系统内建拦截器部分 -->
<interceptor name="alias" class="com.opensymphony.xwork2.interceptor.
AliasInterceptor"/>
    …<!-- 省略其他拦截器的定义 -->
    <!-- 定义 basicStack 拦截器栈 -->
    <interceptor-stack name="basicStack">
        <!-- 引用系统定义的 exception 拦截器 -->
        <interceptor-ref name="exception"/>
        …
    </interceptor-stack>
    …
    <interceptor-stack name="i18nStack">
        <interceptor-ref name="i18n"/>
        <!-- 引用系统定义的 basicStack 拦截器栈 -->
        <interceptor-ref name="basicStack"/>
    </interceptor-stack>
    <!-- 定义 defaultStack 拦截器栈 -->
    <interceptor-stack name="defaultStack">
        <interceptor-ref name="exception"/>
        <interceptor-ref name="alias"/>
        …<!-- 省略其他拦截器 -->
        <interceptor-ref name="params"/>
        <interceptor-ref name="conversionError"/>
        <interceptor-ref name="validation">
            <param name="excludeMethods">input,back,cancel,browse
            </param>
        </interceptor-ref>
        <interceptor-ref name="workflow">
            <param name="excludeMethods">input,back,cancel,browse
            </param>
        </interceptor-ref>
        <interceptor-ref name="debugging"/>
        <interceptor-ref name="deprecation"/>
    </interceptor-stack>
    …<!-- 省略其他拦截器栈 -->
</interceptors>
<!-- 将 defaultStack 拦截器栈配置为系统默认拦截器 -->
<default-interceptor-ref name="defaultStack"/>
```

10.2.2 配置拦截器

在上面的示例中,看到了定义拦截器的部分代码,要使用拦截器,需要以下两个步骤。

- 通过<interceptor…/>元素来定义拦截。

- 通过<interceptor-ref.../>元素来使用拦截器。

使用拦截器需要在 struts.xml 配置文件中进行相应操作，代码如下：

```
<package name="packageName" extends="struts-default" namespace="/">
<interceptors>
        <!-- 定义拦截器 -->
        <interceptor name="interceptorName" class="interceptorClass"/>
        <!-- 定义拦截器栈 -->
        <interceptor-stack name=" interceptorStackName">
            <!-- 指定引用的拦截器 -->
            <interceptor-ref name="interceptorName|interceptorStackName">
            </interceptor-ref>
        </interceptor-stack>
    </interceptors>
    <!-- 定义默认的拦截器引用 -->
<default-interceptor-ref name="interceptorName|interceptorStackName" />
    <action name="actionName" class="actionClass">
        <!-- 为 Action 指定拦截器引用 -->
        <interceptor-ref name="interceptorName|interceptorStackName" />
        <!-- 省略其他配置 -->
    </action>
</package>
```

在示例代码中，我们可以在配置文件的 interceptors 元素中使用 interceptor 元素来定义拦截器，interceptor 元素的 name 属性与 class 属性是必须填写的。前者指定拦截器的名称，后者指定拦截器的全限定类名。然后在 action 元素中使用 interceptor-ref 元素指定引用的拦截器。

如果想要把多个拦截器组合成一个拦截器栈，就需要在 interceptors 元素中使用 interceptor-stack 元素定义拦截器栈，其中 name 属性指定拦截器栈的名称，依然使用 interceptor-ref 元素指定引用的拦截器栈。

引用拦截器时，Struts 2 并不区分拦截器和拦截器栈，所以在定义拦截器栈时，还可以引用其他拦截器栈。

如果配置文件中的大多数 Action 都引用拦截器，建议大家定义默认的拦截器引用。default-interceptor-ref 元素定义默认的拦截器引用，其 name 属性指定引用的拦截器或拦截器栈的名称。Struts 2 为我们提供了如此丰富的拦截器，但是并不意味着我们失去了创建自定义拦截器的能力；相反，自定义拦截器也不是一件难事。

10.2.3 自定义拦截器

在 Struts 2 程序的开发中，如果想要开发自己的拦截器类，所有的 Struts 2 拦截器都直接或间接实现接口 com.opensymphony.xwork2.interceptor.Interceptor。该接口提供三个方法，具体介绍如下。

(1) void init()：拦截器被初始化之后，在该拦截器执行拦截之前，系统回调该方法。对每个拦截器而言，此方法只执行一次。

(2) void destroy()：该方法跟 init()方法对应，在拦截器示例被销毁之前，系统将回调该方法。

(3) String intercept(ActionInvocation invocation) throws Exception：该方法是用户需要实现的拦截动作，该方法会返回一个字符串作为逻辑视图。

除此之外，继承 com.opensymphony.xwork2.interceptor.AbstractInterceptor 类是更简单的一种实现拦截器的方式，AbstractInterceptor 类提供了 init()和 destroy()方法的空实现，这样我们只需要实现 intercept()方法，就可以创建自己的拦截器了，定义如下：

```java
public abstract class AbstractInterceptor implements Interceptor
{
    public void init(){ }
    public void destroy(){ }
    public abstract String intercept(ActionInvocation invocation) throws Exception;
}
```

10.3　自定义权限验证的拦截器

为登录模块开发一个自定义的拦截器来判断用户是否登录。当用户需要请求执行某个受保护的操作时，先检查用户是否已经登录。如果没有登录，则向用户显示登录页面；如果请求的用户已经登录，则继续操作。实现思路如下。

(1) 编写自定义拦截器，继承自 AbstractInterceptor。
(2) 在 struts.xml 配置文件中定义拦截器。
(3) 引用自定义的拦截器。

1．登录页面和主页面

在 WebRoot 中的 ch10 文件夹中，新建 login.jsp 页面和 main.jsp 页面。
login.jsp 页面的代码如下：

```html
<form name="form1" method="post" action="logAction">
    用户名：<input type="text" name="user.loginName">  <br><br>
    密    码：<input type="password" name="user.loginPwd"><br><br>
    <input type="submit" value="登录">
    <input type="reset" value="取消">
</form>
```

main.jsp 页面的代码如下：

```html
<s:if test="#session.user==null">
    <a href="../restaurant/ch10/login.jsp">
    <span class="blue">[登录]</span></a>  您还未登录，请单击登录链接。
</s:if>
<s:if test="#session.user!=null">
    欢迎您：<span class="red">${sessionScope.user.loginName}</span>
    您已经登录！
</s:if>
```

2. LogAction 和 ShowAction 业务类

在 com.restaurant.action 包中，新建 LogAction 和 ShowAction 业务类。LogAction 业务类，做登录请求处理，代码如下：

```java
package com.restaurant.action;
import java.util.Map;
import com.opensymphony.xwork2.ActionContext;
import com.opensymphony.xwork2.ActionSupport;
import com.restaurant.entity.Users;
public class LogAction extends ActionSupport {
    private Users user;
    // 省略属性 user 的 getter、setter 方法
    public String execute() throws Exception{
        if (user.getLoginName().equals("admin") && user.getLoginPwd().equals("123")) {
            Map<String,Object> session=null;
            ActionContext ac=ActionContext.getContext();
            session=ac.getSession();
            session.put("user", user);
            return "success";        //返回 success 字符串
        }else{
            return "login";           //返回 login 字符串
        }
    }
}
```

ShowAction 业务类，只做一个简单的请求处理，代码如下：

```java
package com.restaurant.action;
import com.opensymphony.xwork2.ActionSupport;
public class ShowAction extends ActionSupport {
    @Override
    public String execute() throws Exception {
        return super.execute();      // 返回 success 字符串
    }
}
```

3. 编写 AuthorityInterceptor 拦截器

在 restaurant 项目的 src 路径下的 com.restaurant.interceptor 包中，新建 AuthorityInterceptor 类，并继承自 AbstractInterceptor，代码如下：

```java
package com.restaurant.interceptor;
import java.util.Map;
import com.opensymphony.xwork2.ActionInvocation;
import com.opensymphony.xwork2.interceptor.AbstractInterceptor;
import com.restaurant.entity.Users;
public class AuthorityInterceptor extends AbstractInterceptor {
    @Override
    public String intercept(ActionInvocation invocation) throws Exception {
        // 取得用户会话，获取用户会话信息
        Map session = invocation.getInvocationContext().getSession();
        if (session == null) {
```

```
            return "login";
        } else {
            Users user = (Users) session.get("user");
            if (user == null) {
                return "login";
            } else {
                return invocation.invoke();
            }
        }
    }
}
```

4．配置拦截器

修改 struts.xml 配置文件，添加相应的拦截器配置，代码如下：

```xml
<package name="restaurant" namespace="/" extends="struts-default">
    <interceptors>
        <!-- 定义权限验证拦截器 -->
        <interceptor name="myAuthorization" class=
"com.restaurant.interceptor.AuthorityInterceptor"></interceptor>
        <!-- 定义拦截器栈 -->
        <interceptor-stack name="myStack">
            <!-- 指定引用的拦截器或拦截器栈 -->
            <interceptor-ref name="myAuthorization" />
            <interceptor-ref name="defaultStack" />
        </interceptor-stack>
    </interceptors>
    <!-- 定义默认的拦截器引用，即可去除响应 action 中的 myStack 引用 -->
    <!-- <default-interceptor-ref name="myStack" /> -->
    <action name="logAction" class= "com.restaurant.action.LogAction">
        <result name="login">/ch10/login.jsp</result>
        <result name="success">/ch10/main.jsp</result>
    </action>
    <action name="show" class="com.restaurant.action.ShowAction">
        <interceptor-ref name="myStack"></interceptor-ref>
        <result name="success">/ch10/main.jsp</result>
        <result name="login">/ch10/login.jsp</result>
    </action>
</package>
```

重新部署项目，在浏览器中访问 http://localhost:8080/restaurant/ch10/main.jsp，如图 10-2 所示，显示与登录相关的信息，单击页面上的"登录"链接可跳转到登录页面。在浏览器中访问 http://localhost:8080/restaurant/show，直接跳转到登录页面，说明自定义的权限验证拦截器起了作用，如图 10-3 所示。

图 10-2　未登录的主页面

图 10-3　登录页面

在登录页面输入用户名和密码，如果正确则进入 main.jsp 页面，如图 10-4 所示。如果输入的用户名和密码错误，还是跳转到登录页面。

图 10-4　成功登录后的主页面

在配置中，将自己定义的 myAuthorization 拦截器定义在 myStack 拦截器栈中，并将 myStack 定义为默认的拦截器并引用它。所以，当访问 Action 请求时，会执行该默认的拦截器。如果没有用户登录，返回相应的 login 字符串。在响应请求的 Action 配置中，若有相应 login 的逻辑视图，则跳转到该 login 所映射的页面；若没有，则出现错误信息。

5．定义全局的 results

上面配置默认拦截器的方法是，在 Action 请求中，都要配置相应的逻辑视图。我们也可以将定义的 myStack 拦截器栈，在需要验证拦截的 Action 中引用，去除定义的默认拦截器，定义全局的<global-results>，在其中定义返回的<result>，修改配置如下：

```
<!-- 省略定义的拦截器-->
<global-results>
    <result name="login">/ch10/login.jsp</result>
</global-results>
<!-- 省略前面已配置的部分 -->
<action name="show" class="com.restaurant.action.ShowAction">
    <interceptor-ref name="myStack"></interceptor-ref>
    <result name="success">/ch10/main.jsp</result>
</action>
```

10.4　小　　结

本章主要介绍了拦截器的基础知识；拦截器的配置和使用方法，Struts 2 内建的拦截器，这是 Struts 2 运行机制的核心；自定义拦截器的实现方式。并通过示例讲解了自定义拦截器的实现过程。在学习的过程中，应学会查阅 Struts 2 的 API 文档。

第 11 章
Hibernate 初步

前面章节学习了 Struts 2 框架，Struts 2 技术的应用使得基于 MVC 架构的 Web 项目的开发变得更加快捷。然而，Struts 2 框架和三层架构面对软件需求量越来越大的时候，往往束手无策，程序员仍然需要在数据访问层编写大量重复的代码。为了提高数据访问层的编码效率，Gavin King 领导开发出了当今流行的"对象—关系映射(ORM)"框架：Hibernate。

11.1 Hibernate 概述

目前的主流数据库依然是关系数据库,而 Java 语言则是面向对象的编程语言,当把二者结合在一起使用时相当麻烦,而 Hibernate 则减少了这个问题的困扰,它完成了对象模型和基于 SQL 的关系模型的映射关系,使得开发者可以完全采用面向对象的方式来开发程序。

11.1.1 JDBC 的困扰

我们在做项目的时候,通过 JDBC 访问数据库,发现反复编写数据访问层代码太麻烦了,每个表都少则几个字段,多则几十个字段,对几十张包含几十个字段的数据表进行插入操作,编写的 SQL 语句将会很长,非常烦琐。读取数据时,需要写多条 getString()或 getInt()语句从 ResultSet 数据集中取出各个字段信息,不仅枯燥,而且工作量巨大。这种重复性的编码工作没有任何创造性,而且容易出错。

这些问题都是 JDBC 的劣势,这些烦琐的编码不但困扰着我们,同样也困扰着伟大的软件工程师 Gavin King。他觉得访问数据库的代码开发效率太低了,且觉得可以开发出一套更好的数据库访问框架,把项目开发的时间大大缩短。有了初步的想法之后,他决定开始行动,两年后一个优秀的开源框架——Hibernate 诞生了。

11.1.2 Hibernate 的优势

概括地说,Hibernate 是一个优秀的 Java 持久化层解决方案,是当今主流的对象—关系映射工具。Hibernate 简化了 JDBC 烦琐的编码,例如要将用户添加到 List 集合对象中,只需要短短几行代码。示例代码如下:

```
Session session = HibernateUtil.getSession();
Query query = session.createQuery("from User");
List<User> users = (List<User>)query.list();
```

可见 Hibernate 处理数据库查询时,需要编写的代码非常简洁,Hibernate 直接返回的是一个 List 集合类型的对象,可以直接使用,这样避免了烦琐的重复性的数据转换过程。

Hibernate 将数据库的连接信息都存放在配置文件中,这样不仅有利于项目的实施,而且降低了项目的风险。当数据库连接信息发生变化时,如用户名、密码变化,甚至更换了后台数据库软件,只需要修改配置文件中的连接信息即可,无须重新编译源代码,非常方便。

Hibernate 完全是建立在 JDBC 的基础上的,是对 JDBC 有丰富开发经验的人根据实际使用 JDBC 的经验,对 JDBC 的操作进行了封装。Hibernate 不仅解决了 JDBC 的劣势,还能帮助 JDBC 初学者避免操作数据库时出现一些低级的耗费性能的错误,降低相关项目成本。

11.1.3 持久化和 ORM

程序运行时,有些程序数据保存在内存中,当程序退出后,这些数据就不复存在了,所以，我们称这些数据的状态为瞬时的(Transient)。有些数据，在程序退出后，还以文件等形式保存在存储设备中，我们称这些数据的状态是持久的(Persistent)。持久化是程序中的数据在瞬时状态和持久状态间转换的机制，持久化的概念如图 11-1 所示。

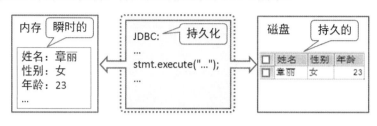

图 11-1　持久化的概念

JDBC 就是一种持久化机制。将程序数据直接保存成文本文件也是持久化机制的一种实现。但常用的是将程序数据保存到数据库中。在分层结构中，DAO 层(数据访问层)有时也被称为持久化层，因为这一层承担的主要工作就是将数据保存到数据库中或把数据从数据库中读取出来，如图 11-2 所示。

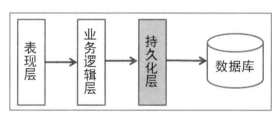

图 11-2　持久化层

以面向对象的方式组织程序，瞬时的数据也以对象的形式存在，而持久的数据多保存在关系型数据库中。所以，在通常的情况下，持久化要完成的操作就是把对象保存到关系型数据库中，或者把关系型数据库中的数据读取出来以对象的形式封装。Hibernate 是在 JDBC 的基础上进行封装，以简化 JDBC 方式烦琐的编码工作。使用 JDBC 将对象保存到数据库中要编写 SQL 语句，并将对象中的属性值取出来赋值给数据表中对应的字段，示例代码如下：

```
String sql="insert into users(loginName,loginPwd,trueName,email,
phone,address,status) values(?,?,?,?,?,?,?)";
try {
    pstmt= getConnection().prepareStatement(sql);
    pstmt.setString(1, user.getLoginName());
    …
    pstmt.executeUpdate();
} catch (Exception e) {    e.printStackTrace();    }
```

Hibernate 的工作原理和 JDBC 编程一样，通过 insert 插入数据、Delete 删除数据、update 更新数据和 select 查询数据，不过 Hibernate 充当 DAO 层，根据 POJO 与实体类的映射配置自动生成 SQL 语句。上面的示例，使用 Hibernate 只要简单地执行 session.save(user)，就可以把 user 对象保存到数据库对应的表中，示例代码如下：

```
Session session = HibernateUtil.getSession();
Transaction transaction=session.beginTransaction();
```

```
Session.save(User);
transaction.commit();
```

Hibernate 是怎么知道 user 对象保存到哪一个表中的，user 对象中的每个属性又对应到数据库表的哪个字段呢？对象—关系映射信息的示例如图 11-3 所示。

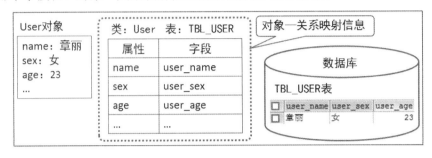

图 11-3 对象—关系映射信息

在编写程序的时候，以面向对象的方式处理数据；保存数据的时候，却以关系型数据库的方式存储。所以说，客观上我们需要一种能在两者间进行转换的机制，这样的机制就是 ORM(对象关系映射)，这个机制需要保存对象和关系数据库表的映射信息，当数据在对象和关系数据库中转换的时候，协助正确地完成转换。简而言之，ORM 就是利用描述对象和数据库之间的映射，自动地把 Java 应用程序中的对象持久化到关系数据库的表中。

11.1.4 Hibernate 的体系架构

Hibernate 作为数据访问层，通过配置文件和映射文件(或持久化注解)将持久化对象映射到数据库的表中，然后通过操作持久化对象，对数据库表进行各种操作。Hibernate 的简要体系架构如图 11-4 所示。

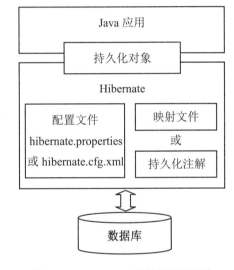

图 11-4 Hibernate 的简要体系架构

从图 11-4 可以看出，Hibernate 使用数据库和配置信息来为应用程序提供持久化服务。Hibernate 比较灵活且支持多种应用方案，一种"全面解决"方案的 Hibernate 体系架构如图 11-5 所示。

下面对图 11-5 中各个对象和接口的含义逐一进行解释。

- Transient Objects(瞬时对象)：由 new 创建，未与 Hibernate Session 关联的对象。
- Persistent Objects(持久化对象)：带有持久化状态、具有业务功能的单线程对象。这些对象是与唯一的 Session 相关联的普通的 JavaBean 或 POJO。
- SessionFactory 接口：生成 Session 的工厂，负责创建 Session 对象，需要使用 ConnectionProvider。

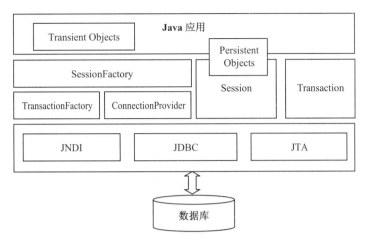

图 11-5 “全面解决”方案的 Hibernate 体系架构

- Session 接口：表示应用程序和持久层之间交互操作的一个单线程对象，隐藏了 JDBC 连接。用于执行被持久化对象的 CRUD 操作，Session 对象是非线程安全的。
- TransactionFactory 接口：生成 Transaction 的工厂。
- Transaction 接口：应用程序用来指定原子操作单元范围的对象。它通过抽象将应用与底层具体的 JDBC、JTA 以及 CORBA 事务隔离开。
- ConnectionProvider：生成 JDBC 的工厂(同时起到连接池的作用)。它通过抽象将应用与底层的 Datasource 或 DriverManager 分离。

11.2 Hibernate 的下载与安装

Hibernate 的用法非常简单，只要在 Java 项目或 Web 项目中引入 Hibernate 框架，就能以面向对象的方式操作关系数据库。读者可以从官方网站 http://www.hibernate.org 下载所需要的版本(这里以 hibernate 5.2.6 版本为例)，具体步骤如下。

(1) 登录 http://hibernate.org/orm/，即可在页面上看到一个绿色的 Download(5.2.6Final)按钮，单击该按钮跳转到 https://sourceforge.net/的相应页面下载 Hibernate 压缩包。

(2) 解压刚下载的压缩包，得到一个名为 hibernate-release-5.2.6.Final 的文件夹，该文件夹的目录结构如图 11-6 所示。

名称	修改日期	类型
documentation	2016/12/20 13:03	文件夹
lib	2016/12/20 13:03	文件夹
project	2016/12/20 13:03	文件夹
changelog.txt	2016/12/20 10:21	文本文档
hibernate_logo.gif	2015/3/27 15:34	看图王 GIF 图片文件
lgpl.txt	2015/9/11 19:30	文本文档

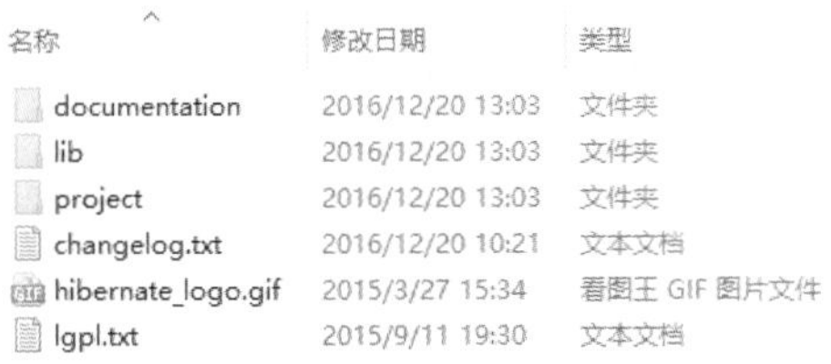

图 11-6 Hibernate 压缩包的文件结构

图 11-6 中各个文件和文件夹的说明如下。

- documentation：该路径下存放 Hibernate 的相关文档，包括参考文档和 API 文档。

- lib：该路径下存放 Hibernate 5.2 的核心类库，以及编译和运行所依赖的第三方类库。其中 required 子目录下保存运行 Hibernate 5.2 项目必需的 jar 包。
- project：存放 Hibernate 各种相关项目的源代码。
- lgpl.txt：logo 等杂项文件。

（3）将 required 子目录下的所有 jar 包添加到应用程序的类加载路径中——既可通过添加环境变量的方式来添加，也可使用 IDE 或 Ant 工具来管理应用程序的类加载路径。required 子目录下包含的 jar 包如图 11-7 所示。

图 11-7　Hibernate 必需的 jar 包

这些 jar 包的作用说明如表 11-1 所示。

表 11-1　Hibernate 必需的 jar 包的作用说明

jar 包名	说　明
antlr-2.7.7.jar	语言转换工具，Hibernate 利用它实现 HQL 到 SQL 的转换
dom4j-1.6.1.jar	一个 Java 的 XML API，类似于 jdom，读写 XML 文件
geronimo-jta_1.1_spec-1.1.1.jar	指定事务、事务处理、分布式事务处理系统之间的标准，如果缺少此文件，运行时会抛出异常
hibernate-commons-annotations-5.0.1.Final.jar	常见的反射代码用于支持注解处理
hibernate-core-5.2.6.Final.jar	hibernate 核心类库
hibernate-jpa-2.1-api-1.0.0.Final.jar	对 JPA(Java 持久化 API)规范的支持
jandex-2.0.3.Final.jar	用来索引 annotation 的包路径及主要类
javassist-3.20.1-GA.jar	一个开源的分析、编辑和创建 Java 字节码的类库
jboss-logging-3.3.0.Final.jar	Jboss 的日志框架

由于 Hibernate 底层依然是基于 JDBC 的，因此在应用程序中使用 Hibernate 执行持久化时同样少不了 JDBC 驱动，将 MySQL 数据库的驱动 mysql-connector-java-5.1.42-bin.jar 添加到应用程序的类加载路径中。

11.3　小　　结

本章介绍了 Hibernate 框架技术，JDBC 的困扰，以及 Hibernate 与 JDBC 相比较的优势所在；还介绍了持久化和 ORM 的相关概念，以及 Hibernate 的体系结构和 Hibernate 的下载与安装。

第 12 章
使用 Hibernate 实现数据的增删改查

上一章我们学习了 Hibernate 的优势、持久化和 ORM、体系架构以及下载和安装。本章我们学习使用 Hibernate 实现数据的增删改查的操作，主要有两种方法：一是基于 XML 映射文件实现；二是基于持久化注解实现。

12.1 基于 XML 映射文件实现数据的增删改查

在所有的 ORM 框架中有一个非常重要的媒介：PO(持久化对象)。持久化对象的作用是完成持久化操作，简单地说，通过该对象可以对数据执行增、删、改、查的操作。在 Java 或 Java Web 项目中添加 Hibernate 框架后，就能以面向对象的方式操作关系型数据库了。

12.1.1 Hibernate 数据操作流程

作为一个优秀的持久层框架，Hibernate 很容易入门。应用程序无须直接访问数据库，甚至无须理会底层采用何种数据库——这一切对应用程序完全透明，应用程序只需要创建、修改、删除持久化对象即可；与此同时，Hibernate 则负责把这种操作转换为对指定数据表的操作。在使用 Hibernate 框架前，先来看看 Hibernate 是如何实现 ORM 框架的，即 Hibernate 的执行流程，如图 12-1 所示。

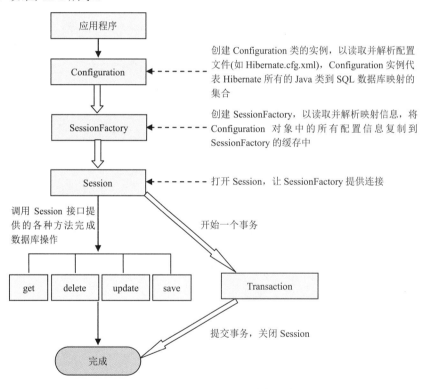

图 12-1 Hibernate 的执行流程

(1) 应用程序先调用 Configuration 类，该类读取 Hibernate 的配置文件及映射文件中的信息，并用这些信息生成一个 SessionFactory 对象。

(2) 用 SessionFactory 对象生成一个 Session 对象，并用 Session 对象生成 Transaction 对象；可通过 Session 对象的 get()、load()、save()、update()、delete()和 saveOrUpdate()等方法对 PO 进行加载、保存、更新、删除等操作；在查询的情况下，可通过 Session 对象生成一个

Query 对象,然后利用 Query 对象执行查询操作;如果没有异常,Transaction 对象将提交这些操作结果到数据库中。

通过 Hibernate 操作数据库需要经过以下步骤。

(1) 读取并解析配置文件。

(2) 读取并解析映射信息,创建 SessionFactory。

(3) 打开 Session。

(4) 开启一个事务。

(5) 执行数据库操作。

(6) 提交事务(回滚事务)。

(7) 关闭 Session 和 SessionFactory。

12.1.2 添加数据

依照图 12-1 所示的 Hibernate 执行流程,下面通过一个简单的实例来体验 Hibernate 的魅力。本实例采用的数据库为 MySQL 5.7,使用 Hibernate 向数据库 restrant 中的 users 表中添加新记录。数据表 users 的部分字段如表 12-1 所示。

表 12-1 数据表 users 的部分字段

字段名	类型	说 明
Id	int	用户编号,主键,自增
LoginName	varchar(20)	登录名称
LoginPwd	varchar(20)	登录密码
TrueName	varchar(20)	真实姓名

实现该功能的具体操作步骤如下。

(1) 在 MyEclipse 中创建 Java 项目,名称为 hibernate-1。在项目中新建文件夹 lib,用于存放项目所需的 jar 包,项目 hibernate-1 最终的目录结构如图 12-2 所示。

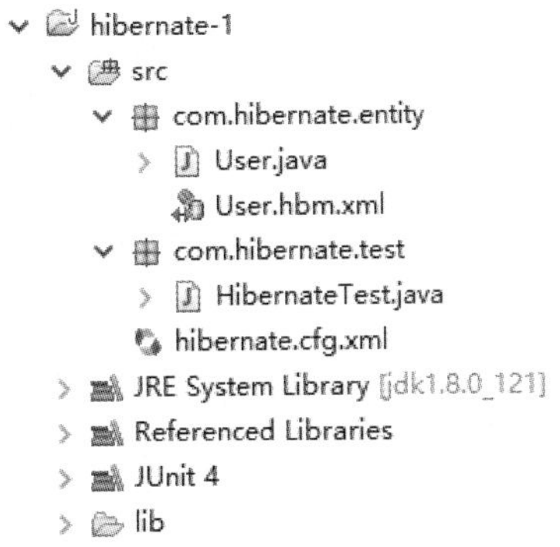

(2) 将第 11 章中图 11-7 所示的 Hibernate 必需的 jar 包,复制到该项目下的 lib 目录中,即完成了 Hibernate 的安装。

(3) 将 MySQL 的 JDBC 驱动包也复制到该项目的 lib 目录中,这里使用的版本为 mysql-connector-java-5.1.42-bin.jar。

图 12-2 项目 hibernate-1 的目录结构

(4) 选中该项目 lib 目录下的所有 jar 包,右击并选择 Build Path→Add to Build Path 命令,将这些 jar 包添加到项目的构建路径中。

(5) 创建实体类。

在 src 目录下新建 com.hibernate.entity 包,并在其中创建实体类 User(对应数据表 users)。

User 类包含一些属性(对应数据表 users 的部分字段)，以及与之对应的 getXxx()和 setXxx()方法，还可以根据需要添加构造方法，其代码如下：

```java
package com.hibernate.entity;
public class User {
    private int id;
    private String loginName;
    private String loginPwd;
    private String trueName;
    //省略属性的getter、setter方法
    //无参构造方法
    public User() {    }
    //有参构造方法
    public User(String loginName, String loginPwd, String trueName) {
        this.loginName = loginName;
        this.loginPwd = loginPwd;
        this.trueName = trueName;
    }
    @Override     //重写toString方法
    public String toString() {
        return "User [Id="+id+", LoginName="+loginName+", LoginPwd"+loginPwd+", TrueName"+trueName+"]";
    }
}
```

(6) 编写映射文件。

实体类 User 目前还不具备持久化操作的能力，为了使其具备这种能力，需要告知 Hibernate 框架将实体类 User 映射到数据库 restrant 中的哪个表，以及类中的哪个属性对应数据库表中的哪个字段，这些都需要在映射文件中配置。在实体类 User 所在的包 com.hibernate.entity 中创建 User.hbm.xml 文件，该文件的具体配置如下：

```xml
<?xml version="1.0" encoding="utf-8"?>
<!DOCTYPE hibernate-mapping PUBLIC "-//Hibernate/Hibernate Mapping DTD 3.0//EN" "http://www.hibernate.org/dtd/hibernate-mapping-3.0.dtd">
<hibernate-mapping package="com.hibernate.entity">
    <class name="User" table="users" catalog="restrant">
        <id name="id" type="java.lang.Integer">
            <column name="Id" />
            <generator class="native"></generator>
        </id>
        <property name="loginName" type="java.lang.String">
            <column name="LoginName" length="20" not-null="true" />
        </property>
        <property name="loginPwd" type="java.lang.String">
            <column name="LoginPwd" length="20" not-null="true" />
        </property>
        <property name="trueName" type="java.lang.String">
            <column name="TrueName" length="20" not-null="true" />
        </property>
    </class>
</hibernate-mapping>
```

上述配置展示了从实体类 User 到数据库表 users 的映射。在映射文件中，每个<class>节

点配置一个实体类的映射信息，<class>节点的 name 属性对应实体类的名字，table 属性对应数据库表的名字，catalog 属性对应数据库的名字。

在<class>节点下，必须有一个<id>节点，用于定义实体的标识属性(对应数据库表的主键)。<id>节点的 name 属性对应实体类的属性，type 属性指定实体类属性的类型。例如，这里的 id 为实体类 Users 中的属性，该属性类型为 Integer。<column>用于指定对应数据库表的主键，<generator>节点用于指定主键的生成器策略。Hibernate 提供的常用主键生成器策略如下。

- increment：对象标识符由 Hibernate 以递增方式生成，如果有多个应用实例向同一张表中插入数据，则会出现重复的主键，应当谨慎使用。
- identity：对象标识符由底层数据库的自增主键生成机制产生，要求底层数据库支持自增字段类型，如 MySQL 的 auto_increment 类型主键和 SQL Server 的 identity 类型主键。还适用于 DB2、Sybase 和 HypersonicSQL。
- sequence：对象标识符由底层数据库的序列生成机制产生，要求底层数据库支持序列，如 Oracle 数据库的序列。还适用于 DB2、PostgreSQL、SAP DB、McKoi 等。
- hilo：对象标识符由 Hibernate 按照高/低位算法生成，该算法从特定表的字段读取高位值，在默认情况下选用 hibernate_unique_key 表中的 next_hi 字段。高/低位算法生成的标识符仅在一个特定的数据库中是唯一的。
- native：根据底层数据库对自动生成标识符的支持能力，选择 identity、sequence 或 hilo。适合于跨数据库平台的开发。
- assigned：对象标识符由应用程序产生，如果不指定<generator>节点，则默认使用该生成器策略。

大部分数据库，如 MySQL、Oracle、DB2 等，都提供了易用的主键生成机制(identity 字段或 sequence)，因此可以在数据库提供的主键生成机制上，采用<generator class="native">的主键生成方式。

<class>节点下除了<id>子节点，还包括<property>子节点，用于映射普通属性。<property>节点与<id>节点类似，只是不能包括<generator>子节点。每个<property>节点指定一对属性和字段的对应关系。

(7) 编写 Hibernate 配置文件。

Hibernate 映射文件反映了持久化类和数据库表的映射信息，而 Hibernate 配置文件则反映了 Hibernate 连接的数据库的相关信息，如数据库用户名、密码、驱动类等。在项目 src 目录下创建 Hibernate 配置文件，文件名为 hibernate.cfg.xml，配置文件的内容如下：

```
<!DOCTYPE hibernate-configuration PUBLIC
    "-//Hibernate/Hibernate Configuration DTD 3.0//EN"
    "http://www.hibernate.org/dtd/hibernate-configuration-3.0.dtd">
<hibernate-configuration>
    <session-factory>
        <!-- Hiberante 连接的基本信息  -->
        <property name="connection.username">root</property>
        <property name="connection.password">123456</property>
        <property name="connection.driver_class">
                    com.mysql.jdbc.Driver</property>
```

```xml
            <property name="connection.url">
                        jdbc:mysql:///restrant</property>
        <!-- Hiberante 方言 -->
        <property name="dialect">
                org.hibernate.dialect.MySQLInnoDBDialect</property>
        <!-- 是否打印 SQL -->
        <property name="show_sql">true</property>
        <!-- 关联 Hibernate 的映射文件 -->
        <mapping resource="com/hibernate/entity/User.hbm.xml"/>
    </session-factory>
</hibernate-configuration>
```

其中，connection.username 属性定义数据库用户名；connection.password 属性定义数据库密码；connection.driver_class 属性定义数据库驱动类；connection.url 属性定义数据库连接 URL。dialect 参数是必须配置的，用于配置 Hibernate 使用的不同数据库类型。Hibernate 支持几乎所有主流数据库，包括 MS SQL Server、MySQL、DB2、Oracle 等。show_sql 参数设置为 true，表示程序运行时在控制台输出执行的 SQL 语句。此外，配置实体类和数据表的映射信息的映射文件需要在 Hibernate 配置文件中声明，代码如下：

```xml
<mapping resource="com/hibernate/entity/User.hbm.xml"/>
```

（8）编写测试类。

在 src 目录下新建 com.hibernate.test 包，并在其中新建 JUnit 测试类 HibernateTest.java，代码如下：

```java
package com.hibernate.test;
import org.hibernate.Session;
import org.hibernate.SessionFactory;
import org.hibernate.Transaction;
import org.hibernate.boot.MetadataSources;
import org.hibernate.boot.registry.StandardServiceRegistry;
import org.hibernate.boot.registry.StandardServiceRegistryBuilder;
import org.junit.After;
import org.junit.Before;
import org.junit.Test;
import com.hibernate.entity.User;
public class HibernateTest {
    private SessionFactory sessionFactory;
    private Session session;
    private Transaction transaction;
    @Before
    public void init(){
        // 加载 hibernate.cfg.xml
        final StandardServiceRegistry registry=new StandardServiceRegistryBuilder().configure().build();
        try {
            // 根据 hibernate.cfg.xml 配置,初始化 SessionFactory
            sessionFactory=new MetadataSources(registry)
                    .buildMetadata().buildSessionFactory();
            //创建 session
            session=sessionFactory.openSession();
            //通过 session 开始事务
```

```java
            transaction=session.beginTransaction();
        } catch (Exception e) {
            StandardServiceRegistryBuilder.destroy(registry);
        }
    }
    //添加数据
    @Test
    public void testSaveUser() {
        User user=new User("hiberUser1","123456","用户1");
        session.save(user);
    }
    @After
    public void destroy(){
        transaction.commit();              //提交事务
        session.close();                   //关闭session
        sessionFactory.close();            //关闭sessionFactory
    }
}
```

在测试类 HibernateTest 中，首先添加 init()方法，并在方法前面添加@Before 注解。JUnit4 使用 Java 5 中的@Before 注解，用于进行初始化。init()方法对于每一个测试方法都要执行一次，方法中的代码根据 hibernate.cfg.xml 配置初始化 SessionFactory，获取 Session，开始事务。

然后添加 destroy()方法，并在方法前面添加@After 注解。JUnit4 使用 Java 5 中的@After 注解，用于释放资源。destroy()方法对于每一个测试方法都要执行一次。方法中的代码会执行事务提交，释放 session 和 sessionFactory 资源。

接下来编写 testSaveUser()方法，并在方法前面添加@Test 注解。JUnit4 使用 Java 5 中的@Test 注解，用于测试方法，可以测试期望异常和超时时间。

在 testSaveUser()方法中，通过构造方法实例化 User 类得到对象 user，将需要添加到数据表 users 中的用户信息封装到该对象中，然后调用 session 的 save 方法。

(9) 运行测试方法 testSaveUser。

在测试类 HibernateTest 中，选中 testSaveUser()方法，右击，在弹出的快捷菜单中选择 Run As→JUnit Test 命令，执行结束后，可以看到数据表 users 中添加了一条新用户记录，如图 12-3 所示。

Id	LoginName	LoginPwd	TrueName	Email	Phone	Address	Status
1	zhangsan	123456	张三	user@163.com	13200000001	江苏南京X区	1
2	zs	zs	zs	zs	zs	zs	1
3	lisi	lisi	lisi	lisi	lisi	lisi	1
4	sj	123	shijun	shijun@126.com	12345678901	江苏扬州	1
5	shi	123	shijun	dd@123.com	0987654321	江苏扬州	1
6	hiberUser1	123456	用户1	(NULL)	(NULL)	(NULL)	(NULL)

图 12-3 使用 Hibernate 添加数据

由于在 Hibernate 配置文件中将 show_sql 参数设置为 true，因此程序运行时会将 Session 的 save 方法所封装的 SQL 语句输出到控制台，SQL 语句如下：

```
Hibernate: insert into restrant.users (LoginName, LoginPwd, TrueName) values (?, ?, ?)
```

在测试类 HibernateTest 中，使用@Test 注解修饰的方法的调用顺序为：@Before→@Test→@After，因此执行 testSaveUser()方法前先调用@Before 注解修饰的方法 init()，执行结束后再调用@After 注解修饰的方法 destroy()。

Session 的 save 方法必须在事务环境中完成，并需要使用 commit 方法提交事务，记录才能成功添加到数据表中。使用 Session 对象的 save 方法虽然可以完成对象的持久化操作，但有时会出现问题，如一个对象已经被持久化了，此时如果再次调用 save()方法将会出现异常。使用 saveOrUpdate()方法可以很好地解决这一问题，因为它会自动判断该对象是否已经持久化，如果已经持久化将执行更新操作，否则执行添加操作。如果标识(主键)的生成策略是自增型的，则使用 Session 对象的 save()和 saveOrUpdate()方法是完全相同的。

可以看出，Hibernate 以面向对象的方式实现对数据库的操作，即将对数据表和字段的操作转变为对实体类和属性的操作。在这一过程中，Hibernate 对象经历了状态的变迁。Hibernate 的对象有三种状态，分别为瞬态(Transient)、持久态(Persistent)和脱管态(Detached)。处于持久态的对象也称为 PO(Persistence Object)，瞬时对象和脱管对象也称为 VO(Value Object)。

由 new 关键字创建的对象，如果它与数据库中的数据没有任何关联，也没有通过 Session 实例进行任何持久化操作，则该对象处于瞬态。瞬态对象一旦不再被其他对象引用，那么将很快被 Java 虚拟机回收。例如，测试类中通过 new 关键字创建的实体类 user，其状态为瞬态。

在 Hibernate 中通过 Session 的 save()和 saveOrUpdate()方法，可以将瞬时对象转变成持久态对象，同时将对象中携带的数据插入数据库表中。处于持久态的对象在数据库中具有相应的记录，并拥有一个持久化标识。持久态对象位于一个 Session 实例的缓存中，即总是与一个 Session 实例相关联。当 Session 清理缓存时，会根据持久态对象的属性的变化，同步更新数据库。例如，测试类中调用 Session 实例的 save 方法后，user 对象的状态由瞬态转变为持久态。

持久态对象相关联的 Session 实例执行 delete()方法之后，持久态对象将转变为瞬态，同时删除数据库中相应的记录，该对象不再与数据库的记录相关联。

持久态对象相关联的 Session 实例执行 close 方法、clear 方法或者 evict 方法之后，持久态对象将转变成脱管态。例如，测试类中调用 session.close()方法关闭 Session 后，user 对象状态由持久态转为脱管态。此后，如果 user 对象中的属性值发生变化，Hibernate 不会再将变化同步到数据库中。

脱管态对象如果不再被任何对象引用，将很快被垃圾回收。如果被重新关联到 Session 上，脱管态对象将再次转变为持久态。脱管态对象具有数据库记录标识，可以使用 Session 的 update()或者 saveOrUpdate()方法将脱管态对象转变为持久态，即对象与数据库记录同步。

脱管态对象与瞬态对象的相同之处在于：如果不再被任何对象引用，将很快被垃圾回收；不同之处在于：脱管态对象有数据库记录标识，瞬时对象没有。

Hibernate 的对象三种状态的转变关系如图 12-4 所示。

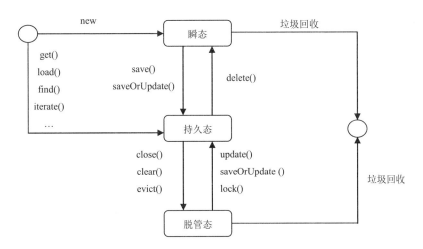

图 12-4　Hibernate 的对象三种状态的转变关系

从图 12-4 可以看出，通过 Sesssion 实例调用一系列方法后会引起 Hibernate 的对象状态转变。其中，能够使 Hibernate 的对象由瞬态或脱管态转变为持久态的方法如下。

① save()方法将对象由瞬态转变为持久态。
② load()或 get()方法获得的对象的状态处于持久态。
③ find()方法获得的 List 集合中的对象状态处于持久态。
④ update()、saveOrUpdate()和 lock()方法可将脱管态对象转变为持久态。

能够使 Hibernate 的对象由持久态转变为脱管态的方法如下。

① close()方法调用后，Session 的缓存会被清空，缓存中所有持久态对象都转变为脱管态。处于脱管状态的对象称为游离对象，当游离对象不再被引用时，将被 Java 虚拟机垃圾回收机制清除。
② evict()方法可将 Session 缓存中一个指定的持久态对象删除，使其转变为脱管态对象。当缓存中保存了大量处于持久态的对象时，为了节省内存空间，可以调用 evict()方法删除一些持久态对象。

12.1.3　加载数据

加载数据是指通过标识符得到指定类的持久化对象，可以通过 Session 实例加载数据。Session 提供了两种方法来加载数据，分别如下。

(1) Object get(Class class, Serializable id)：通过实体类 class 对象和 id 加载数据。
(2) Object load(Class class, Serializable id)：通过实体类 class 对象和 id 加载数据。

1．使用 get()方法

在测试类 HibernateTest 中，添加 testGetUser()方法，并使用@Test 注解加以修饰，实现从数据表 users 中加载编号 Id 为 1 的用户对象，并将用户信息输出到控制台，代码如下：

```
@Test
public void testGetUser(){
    // 从数据表 users 中加载编号 Id 为 1 的用户对象
```

```
    User user = (User)session.get(User.class, 1);
    // 在控制台输出用户对象信息
    System.out.println(user.toString());
}
```

在该示例中，使用 Session 的 get()方法加载数据，只需要一行代码，不再需要烦琐地从 ResultSet 中取数据封装到实体的代码。执行 testGetUser ()方法，控制台输出结果如下：

```
…
User [Id=1, LoginName=zhangsan, LoginPwd=123456, TrueName=张三]
```

如果加载的数据不存在(如 Id=8)，get()方法会返回一个 null 对象。

2. 使用 load()方法

在测试类 HibernateTest 中，添加 testLoadUser()方法，并使用@Test 注解加以修饰，实现从数据表 users 中加载编号 Id 为 2 的用户对象，并将用户信息输出到控制台，代码如下：

```
@Test
public void testLoadUser(){
    try {
        // 从数据表users 中加载编号Id 为2 的用户对象
        User user = (User) session.load(User.class,2);
        // 在控制台输出用户对象信息
        System.out.println(user.toString());
    } catch (Exception e) {
        e.printStackTrace();
    }
}
```

执行 testLoadUser()方法，控制台输出结果如下：

```
…
User [Id=2, LoginName=zs, LoginPwd=zs, TrueName=zs]
```

如果获取的 id 号不存在(如 id=8)，则会抛出错误，具体如下：

```
org.hibernate.ObjectNotFoundException: No row with the given identifier exists: [com.hibernate.entity.User#8]
    at org.hibernate.boot.internal.StandardEntityNotFoundDelegate.handleEntityNotFound(StandardEntityNotFoundDelegate.java:28)
    at org.hibernate.proxy.AbstractLazyInitializer.checkTargetState(AbstractLazyInitializer.java:235)
    at org.hibernate.proxy.AbstractLazyInitializer.initialize(AbstractLazyInitializer.java:157)
    at org.hibernate.proxy.AbstractLazyInitializer.getImplementation(AbstractLazyInitializer.java:259)
    at org.hibernate.proxy.pojo.javassist.JavassistLazyInitializer.invoke(JavassistLazyInitializer.java:73)
    at com.hibernate.entity.User_$$_jvstd5a_0.toString(User_$$_jvstd5a_0.java)
    at com.hibernate.test.HibernateTest.testLoadUser(HibernateTest.java:53)
```

运行结果出现了 ObjectNotFoundException 异常，表示对象没有发现。这一异常说明使用 load()方法加载数据时，要求记录必须存在，这一点与 get()方法是不同的。

12.1.4 删除数据

删除数据是指根据主键值将一条记录从数据表中删除，可以通过 Session 实例的 delete(Object obj)方法来删除数据库中的记录。delete 方法的参数 obj 表示要删除的持久态对象。因此，在调用 delete 方法前，需要通过 Session 的 get 方法获得指定标识的持久态对象。

在测试类 HibernateTest 中，添加 testDeleteUser()方法，并使用@Test 注解加以修饰，实

现将数据表 users 中编号 id 为 6 的记录删除。代码如下：

```
@Test
public void testDeleteUser () {
    // 从数据表 users 中加载编号 id 为 6 的用户对象，数据表中要有 id 为 6 的记录
    User user = (User) session.get(User.class,6);
    // 删除对象
    session.delete(user);
}
```

执行 testDeleteUser()方法，控制台输出结果如下：

```
…
Hibernate: delete from restrant.users where Id=?
```

执行 testDeleteUser()方法，打开数据表 users，可以看到编号为 6 的记录已被删除。

12.1.5 修改数据

通过 Session 实例的 update(Object obj)方法可以修改数据库中的记录，参数 obj 表示要修改的对象。update 方法可将一个处于脱管态的对象加载到 Session 缓存中，与一个具体的 Session 实例关联，使其状态转变为持久态。在调用 update 方法前，需要通过 Session 的 get 方法获得指定标识的持久态对象。

在测试类 HibernateTest 中，添加 testUpdateUser()方法，并使用@Test 注解加以修饰，实现将数据表 users 中编号 Id 为 2 的记录中的登录名由 zs 修改为 zhang。代码如下：

```
@Test
public void testUpdateUser() {
    // 从数据表 User 中加载编号 id 为 2 的用户对象
    User user = (User) session.get(User.class,2);
    // 修改数据
    user.setLoginName("zhang");
    // 更新对象
    session.update(user);
}
```

执行 testUpdateUser()方法，打开数据表 users，可看到编号 Id 为 2 的记录中的登录名被修改。除了 update 方法，也可以通过 Session 实例 saveOrUpdate(Object obj)方法修改数据库记录。在使用 Hibernate 编写持久化代码时，不需要再有数据库表和字段等概念，取而代之的是对象和属性。以面向对象的思维编写代码是 Hibernate 持久化操作的一个理念。

12.2 基于 Annotation 注解实现数据的增删改查

从 JDK 1.5 开始，Java 增加了 Annotation 注解技术解决方案，将原来通过 XML 配置文件管理的信息改为通过 Annotation 进行管理，从而实现 Hibernate 的零配置。Hibernate 的 Annotation 方案是以 Java 持久化(Java Persistence API，JPA)为基础，进一步扩展而来的。

使用 Annotation 注解实现数据的增、删、改、查操作步骤如下：

(1) 先将项目 hibernate-1 复制并命名为 hibernate-2，再导入 MyEclipse 开发环境中。
(2) 修改实体类 User.java。

通过 Annotation 注解将数据表与实体类之间的映射在实体类中完成，无须使用映射文件，因此需要先将项目 hibernate-2 的 com.hibernate.entity 包中的映射文件 User.hbm.xml 删除，然后修改实体类 User.java，代码如下：

```java
package com.hibernate.entity;
import javax.persistence.*;
// 使用@Entity注解，表示当前类为实体Bean，需要进行持久化
@Entity
// 使用@Table注解实现数据表users与持久化类User之间的映射，
// catalog指定数据库名，name指定表名
@Table(name="users",catalog="restrant")
public class User {
    private int id;
    private String loginName;
    private String loginPwd;
    private String trueName;
    @Id    // 使用@Id注解指定当前持久化类的ID标识属性
    // 使用@GeneratedValue注解指定主键生成策略为IDENTITY
    @GeneratedValue(strategy=GenerationType.IDENTITY)
    // 使用@Column注解指定当前属性所对应的数据表中的字段，name指定字段名，
    // unique指定是否唯一，nullable指定是否可为null
    @Column(name="id",unique=true,nullable=false)
    public int getId() {
        return id;
    }
    public void setId(int id) {
        this.id = id;
    }
    // 使用@Column注解指定当前属性所对应的数据表的字段，
    // name指定字段名，length指定字段长度
    @Column(name = "loginName", length = 20)
    public String getLoginName() {
        return loginName;
    }
    public void setLoginName(String loginName) {
        this.loginName = loginName;
    }
    @Column(name="loginPwd",length = 20)
    public String getLoginPwd() {
        return loginPwd;
    }
    public void setLoginPwd(String loginPwd) {
        this.loginPwd = loginPwd;
    }
    @Column(name="trueName",length = 20)
    public String getTrueName() {
        return trueName;
    }
    public void setTrueName(String trueName) {
        this.trueName = trueName;
    }
```

```
    //省略无参构造方法和有参构造方法
    @Override    //重写toString方法
    public String toString() {
        return "User [Id="+id+", LoginName="+loginName+",
LoginPwd="+loginPwd+", TrueName="+trueName+"]";
    }
}
```

JPA (Java Persistence API)规范推荐使用 Annotation 来管理实体类与数据表之间的映射关系，可以避免同时维护两份文件(Java 实体类和 XML 映射文件)，而是将映射信息(写在 Annotation 中)与实体类集中在一起。在实体类 User.java 代码中，使用了@Entity 注解、@Table 注解、@Id 注解、@GeneratedValue 注解和@Column 注解，这些注解的含义如表 12-2 所示。

表 12-2 实体类 User 中 Annotation 注解的含义

Annotation 名称	功能描述
@Entity	表示当前类为实体 Bean，需要进行持久化。将一个 JavaBean 声明为持久化类时，在默认情况下，该类的所有属性都将映射到数据表的字段。如果在该类中添加了无须映射的属性，则需要使用@Transient 注解声明
@Table	实现数据表与持久化类之间的映射，catalog 指定数据库名，name 指定表名。@Table 注解的位置在@Entity 注解之下
@Id	指定当前持久化类的 ID 标识属性，与@GeneratedValue 配合使用
@GeneratedValue	指定 ID 标识生成器，即主键生成策略，与@Id 配合使用
@Column	指定当前属性所对应的数据库表中的字段，name 指定字段名，unique 指定是否唯一，nullable 指定是否可为 null

主键生成策略通过 GenerationType 来指定，GenerationType 是一个枚举，它定义了主键生成策略的类型。在实体类 User 中，使用@GeneratedValue 注解指定主键生成策略为 GenerationType.IDENTITY。该策略用于 MySQL 数据库，特点是递增。对于 MySQL 数据库，使用递增序列时需要在建表时为主键指定 auto_increment 属性。GenerationType 枚举值还有以下三种。

① GenerationType.AUTO。

自动选择一个最适合底层数据库的主键生成策略，这个是默认选项，即如果只写@GeneratedValue，等价于@GeneratedValue(strategy=GenerationType.AUTO)。在本例中，如果将主键生成策略指定为 GenerationType.AUTO，运行测试类中的 testSaveUser()方法时，控制台抛出如下异常信息：

```
com.mysql.jdbc.exceptions.jdbc4.MySQLSyntaxErrorException: Table
'restrant.hibernate_sequence' doesn't exist
```

说明需要通过 hibernate_sequence 来生成主键，而数据库 restrant 中并不存在该主键。

② GenerationType.SEQUENCE。

根据底层数据库的序列来生成主键，条件是数据库支持序列。用于 Oracle 数据库，MySQL 不支持 GenerationType.SEQUENCE。

③ GenerationType.TABLE。

使用一个特定的数据库表格来保存主键，框架借由表模拟序列产生主键，使用该策略可以使应用更易于数据库移植。不同的 JPA 实现上生成的表名是不同的，如 OpenJPA 生成 openjpa_sequence_table 表，Hibernate 生成 hibernate_sequences 表，而 TopLink 则生成 sequence 表。这些表都具有一个序列名和对应值两个字段，如 SEQ_NAME 和 SEQ_COUNT。

(3) 修改 Hibernate 配置文件。

由于不再使用映射文件，因此需要将 Hibernate 配置文件 hibernate.cfg.xml 中使用的映射文件由原来的*.hbm.xml 文件转变成持久化类文件，代码如下：

```
<!-- 关联 Hibernate 的持久化类 -->
<mapping class="com.hibernate.entity.User" />
```

这样原来大量的*.hbm.xml 文件不再需要了，所有的配置都通过 Annotation 注解直接在持久化类中进行配置完成。

使用与 hibernate-1 相同的测试类 HibernateTest 进行测试，同样可以完成数据的增、删、改、查操作。

12.3 小　　结

本章讲解了基于 XML 映射文件实现数据的增、删、改、查操作，以及基于 Annotation 注解结束实现 Hibernate 的零配置。Hibernate 为开发带来的便捷还有很多，通过后面的学习，读者将逐步领略 Hibernate 框架的魅力所在。

第 13 章
使用 Hibernate 实现关联映射和继承映射

上一章介绍了如何配置并使用 Hibernate 对数据库的增、删、改、查操作。在学习面向对象时，曾经学过对象间存在关联的关系，学习数据库时，也学习过表与表间可以通过外键关联起来。本章学习怎样映射面向对象领域的关联关系和数据库关系模型中的外键关联。

13.1 基于 XML 映射文件实现关联映射

上一节实现数据的 CRUD 操作针对的是单个对象(映射到数据库中的单个表)，由于数据库中表之间可以通过外键进行关联，因此在使用 Hibernate 操作映射到存在关联关系的数据表的对象时，需要将对象的关联关系和数据表的外键关联进行映射。本节将基于 XML 映射文件讲述 Hibernate 的关联映射，包括单向多对一关联、单向一对多关联、双向多对一关联、多对多关联、双向一对一关联。

13.1.1 单向多对一关联

单向多对一关联是最为常见的单向关联关系。单向多对一映射关系是由"多"的一方指向"一"的一方。在表示"多"的一方数据表中增加一个外键来指向表示"一"的一方数据表，"一"的一方作为主表，"多"的一方作为从表。

例如，数据库 restrant 中餐品表 meal 和菜系表 mealseries 的对应关系就是一种多对一关系，因为在 meal 表中有多条餐品记录对应 mealseries 表中同一个菜系记录，如图 13-1 所示。

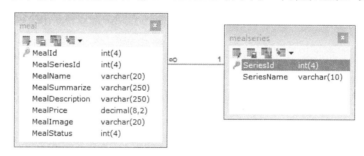

图 13-1 数据表 meal 和 mealseries 间的多对一关系

单向多对一关联只需要从"多"的一端访问"一"的一端，所以只需要在"多"的一方的实体类和映射文件中进行配置，而不用考虑"一"的一方。实现数据表 meal 和 mealseries 之间单向多对一关联映射的步骤如下。

(1) 将项目 hibernate-1 复制并命名为 hibernate-3，再导入 MyEclipse 开发环境中。

(2) 创建实体类。

在项目 hibernate-3 的 com.hibernate.entity 包中创建实体类 Mealseries.java 和 Meal.java，分别对应数据表 mealseries 和 meal。

其中，菜系实体 Mealseries.java 代码如下：

```
package com.hibernate.entity;
public class Mealseries {
    private int seriesId;           //菜系编号
    private String seriesName;      //菜系名称
    //省略属性的getter、setter方法
    //省略无参构造方法和有参构造方法
}
```

餐品实体 Meal.java 代码如下：

```
package com.hibernate.entity;
public class Meal {
    //餐品基本信息
    private int mealId;                     //餐品编号
    private Mealseries mealseries;          //餐品菜系关联属性
    private String mealName;                //餐品名称
    private String mealSummarize;           //餐品摘要
    private String mealDescription;         //餐品详细描述信息
    private Double mealPrice;               //餐品价格
    private String mealImage;               //餐品图片文件名
    //省略属性的 getter、setter 方法
    //省略无参构造方法和有参构造方法
}
```

在实体类 Meal 中使用 Mealseries 类声明 mealseries 属性，并添加该属性的 getter 和 setter 方法，以体现实体类 Meal 对 Mealseries 的关联关系。

（3）创建映射文件。

在 com.hibernate.entity 包中，创建映射文件 Meal.hbm.xml 和 Mealseries.hbm.xml，分别对应实体类 Meal.java 和 Mealseries.java。

其中，映射文件 Meal.hbm.xml 内容如下：

```xml
<?xml version="1.0" encoding="utf-8"?>
<!DOCTYPE hibernate-mapping PUBLIC "-//Hibernate/Hibernate Mapping DTD
3.0//EN" "http://www.hibernate.org/dtd/hibernate-mapping-3.0.dtd">
<hibernate-mapping package="com.hibernate.entity">
    <class name="Meal" table="meal" catalog="restrant">
        <id name="mealId" type="java.lang.Integer">
            <column name="MealId" />
            <generator class="native"></generator>
        </id>
        <!-- 映射实体类 Meal 到 Mealseries 的单向多对一的关联 -->
        <many-to-one name="mealseries" column="MealSeriesId" class="Mealseries">
        </many-to-one>
        <property name="mealName" type="java.lang.String">
            <column name="MealName" length="20" not-null="true"/>
        </property>
        <!-- 省略 mealSummarize、mealDescription 字段的映射配置 -->
        <property name="mealPrice" type="java.lang.Double">
            <column name="MealPrice" precision="8" />
        </property>
        <property name="mealImage" type="java.lang.String">
            <column name="MealImage" length="20" />
        </property>
    </class>
</hibernate-mapping>
```

<many-to-one>元素用来映射从实体类 Meal 到 Mealseries 的单向多对一的关联关系，class 属性指定关联类的名字，这里为 Mealseries；name 属性指定在实体类 Meal 中关联的 Mealseries 类的属性名，这里为 mealseries；column 属性指定数据表关联的外键，这里为

MealseriesId。

换句话说，实体类 Meal 对 Mealseries 的多对一关联在本质上是通过数据表 Meal 中的外键 MealseriesId 与数据表 Mealseries 关联实现的，但 Hibernate 将表之间的关联通过<many-to-one>元素进行了封装。在读取餐品对象时，通过关联可以获取该餐品所关联的菜系对象，并将其赋值给实体类 Meal 中所定义的 Mealseries 菜系的属性 mealseries。

映射文件 Mealseries.hbm.xml 内容如下：

```xml
<hibernate-mapping package="com.hibernate.entity">
    <class name="Mealseries" table="mealseries" catalog="restrant">
        <id name="seriesId" type="java.lang.Integer">
            <column name="SeriesId" />
            <generator class="native"></generator>
        </id>
        <property name="seriesName" type="java.lang.String">
            <column name="SeriesName" length="10" not-null="true" />
        </property>
    </class>
</hibernate-mapping>
```

将上述两个映射文件通过<mapping>标签添加到 Hibernate 配置文件 hibernate.cfg.xml 中，如下：

```xml
<mapping resource="com/hibernate/entity/Meal.hbm.xml"/>
<mapping resource="com/hibernate/entity/Mealseries.hbm.xml"/>
```

在项目 hibernate-3 的测试类 HibernateTest 中添加测试方法 testM2OGet()，并使用@Test 注解修饰。在 testM2OGet()方法中获取指定编号的餐品对象，同时获取关联的菜系对象，代码如下：

```java
@Test
public void testM2OGet(){
    //加载餐品对象
    Meal meal=(Meal)session.get(Meal.class, 1);
    System.out.println("餐品名称为: "+ meal.getMealName());
    //获取关联的菜系对象信息
    System.out.println("菜系是: "+ meal.getMealseries().getSeriesName());
}
```

执行 testM2OGet()方法，控制台输出结果如下：

```
Hibernate: select … from restrant.meal meal0_ where meal0_.MealId=?
餐品名称为：雪梨肉肘棒
Hibernate: select … from restrant.mealseries mealseries0_ where mealseries0_.SeriesId=?
菜系是：鲁菜
```

从控制台输出可以看出，第一条 SQL 语句只从数据表 meal 中获取数据，并没有立即获取关联的 mealseries 数据。只有执行 meal.getMealseries()时才发出第二条 SQL 语句从数据表 mealseries 中获取关联的菜系，这种数据加载的策略称为"懒加载"或"延迟加载"。懒加载可以提高系统性能，但如果在加载关联数据前关闭 session，会抛出"懒加载异常"。

如果修改映射文件 Meal.hbm.xml，在<many-to-one>元素中添加属性 lazy=false。

```xml
<many-to-one name="mealseries" column="MealSeriesId" class="Mealseries"
lazy="false" >
</many-to-one>
```

再次执行 testM2OGet()方法，控制台输出结果如下：

```
Hibernate: select … from restrant.meal meal0_ where meal0_.MealId=?
Hibernate: select … from restrant.mealseries mealseries0_ where
mealseries0_.SeriesId=?
餐品名称为：雪梨肉肘棒
菜系是：鲁菜
```

由于设置了 lazy=false，加载策略变成了"立即加载"，即在执行 meal.getMealseries()获取关联的数据前，也会发出 SQL 语句。所以发出第一条 SQL 语句后，立即发出了第二条 SQL 语句。

13.1.2 单向一对多映射

meal 对 mealseries 是单向多对一关联，反过来看，mealseries 对 meal 便是单向一对多关联，即一个菜系可以包含多个餐品。这也意味着每个 Mealseries 对象会引用一组 Meal 对象，因此需要在 Mealseries 类中定义一个集合类型的属性，在访问"一"的一方 Mealseries 对象时，关联的"多"的一方 Meal 的多个对象将保存到该集合类型的属性中。

实现数据表 mealseries 和 meal 之间单向一对多关联映射的步骤如下。

(1) 修改实体类 Mealseries.java。

在实体类 Mealseries 中，添加一个集合属性，并添加该属性的 getter()和 setter()方法，添加的代码如下：

```java
private Set mealSet=new HashSet();      // 菜系关联的集合属性
// 省略 mealSet 属性的 getter 方法和 setter 方法
```

(2) 修改映射文件 Mealseries.hbm.xml。

在实体类 Mealseries 中添加集合属性 mealSet 后，需要在映射文件 Mealseries.hbm.xml 中映射集合类型的 mealSet 属性。由于在数据表 mealseries 中没有直接与 mealSet 属性对应的字段，因此不能直接使用<property>元素来映射 mealSet 属性，因此需要使用<set>元素。在 Mealseries.hbm.xml 中通过<set>元素添加一对多的配置，代码如下：

```xml
<!-- 配置一对多关联映射 -->
<set name="mealSet">
    <key column="MealSeriesId" />
    <one-to-many class="Meal" />
</set>
```

在<set>元素中，使用 name 属性设定为 Mealseries 类中定义的 mealSet 属性。<set>元素包含两个子元素，<key>子元素的 column 属性设定为数据表 meal 的外键，这里为 MealSeriesId，<one-to-many>子元素用于映射关联实体，其 class 属性设定为"多"的一方的类名，这里为 Meal。

通过上述单向一对多的配置，当加载"一"的一方的 Mealseries 对象时，底层通过数据表 meal 中的外键 MealSeriesId 与数据表 mealseries 关联，从而实现加载关联的"多"的一方的多个 Meal 对象，并将这些对象存放在集合 mealSet 中。

在测试类 HibernateTest 中添加测试方法 testO2MGet()，并使用@Test 注解修饰。在 testO2MGet()方法中获取指定编号的菜系对象，同时获取关联的餐品对象集合，代码如下：

```java
@Test
public void testO2MGet(){
    // 加载菜系对象
    Mealseries ms = (Mealseries) session.get(Mealseries.class, 1);
    System.out.println(ms.getSeriesName() + "菜系的餐品有：");
    // 获取关联的餐品对象集合并通过迭代输出到控制台
    Iterator iterator = ms.getMealSet().iterator();
    while (iterator.hasNext()) {
        Meal meal = (Meal) iterator.next();
        System.out.println(meal.getMealName());
    }
}
```

执行 testO2MGet()方法，控制台输出结果如下：

```
Hibernate: ……
鲁菜菜系的餐品有：
Hibernate: ……
肉冬菜肉末
糖醋红柿椒
烤花肉揽桂鱼
雪梨肉肘棒
素锅烤鸭肉
……
```

13.1.3 双向多对一映射

13.1.1 和 13.1.2 小节分别对数据表 meal 与 mealseries 进行了单向多对一和单向一对多关联，如果将两者结合起来便形成了双向关联。双向多对一关联也可称为双向一对多关联。13.1.1 和 13.1.2 小节中测试单向多对一和单向一对多关联时，只是以数据加载为例。下面介绍如何实现双向多对一关联中数据的增、删、改的操作。

在 Hibernate 中通过设置 inverse 属性来决定由双向关联的哪一方来维护表和表之间的关联关系，当其中的一方设置 inverse=true 时，表示将控制权反转，此时由对方(主动方)负责维护关联关系。inverse 属性的默认值为 false，在双方都没有设置 inverse=true 的情况下，双方都维护关联关系，会影响性能。通常，由"多"的一方作为主动方维护关联关系有助于性能的改善。

1．添加数据

在数据表 mealseries 中添加一个新的菜系"淮扬菜"，并在数据表 meal 中添加两个菜系为淮扬菜的餐品。

修改映射文件 Mealseries.hbm.xml，给<set>元素添加 inverse 属性，并设置 inverse=true，将控制权反转，此时由"多"的一方(Meal)作为主动方来维护双方的关联关系，即"一"的一方(Mealseries)放弃维护关联关系。代码如下：

```
<set name="mealSet" inverse="true">
```

在测试类 HibernateTest 中添加测试方法 testM2OAndO2MSave()，并使用@Test 注解修饰。testM2OAndO2MSave()方法代码如下：

```
    @Test
    public void testM2OAndO2MSave() {
        // 新建菜系对象
        Mealseries ms = new Mealseries("淮扬菜");
        // 新建餐品对象
        Meal meal1=new Meal();
        meal1.setMealName("大煮干丝");
        meal1.setMealSummarize("又称鸡汁煮干丝，传统名菜，吃起来爽口开胃。");
        meal1.setMealDescription("一道既清爽，又有营养的佳肴，其风味之美，历来被推为
席上美馔，淮扬菜系中的看家菜。");
        meal1.setMealPrice(15.00);
        Meal meal2=new Meal();
        meal2.setMealName("狮子头");
        meal2.setMealSummarize("口感松软，肥而不腻，营养丰富，红烧、清蒸皆可");
        meal2.setMealDescription("千百年来盛誉不衰，成功之举在于保持基本格调传统烹调
方法，随候用料因物而异，富于变化，成为系列佳肴。");
        meal2.setMealPrice(25.00);
        // 设置关联关系
        meal1.setMealseries(ms);
        meal2.setMealseries(ms);
        ms.getMealSet().add(meal1);
        ms.getMealSet().add(meal2);
        // 先插入一的一端
        session.save(ms);
        // 再插入多的一端
        session.save(meal1);
        session.save(meal2);
    }
```

执行 testM2OAndO2MSave()方法，数据表 mealseries 和 meal 添加的数据分别如图 13-2 和图 13-3 所示。

SeriesId	SeriesName
7	湘菜
8	徽菜
9	西餐
10	西点
11	药膳
12	私房菜
13	淮扬菜

图 13-2　数据表 mealseries 中的记录

MealId	MealSeriesId	MealName	MealSummarize	MealDescription	MealPrice
7	1	糖醋红柿椒	色红美，味鲜香。	色红美，味鲜香。	8.00
8	2	芹黄烧鱼条	色泽金黄润亮，鱼	色泽金黄润亮，鱼	15.00
9	2	巴国玉米糕	风味浓、口感奇、	风味浓、口感奇、	13.00
10	2	酥皮龙虾	色泽黄绿相衬，协	酥皮龙虾成菜配以	20.00
11	1	香煎茄片	色泽红亮，口感似	色泽红亮，口感似	9.00
12	1	金陵片皮鸭	此菜深红明亮，皮	此菜深红明亮，皮	10.00
13	13	大煮干丝	又称鸡汁煮干丝，	一道既清爽，又有	15.00
14	13	狮子头	口感松软，肥而不	千百年来盛誉不衰	25.00

图 13-3　数据表 meal 中的记录

此时，控制台输出的 SQL 语句如下：

```
Hibernate: insert into restrant.mealseries (SeriesName) values (?)
Hibernate: insert into restrant.meal (MealSeriesId, MealName, MealSummarize,
MealDescription, MealPrice, MealImage) values (?, ?, ?, ?, ?, ?)
Hibernate: insert into restrant.meal (MealSeriesId, MealName, MealSummarize,
```

MealDescription, MealPrice, MealImage) values (?, ?, ?, ?, ?, ?)

可以看出仅仅使用最少的 3 条 insert 语句就完成了 3 条记录的添加，因此性能最好。

2. 修改数据

在测试类 HibernateTest 中添加测试方法 testM2OAndO2MUpdate()，并使用@Test 注解修饰。在 testM2OAndO2MUpdate()方法中，将数据表 meal 中编号 id=7 的餐品的菜系由"鲁菜"(id 为 1)修改为"川菜"(id 为 2)，代码如下：

```java
@Test
public void testM2OAndO2MUpdate() {
    // 加载编号 id=7 的餐品对象
    Meal meal= (Meal) session.get(Meal.class, 7);
    // 加载编号 id=2(川菜)的菜系对象
    Mealseries ms = (Mealseries)session.get(Mealseries.class, 2);
    // 修改关联关系
    meal.setMealseries(ms);
    // 更新 meal 餐品对象
    session.update(meal);
}
```

执行 testM2OAndO2MUpdate()方法后，在控制台输出 3 条 select 的 SQL 和 1 条 update 的 SQL 语句，打开数据表 meal，可以看到编号 id=7 的餐品的菜系由原先的 1 修改成了 2。

3. 删除数据

在测试类 HibernateTest 中添加测试方法 testM2OAndO2MDelete()，并使用@Test 注解修饰。在 testM2OAndO2MDelete()方法中，将数据表 mealseries 中编号 id=13 的菜系记录删除。代码如下：

```java
@Test
public void testM2OAndO2MDelete() {
    // 加载编号 id=13 的菜系对象
    Mealseries ms = (Mealseries) session.get(Mealseries.class, 13);
    // 删除对象 ms
    session.delete(ms);
}
```

执行 testM2OAndO2MDelete()方法，在控制台输出一条 select 的 SQL 语句和一条 delete 的 SQL 语句，紧接着抛出如下异常信息：

```
org.hibernate.engine.jdbc.spi.SqlExceptionHelper logExceptions
ERROR: Cannot delete or update a parent row: a foreign key constraint fails
(`restrant`.`meal`, CONSTRAINT `meal_ibfk_1` FOREIGN KEY (`MealSeriesId`)
REFERENCES `mealseries` (`SeriesId`))
org.hibernate.internal.ExceptionMapperStandardImpl
mapManagedFlushFailure
ERROR: HHH000346: Error during managed flush
[org.hibernate.exception.ConstraintViolationException: could not execute
statement]
```

发生异常的原因在于：数据表 meal 中 MealSeriesId 外键字段引用了数据表 mealseries 的

SeriesId 字段，当准备从 mealseries 表中删除"淮扬菜"时，该菜系的 SeriesId 被 meal 表中的两条相关餐品记录所引用，只有先将 meal 表中参考该 SeriesId 的两条餐品记录删除。当然，没有必要这么麻烦，可以采用级联删除的方法，在删除 mealseries 表中记录的同时，会将 meal 表中关联的记录一同删除。修改映射文件 Mealseries.hbm.xml，在<set>元素中添加属性 cascade，并将值设置为 delete，代码如下：

```xml
<set name="mealSet" inverse="true" cascade="delete">
```

再次执行 testM2OAndO2MDelete()方法，在控制台输出两条 select 的 SQL 语句和三条 delete 的 SQL 语句，再打开数据表 meal 和 mealseries，可以看到数据被成功删除。

13.1.4 双向多对多映射

在数据库 restrant 中，数据表 admin 和 functions 之间存在多对多关联关系，因为一个系统管理员可以使用系统的多个功能，一个系统功能也可能被多个管理员使用。

在程序设计时，一般不建议直接在 admin 和 functions 之间建立多对多关联，这会造成两者之间的相互依赖。可以通过一个中间表来维护两者之间的多对多关联，这个中间表分别与 admin 和 functions 构成多对一关联。在数据库 restrant 中，管理员信息表为 admin，系统功能表为 functions，中间表为 powers，它同时参照 admin 和 functions 表。这三张表之间的关系如图 13-4 所示。

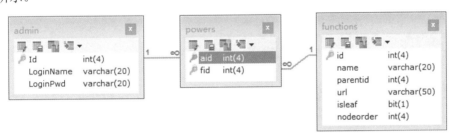

图 13-4 多对多关联关系

powers 表以 aid 和 fid 作为联合主键，其中，aid 字段作为外键参照 admin 表的 Id 字段，fid 字段作为外键参照 functions 表的 id 字段。

实现数据表 admin 和 functions 双向多对多关联映射的步骤如下。

(1) 将项目 hibernate-1 复制并命名为 hibernate-4，再导入 MyEclipse 开发环境中。

(2) 在项目 hibernate-4 的 com.hibernate.entity 包中新建两个实体类 Admin.java 和 Functions.java。其中，实体类 Admin.java 代码如下：

```java
package com.hibernate.entity;
import java.util.HashSet;
import java.util.Set;
public class Admin {
    private int id;                          //id 号
    private String loginName;                //登录名
    private String loginPwd;                 //登录密码
    private Set fs=new HashSet();            // 关联的属性
```

```
    // 省略属性的getter、setter方法
    // 省略无参构造方法和有参构造方法
}
```

在实体类 Admin.java 中添加 Set 类型的属性 fs，以体现与 Functions 的关联。

实体类 Functions.java 代码如下：

```
package com.hibernate.entity;
import java.util.HashSet;
import java.util.Set;
public class Functions {
    private int id;                         //id号
    private String name;                    //功能名称
    private Set as=new HashSet();           //关联的属性
    // 省略属性的getter、setter方法
    // 省略无参构造方法和有参构造方法
}
```

同样，在实体类 Functions.java 中添加 Set 类型的属性 as，以体现与 Admin 的关联。

(3) 在项目 hibernate-4 的 com.hibernate.entity 包中创建映射文件 Admin.hbm.xml 和 Functions.hbm.xml。其中，Admin.hbm.xml 文件内容如下：

```xml
<?xml version="1.0" encoding="utf-8"?>
<!DOCTYPE hibernate-mapping PUBLIC "-//Hibernate/Hibernate Mapping DTD 
3.0//EN" "http://www.hibernate.org/dtd/hibernate-mapping-3.0.dtd">
<hibernate-mapping package="com.hibernate.entity">
    <class name="Admin" table="admin" catalog="restrant">
        <id name="id" type="java.lang.Integer">
            <column name="Id" />
            <generator class="native"></generator>
        </id>
        <property name="loginName" type="java.lang.String">
            <column name="LoginName" length="20" not-null="true" />
        </property>
        <property name="loginPwd" type="java.lang.String">
            <column name="LoginPwd" length="20" not-null="true" />
        </property>
        <!-- 配置多对多关联 -->
        <set name="fs" table="powers">
            <key column="aid" not-null="true" />
            <many-to-many column="fid" class="Functions"/>
        </set>
    </class>
</hibernate-mapping>
```

首先给<set>元素添加 name 属性，值设定为 Admin 实体类中 Set 类型的属性 fs；然后添加一个 table 属性，值为中间表的名称，这里为数据表 powers。再给<set>元素添加两个子元素。<key>子元素的 column 属性指定中间表 powers 中参照 admin 表的外键名字，这里为 aid。<many-to-many>子元素中需要设定两个属性，class 属性设定为多对多关联中另一方的类，这里为 Functions；column 属性指定中间表 powers 中参照 functions 表的外键名字，这里为 fid。

Functions.hbm.xml 文件内容如下：

```xml
<?xml version="1.0" encoding="utf-8"?>
<!DOCTYPE hibernate-mapping PUBLIC "-//Hibernate/Hibernate Mapping DTD
3.0//EN" "http://www.hibernate.org/dtd/hibernate-mapping-3.0.dtd">
<hibernate-mapping package="com.hibernate.entity">
    <class name="Functions" table="functions" catalog="restrant">
        <id name="id" type="java.lang.Integer">
            <column name="id" />
            <generator class="native"></generator>
        </id>
        <property name="name" type="java.lang.String">
            <column name="name" length="20" not-null="true" />
        </property>
        <!-- 配置多对多关联 -->
        <set name="as" table="powers" inverse="true">
            <key column="fid" not-null="true" />
            <many-to-many column="aid" class="Admin"/>
        </set>
    </class>
</hibernate-mapping>
```

在 Functions.hbm.xml 映射文件的<set>元素中设置了 inverse="true"，表示将关联关系的控制权反转，即由对方(Admin)管理关联关系。

最后将这两个映射文件通过<mapping>标签添加到 Hibernate 配置文件 hibernate.cfg.xml 中，代码如下：

```xml
<mapping resource="com/hibernate/entity/Admin.hbm.xml"/>
<mapping resource="com/hibernate/entity/Functions.hbm.xml"/>
```

(4) 添加数据。

在测试类 HibernateTest 中添加测试方法 testM2MSave()，并使用@Test 注解修饰。在 testM2MSave()方法中新建两个管理员，新建三个系统功能，并设置关联关系，最后执行保存操作。

代码如下：

```java
@Test
public void testM2MSave(){
    // 新建两个管理员对象
    Admin admin1=new Admin("admin1","123");
    Admin admin2=new Admin("admin2","123");
    // 新建三个系统功能
    Functions f1=new Functions("测试餐品管理");
    Functions f2=new Functions("测试订单管理");
    Functions f3=new Functions("测试用户管理");
    // 设置关联关系
    admin1.getFs().add(f1);
    admin1.getFs().add(f2);
    admin1.getFs().add(f3);
    admin2.getFs().add(f1);
    admin2.getFs().add(f2);
    // 保存
    session.save(admin1);
    session.save(admin2);
```

```
        session.save(f1);
        session.save(f2);
        session.save(f3);
    }
```

执行 testM2MSave()方法，数据表 admin、functions 和 powers 中记录分别如图 13-5、图 13-6 和图 13-7 所示。

Id	LoginName	LoginPwd
1	admin	123456
2	admin1	123
3	admin2	123

id	name
13	测试餐品管理
14	测试订单管理
15	测试用户管理

aid	fid
2	13
3	13
2	14
3	14
2	15

图 13-5　数据表 admin 中的记录　　图 13-6　数据表 functions 中的记录　　图 13-7　数据表 powers 中的记录

(5) 加载数据。

在测试类 HibernateTest 中添加测试方法 testM2MGet()，并使用@Test 注解修饰。在 testM2MGet()方法中加载管理员及关联的系统功能属性，代码如下：

```
    @Test
    public void testM2MGet() {
        // 加载管理员对象
        Admin admin = (Admin) session.get(Admin.class, 2);
        System.out.println(admin.getLoginName());
        // 加载关联的系统功能属性
        Set fs = admin.getFs();
        System.out.println(fs.size());
    }
```

执行 testM2MGet()方法，控制台输出结果如下：

```
Hibernate: select … from restrant.admin admin0_ where admin0_.Id=?
admin1
Hibernate: select … from powers fs0_ inner join restrant.functions functions1_ on fs0_.fid=functions1_.id where fs0_.aid=?
3
```

(6) 删除数据。

在测试类 HibernateTest 中添加测试方法 testM2MDelete()，并使用@Test 注解修饰。在 testM2MDelete()方法中删除指定管理员，代码如下：

```
    @Test
    public void testM2MDelete() {
        // 加载编号 id=3 管理员对象，根据数据表中的实际情况选择 id
        Admin admin = (Admin) session.get(Admin.class, 3);
        // 执行删除操作
        session.delete(admin);
    }
```

执行 testM2MDelete()方法，数据表 admin 和 powers 中相关记录被删除。控制台输出结果如下：

```
Hibernate: select admin0_.Id as Id1_0_0_, admin0_.LoginName as
LoginNam2_0_0_, admin0_.LoginPwd as LoginPwd3_0_0_ from restrant.admin
admin0_ where admin0_.Id=?
Hibernate: delete from powers where aid=?
Hibernate: delete from restrant.admin where Id=?
```

13.1.5 双向一对一映射

双向一对一关联映射可以通过基于外键和基于主键两种方式实现。

1. 基于外键的一对一关联映射

基于外键的一对一关联与多对一关联实质相同,是多对一关联的一个特例。外键可以存放在任意一端,在存放外键的一端,增加<many-to-one>元素,并在该元素中增加 unique="true"属性,表示多的一方也必须唯一,并使用 name 属性来指定关联属性的属性名。在另一端需要使用<one-to-one>元素,同样使用 name 属性来指定关联属性的属性名。

在数据库 restrant 中,新建数据表 admin_detail,用于存储管理员的详细信息,如图 13-8 所示。在数据表 admin 中添加一个字段 Did,然后设置其与数据表 admin_detail 的 Id 字段关联,如图 13-9 所示。

图 13-8 数据表 admin_detail 的结构　　图 13-9 数据表 admin 的结构

管理员信息表 admin 的 Did 字段作为该表的外键,需要保证该字段的唯一性,否则就不是一对一映射关系,而是多对一映射关系。

实现数据表 admin 和 admin_detail 双向一对一关联映射的步骤如下。

(1) 将项目 hibernate-1 复制并命名为 hibernate-5,再导入 MyEclipse 开发环境中。
(2) 创建实体类。

在项目 hibernate-5 的 com.hibernate.entity 包中创建实体类 Admin.java 和 AdminDetail.java,分别对应数据表 admin 和 admin_detail。

其中,实体类 Admin.java 的代码如下:

```
package com.hibernate.entity;
public class Admin {
    private int id;                      //id 号
    private String loginName;            //登录名
    private String loginPwd;             //登录密码
    private AdminDetail ad;              //关联的属性
    // 省略属性的 getter、setter 方法
    // 省略无参构造方法和有参构造方法
}
```

实体类 AdminDetail.java 的代码如下:

```
package com.hibernate.entity;
public class AdminDetail {
    private int id;                    //id号
    private String address;            //地址
    private String realName;           //真实姓名
    private Admin admin;               //关联的属性
    // 省略属性的 getter、setter 方法
    // 省略无参构造方法和有参构造方法
}
```

在实体类 Admin.java 和 AdminDetail.java 中都添加了与对方关联的属性。

(3) 创建映射文件。

在项目 hibernate-5 的 com.hibernate.entity 包中创建映射文件 Admin.hbm.xml 和 AdminDetail.hbm.xml。其中，Admin.hbm.xml 文件内容如下：

```xml
<?xml version="1.0" encoding="utf-8"?>
<!DOCTYPE hibernate-mapping PUBLIC "-//Hibernate/Hibernate Mapping DTD 3.0//EN" "http://www.hibernate.org/dtd/hibernate-mapping-3.0.dtd">
<hibernate-mapping package="com.hibernate.entity">
    <class name="Admin" table="admin" catalog="restrant">
        <id name="id" type="java.lang.Integer">
            <column name="Id" />
            <generator class="native"></generator>
        </id>
        <property name="loginName" type="java.lang.String">
            <column name="LoginName" length="20" not-null="true" />
        </property>
        <property name="loginPwd" type="java.lang.String">
            <column name="LoginPwd" length="20" not-null="true" />
        </property>
        <!-- 使用many-to-one的方式来映射一对一关联关系 -->
        <many-to-one name="ad" class="AdminDetail" column="Did" unique="true" />
    </class>
</hibernate-mapping>
```

在映射文件 Admin.hbm.xml 中需要使用<many-to-one>元素而不是<one-to-one>元素来映射在 Admin 类中定义的 AdminDetail 类型的 ad 属性，但必须使用 unique="true"指定多的一端唯一，即满足唯一性约束，以实现一对一关联。

AdminDetail.hbm.xml 文件内容如下：

```xml
<?xml version="1.0" encoding="utf-8"?>
<!DOCTYPE hibernate-mapping PUBLIC "-//Hibernate/Hibernate Mapping DTD 3.0//EN" "http://www.hibernate.org/dtd/hibernate-mapping-3.0.dtd">
<hibernate-mapping package="com.hibernate.entity">
    <class name="AdminDetail" table="admin_detail" catalog="restrant">
        <id name="id" type="java.lang.Integer">
            <column name="Id" />
            <generator class="native"></generator>
        </id>
        <property name="address" type="java.lang.String">
            <column name="Address" length="255" not-null="true" />
        </property>
```

```xml
        <property name="realName" type="java.lang.String">
            <column name="RealName" length="10" not-null="true" />
        </property>
        <!-- 映射一对一关联关系 -->
        <one-to-one name="admin" class="Admin" property-ref="ad" />
    </class>
</hibernate-mapping>
```

在映射文件 AdminDetail.hbm.xml 中，需要通过<one-to-one>元素来映射从 AdminDetail 到 Admin 的一对一关联。使用 property-ref="ad"表明建立了从 AdminDetail 对象到 Admin 对象的关联，因此只需要调用 AdminDetail 对象的 getAdmin()方法就可以访问到 Admin 对象。

将这两个映射文件通过<mapping>标签添加到 Hibernate 配置文件 hibernate.cfg.xml 中，代码如下：

```xml
<mapping resource="com/hibernate/entity/Admin.hbm.xml"/>
<mapping resource="com/hibernate/entity/AdminDetail.hbm.xml"/>
```

（4）添加数据。

在测试类 HibernateTest 中添加测试方法 testO2OSave()，并使用@Test 注解修饰。在 testO2OSave()方法中新建管理员信息对象，新建管理员信息详情对象，并设置关联关系，最后执行保存操作。代码如下：

```java
@Test
public void testO2OSave(){
    // 新建管理员信息对象
    Admin admin = new Admin("admin", "123456");
    // 新建管理员信息详情对象
    AdminDetail ad = new AdminDetail("江苏扬州", "管理员");
    // 设置关联关系
    admin.setAd(ad);
    ad.setAdmin(admin);
    // 保存操作
    session.save(ad);
    session.save(admin);
}
```

执行 testM2MSave()方法，数据表 admin_detail 和 admin 中的记录分别如图 13-10 和图 13-11 所示。

Id	Address	RealName
1	江苏扬州	管理员

Id	LoginName	LoginPwd	Did
2	admin	admin	1

图 13-10　数据表 admin_detail 中的记录　　　　图 13-11　数据表 admin 中的记录

控制台输出的 SQL 语句如下：

```
Hibernate: insert into restrant.admin_detail (Address, RealName) values (?, ?)
Hibernate: insert into restrant.admin (LoginName, LoginPwd, Did) values (?, ?, ?)
```

(5) 加载数据。

在测试类 HibernateTest 中添加测试方法 testO2OGet_1()，并使用@Test 注解修饰。在 testO2OGet_1()方法中先加载编号 id=2 的 Admin 对象，再加载关联的 AdminDetail 对象。代码如下：

```java
@Test
public void testO2OGet_1() {
    // 加载编号 id=2 的 Admin 对象，根据数据表中的实际情况选择 id
    Admin admin = (Admin)session.get(Admin.class, 2);
    System.out.println(admin.getLoginName());
    // 加载关联的 AdminDetail 对象
    System.out.println(admin.getAd().getAddress());
}
```

执行 testO2OGet_1()方法，控制台输出结果如下：

```
Hibernate: select ……
admin
Hibernate: select ……
Hibernate: select ……
江苏扬州
```

在测试类 HibernateTest 中添加测试方法 testO2OGet_2()，并使用@Test 注解修饰。在 testO2OGet_2()方法中先加载编号 id=1 的 AdminDetail 对象，再获取关联的 Admin 对象信息。代码如下：

```java
@Test
public void testO2OGet_2() {
    // 加载编号 id=1 的 AdminDetail 对象
    AdminDetail ad = (AdminDetail) session.get(AdminDetail.class, 1);
    System.out.println(ad.getRealName());
    // 获取 Admin 对象
    System.out.println(ad.getAdmin().getLoginName());
}
```

执行 testO2OGet_2()方法，控制台输出结果如下：

```
Hibernate: select …… from restrant.admin_detail admindetai0_ left outer join restrant.admin admin1_ on admindetai0_.Id=admin1_.Did where admindetai0_.Id=?
管理员
admin
```

由于在映射文件 AdminDetail.hbm.xml 的<one-to-one>元素中没有通过 Column 属性设置与 Admin 对象的关联，从控制台输出可以看出，通过 AdminDetail 对象获取 Admin 对象是通过左外连接(left outer join)实现的。

(6) 删除数据。

在测试类 HibernateTest 中添加测试方法 testO2ODelete()，并使用@Test 注解修饰。在 testO2ODelete()方法中删除编号 id=1 的 AdminDetail 对象，代码如下：

```java
@Test
public void testO2ODelete() {
```

```
        // 删除编号 id=1 的 AdminDetail 对象
        AdminDetail ad = (AdminDetail) session.get(AdminDetail.class,1);
        // 执行删除操作
        session.delete(ad);
    }
```

执行 testO2ODelete()方法，在控制台抛出外键约束异常 ConstraintViolationException。代码如下：

```
ERROR: HHH000346: Error during managed flush
[org.hibernate.exception.ConstraintViolationException: could not execute
statement]
```

只需要修改映射文件 AdminDetail.hbm.xml，在其<one-to-one>元素中添加 cascade 属性，并设置为 delete，表示采用删除级联操作。代码如下：

```
<one-to-one name="admin" class="Admin" property-ref="ad" cascade="delete"/>
```

再次执行 testO2ODelete()方法，数据表 admin_detail 和 admin 中记录成功删除。

2．基于主键的一对一关联映射

基于主键的一对一关联就是限制两个数据表的主键使用相同的值，通过主键形成一对一映射关系。在实现基于外键的双向一对一关联映射时，数据表 admin 中添加了一个外键字段 Did，与数据表 admin_detail 的 Id 字段关联。在实现基于主键的双向一对一关联时，需要将该外键字段 Did 删除，并设置 admin 表的主键 Id 与 admin_detail 表主键 Id 关联。由于两个表主键相关联，因此需要保持两个数据表的主键 Id 一致，最好将两个表的数据全部清空后，再设置关联，如图 13-12 所示。

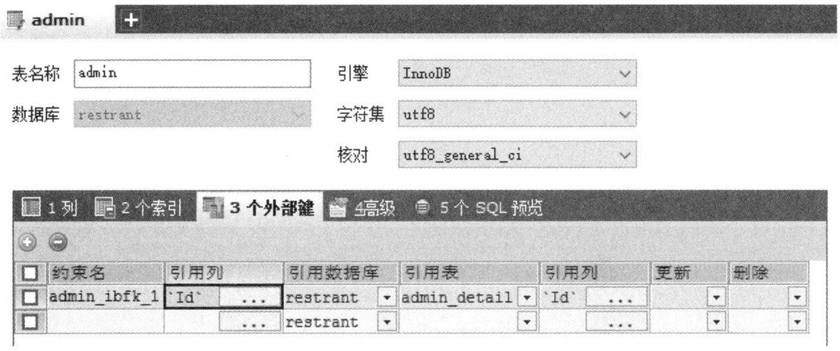

图 13-12 设置 admin 表的主键 Id 与 admin_detail 表主键 Id 关联

基于主键的双向一对一关联映射，需要使用 foreign 策略生成主键，任何一方都可以使用 foreign 策略，表明根据对方主键生成自己的主键。使用 foreign 策略的一方增加<one-to-one>元素映射关联关系，还必须将其 constrained 属性设置为 true，而另一方只需要增加<one-to-one>元素映射关联属性即可。

基于主键实现数据表 admin_detail 和 admin 之间双向一对一关联映射的步骤如下。

(1) 将项目 hibernate-5 复制并命名为 hibernate-6，再导入 MyEclipse 开发环境中。
(2) 修改映射文件 Admin.hbm.xml 和 AdminDetail.hbm.xml。

在映射文件 Admin.hbm.xml 中，先修改<generator>元素，再删除原先的<many-to-one>元素，并配置<one-to-one>元素，代码如下：

```xml
<?xml version="1.0" encoding="utf-8"?>
<!DOCTYPE hibernate-mapping PUBLIC "-//Hibernate/Hibernate Mapping DTD
3.0//EN" "http://www.hibernate.org/dtd/hibernate-mapping-3.0.dtd">
<hibernate-mapping package="com.hibernate.entity">
    <class name="Admin" table="admin" catalog="restrant">
        <id name="id" type="java.lang.Integer">
            <column name="Id" />
            <!-- 采用 foreign 策略，直接使用另一个关联的实体的标识属性 -->
            <generator class="foreign">
                <param name="property">ad</param>
            </generator>
        </id>
        <!-- 省略未修改的 loginName 和 loginPwd 的配置 -->
        <!-- 使用 one-to-one 的方式来映射一对一关联关系 -->
        <one-to-one name="ad" class="AdminDetail" constrained="true" />
    </class>
</hibernate-mapping>
```

由于 admin_detail 表中的主键是引用 admin 表主键的外键，所以 Admin 类的主键在映射时的生成策略，需要由关联类 AdminDetail 来指定。在映射文件 Admin.hbm.xml 中通过<generator>元素指定属性 class="foreign"，表明根据关联类 AdminDetail 生成 Admin 类的主键。子元素<param>设置 property 指定关联类 AdminDetail 的属性，即在 Admin 类中所定义的 AdminDetail 类型的属性 ad。通过<one-to-one>元素的 name 属性指定关联类属性为 ad，constrained="true"表明 Admin 类的主键由关联类 AdminDetail 生成。

在映射文件 AdminDetail.hbm.xml 中，只需要修改<one-to-one>元素，代码如下：

```xml
<!-- 修改映射一对一关联关系 -->
<one-to-one name="admin" class="Admin" cascade="delete" />
```

(3) 添加数据。

执行测试类 HibernateTest 中的 testO2OSave()方法，数据表中 admin_detail 和 admin 成功添加记录，分别如图 13-13 和图 13-14 所示。

☐	Id	Address	RealName
☐	2	江苏扬州	管理员
*	(Auto)	(NULL)	(NULL)

☐	Id	LoginName	LoginPwd	Did
☐	2	admin	admin	(NULL)
*	(Auto)	(NULL)	(NULL)	(NULL)

图 13-13　数据表 admin_detail 中的记录　　图 13-14　数据表 admin 中的记录

控制台输出的 SQL 语句如下：

```
Hibernate: insert into restrant.admin_detail (Address, RealName) values (?, ?)
Hibernate: insert into restrant.admin (LoginName, LoginPwd, Id) values (?, ?, ?)
```

依次执行测试类 HibernateTest 中的 testO2OGet_1()、testO2OGet_2()和 testO2ODelete()等方法，并根据情况修改这些方法中所加载记录的编号 Id，执行结果与基于外键的一对一关联映射时相同。

13.2 基于 Annotation 注解实现关联映射

在上一章的 12.2 小节针对单张数据表 users，介绍过如何使用 Annotation 注解实现 Hibernate 零配置。下面针对多对一、多对多和一对一双向关联映射介绍使用 Annotation 注解实现关联映射的零配置。

13.2.1 双向多对一映射

基于 Annotation 注解实现数据表 meal 和 mealseries 双向多对一关联映射的步骤如下。
(1) 先将项目 hibernate-2 复制并命名为 hibernate-7，再导入 MyEclipse 开发环境中。
(2) 在项目 hibernate-7 的 com.hibernate.entity 包中创建实体类 Meal.java 和 Mealseries.java。其中，基于 Annotation 注解实现的持久化类 Meal 代码如下：

```java
package com.hibernate.entity;
import javax.persistence.*;
@Entity
@Table(name="meal",catalog="restrant")
public class Meal {
    // 餐品基本信息(部分)
    private int mealId;                       //餐品编号
    private Mealseries mealseries;            //餐品的菜系关联属性
    private String mealName;                  //餐品名称
    private String mealSummarize;             //餐品摘要
    private String mealDescription;           //餐品详细描述信息
    private Double mealPrice;                 //餐品价格
    private String mealImage;                 //餐品图片
    @Id
    @GeneratedValue(strategy=GenerationType.IDENTITY)
    @Column(name="MealId", unique=true, nullable=false)
    public int getMealId() {
        return mealId;
    }
    public void setMealId(int mealId) {
        this.mealId = mealId;
    }
    //使用@ManyToOne 和@JoinColumn 注解实现 Meal 到 Mealseries 的多对一关联
    @ManyToOne
    @JoinColumn(name="MealSeriesId")
    public Mealseries getMealseries() {
        return mealseries;
    }
    public void setMealseries(Mealseries mealseries) {
        this.mealseries = mealseries;
    }
    @Column(name="MealName", nullable=false, length=20)
    public String getMealName() {
        return mealName;
    }
    public void setMealName(String mealName) {
```

```java
        this.mealName = mealName;
    }
    @Column(name="MealSummarize", nullable=false, length=250)
    public String getMealSummarize() {
        return mealSummarize;
    }
    public void setMealSummarize(String mealSummarize) {
        this.mealSummarize = mealSummarize;
    }
    @Column(name="MealDescription", nullable=false)
    public String getMealDescription() {
        return mealDescription;
    }
    public void setMealDescription(String mealDescription) {
        this.mealDescription = mealDescription;
    }
    @Column(name="MealPrice", nullable=false)
    public Double getMealPrice() {
        return mealPrice;
    }
    public void setMealPrice(Double mealPrice) {
        this.mealPrice = mealPrice;
    }
    @Column(name="MealImage")
    public String getMealImage() {
        return mealImage;
    }
    public void setMealImage(String mealImage) {
        this.mealImage = mealImage;
    }
    // 省略无参构造方法
    // 省略有参构造方法
}
```

在持久化类 Meal 中，需要定义一个 Mealseries 类型的关联属性 mealseries，再使用 @ManyToOne 和 @JoinColumn 注解实现 Meal 到 Mealseries 的多对一关联。@ManyToOne 注解的 fetch 属性可选择项包括：FetchType.EAGER 和 FetchType.LAZY，前者表示关联类在主类加载的时候同时加载(立即加载)，后者表示关联类在被访问时才加载(懒加载或延迟加载)，在多对一时默认值是 FetchType.EAGER，在一对多时默认值是 FetchType.LAZY。@JoinColumn(name="MealSeriesId")指定数据表 meal 的 MealSeriesId 字段作为外键与数据表 mealseries 的主键关联。

基于 Annotation 注解实现的持久化类 Mealseries 如下：

```java
package com.hibernate.entity;
import java.util.HashSet;
import java.util.Set;
import javax.persistence.*;
@Entity
@Table(name="mealseries",catalog="restrant")
public class Mealseries {
    private int seriesId;              //菜系 Id
    private String seriesName;         //菜系名称
```

```java
    private Set<Meal> mealSet=new HashSet<Meal>();         //关联的集合属性
    @Id
    @GeneratedValue(strategy=GenerationType.IDENTITY)
    @Column(name="SeriesId", unique=true, nullable=false)
    public int getSeriesId() {
        return seriesId;
    }
    public void setSeriesId(int seriesId) {
        this.seriesId = seriesId;
    }
    @Column(name="SeriesName",nullable=false, length=10)
    public String getSeriesName() {
        return seriesName;
    }
    public void setSeriesName(String seriesName) {
        this.seriesName = seriesName;
    }
    // 使用@OneToMany注解实现Mealseries到Meal的一对多关联
    @OneToMany(mappedBy="mealseries", cascade={CascadeType.REMOVE})
    public Set<Meal> getMealSet() {
        return mealSet;
    }
    public void setMealSet(Set<Meal> mealSet) {
        this.mealSet = mealSet;
    }
    // 无参构造方法
    public Mealseries() {
    }
    // 有参构造方法
    public Mealseries(String seriesName) {
        this.seriesName = seriesName;
    }
}
```

在持久化类 Mealseries 中，需要定义元素类型为 Meal 的关联集合属性 mealSet，再使用 @OneToMany 注解实现 Mealseries 到 Meal 的一对多关联。@OneToMany 注解的 mappedBy 属性作用相当于设置 inverse=true，表示将关联关系的主管权反转，即由 Meal 管理双方的关联关系。mappedBy 属性值为关联的多的一方(Meal 类)所定义 Mealseries 类型的属性 mealseries。cascade = { CascadeType.REMOVE }指定级联删除。

基于 Annotation 注解的持久化类 Meal 和 Mealseries 配置完成后，还需要在 Hibernate 配置文件 hibernate.cfg.xml 中添加对持久化类 Meal 和 Mealseries 类的引用，代码如下：

```xml
<mapping class="com.hibernate.entity.Meal" />
<mapping class="com.hibernate.entity.Mealseries" />
```

将项目 hibernate-3 的测试类 HibernateTest 中 testM2OGet、testO2MGet、testM2OAndO2MSave、testM2OAndO2MUpdate 和 testM2OAndO2MDelete 等测试方法复制到项目 hibernate-7 的测试类 HibernateTest 中，依次测试这些方法，测试效果与 hibernate-3 相同。

13.2.2 双向多对多映射

13.1.4 节介绍过基于 XML 映射文件实现数据表 admin 和 functions 之间的双向多对多关联映射。下面介绍基于 Annotation 注解实现双向多对多映射。具体步骤如下。

在项目 hibernate-7 的 com.hibernate.entity 包中创建实体类 Admin.java 和 Functions.java。其中，基于 Annotation 注解实现的持久化类 Functions 代码如下：

```java
package com.hibernate.entity;
import java.util.HashSet;
import java.util.Set;
import javax.persistence.*;
@Entity
@Table(name="functions",catalog="restrant")
public class Functions {
    private int id;           //id号
    private String name;      //功能名称
    private Set<Admin> adminSet=new HashSet<Admin>();// 关联的属性
    @Id
    @GeneratedValue(strategy=GenerationType.IDENTITY)
    @Column(name="id",unique=true, nullable=false)
    public int getId() {
        return id;
    }
    public void setId(int id) {
        this.id = id;
    }
    @Column(name="name",length=20)
    public String getName() {
        return name;
    }
    public void setName(String name) {
        this.name = name;
    }
    // 使用@ManyToMany注解实现Functions到Admin的多对多关联
    @ManyToMany(mappedBy="fs")
    public Set<Admin> getAdminSet() {
        return adminSet;
    }
    public void setAdminSet(Set<Admin> adminSet) {
        this.adminSet = adminSet;
    }
    // 无参构造方法
    public Functions() {
    }
    // 有参构造方法
    public Functions(String name) {
        this.name = name;
    }
}
```

在持久化类 Functions 中，定义了一个元素类型为 Admin 的关联集合 adminSet，再使用

@ManyToMany 注解实现 Functions 到 Admin 的多对多关联映射。在@ManyToMany 注解中，设置属性 mappedBy="fs"，作用相当于 inverse="true"，将关联关系控制权反转，即由 Admin 管理双方关联关系。fs 是 Admin 类中定义的元素类型为 Functions 的集合。由于 Admin 是关联关系的主管方，因此 Admin 类和 Functions 类的多对多关联映射是在 Admin 类中实现的。

基于 Annotation 注解实现的持久化类 Admin 代码如下：

```java
package com.hibernate.entity;
import java.util.HashSet;
import java.util.Set;
import javax.persistence.*;
@Entity
@Table(name="admin",catalog="restrant")
public class Admin {
    private int id;                         //id 号
    private String loginName;               //登录名
    private String loginPwd;                //登录密码
    private Set<Functions> fs=new HashSet<Functions>();// 关联的属性
    @Id
    @GeneratedValue(strategy=GenerationType.IDENTITY)
    @Column(name="Id",unique=true, nullable=false)
    public int getId() {
        return id;
    }
    public void setId(int id) {
        this.id = id;
    }
    @Column(name="LoginName",length=20)
    public String getLoginName() {
        return loginName;
    }
    public void setLoginName(String loginName) {
        this.loginName = loginName;
    }
    @Column(name="LoginPwd",length=20)
    public String getLoginPwd() {
        return loginPwd;
    }
    public void setLoginPwd(String loginPwd) {
        this.loginPwd = loginPwd;
    }
    // 使用@ManyToMany 注解实现 Admin 到 Functions 的多对多关联
    @ManyToMany()
@JoinTable(name="powers",joinColumns={@JoinColumn(name="aid")},inverseJoinColumns={@JoinColumn(name="fid")})
    public Set<Functions> getFs() {
        return fs;
    }
    public void setFs(Set<Functions> fs) {
        this.fs = fs;
    }
    // 无参构造方法
    public Admin() {
```

```
    }
    // 有参构造方法
    public Admin(String loginName, String loginPwd) {
        this.loginName = loginName;
        this.loginPwd = loginPwd;
    }
}
```

在持久化类 Admin 中，定义一个元素类型为 Functions 的关联集合 fs，再使用 @ManyToMany 注解和@JoinTable 注解实现 Admin 到 Functions 的多对多关联。@JoinTable 注解描述了多对多关系的数据表关系，name 属性指定中间表的名称，这里为 powers；joinColumns 属性定义中间表 powers 与管理员表 admin 关联的外键列，这里为 aid；inverseJoinColumns 属性定义中间表 powers 与另一端系统功能表 functions 关联的外键列，这里为 fid。

基于 Annotation 注解的持久化类 Admin 和 Functions 的双向多对多关联配置完成后，还需要在 Hibernate 配置文件 hibernate.cfg.xml 中添加对持久化类的引用，代码如下：

```
<mapping class="com.hibernate.entity.Admin" />
<mapping class="com.hibernate.entity.Functions" />
```

将项目 hibernate-4 的测试类 HibernateTest 中 testM2MSave、testM2MGet 和 testM2MDelete 等测试方法复制到项目 hibernate-7 的测试类 HibernateTest 中，依次测试这些方法，测试效果与 hibernate-4 相同。

13.2.3　双向一对一映射

13.1.5 节介绍过使用 XML 映射文件实现数据表 admin 和 admin_detail 之间基于外键和基于主键的双向一对一关联映射。下面介绍使用 Annotation 注解实现基于主键的双向一对一关联映射。具体步骤如下。

（1）在数据库 restrant 中，数据表 admin 和 admin_detail 的结构分别如图 13-15 和图 13-16 所示。

图 13-15　数据表 admin 的结构　　　图 13-16　数据表 admin_detail 的结构

设置 admin 表的主键 Id 与 admin_detail 表的主键 Id 关联。由于两个表主键相关联，因此需要保持两个数据表的主键 Id 一致，最好将两个表的数据全部清空后，再设置关联。

（2）将项目 hibernate-2 复制并命名为 hibernate-8，再导入 MyEclipse 开发环境中。

（3）创建实体类。

在项目 hibernate-8 的 com.hibernate.entity 包中创建实体类 Admin.java 和 AdminDetail.java，分别对应数据表 admin 和 admin_detail。

其中，实体类 Admin.java 代码如下：

```java
package com.hibernate.entity;
import javax.persistence.*;
import org.hibernate.annotations.GenericGenerator;
import org.hibernate.annotations.Parameter;
@Entity
@Table(name="admin",catalog="restrant")
public class Admin {
    private int id;                         //id号
    private String loginName;               //登录名
    private String loginPwd;                //登录密码
    private AdminDetail ad;                 //关联的属性
    // 这组注解功能是将当前对象中 ad 属性的主键来作为本对象的主键

@GenericGenerator(name="generator",strategy="foreign",parameters=@Parameter(name="property",value="ad"))
    @Id
    @GeneratedValue(generator="generator")
    @Column(name="Id",unique=true,nullable=false)
    public int getId() {
        return id;
    }
    public void setId(int id) {
        this.id = id;
    }
    @Column(name="LoginName",length=20)
    public String getLoginName() {
        return loginName;
    }
    public void setLoginName(String loginName) {
        this.loginName = loginName;
    }
    @Column(name="LoginPwd",length=20)
    public String getLoginPwd() {
        return loginPwd;
    }
    public void setLoginPwd(String loginPwd) {
        this.loginPwd = loginPwd;
    }
    // 使用@OneToOne 注解实现 Admin 与 AdminDetail
    // 的基于主键的一对一关联
    @OneToOne(mappedBy="admin",optional=false)
    public AdminDetail getAd() {
        return ad;
    }
    public void setAd(AdminDetail ad) {
        this.ad = ad;
    }
    public Admin() {
    }
    public Admin(String loginName, String loginPwd) {
```

```java
        this.loginName = loginName;
        this.loginPwd = loginPwd;
    }
}
```

在持久化类 Admin 中，首先定义了 AdminDetail 类型关联属性 ad，然后使用@GenericGenerator、@Id、@GeneratedValue 和@Column 这一组注解将 Admin 类中定义的 AdminDetail 类型的属性 ad 的主键作为 Admin 类对象的主键。

其中，@GenericGenerator 注解声明了一个 Hibernate 的主键生成策略，支持 13 种策略。该注解的 name 属性指定生成器名称，strategy 属性指定具体生成器的类名(即生成策略)，这里选择 foreign 策略，表示使用另一个关联对象的主键，通常和<one-to-one>联合起来使用。parameters 属性得到 strategy 指定的具体生成器所用到的参数，设置 value="ad"表示将当前类 Admin 中定义的 AdminDetail 类型的 ad 属性的主键作为 Admin 类对象的主键。

再使用@OneToOne 注解实现 Admin 与 AdminDetail 的基于主键的一对一关联，设置属性 mappedBy="admin" 作用相当于 inverse=true，表示将关联关系的控制权反转，即由 AdminDetail 方管理关联关系，admin 为 AdminDetail 类中定义的 Admin 类型的关联属性。设置属性 optional=false 指定关联属性 ad 不能为空。

实体类 AdminDetail.java 代码如下：

```java
package com.hibernate.entity;
import javax.persistence.*;
@Entity
@Table(name="admin_detail",catalog="restrant")
public class AdminDetail {
    private int id;              //id号
    private String address;      //地址
    private String realName;     //真实姓名
    private Admin admin;         //关联的属性
    @Id
    @GeneratedValue(strategy=GenerationType.IDENTITY)
    @Column(name="Id",unique=true,nullable=false)
    public int getId() {
        return id;
    }
    public void setId(int id) {
        this.id = id;
    }
    @Column(name="Address",length=255)
    public String getAddress() {
        return address;
    }
    public void setAddress(String address) {
        this.address = address;
    }
    @Column(name="RealName",length=10)
    public String getRealName() {
        return realName;
    }
    public void setRealName(String realName) {
        this.realName = realName;
```

```java
    }
    // 使用@OneToOne 和@PrimaryKeyJoinColumn 注解
    // 实现 AdminDetail 与 Admin 基于主键的一对一关联
    @OneToOne(cascade=CascadeType.REMOVE)
    @PrimaryKeyJoinColumn
    public Admin getAdmin() {
        return admin;
    }
    public void setAdmin(Admin admin) {
        this.admin = admin;
    }
    public AdminDetail() {
    }
    public AdminDetail(String address, String realName) {
        this.address = address;
        this.realName = realName;
    }
}
```

在持久化类 AdminDetail 中，首先定义了 Admin 类型的关联属性 admin，再使用 @OneToOne 和@PrimaryKeyJoinColumn 注解实现 AdminDetail 与 Admin 的基于主键的一对一关联。设置 cascade=CascadeType.REMOVE 表示级联删除。@PrimaryKeyJoinColumn 注解表示两个实体通过主键关联。

使用 Annotation 注解完成持久化类 Admin 和 AdminDetail 的配置后，还需要在 Hibernate 配置文件 hibernate.cfg.xml 中添加对持久化类的引用，代码如下：

```xml
<mapping class="com.hibernate.entity.Admin"/>
<mapping class="com.hibernate.entity.AdminDetail"/>
```

将项目 hibernate-6 的测试类 HibernateTest 中 testO2OSave、testO2OGet_1 和 testO2OGet_2 和 testO2ODelete 等测试方法复制到项目 hibernate-8 的测试类 HibernateTest 中，并根据情况修改这些方法中所加载记录的编号 Id。依次测试这些方法，测试效果与 hibernate-6 相同。

13.3 基于 XML 映射文件实现继承映射

面向对象程序的程序设计语言，继承和多态是两个最基本最重要的概念，它实现了代码的重用。Hibernate 的继承映射可以理解为两个持久化类之间的继承关系。例如：教师、学生和人之间的关系。老师继承了人，可以认为老师是一个特殊的人，如果对人进行查询，老师的实例也将被得到，而无须关注到人的实例、老师的实例在底层数据库的存储。Hibernate 支持三种继承映射策略。

13.3.1 使用 subclass 进行映射

将域模型中的每一个实体对象映射到一个独立的表中，也就是说，不用在关系数据模型中考虑域模型中的继承关系和多态。使用 subclass 的继承映射可以实现对于继承关系中的父类和子类使用相同的一张表。因为父类和子类的实例全部都保存在同一个表中，因此需要在该表内增加一列，使用该列来区分每行记录到底是哪个类的实例，这个列被称为辨别者列(discriminator)。在这种映射策略下，使用 subclass 来映射子类，使用 class 或 subclass 的 discriminator-vlaue 属性指定辨别者列的值。所有子类定义的字段都不能有非空约束。如果为这些字段添加非空约束，那么父类的实例在那些列其实并没有值，这将引起数据库完整冲突，导致父类的实例无法保存到数据库中。

在 restrant 数据库中，新建 persons 表，表结构如图 13-17 所示。

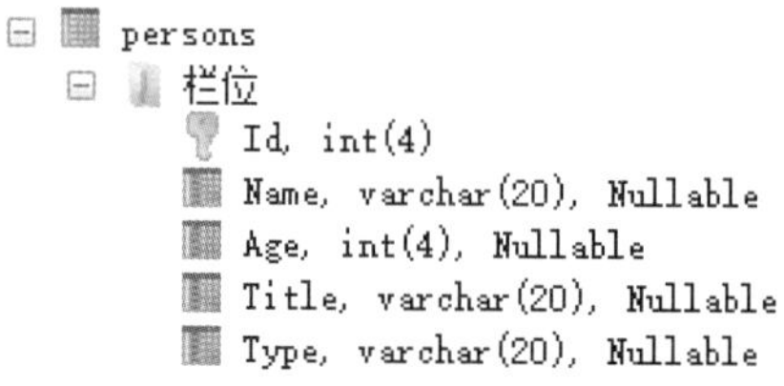

将项目 hibernate-1 复制并重命名为 hibernate-9，再导入 MyEclipse 开发环境中。

在 hibernate-9 项目的 com.hibernate.entity 包中新建 Person 类，代码如下：

图 13-17　persons 表的结构

```java
package com.hibernate.entity;
public class Person {
    private Integer id;
    private String name;
    private int age;
    // 省略属性的 getter、setter 方法
    // 省略无参构造方法和有参构造方法
}
```

新建 Teacher 类，并继承 Person 类，代码如下：

```java
package com.hibernate.entity;
public class Teacher extends Person {
    private String title;
    // 省略属性的 getter、setter 方法
    // 省略无参构造方法
    public Teacher(String name, int age, String title) {
        super(name, age);
        this.title = title;
    }
}
```

在 com.hibernate.entity 包中，创建映射文件 Person.hbm.xml，对应实体类 Person，代码如下：

```xml
<?xml version="1.0" encoding="utf-8"?>
<!DOCTYPE hibernate-mapping PUBLIC "-//Hibernate/Hibernate Mapping DTD 3.0//EN" "http://www.hibernate.org/dtd/hibernate-mapping-3.0.dtd">
<hibernate-mapping package="com.hibernate.entity">
    <class name="Person" table="persons" discriminator-value="PERSON" catalog="restrant" >
        <id name="id" type="java.lang.Integer">
            <column name="Id" />
            <generator class="native"></generator>
        </id>
        <!-- 配置辨别者列 -->
        <discriminator column="Type" type="java.lang.String">
        </discriminator>
        <property name="name" type="java.lang.String">
```

```xml
            <column name="Name" length="20" />
        </property>
        <property name="age" type="java.lang.Integer">
            <column name="Age" />
        </property>
        <!-- 映射子类Teacher，使用subclass进行映射   -->
        <subclass name="Teacher" discriminator-value="TEACHER">
            <property name="title" type="java.lang.String" column="Title"></property>
        </subclass>
    </class>
</hibernate-mapping>
```

将 Person.hbm.xml 映射文件通过<mapping>添加到 Hibernate 配置文件 hibernate.cfg.xml 中，代码如下：

```xml
<mapping resource="com/hibernate/entity/Person.hbm.xml"/>
```

在项目 hibernate-9 的测试类 HibernateTest 中添加测试方法 testExtendsSave()，并使用 @Test 注解修饰，在该方法中新建 Person 对象和 Teacher 对象，添加数据，代码如下：

```java
// 插入数据
@Test
public void testExtendsSave(){
    Person person=new Person("张瑜涵",20);
    session.save(person);
    Teacher teacher=new Teacher("王小龙",46,"副教授");
    session.save(teacher);
}
```

执行 testExtendsSave()方法，控制台输出两条插入语句，结果如下：

```
Hibernate: insert into restrant.persons (Name, Age, Type) values (?, ?, 'PERSON')
Hibernate: insert into restrant.persons (Name, Age, Title, Type) values (?, ?, ?, 'TEACHER')
```

查看数据库中的 persons 表，如图 13-18 所示。

Id	Name	Age	Title	Type
1	张瑜涵	20	(NULL)	PERSON
2	王小龙	46	副教授	TEACHER

图 13-18 插入记录后的 persons 表

从结果来看，Type 列中的 PERSON 和 TEACHER 是我们配置的辨别者列。

在 HibernateTest 中添加测试方法 testExtendsQuery()查询数据，并使用@Test 注解修饰，代码如下：

```java
// 查询父类记录，只需要查询一张数据表
// 查询子类记录，也只需要查询一张数据表
@Test
public void testExtendsQuery(){
    List<Person> persons=session.createQuery("From Person").getResultList();
    System.out.println(persons.size());
```

```
        List<Teacher> teachers=session.createQuery("From
Teacher").getResultList();
        System.out.println(teachers.size());
    }
```

执行 testExtendsQuery()方法，控制台输出两条查询语句，结果如下：

```
Hibernate: select … from restrant.persons person0_
2
Hibernate: select … from restrant.persons teacher0_ where
teacher0_.Type='TEACHER'
1
```

13.3.2 使用 joined-subclass 进行映射

对于继承关系中的子类使用同一个表，这就需要在数据库表中增加额外的区分子类类型的字段。采用 joined-subclass 元素继承映射可以实现每一个子类一张表。

采用这种映射策略时，父类实例保存在父类表中，子类实例由父类表和子类表共同存储。因为子类实例也是一个特殊的父类实例，因此必然也包含了父类实例的属性。于是将子类和父类公有的属性保存在父类表中，子类增加的属性，则保存在子类表中。

在这种映射策略下，无须使用辨别者列，但需要为每个子类使用 key 元素映射共有主键。子类增加的属性可以添加非空约束，因为子类的属性和父类的属性没有保存在同一个表中。

在 restrant 数据库中，修改 persons 表结构，去除 Title 和 Type 字段，结构如图 13-19 所示。新建 teachers 表，结构如图 13-20 所示。

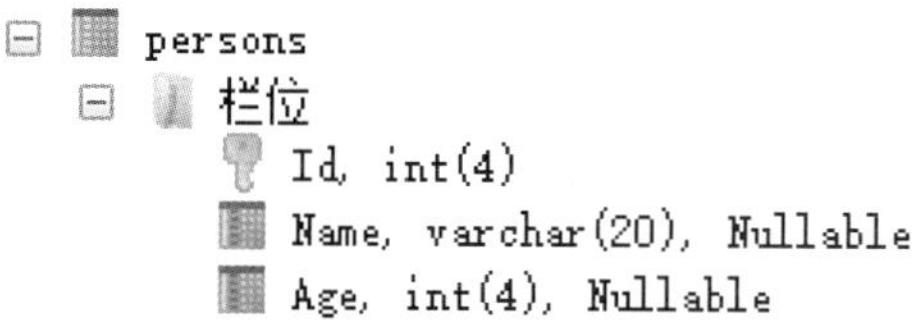

图 13-19 修改后 persons 表的结构 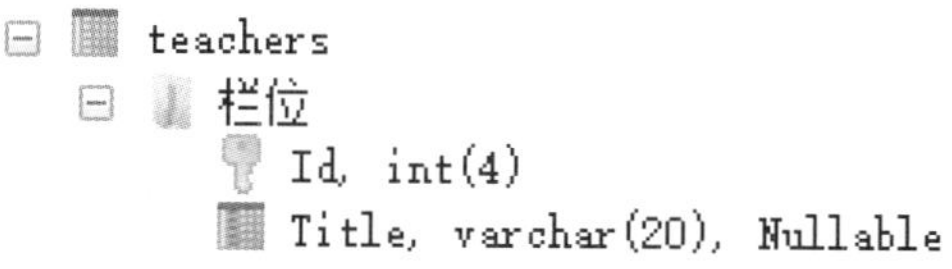　　图 13-20 teachers 表的结构

修改 com.hibernate.entity 包中的映射文件 Person.hbm.xml，去除辨别者列，代码如下：

```xml
<hibernate-mapping package="com.hibernate.entity">
   <class name="Person" table="persons" catalog="restrant" >
      <id name="id" type="java.lang.Integer">
         <column name="Id" />
         <generator class="native"></generator>
      </id>
      <!-- 省略未修改的 name、age 字段映射 -->
      <!-- 添加 joined-subclass 配置 -->
      <joined-subclass name="Teacher" table="teachers">
          <key column="Id"></key>   <!-- 参照父表的主键值 -->
          <property name="title" type="java.lang.String" column="Title">
          </property>
      </joined-subclass>
   </class>
</hibernate-mapping>
```

执行 testExtendsSave()方法，控制台输出三条插入语句，结果如下：

```
Hibernate: insert into restrant.persons (Name, Age) values (?, ?)
Hibernate: insert into restrant.persons (Name, Age) values (?, ?)
Hibernate: insert into teachers (Title, Id) values (?, ?)
```

数据库中的 persons 表和 teachers 表插入数据后的情况如图 13-21 和图 13-22 所示。

图 13-21　persons 表中的记录　　　　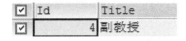
　　　　　　　　　　　　　　　　　　图 13-22　teachers 表中的记录

在这种情况下，插入父类记录的话，直接插入即可；插入子类记录的话，需要插入两张表。

13.3.3　使用 union-subclass 进行映射

域模型中的每个类映射到一个表，通过关系数据模型中的外键来描述表之间的继承关系。这也就相当于按照域模型的结构来建立数据库中的表，并通过外键来建立表之间的继承关系。

采用 union-subclass 元素可以实现将每一个实体对象映射到一个独立的表中。子类增加的属性可以有非空约束，即父类实例的数据保存在父类表中，而子类实例的数据保存在子类表中。在这种映射策略下，子类表的字段会比父类表的字段要多，因为子类表的字段等于父类表的字段加上子类增加属性的总和。在该策略下，既不需要使用辨别者列，也无须使用 key 元素来映射共有主键。

使用 union-subclass 映射策略时不可使用 identity 的主键生成策略，因为同一类继承层次中所有实体类都需要使用同一个主键种子，即多个持久化实体对应的记录的主键应该是连续的，受此影响，也不该使用 native 主键生成策略，因为 native 会根据数据库来选择使用 identity 或 sequence。

在 restrant 数据库中，persons 表结构不变，修改 teachers 表结构，结构如图 13-23 所示。

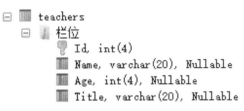

图 13-23　修改后 teachers 表的结构

修改 com.hibernate.entity 包中的映射文件 Person.hbm.xml，代码如下：

```
<?xml version="1.0" encoding="utf-8"?>
<!DOCTYPE hibernate-mapping PUBLIC "-//Hibernate/Hibernate Mapping DTD
3.0//EN" "http://www.hibernate.org/dtd/hibernate-mapping-3.0.dtd">
```

```xml
<hibernate-mapping package="com.hibernate.entity">
    <class name="Person" table="persons" catalog="restrant" >
        <id name="id" type="java.lang.Integer">
            <column name="Id" />
            <generator class="hilo"></generator>
        </id>
        <!-- 省略未修改的 name、age 字段映射 -->
        <union-subclass name="Teacher" table="teachers">
            <property name="title" type="java.lang.String" column="Title">
            </property>
        </union-subclass>
    </class>
</hibernate-mapping>
```

再次执行 testExtendsSave()方法，观察控制台的输出结果，并查看数据表。

13.4 小　　结

本章主要介绍了 Hibernate 中的关联映射，分别采用基于 XML 映射文件实现关联映射和基于 Annotation 注解实现关联映射。包括多对一、一对多、双向多对一关联、双向一对一和多对多关联。其中，常用的就是一对多和多对一，并且在能不用中间表的时候尽量不用中间表。多对多关联会用到，如果用到了，应该首先考虑底层数据库设计是否合理。还讲解了基于 XML 映射分别使用 subclass、joined-subclass 和 union-subclass 实现继承映射。

第 14 章
使用 Hibernate 查询数据

前面学习了 Hibernate 的基础知识,以及如何使用 Hibernate 管理对象间的关联关系,掌握了如何使用 Hibernate 完成增、删、改以及加载对象数据的方法,还没有介绍如何使用 Hibernate 进行查询操作,本章将完成这部分知识的学习。

14.1 使用 HQL 查询数据

HQL(Hibernate Query Language)是 Hibernate 提供的一种面向对象的查询语言。HQL 提供了更加丰富灵活的特性，提供了强大的查询能力。在 Hibernate 中，将 HQL 作为推荐的查询模式，使用类、对象和属性概念，没有表和字段的概念。HQL 提供了更接近传统 SQL 语句的查询语法，也提供了更全面的特性。

使用传统的 JDBC API 来查询数据，需要编写复杂的 SQL 语句，然后还要将查询结果以对象的形式进行封装，放到集合对象中保存。这种查询方式不仅麻烦，而且容易出错。

HQL 查询与 JDBC 查询相比，具体有以下几个优点。

(1) 直接针对实体类和属性进行查询，不需要再编写烦琐的 SQL 语句。

(2) 查询结果直接保存在 List 中的对象，不需要再次封装。

(3) 可以通过配置 dialect 属性，对不同的数据库自动生成不同的用于执行的 SQL 语句。

在 Hibernate 提供的各种查询方式中，HQL 应用最为广泛。HQL 支持属性查询、参数查询、关联查询、分页查询，提供内置聚集函数。

14.1.1 简单查询

从数据表 meal 中查询所有的餐品对象，按照名称升序排序，将查询结果输出到控制台。将项目 hibernate-3 复制并命名为 hibernate-10，再导入 MyEclipse 开发环境中。在测试类 HibernateTest 中添加 testHql_1()方法，并使用@Test 注解加以修饰，代码如下：

```java
import org.hibernate.query.Query;
……
@Test
public void testHql_1(){
    // 编写 HQL 语句
    String hql = "from Meal as m order by m.mealName asc";
    // 创建 Query 对象
    Query<Meal> query = session.createQuery(hql,Meal.class);
    // 执行查询，获得结果
    List<Meal> list=query.getResultList() ;
    // 遍历查找结果
    Iterator<Meal> iterator=list.iterator();
    while (iterator.hasNext()) {
        Meal meal = iterator.next();
        System.out.println(meal.getMealId()+". " +meal.getMealName() + ": \t" + meal.getMealSummarize());
    }
}
```

HQL 语句 from Meal as m 中的 Meal 是类名，而不是表名，因此需要区分大小写，关键字 from 不区分大小写。在 HQL 语句中可以使用别名，例如 m 是 Meal 类的别名，别名可以使用关键字 as 指定，as 关键字也可以省略。通过 order by 子句将查询结果按照餐品名称升序排序。升序排序的关键字是 asc，降序排序的关键字是 desc，查询语句中默认为升序。

执行 testHql_1()方法，控制台输出结果如下：

```
Hibernate: select meal0_.MealId as MealId1_0_, meal0_.MealSeriesId as
MealSeri2_0_, meal0_.MealName as MealName3_0_, meal0_.MealSummarize as
MealSumm4_0_, meal0_.MealDescription as MealDesc5_0_, meal0_.MealPrice as
MealPric6_0_, meal0_.MealImage as MealImag7_0_ from restrant.meal meal0_
order by meal0_.MealName asc
Hibernate: ……
Hibernate: ……
9. 巴国玉米糕肉：    风味浓、口感奇、品种多
4. 泰安肉三美豆腐：  汤汁乳白而鲜，豆腐软滑，白菜鲜嫩，清淡爽口。
3. 烤花肉揽桂鱼：    味道特鲜，白中泛红，佐以姜末、香醋，尤胜一等。
7. 糖醋红柿椒：      色红美，味鲜香。
2. 素锅烤鸭肉：      颜色鲜艳，酷似鸭肉，鲜香不腻。
……
```

14.1.2 属性查询

以上查询的结果是对象的所有属性，如果只查询对象的部分属性，则称为属性查询，也称为投影查询。在测试类 HibernateTest 中添加 testHql_2()方法，并使用@Test 注解加以修饰，代码如下：

```
@Test
public void testHql_2(){
    // 编写HQL语句，使用属性查询
    String hql = "select m.mealId,m.mealName from Meal as m";
    // 创建Query对象，此处不使用泛型
    Query query= session.createQuery(hql);
    // 执行查询，获得结果
    List list=query.getResultList() ;
    // 遍历查找结果
    Iterator iterator=list.iterator();
    while (iterator.hasNext()) {
        Object[] object=(Object[])iterator.next();
        System.out.println(object[0] +". "+object[1] );
    }
}
```

执行 testHql_2()方法，控制台输出结果如下：

```
Hibernate: select meal0_.MealId as col_0_0_, meal0_.MealName as col_1_0_,
meal0_.MealSummarize as col_2_0_ from restrant.meal meal0_
1. 雪梨肉肘棒
2. 素锅烤鸭肉
3. 烤花肉揽桂鱼
4. 泰安肉三美豆腐
5. 落叶琵琶肉虾
……
```

14.1.3 聚集函数

在 HQL 语句中可以使用的聚集函数包括统计记录总数(count)、计算最小值(min)、计算最

大值(max)、计算和(sum)、计算平均值(avg)。

在实体类 Meal.java 中有一个属性 mealPrice，在映射文件 Meal.hbm.xml 中已经为 Meal.java 中的属性 mealPrice 配置了映射，对应数据表 meal 中的 MealPrice 字段。在测试类 HibernateTest 中添加 testHql_3()方法，并使用@Test 注解加以修饰，代码如下：

```java
@Test
public void testHql_3(){
    // 使用count 统计餐品的记录总数
    String hql1 = "select count(m) from Meal m";
    Query query1= session.createQuery(hql1);
    Long count=(Long)query1.uniqueResult();
    // 使用avg 统计餐品的平均价格
    String hql2 = "select avg(m.mealPrice) from Meal m";
    Query query2= session.createQuery(hql2);
    Double money=(Double)query2.uniqueResult();
    // 使用max 和min 统计最贵和最便宜的餐品
    String hql3= "select max(m.mealPrice),min(m.mealPrice) from Meal m";
    Query query3= session.createQuery(hql3);
    Object[] price=(Object[])query3.uniqueResult();
    System.out.println("记录总数："+count+"，平均金额："+money+"，最低价格为："+price[0] +"，最高价格为："+price[1]);
}
```

执行 testHql_2()方法，控制台输出结果如下：

记录总数：12，平均金额：12.83，最低价格为：20.0，最高价格为：8.0

14.1.4 分组查询

在测试类 HibernateTest 中添加 testHql_4()方法，并使用@Test 注解修饰，以菜系为分组依据对所有餐品进行分组，查询数据表 meal 中各种菜系的餐品总数。代码如下：

```java
@Test
public void testHql_4(){
    // 分组统计餐品的菜系总数
    String hql = "select m.mealseries.seriesName,count(*) from Meal m group by m.mealseries";
    Query query= session.createQuery(hql);
    List list=query.getResultList() ;
    Iterator iterator=list.iterator();
    // 每条记录封装成一个Object 数组
    while (iterator.hasNext()) {
        Object[] object=(Object[]) iterator.next();
        System.out.println("菜系："+ object[0] +"，餐品总数："+ object[1]);
    }
}
```

执行 testHql_4()方法，控制台输出结果如下：

Hibernate: select ……
菜系：鲁菜，餐品总数：9
菜系：川菜，餐品总数：3

14.1.5 动态实例查询

在属性查询(或投影查询)时,返回的查询结果是一个对象数组,不易操作。为了提高检索效率,可将检索出来的属性封装到一个实体类的对象中,这种方式就是动态实例查询。

在测试类 HibernateTest 中添加 testHql_5()方法,并使用@Test 注解加以修饰,只查询餐品中的名称和 Id 号,将检索出来的属性封装到一个实体类的对象中,代码如下:

```
@Test
public void testHql_5(){
    // 编写 Hql 语句,使用动态实例查询
    String hql = "select new Meal(m.mealId,m.mealName) from Meal m";
    Query<Meal> query= session.createQuery(hql,Meal.class);
    List<Meal> list=query.getResultList() ;
    for(Meal m : list){
        System.out.println(m.getMealId() + ". " +m.getMealName());
    }
}
```

在 HQL 语句中使用了 Meal 类的带餐品 Id 号和餐品名称两个参数的构造方法,因此需要在实体类 Meal 类中添加这个构造方法,代码如下:

```
public Meal(int mealId, String mealName) {
    this.mealId = mealId;
    this.mealName = mealName;
}
```

执行 testHql_5()方法,控制台输出结果如下:

```
Hibernate: select meal0_.MealId as col_0_0_, meal0_.MealName as col_1_0_
from restrant.meal meal0_
1. 雪梨肉肘棒
2. 素锅烤鸭肉
3. 烤花肉揽桂鱼
......
```

14.1.6 分页查询

批量查询数据时,在单个页面上显示所有的查询结果会存在一定的问题,因此需要对查询结果进行分页显示。Query 接口提供了用于分页显示查询结果的方法,具体介绍如下。

(1) setFirstResult(int firstResult)方法设定从哪个对象开始查询,参数 firstResult 表示这个对象在查询结果中的索引(索引的起始值为 0)。

(2) setMaxResult(int maxResult)方法设定一次返回多少个对象。默认时,返回查询结果中的所有对象。

分页查询是系统中常用的一个功能,为了方便调用,先在测试类 HibernateTest 中添加方法 pagedSearch(int pageIndex, int pageSize),根据页码和每页显示记录数从数据表 meal 中获取相应的记录。代码如下:

```
public void pagedSearch(int pageIndex, int pageSize) {
```

```
        String hql = "from Meal m order by m.mealName asc";
        Query<Meal> query = session.createQuery(hql,Meal.class);
        int startIndex = (pageIndex - 1) * pageSize;
        query.setFirstResult(startIndex);
        query.setMaxResults(pageSize);
        List<Meal> list = query.getResultList() ;
        for(Meal m : list){
            System.out.println(m.getMealId() + ". " +m.getMealName());
        }
    }
```

在 pagedSearch(int pageIndex, int pageSize)方法中，第一个参数表示当前页码，第二个参数表示每页显示多少个对象。

然后在测试类 HibernateTest 中添加 testHql_6()方法，并使用@Test 注解加以修饰，调用 pagedSearch 方法。代码如下：

```
@Test
public void testHql_6(){
    pagedSearch(2, 4);
}
```

执行 testHql_6()方法，控制台输出结果如下：

```
2．素锅烤鸭肉
6．肉冬菜肉末
8．芹黄烧鱼条
5．落叶琵琶肉虾
```

14.1.7 条件查询

在实际应用中，常常需要根据指定的条件进行查询。此时，可以使用 HQL 语句提供的 where 子句进行查询，或者使用 like 关键字进行模糊查询。

根据提供的参数形式，条件查询有两种：按参数位置查询和按参数名字查询。

1．按参数位置查询

按参数位置查询时，在 HQL 语句中需要使用"？"来定义参数的位置。在测试类 HibernateTest 中添加 testHql_7()方法，并使用@Test 注解加以修饰，按照参数位置查询的方式，查询餐品名称包含"鱼"的餐品信息。代码如下：

```
@Test
public void testHql_7(){
    // 编写 Hql 语句，使用参数查询
    String hql = "from Meal m where m.mealName like ?";
    Query<Meal> query= session.createQuery(hql,Meal.class);
    // 为HQL语句中"?"代表的参数设置值
    query.setParameter(0, "%鱼%");
    List<Meal> list=query.getResultList() ;
    for(Meal m : list){
        System.out.println(m.getMealId() + ". " +m.getMealName());
    }
}
```

在 testHql_7()方法中，HQL 语句使用了"?"来定义参数的位置，这里的 HQL 语句中定义了一个参数，第一个参数的位置为零。接下来使用 Query 提供 query.setParameter (0, "%鱼%")方法设置参数的值。在 setParameter()方法中，第一个参数表示 HQL 语句中参数的位置，第二个参数表示 HQL 语句中参数的值。这里给参数赋值时，使用了"%"通配符，以匹配任意类型和任意长度的字符串。如果 HQL 语句中有多个参数，可以依次进行赋值。

执行 testHql_7()方法，控制台输出结果如下：

```
Hibernate: ……
3. 烤花肉揽桂鱼
8. 芹黄烧鱼条
```

2．按参数名字查询

按参数名字查询时，需要在 HQL 语句中定义命名参数，且命名参数需要以"："开头。在测试类 HibernateTest 中添加 testHql_8()方法，并使用@Test 修饰，按照参数名字查询的方式，查询餐品名称包含"鱼"的餐品信息。代码如下：

```
@Test
public void testHql_8(){
    // 编写 Hql 语句，使用参数查询
    String hql = "from Meal m where m.mealName like :mname";
    Query<Meal> query= session.createQuery(hql,Meal.class);
    // 为 HQL 语句中 mname 代表的参数设置值
    query.setParameter("mname", "%鱼%");
    List<Meal> list=query.getResultList() ;
    for(Meal m : list){
        System.out.println(m.getMealId() + ". " +m.getMealName());
    }
}
```

执行 testHql_8()方法，控制台输出结果同 testHql_7()方法。

在 HQL 语句中设定查询条件时，可以使用如表 14-1 所示的各种运算。

表 14-1　HQL 支持的各种运算

类　型	HQL 运算符
比较运算	=、<>、>、>=、<、<=、is null、is not null
范围运算	in、not in、between、not between
逻辑运算	and(逻辑与)、or(逻辑或)、not(逻辑非)
模式匹配	like

14.1.8　连接查询

HQL 支持各种连接查询，如内连接、隐式内连接等。

1．内连接

内连接是指两个表中指定的关键字相等的值才会出现在查询结果集中的一种查询方式。

在 HQL 中，使用关键字 inner join 进行内连接。在测试类 HibernateTest 中添加 testHql_9()方法，并使用@Test 注解加以修饰，从数据表 meal 中查询菜系为"川菜"的餐品信息。代码如下：

```java
@Test
public void testHql_9(){
    // 编写 Hql 语句，使用内连接查询
    String hql = "from Meal as m inner join m.mealseries as ms where ms.seriesName='川菜'";
    Query query= session.createQuery(hql);
    List list=query.getResultList() ;
    Iterator iterator=list.iterator();
    Object[] obj=null;
    Meal m=null;
    Mealseries ms=null;
    while (iterator.hasNext()) {
        obj= (Object[]) iterator.next();
        m=(Meal)obj[0];
        ms=(Mealseries)obj[1];
        System.out.println("餐品名称："+m.getMealName()+" \t 价格："+m.getMealPrice()+"\t 菜系："+ms.getSeriesName());
    }
}
```

HQL 语句中使用 inner join 进行内连接，查询返回的结果并不是 Meal 对象(虽然 from 关键字后面只有 Meal)，而是一个对象数组，对象数组中的第一列是 Meal 对象，第二列是 Mealseries 对象。

执行 testHql_9()方法，控制台输出结果如下：

```
Hibernate: select … from restrant.meal meal0_ inner join restrant.mealseries mealseries1_ on meal0_.MealSeriesId=mealseries1_.SeriesId where mealseries1_.SeriesName='川菜'
餐品名称：芹黄烧鱼条        价格：15.0       菜系：川菜
餐品名称：巴国玉米糕肉      价格：13.0       菜系：川菜
餐品名称：酥皮龙虾          价格：20.0       菜系：川菜
```

从上述 SQL 语句可以看出在查询过程中 meal 和 mealseries 两张表的所有列都被查询。因此，查询结果中同时包含了 Meal 和 Mealseries 的所有对象，而不是只有 Meal 对象。

2．隐式内连接

隐式内连接是指 HQL 语句中看不到 join 关键字，好像没有连接一样，但实际上已经发生了内连接。在测试类 HibernateTest 中添加 testHql_10()方法，并使用@Test 注解加以修饰，从数据表 meal 中查询菜系为"川菜"的餐品信息。代码如下：

```java
@Test
public void testHql_10(){
    // 编写 Hql 语句，使用隐式内连接查询
    String hql = "from Meal m, Mealseries ms where m.mealseries=ms and ms.seriesName='川菜'";
    // 以下代码与 testHql_9()中的相同
    Query query= session.createQuery(hql);
```

```
        // 省略和testHql_9()测试方法中相同部分的代码
        ……
    }
```

HQL 语句中没有使用内连接关键字，在 where 子句中 m.mealseries 引用了 Meal 对象的 mealseries 属性，实际上 Meal 与 Mealseries 已经发生了内连接。

执行 testHql_10()方法，控制台输出结果同方法 testHql_9()。

14.1.9 子查询

Hibernate 支持在查询中嵌套子查询，一个子查询必须放在圆括号内。HQL 中的子查询分为相关子查询和无关子查询。

1．相关子查询

相关子查询是指子查询使用外层查询中的对象别名。在测试类 HibernateTest 中添加 testHql_11()方法，并使用@Test 注解加以修饰，使用相关子查询检索数据表 Meal 中所有菜系记录数超过 2 的菜系。代码如下：

```
@Test
public void testHql_11(){
    // 编写Hql语句，使用相关子查询
    String hql = "from Mealseries ms where (select count(*) from ms.mealSet)>2 ";
    Query query= session.createQuery(hql);
    List list=query.getResultList() ;
    Iterator iterator=list.iterator();
    while (iterator.hasNext()) {
        Mealseries ms=(Mealseries) iterator.next();
        System.out.println("菜系名称："+ms.getSeriesName()+"\t菜系记录数："+ms.getMealSet().size());
    }
}
```

在 HQL 语句中，子查询中引用了外层语句中的别名 ms，它是 Mealseries 类的别名。每个菜系包含多个餐品记录，即在 meal 表中有多条餐品记录对应着 mealseries 表中的同一条记录，在 Mealseries 类中创建了 Set 类型的属性 mealSet。

执行方法 testHql_11()，控制台输出结果如下：

```
Hibernate: ……
Hibernate: ……
菜系名称：鲁菜        菜系记录数：9
Hibernate: ……
菜系名称：川菜        菜系记录数：3
```

2．无关子查询

无关子查询是指子查询语句与外层查询语句无关。在测试类 HibernateTest 中添加 testHql_12()方法，并使用@Test 注解加以修饰，使用无关子查询检索所有低于平均价的餐品对象。代码如下：

```java
@Test
public void testHql_12(){
    // 编写 Hql 语句,使用无关子查询检索所有低于平均价的餐品对象
    String hql = "from Meal m where m.mealPrice<(select avg(m1.mealPrice) from Meal m1) ";
    Query query= session.createQuery(hql);
    List list=query.getResultList() ;
    Iterator iterator=list.iterator();
    while (iterator.hasNext()) {
        Meal m=(Meal)iterator.next();
        System.out.println("餐品名称: "+m.getMealName()+" \t 价格: "+m.getMealPrice());
    }
}
```

执行方法 testHql_12(),控制台输出结果如下:

```
餐品名称: 雪梨肉肘棒        价格: 10.0
餐品名称: 泰安肉三美豆腐     价格: 8.0
餐品名称: 肉冬菜肉末        价格: 12.0
餐品名称: 糖醋红柿椒        价格: 8.0
餐品名称: 香煎茄片          价格: 9.0
餐品名称: 金陵片皮鸭        价格: 10.0
```

当子查询结果为多条记录时,Hibernate 提供了相应的关键字,如表 14-2 所示。

表 14-2 与子查询相关的关键字

关 键 字	含 义
all	表示子查询语句返回的所有记录
any	表示子查询语句返回的任意一条记录
some	与 any 关键字相同
in	表示是否出现在子查询返回的所有记录中
exists	表示子查询是否至少返回一条记录

下面主要介绍如何在子查询中使用 exists、in 关键字进行检索。

1) 使用 exists 关键字的子查询

在测试类 HibernateTest 中添加 testHql_13()方法,并使用@Test 注解加以修饰,使用 exists 关键字检索餐品价格大于 15 元的菜系名称。代码如下:

```java
@Test
public void testHql_13(){
    // 编写 Hql 语句,检索餐品价格大于15元的菜系名称
    String hql = "from Mealseries ms where exists(select m from ms.mealSet as m where m.mealPrice>15)";
    Query query= session.createQuery(hql);
    List list=query.getResultList() ;
    Iterator iterator=list.iterator();
    while (iterator.hasNext()) {
        Mealseries ms=(Mealseries) iterator.next();
        System.out.println("菜系名称: "+ms.getSeriesName());
    }
}
```

执行方法 testHql_13()，控制台输出结果如下：

```
Hibernate: ……
菜系名称：鲁菜
菜系名称：川菜
```

2）使用 in 关键字的子查询

在测试类 HibernateTest 中添加 testHql_14()方法，并使用@Test 注解加以修饰，使用 in 关键字查询数据表 meal 中包含 3 条餐品记录的菜系名称。代码如下：

```java
@Test
public void testHql_14(){
    // 编写 Hql 语句，查询数据表 meal 中包含 3 条餐品记录的菜系名称
    String hql = "from Mealseries ms where 3 in(select count(m) from ms.mealSet as m) ";
    Query query= session.createQuery(hql);
    List list=query.getResultList() ;
    Iterator iterator=list.iterator();
    while (iterator.hasNext()) {
        Mealseries ms=(Mealseries) iterator.next();
        System.out.println("菜系名称："+ms.getSeriesName());
    }
}
```

执行方法 testHql_14()，控制台输出结果如下：

```
Hibernate: ……
菜系名称：川菜
```

14.2 使用 QBC 查询数据

QBC 是 Query By Criteria 首字母缩写，Criteria 是 Hibernate API 提供的一个查询接口，位于 org.hibernate 包下。Criteria 查询又称为对象查询，它使用一种封装了基于字符串形式的查询语句的 API 来查询对象。

QBC 查询主要由 Criteria 接口来完成，该接口由 Hibernate Session 创建，Criterion 是 Criteria 的查询条件。Criteria 提供了 add(Criterion criterion)方法来添加查询条件。

Criterion 接口的主要实现类包括 Example、Junction 和 SimpleExpression。Example 主要用来提供 QBE(Query By Example)检索方式，是 QBC 的子功能。

Criterion 接口的实现类一般通过 Restrictions 工具类来创建。使用工具类 Order 相关方法设置排序方式，如 Order.asc 表示升序，Order.desc 表示降序。

14.2.1 简单查询

在使用 HQL 查询方式时，需要定义基于字符串形式的 HQL 语句，虽然比 JDBC 代码有所进步，但仍然烦琐且不方便使用参数查询。Criteria 采用面向对象的方式封装查询条件，Criteria API 提供了查询对象的另一种方式，提供了 Criteria 接口、Criterion 接口、Expression

类，以及 Restrictions 类作为辅助，从而使得查询代码的编写更加方便。

使用 Restrictions 辅助类，进行 Criteria 查询的基本步骤如下。

(1) 创建 Criteria 对象。

(2) 使用 Restrictions 对象编写查询条件，并将查询条件加入 Criteria 对象。

(3) 执行查询，获得结果。

下面通过示例演示如何使用 Criteria 查询，在测试类 HibernateTest 中添加 testCriteria_1() 方法，使用@Test 注解加以修饰，使用 Criteria 方式从数据表 meal 中查询所有餐品对象。代码如下：

```java
@Test
public void testCriteria_1(){
    // 创建查询所有餐品的 Criteria 对象
    Criteria c=session.createCriteria(Meal.class);
    // 对查询结果按 mealName 升序排序
    c.addOrder(Order.asc("mealName"));
    // 执行查询，获取结果
    List<Meal> list=c.list();
    // 循环输出查询结果
    for(Meal m : list){
        System.out.println(m.getMealId() + ". " +m.getMealName());
    }
}
```

执行 testCriteria_1()方法，控制台输出结果如下：

```
Hibernate: select … from restrant.meal this_ order by this_.MealName asc
Hibernate: ……
Hibernate: ……
9. 巴国玉米糕肉
4. 泰安肉三美豆腐
3. 烤花肉揽桂鱼
7. 糖醋红柿椒
……
```

14.2.2 分组查询

根据所属菜系对餐品记录进行分组，在测试类 HibernateTest 中添加 testCriteria_2()方法，使用@Test 注解加以修饰，查询 meal 表中各个菜系的餐品总记录数及总金额。代码如下：

```java
@Test
public void testCriteria_2(){
    // 创建 Criteria 对象
    Criteria c=session.createCriteria(Meal.class);
    // 构建 ProjectionList 对象
    ProjectionList pList=Projections.projectionList();
    // 创建分组依据，对菜系进行分组
    pList.add(Projections.groupProperty("mealseries"));
    // 统计各分组中的记录数
    pList.add(Projections.rowCount());
    // 统计各分组中的餐品价格总和
    pList.add(Projections.sum("mealPrice"));
```

```
        // 为 Criteria 对象设置 Projection
        c.setProjection(pList);
        List list=c.list();              // 执行查询,获取结果
        Iterator iterator=list.iterator();    // 遍历查询结果
        while(iterator.hasNext()){
            Object[] obj=(Object[]) iterator.next();
            Mealseries ms=(Mealseries)obj[0];
            System.out.println("菜系名称: " + ms.getSeriesName() + "\t 餐品记录总
数: " +obj[1]+ "\t 价格总和: "+ obj[2]);
        }
    }
```

执行 testCriteria_2()方法,控制台输出结果如下:

```
Hibernate: select … from restrant.meal this_ group by this_.MealSeriesId
Hibernate: ……
Hibernate: ……
菜系名称: 鲁菜        餐品记录总数: 9        价格总和: 106.0
菜系名称: 川菜        餐品记录总数: 3        价格总和: 48.0
```

14.2.3 聚集函数

在测试类 HibernateTest 中添加 testCriteria_3()方法,并使用@Test 注解加以修饰,使用内置聚集函数统计 meal 表中所有餐品价格总和、平均价格、最大价格和最小价格。代码如下:

```
@Test
public void testCriteria_3(){
    // 创建 Criteria 对象
    Criteria c=session.createCriteria(Meal.class);
    // 构建 ProjectionList 对象
    ProjectionList pList=Projections.projectionList();
    // 统计餐品价格总和
    pList.add(Projections.sum("mealPrice"));
    // 统计餐品平均价格
    pList.add(Projections.avg("mealPrice"));
    // 统计餐品最高价格
    pList.add(Projections.max("mealPrice"));
    // 统计餐品最低价格
    pList.add(Projections.min("mealPrice"));
    // 为 Criteria 对象设置 Projection
    c.setProjection(pList);
    // 执行查询,获取结果
    List list=c.list();
    // 遍历查询结果
    Iterator iterator=list.iterator();
    while(iterator.hasNext()){
        Object[] obj=(Object[]) iterator.next();
        System.out.println("餐品价格总和: "+obj[0]+"\t 平均价格: "+obj[1]
+"\t 最高价格: "+obj[2]+"\t 最低价格: "+obj[3]);
    }
}
```

执行 testCriteria_3()方法,控制台输出结果如下:

```
Hibernate: select sum(this_.MealPrice) ……
餐品价格总和：154.0    平均价格：12.833333    最高价格：20.0    最低价格：8.0
```

14.2.4 组合查询

组合查询是指通过 Restrictions 工具类的相应方法动态构造查询条件，并将查询条件加入 Criteria 对象，从而实现查询功能。在测试类 HibernateTest 中编写 testCriteria_4()方法，并使用@Test 注解加以修饰，按餐品名称和餐品价格查询餐品对象。代码如下：

```java
@Test
public void testCriteria_4(){
    // 封装查询条件
    Meal condition=new Meal();
    condition.setMealName("虾");
    condition.setMealPrice(18.00);
    // 创建 Criteria 对象
    Criteria c=session.createCriteria(Meal.class);
    // 使用 Restrictions 对象编写查询条件，并将查询条件加入 Criteria 对象
    if (condition!=null) {
        if (condition.getMealName()!=null
&& !condition.getMealName().equals("")) {  // 按餐品名称进行筛选
            c.add(Restrictions.like("mealName",
condition.getMealName(),MatchMode.ANYWHERE));
        }
        if (condition.getMealPrice()>0) {// 按餐品价格进行筛选
            c.add(Restrictions.le("mealPrice", condition.getMealPrice()));
        }
    }
    List<Meal> list=c.list();           // 执行查询，获取结果
    // 循环输出查询结果
    for(Meal m : list){
        System.out.println("餐品名称： " +m.getMealName()+"\t 价格： "+m.getMealPrice());
    }
}
```

在 testCriteria_4()方法中使用 Restrictions 对象编写查询条件，并将查询条件加入 Criteria 对象。Restrictions 提供了大量的静态方法，来创建查询条件，如表 14-3 所示。

表 14-3 Restrictions 提供的方法

方 法 名	说 明
Restrictions.eq	等于
Restrictions.allEq	使用 Map，使用 key/value 进行多个等于的比较
Restrictions.gt	大于 >
Restrictions.ge	大于等于 >=
Restrictions.lt	小于 <
Restrictions.le	小于等于 <=
Restrictions.between	对应 SQL 的 BETWEEN 子句
Restrictions.like	对应 SQL 的 LIKE 子句

续表

方 法 名	说 明
Restrictions.in	对应 SQL 的 in 子句
Restrictions.and	and 关系
Restrictions.or	or 关系
Restrictions.sqlRestriction	SQL 限定查询

MatchMode 表示匹配模式，包含的静态常量如表 14-4 所示。

表 14-4　MatchMode 包含的静态常量

匹配模式	说 明
MatchMode.ANYWHERE	模糊匹配
MatchMode.EXACT	精确匹配
MatchMode.START	以某个字符为开头进行匹配
MatchMode.END	以某个字符为结尾进行匹配

执行 testCriteria_4()方法，输出两条查询语句和响应结果，控制台输出结果如下：

餐品名称：落叶琵琶肉虾　　　价格：14.0

14.2.5　关联查询

使用 Criteria 并通过 Restrictions 工具类，可以实现关联查询。在测试类 HibernateTest 中编写 testCriteria_5()方法，并使用@Test 注解加以修饰，实现从数据表 meal 中查询菜系为"川菜"的餐品名称包含"鱼"的餐品。代码如下：

```java
@Test
public void testCriteria_5(){
    // 创建 Criteria 对象
    Criteria mCriteria=session.createCriteria(Meal.class);
    // 设置从 Meal 类中查询的条件
    mCriteria.add(Restrictions.like("mealName","鱼",
MatchMode.ANYWHERE));
    // 创建一个新的 Criteria 实例，以引用 mealseries
    Criteria msCriteria=mCriteria.createCriteria("mealseries");
    // 设置从关联的 Mealseries 类中查询的条件
    msCriteria.add(Restrictions.like("seriesName", "川菜"));
    List<Meal> list=mCriteria.list();    // 执行查询，获取结果
    // 循环输出查询结果
    for(Meal m : list){
        System.out.println("餐品名称：" + m.getMealName()+"\t 价格："
+m.getMealPrice());
    }
}
```

执行 testCriteria_5()方法，控制台输出结果如下：

Hibernate: ……

餐品名称：芹黄烧鱼条   价格：15.0

创建 Criteria 对象和使用 Restrictions 对象编写查询条件，可以采用方法链编程风格。

```java
@Test
public void testCriteria_5_1(){
    List<Meal> list=session.createCriteria(Meal.class)
        .add(Restrictions.like("mealName","鱼",MatchMode.ANYWHERE))
        .createCriteria("mealseries").add(Restrictions.like("seriesName", "川菜")).list();
        for(Meal m : list){            // 循环输出查询结果
            System.out.println("餐品名称: " +m.getMealName()+"\t 价格: "+m.getMealPrice());
        }
}
```

14.2.6 分页查询

使用 Criteria 并通过 Restrictions 工具类，可以实现分页查询。Hibernate 的 Criteria 也提供了两个用于实现分页的方法：setFirstResult(int firstResult)和 setMaxResults(int maxResults)。其中，setFirstResult(int firstResult)方法用于指定从哪个对象开始检索，默认为第一个对象(序号为 0)；setMaxResults(int maxResults)方法用于指定一次最多检索的对象数，默认为所有对象。在测试类 HibernateTest 中添加 testCriteria_6()方法，使用@Test 注解加以修饰。代码如下：

```java
@Test
public void testCriteria_6(){
    Criteria c=session.createCriteria(Meal.class);
    // 从第一个对象开始查询
    c.setFirstResult(0);
    // 每次从查询结果中返回 4 个对象
    c.setMaxResults(4);
    List<Meal> list=c.list();
    for(Meal m : list){  // 循环输出查询结果
        System.out.println("餐品名称: " +m.getMealName()+"\t 价格: "+m.getMealPrice());
    }
}
```

执行 testCriteria_6()方法，控制台输出结果如下：

```
Hibernate: ……
Hibernate: ……
餐品名称：雪梨肉肘棒         价格：10.0
餐品名称：素锅烤鸭肉         价格：20.0
餐品名称：烤花肉揽桂鱼       价格：15.0
餐品名称：泰安肉三美豆腐     价格：8.0
```

14.2.7 QBE 查询

QBE(Query By Example)查询为举例查询，也称示例查询。由于 QBE 查询检索与指定示例对象具有相同属性值的对象，因此示例对象的创建是 QBE 查询的关键。示例对象中的所有非空属性都作为查询的条件。

以 testCriteria_4()方法中实现的组合查询为例，组合的条件越多，需要的 if 语句就越多，相当烦琐，此时使用 QBE 查询最方便。在测试类 HibernateTest 中编写 testCriteria_7()方法，使用@Test 注解加以修饰，按餐品名称和餐品价格查询餐品对象。代码如下：

```java
@Test
public void testCriteria_7(){
    // 封装查询条件
    Meal condition=new Meal();
    condition.setMealName("虾");
    condition.setMealPrice(18.00);
    // 创建 Criteria 对象
    Criteria c=session.createCriteria(Meal.class);
    // 使用 Example 工具类创建示例对象，将属性 mealPrice 排除在示例查询外
    Example example=
        Example.create(condition).excludeProperty("mealPrice");
    // 设置不区分大小写
    example.ignoreCase();
    // 设置匹配模式为 ANYWHERE
    example.enableLike(MatchMode.ANYWHERE);
    // 为 Criteria 对象指定示例对象 example 作为查询条件
    c.add(example);
    // 将 mealPrice 作为一个额外的条件加入查询
    if (condition.getMealPrice()>0) {
      c.add(Restrictions.le("mealPrice", condition.getMealPrice()));
    }
    List<Meal> list=c.list();       // 执行查询，获取结果
    // 循环输出查询结果
    for(Meal m : list){
        System.out.println("餐品名称：" +m.getMealName()+" \t 价格：" +
m.getMealPrice());
    }
}
```

testCriteria_7()方法查询餐品名称包含"虾"且餐品价格小于等于 18 的餐品对象，这种查询需要将查询的条件封装到一个 Meal 对象，然后以该对象为参数，Example 工具类调用其静态方法 create()创建一个 Example 对象。由于示例查询针对对象的属性进行模糊匹配、精确匹配、开头匹配和结尾匹配，而 mealPrice 属性查询条件是"小于等于"。因此在创建 Example 对象时，需要先将属性 mealPrice 排除在示例查询外，Example 对象的 enableLike()方法与 Restrictions.like()方法类似，表示模糊查询。ignoreCase()方法表示在查询过程中不区分大小写。再为 Criteria 对象指定示例对象 example 作为查询条件。最后将 mealPrice 作为一个额外的条件加入查询。

执行 testCriteria_7()方法，控制台输出结果如下：

```
Hibernate: select …… from restrant.meal this_ where (lower(this_.MealName)
like ?) and this_.MealPrice<=?
Hibernate: ……
餐品名称：落叶琵琶肉虾        价格：14.0
```

从 SQL 语句可以看出，在 where 子句中调用 lower 函数把 name 字段的值转为小写。Hibernate 在 like 关键字后的问号参数位置绑定参数值时，绑定的参数值为"%虾%"，mealPrice 字段的查询条件是<=18.00。

14.2.8 离线查询

Criteria 查询是一种在线查询方式，它是通过 Hibernate Session 进行创建的。而 DetachedCriteria 查询是一种离线查询方式，创建查询时无须使用 Session，可以在 Session 范围之外创建一个查询，并且可以使用任意的 Session 执行它。DetachedCriteria 提供了两个静态方法：forClass(Class)和 forEntityName(Name)，可以通过这两个方法创建 DetachedCriteria 实例。

在测试类 HibernateTest 中添加 testDetachedCriteria()方法，使用@Test 注解加以修饰，查询餐品名称包含"鱼"的餐品信息。代码如下：

```java
@Test
public void testDetachedCriteria(){
    // 创建离线查询 DetachedCriteria 实例
    DetachedCriteria query=DetachedCriteria.forClass(Meal.class)
.add(Property.forName("mealName").like("鱼",MatchMode.ANYWHERE));
    // 执行查询，获取结果
    List<Meal> list= query.getExecutableCriteria(session).list();
    // 循环输出查询结果
    for(Meal m : list){
        System.out.println("餐品名称: " + m.getMealName()+" \t 价格: "+m.getMealPrice());
    }
}
```

通过 DetachedCriteria 提供的 forClass(Class)方法创建 DetachedCriteria 实例，即在 Session 范围之外创建了一个查询。其中，Property 可以对某个属性进行查询条件的设置，如 Property.forName("name").like("鱼", MatchMode.ANYWHERE)设置的查询条件为：name 属性值包含"鱼"，然后通过 Session 执行查询。

执行 testDetachedCriteria()方法，控制台输出结果如下：

```
Hibernate: ……
Hibernate: ……
餐品名称: 烤花肉揽桂鱼      价格: 15.0
餐品名称: 芹黄烧鱼条        价格: 15.0
```

14.3 小 结

本章主要介绍了 Hibernate 中两个重要的查询：HQL 查询和 QBC 查询。HQL 查询时 Hibernate 提供的一种面向对象的查询方式，其优点在于：直接针对实体类和属性进行查询，不再编写烦琐的 SQL 语句，查询结果是直接保存在 List 中的对象，不用再次封装。QBC 下的查询则采用面向对象的方式封装插入查询条件。

第 15 章
使用 Hibernate 缓存数据

缓存介于应用程序和物理数据源之间,其作用是为了降低应用程序对物理数据源访问的频次,从而提高应用的运行性能。缓存内的数据是对物理数据源中的数据复制,应用程序在运行时从缓存读写数据,可以减少应用程序对永久性数据存储源的访问,使应用程序的运行性能得以提高。

15.1 缓存的概念和范围

缓存的介质一般是内存，所以读写速度很快。但如果缓存中存放的数据量非常大，也会用硬盘作为缓存介质。缓存的实现不仅仅要考虑存储的介质，还要考虑到管理缓存的并发访问和缓存数据的生命周期。

Hibernate 的缓存包括 Session 的缓存和 SessionFactory 的缓存，其中 SessionFactory 的缓存又可以分为两类：内置缓存和外置缓存。Session 的缓存是内置的，不能被卸载，也被称为 Hibernate 的第一级缓存。SessionFactory 的缓存又被称为 Hibernate 二级缓存。SessionFactory 的内置缓存和 Session 的缓存在实现方式上比较相似。SessionFactory 的外置缓存是一个可配置的插件。在默认情况下，SessionFactory 不会启用这个插件。

Hibernate 提供的一级缓存是一个线程对应一个 session，一个线程可以看成一个用户。也就是说，session 级缓存(一级缓存)只能给一个线程用，别的线程用不了，一级缓存就是和线程绑定了。Hibernate 一级缓存的生命周期很短，和 session 生命周期一样，一级缓存也称 session 级的缓存或事务级缓存。如果事务提交或回滚了，我们称 session 就关闭了，生命周期结束。

Hibernate 二级缓存需要 sessionFactory 来管理，它是初级的缓存，所有人都可以使用，它是共享的。使用缓存，肯定是长时间不改变的数据，如果经常变化的数据放到缓存里就没有太大意义了。因为经常变化，还是需要经常到数据库里查询，那就没有必要用缓存了。

Hibernate 缓存的范围包括事务范围、进程范围和集群范围，具体介绍如下。

(1) 事务范围。缓存只能被当前事务访问。缓存的生命周期依赖于事务的生命周期，当事务结束时，缓存也就结束生命周期。在此范围下，缓存的介质是内存。

(2) 进程范围。缓存被进程内的所有事务共享。这些事务有可能是并发访问缓存，因此必须对缓存采取必要的事务隔离机制。

(3) 集群范围。在集群环境中，缓存被一个机器或者多个机器的进程共享。缓存中的数据被复制到集群环境中的每个进程节点，进程间通过远程通信来保证缓存中的数据一致性，缓存中的数据通常采用对象的松散数据形式。

15.2 一 级 缓 存

Hibernate 的一级缓存由 Session 提供，只存在于 Session 的生命周期中。当应用程序调用 Session 接口的 save()、update()、saveOrUpdate()、get()、load()或者 Query 和 Criteria 实例的 list()、iterate()等方法时，如果 Session 缓存中没有相应对象，Hibernate 就会把对象加入一级缓存中。当 Session 关闭时，该 Session 所管理的一级缓存也会立即被清除。

1. 一个 session 中发出两次 get 查询

项目 hibernate-10 复制并命名为 hibernate-11，再导入 MyEclipse 开发环境中。在项目 hibernate-11 的测试类 HibernateTest 中添加 testSessionCache_1()方法，并使用@Test 注解加以

修饰，在同一个 session 中发出两次 get 查询。代码如下：

```
@Test
public void testSessionCache_1(){
    User u1=(User)session.get(User.class, 1);
    System.out.println("用户名："+u1.getLoginName());
    User u2=(User)session.get(User.class, 1);
    System.out.println("用户名："+u2.getLoginName());
}
```

执行 testSessionCache_1()方法，控制台输出结果如下：

```
Hibernate: select … from restrant.users user0_ where user0_.Id=?
用户名：zhangsan
用户名：zhangsan
```

从控制台输出结果可以看出，第一次执行 get 方法时查询了数据库，产生一条 SQL 语句；第二次执行 get 方法时，由于在一级缓存中找到该对象，因此不会查询数据库，不再发出 SQL 语句。

2. 开启两个 session 中发出两次 get 查询

在测试类 HibernateTest 中添加 testSessionCache_2()方法，并使用@Test 注解加以修饰，开启两个 session 中发出两次 get 查询。代码如下：

```
@Test
public void testSessionCache_2(){
    User u1=(User)session.get(User.class, 1);
    System.out.println("用户名："+u1.getLoginName());
    transaction.commit();
    session.close();
    session=sessionFactory.openSession();
    transaction=session.beginTransaction();
    User u2=(User)session.get(User.class, 1);
    System.out.println("用户名："+u2.getLoginName());
}
```

执行 testSessionCache_2()方法，控制台输出结果如下：

```
Hibernate: select … from restrant.users user0_ where user0_.Id=?
用户名：zhangsan
Hibernate: select … from restrant.users user0_ where user0_.Id=?
用户名：zhangsan
```

从控制台输出结果可以看出，两次执行 get 方法时都查询了数据库，产生了两条 SQL 语句。原因在于：第一次执行 get 方法查询出结果后，关闭了 Session，缓存被清除了；第二次执行 get 方法时，从缓存中找不到结果，只能到数据库查询。

3. 在一个 session 中先 save，再执行 load 查询

在测试类 HibernateTest 中添加 testSessionCache_3()方法，并使用@Test 注解加以修饰，在一个 session 中先执行 save 操作，再执行 load 查询。代码如下：

```
@Test
```

```
public void testSessionCache_3(){
    User user=new User("NewUser","123456","新用户");
    Serializable id=session.save(user);
    User u=(User)session.load(User.class, id);
    System.out.println("用户名: "+u.getLoginName());
}
```

执行 testSessionCache_3()方法，控制台输出结果如下：

```
Hibernate: insert into restrant.users (LoginName, LoginPwd, TrueName) values (?, ?, ?)
用户名: NewUser
```

执行 save 操作时，它会在缓存里放一份。执行 load 操作时不会发出 SQL 语句，因为 save 使用了缓存。

session 接口为应用程序提供了两个管理缓存的方法：evict()方法和 clear()方法。其中，evict()方法用于将某个对象从 session 的一级缓存中清除；clear()方法用于将一级缓存中的所有对象全部清除。

在测试类 HibernateTest 中添加 testSessionCacheClear()方法，并使用@Test 注解加以修饰，在同一个 session 中先调用 load 查询，然后执行 clear()方法，最后再调用 load 查询。代码如下：

```
@Test
public void testSessionCacheClear(){
    User u1=(User)session.load(User.class, 1);
    System.out.println("用户名: "+u1.getLoginName());
    session.clear();
    User u2=(User)session.load(User.class, 1);
    System.out.println("用户名: "+u2.getLoginName());
}
```

执行 testSessionCacheClear()方法，控制台输出结果如下：

```
Hibernate: select … from restrant.users user0_ where user0_.Id=?
用户名: zhangsan
Hibernate: select … from restrant.users user0_ where user0_.Id=?
用户名: zhangsan
```

第一次执行 load 操作时发出 SQL 语句，接着由于一级缓存中的实体被清除了，因此第二次执行 load 操作时也会发出 SQL 语句。clear()方法可以管理一级缓存，一级缓存无法取消，但可以管理。

15.3 二级缓存

二级缓存是一个可插拔的缓存插件，由 SessionFactory 负责管理。由于 SessionFactory 对象的生命周期和应用程序的整个过程对应，因此二级缓存是进程范围或者集群范围的缓存。

与一级缓存一样，二级缓存也根据对象的 ID 来加载缓存。当执行某个查询获得的结果集为实体对象集时，Hibernate 就会把它们按照对象 ID 加载到二级缓存中。在访问指定 ID 的对象时，首先从一级缓存查找，找到直接使用，找不到转到二级缓存查找(必须配置且启用二级缓存)。如果二级缓存中找到，则直接使用，否则会查询数据库，并将查询结果根据对象的 ID 放到缓存中。

1．常用的二级缓存插件

Hibernate 的二级缓存功能是通过配置二级缓存插件来实现的。常用的二级缓存插件包括 EHCache、OSCache、SwarmCache 和 JBossCache。其中，EHCache 缓存插件是理想的进程范围的缓存实现，此处以使用 EHCache 缓存插件为例来介绍如何使用 Hibernate 的二级缓存。

2．Hibernate 使用 EHCache 的步骤

1) 引入 EHCache 相关的 jar 包

在 Hibernate 官网下载 Hibernate5.2.6 压缩包 hibernate-release-5.2.6.Final.zip 并解压，将解压后 hibernate-release-5.2.6.Final\lib\optional\ehcache 目录下的 ehcache-2.10.3.jar、hibernate-ehcache-5.2.6.Final.jar、slf4j-api-1.7.7.jar 三个 jar 包复制到项目 hibernate-11 的 lib 目录中，并将它们添加到项目的构建路径。

2) 创建 EHCache 的配置文件 ehcache.xml

可以直接将解压后的 hibernate-release-5.2.6.Final\project\etc\ehcache.xml 复制到项目 hibernate-11 的 src 目录下。ehcache.xml 的主要代码如下：

```xml
<ehcache>
   <diskStore path="java.io.tmpdir"/>
   <defaultCache maxElementsInMemory="10000"
      eternal="false" timeToIdleSeconds="120"
      timeToLiveSeconds="120" overflowToDisk="true"/>
   <cache name="sampleCache1" maxElementsInMemory="10000"
      eternal="false" timeToIdleSeconds="300"
      timeToLiveSeconds="600" overflowToDisk="true"/>
   <cache name="sampleCache2" maxElementsInMemory="1000"
      eternal="true" timeToIdleSeconds="0"
      timeToLiveSeconds="0" overflowToDisk="false"/>
</ehcache>
```

在 ehcache.xml 配置文件中，diskStore 元素设置缓存数据文件的存储目录；defaultCache 元素设置缓存的默认数据过期策略；cache 元素设置具体的命名缓存的数据过期策略。每个命名缓存代表一个缓存区域，命名缓存机制允许用户在每个类以及类的每个集合的粒度上设置数据过期策略。

在 defaultCache 元素中，maxElementsInMemory 属性设置缓存对象的最大数目；eternal 属性指定是否永不过期，true 为不过期，false 为过期；timeToIdleSeconds 属性设置对象处于空闲状态的最大秒数；timeToLiveSeconds 属性设置对象处于缓存状态的最大秒数；overflowToDisk 属性设置内存溢出时是否将溢出对象写入硬盘。

3) 在 Hibernate 配置文件中启用 EHCache

在 hibernate.cfg.xml 配置文件中，启用 EHCache 的配置如下：

```xml
<!-- 启用二级缓存 -->
<property name="hibernate.cache.use_second_level_cache">true
```

```
        </property>
        <!-- 设置二级缓存插件 EHCache 的 Provider 类  -->
        <property name="hibernate.cache.region.factory_class">
        org.hibernate.cache.ehcache.EhCacheRegionFactory</property>
```

4）配置哪些实体类的对象需要二级缓存

有以下两种方式。

(1) 在实体类的映射文件中配置并发访问策略。

在需要进行缓存的持久化对象的映射文件中配置相应的二级缓存策略，如持久化对象 User 的映射文件 User.hbm.xml：

```
<hibernate-mapping package="com.hibernate.entity">
    <class name="User" table="users" catalog="restrant">
        <cache usage="read-write"/>
        <id name="id" type="java.lang.Integer">
            <column name="Id" />
            <generator class="native"></generator>
        </id>
        ……
    </class>
</hibernate-mapping>
```

映射文件中使用<cache>元素设置持久化类 User 的二级缓存并发访问策略，usage 属性取值为 read-only 时表示只读型并发访问策略；read-write 表示读写型并发访问策略；nonstrict-read-write 表示非严格读写型并发访问策略；Ehcache 插件不支持 transactional(事务型并发访问策略)。

注意，<cache>元素只能放在<class>元素的内部，而且必须处在<id>元素的前面。<cache>元素放在哪些<class>元素下面，就说明会对这些类的对象进行缓存。

(2) 在 Hibernate 配置文件中统一配置(推荐使用)。

在 hibernate.cfg.xml 文件中使用<class-cache>元素来配置哪些实体类的对象需要二级缓存，代码如下：

```
<!-- 二级缓存的要求必须放在所有<mapping>元素的后面 -->
<class-cache usage="read-write" class="com.hibernate.entity.User" />
```

在<class-cache>元素中，usage 属性指定缓存策略，需要注意<class-cache>元素必须放在所有<mapping>元素的后面。至此，Hibernate 的二级缓存 EHCache 就配置并启用完成了。

3．实体对象级别的二级缓存测试

再次执行测试类 HibernateTest 中的 testSessionCache_2()方法，控制台输出结果如下：

```
Hibernate: select … from restrant.users user0_ where user0_.Id=?
用户名：zhangsan
用户名：zhangsan
```

第一次执行 get 方法查询出结果后，关闭了 session，一级缓存被清除了，由于配置并启用了二级缓存，查询出的结果会放入二级缓存。第二次执行 get 方法时，首先从一级缓存中查找，没有找到；然后转到二级缓存查找，二级缓存中找到结果，不需要从数据库查询。

4．使用 collection-cache 配置集合级别的二级缓存

在测试类 HibernateTest 中添加 testSecondCache()方法，并使用@Test 注解加以修饰。代码如下：

```java
@Test
public void testSecondCache(){
    Mealseries ms1 = (Mealseries) session.get(Mealseries.class, 1);
    System.out.println("菜系: "+ms1.getSeriesName());
    System.out.println("数量: "+ms1.getMealSet().size());
    transaction.commit();
    session.close();
    session=sessionFactory.openSession();
    transaction=session.beginTransaction();
    Mealseries ms2 = (Mealseries) session.get(Mealseries.class, 1);
    System.out.println("菜系: "+ms2.getSeriesName());
    System.out.println("数量: "+ms2.getMealSet().size());
}
```

执行 testSecondCache()方法，控制台输出结果如下：

```
Hibernate: select … from restrant.mealseries mealseries0_ where
mealseries0_.SeriesId=?
菜系: 鲁菜
Hibernate: select … from restrant.meal mealset0_ where
mealset0_.MealSeriesId=?
数量: 9
Hibernate: select … from restrant.mealseries mealseries0_ where
mealseries0_.SeriesId=?
菜系: 鲁菜
Hibernate: select … from restrant.meal mealset0_ where
mealset0_.MealSeriesId=?
数量: 9
```

第一次执行 get 方法加载 SeriesId=1 的 Mealseries 对象 ms1 时发出第一条 SQL 语句；执行 ms1.getMealSet()时发出第二条 SQL 语句，之后关闭 session，一级缓存被清除了。由于此时没有针对 Mealseries 类配置二级缓存，因此第二次加载 SeriesId=1 的 Mealseries 对象 ms2，执行 ms2.getMealSet()时会发出第三、第四条 SQL 语句重新获取数据。

在 hibernate.cfg.xml 文件中使用<class-cache>元素来配置 Mealseries 类的对象的二级缓存，代码如下：

```xml
<class-cache usage="read-write" class=
        "com.hibernate.entity.Mealseries" />
```

再执行 testSecondCache()方法，控制台输出结果如下：

```
Hibernate: select … from restrant.mealseries mealseries0_ where
mealseries0_.SeriesId=?
菜系: 鲁菜
Hibernate: select … from restrant.meal mealset0_ where
mealset0_.MealSeriesId=?
数量: 9
菜系: 鲁菜
```

```
Hibernate: select … from restrant.meal mealset0_ where
mealset0_.MealSeriesId=?
数量：9
```

由于针对 Mealseries 类的对象配置了二级缓存，关闭 session 后，第二次加载 SeriesId=1 的 Mealseries 对象 ms2 时，可以从二级缓存中获取，因此少发出一条 SQL 语句。但是在执行 ms2.getMealSet()时还是会发出 SQL 语句重新获取数据，这是因为 Mealseries 对象中 Meal 类型的集合 mealSet 没有配置二级缓存。

在 hibernate.cfg.xml 文件中使用<collection-cache>元素来配置集合 mealSet 的二级缓存，代码如下：

```
<class-cache usage="read-write" class=
 "com.hibernate.entity.Mealseries" />
<collection-cache usage="read-write" collection=
 "com.hibernate.entity.Mealseries.mealSet"/>
```

再执行 testSecondCache()方法，控制台输出结果如下：

```
Hibernate: select … from restrant.mealseries mealseries0_ where
mealseries0_.SeriesId=?
菜系：鲁菜
Hibernate: select … from restrant.meal mealset0_ where
mealset0_.MealSeriesId=?
数量：9
菜系：鲁菜
Hibernate: select … from restrant.meal meal0_ where meal0_.MealId=?
Hibernate: //省略其他 8 条语句
数量：9
```

在执行 ms2.getMealSet()时发出了 9 条 SQL 语句，这是因为针对 MealSeries 对象中 Meal 类型的集合 mealSet 缓存时，没有针对 Meal 实体对象配置二级缓存，因此二级缓存中保存的是 Meal 实体对象的 mealId 列表，而不是 Meal 实体对象。因此执行 getMealSet()时，会从二级缓存中获得 mealId，再根据 mealId 从数据表 meal 中获取 Meal 对象。

为了解决这一问题，只需要在 hibernate.cfg.xml 文件中<class-cache>元素来配置 Meal 类的对象的二级缓存，代码如下：

```
<class-cache usage="read-write" class="com.hibernate.entity.Meal" />
```

再执行 testSecondCache()方法，控制台输出结果如下：

```
Hibernate: select … from restrant.mealseries mealseries0_ where
mealseries0_.SeriesId=?
菜系：鲁菜
Hibernate: select … from restrant.meal mealset0_ where
mealset0_.MealSeriesId=?
数量：9
菜系：鲁菜
数量：9
```

第一次加载 SeriesId=1 的 MealSeries 对象 ms1、执行 ms1.getMealSet()时，发出两条 SQL 语句。关闭 session 后，第二次加载 SeriesId=1 的 MealSeries 对象 ms2、执行 ms2.getMealSet()

时，可以从二级缓存中获取数据，不用再重新查找，因此不再发出 SQL 语句。

5. 基于硬盘的二级缓存

在配置文件 ehcache.xml 中，首先修改<diskStore>元素配置，指定 EHCache 把数据写入磁盘时的目录：

```
<diskStore path="E:\\EHCache"/> <!-- 此处路径不要设置在系统盘下 -->
```

修改 ehcache.xml 中原先的命名缓存区域 sampleCache1 和 sampleCache2，为每个缓存区域设置不同的缓存策略。代码如下：

```
<cache name="com.hibernate.entity.Meal" maxElementsInMemory="1"
    eternal="false" timeToIdleSeconds="300" timeToLiveSeconds="600"
    overflowToDisk="true" />
<cache name="com.hibernate.entity.Mealseries.mealSet"
    maxElementsInMemory="1000" eternal="true" timeToIdleSeconds="0"
    timeToLiveSeconds="0" overflowToDisk="false" />
```

为了便于测试，将命名缓存 com.hibernate.entity.Meal 的 maxElementsInMemory 属性设置为 1，即设置基于内存的缓存中可存放 Meal 对象的最大数目为 1。

给测试类 HibernateTest 中用@After 注解修饰的 destroy()方法中的"sessionFactory.close();"语句添加断点，选中 testSecondCache()方法，右击选择"Debug As→1 JUnit Test"，以调试的方式执行测试类中的 testSecondCache()方法。当程序暂停在"sessionFactory.close();"语句时，打开 E:\EHCache，可以看到缓存到磁盘的实体类对象文件。程序执行结束后，sessionFactory 关闭了，缓存到磁盘的文件就会自动被删除。

15.4 查询缓存

对经常使用的查询语句，如果启用了查询缓存，当第一次执行查询语句时，Hibernate 会将查询结果存储在第二级缓存中。以后再次执行该查询语句时，从缓存中获取查询结果，从而提高查询性能。

Hibernate 的查询缓存主要是针对普通属性结果集的缓存，而对于实体对象的结果集只缓存 ID。如果当前关联的表发生修改，则查询缓存生命周期结束。

在测试类 HibernateTest 中添加 testQueryCache()方法，并使用@Test 注解加以修饰，代码如下：

```
@Test
public void testQueryCache(){
    Query query=session.createQuery("From User");
    List<User> us=query.getResultList();
    System.out.println("用户数: "+us.size());
    us=query.getResultList();
    System.out.println("用户数: "+us.size());
}
```

执行 testQueryCache()方法，控制台输出结果如下：

```
Hibernate: select … from restrant.users user0_
用户数：5
Hibernate: select … from restrant.users user0_
用户数：5
```

在默认情况下，设置的缓存对 HQL 和 QBC 查询是无效的。查询缓存基于二级缓存，使用查询缓存前，必须首先配置二级缓存。在配置了二级缓存的基础上，在 Hibernate 的配置文件 hibernate.cfg.xml 中添加如下配置，可以启用查询缓存：

```
<property name="hibernate.cache.use_query_cache">true</property>
```

此外，还需要在 testQueryCache()方法中调用 Query 或 Criteria 的 setCacheable(true)方法。

```
@Test
public void testQueryCache(){
    Query query=session.createQuery("From User");
    query.setCacheable(true);
    ……
}
```

再次执行 testQueryCache()方法，查询缓存生效，控制台输出结果如下：

```
Hibernate: select … from restrant.users user0_
用户数：5
用户数：5
```

15.5 小　　结

本章主要讲解了缓存的概念和范围，一级缓存、二级缓存，以及查询缓存等，并通过具体的方法示例进行验证，让读者进一步加深了对 Hibernate 框架的了解。

第 16 章
MyBatis 框架

前面已介绍了 Hibernate 框架，它是一个开放源代码的对象关系映射框架，且对 JDBC 进行了非常轻量级的对象封装，使得 Java 程序员可以随心所欲地使用对象编程思维来操纵数据库。本章介绍另一个流行的持久层框架——MyBatis，它本是 Apache 的一个开源项目 iBatis，2010 年这个项目由 Apache Software Foundation 迁移到了 Google Code，并且改名为 MyBatis。

16.1 MyBatis 概念与安装

MyBatis 是支持普通 SQL 查询、存储过程和高级映射的优秀持久层框架。MyBatis 消除了几乎所有的 JDBC 代码和参数的手工设置以及对结果集的检索。MyBatis 可以使用简单的 XML 或注解来配置和映射基本数据类型，将接口和 Java 的 POJO(Plain Old Java Objects，普通的 Java 对象)映射成数据库中的记录。

读者可以从官方网站 https://github.com/mybatis 下载所需要的 MyBatis 版本。这里以 mybatis-3.3.0.zip 版本为例，mybatis-3.3.0.zip 解压后的目录如图 16-1 所示。

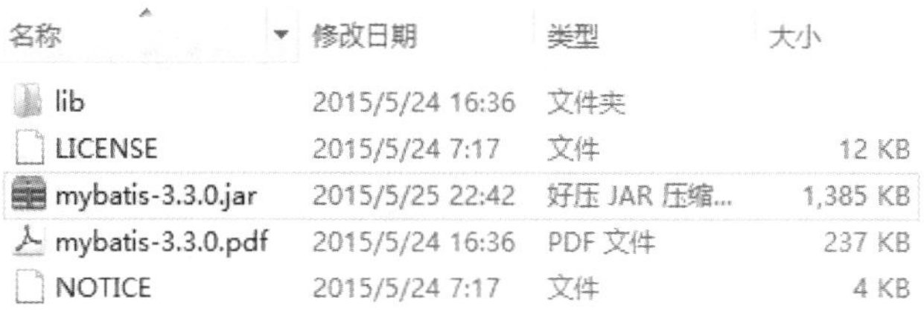

名称	修改日期	类型	大小
lib	2015/5/24 16:36	文件夹	
LICENSE	2015/5/24 7:17	文件	12 KB
mybatis-3.3.0.jar	2015/5/25 22:42	好压 JAR 压缩...	1,385 KB
mybatis-3.3.0.pdf	2015/5/24 16:36	PDF 文件	237 KB
NOTICE	2015/5/24 7:17	文件	4 KB

图 16-1　mybatis-3.3.0.zip 解压后的目录

lib 文件夹中存放 mybatis-3.3.0 所依赖的 jar 包(如日志包 log4j-1.2.17.jar)，mybatis-3.3.0.jar 是 mybatis-3.3.0 必需的核心 jar 包。

16.2 MyBatis 的增删改查

在 Java 或 Java Web 项目中添加 MyBatis 框架后，就能对数据表执行增删改查操作了。下面以数据表 users 为例，使用 MyBatis 实现数据的增删改查，实现过程如下。

(1) 新建一个名为 mybatis_1 的 Java 项目，在项目中新建文件夹 lib，用于存放项目所需的 jar 包。

(2) 将 MyBatis 必需的 jar 包 mybatis-3.3.0.jar，以及日志包 log4j-1.2.17.jar 复制到该项目的 lib 目录中，即完成了 MyBatis 的安装。

(3) 将 MySQL 的驱动包也复制到该项目的 lib 目录中，这里使用的版本为 mysql-connector-java-5.1.18-bin.jar。

(4) 选中该项目 lib 目录下的所有 jar 包，右击并选择 Build Path→Add to Build Path 命令，将这些 jar 包添加到项目的构建路径中。

(5) 创建实体类。

在 src 目录下新建 com.mybatis.pojo 包，并在其中创建实体类 Users(对应数据库 restrant 中的数据表 users)。Users 类包含一些属性(对应数据表 users 的部分字段)，以及与之对应的 getXXX()和 setXXX()方法，还可以根据需要添加构造方法。代码如下：

```
package com.mybatis.pojo;
public class Users {
   private int id;
```

```
    private String loginName;
    private String loginPwd;
// 此处省略属性的getter和setter方法
    // 此处省略有参构造方法和无参构造方法
// 此处省略toString()方法
}
```

(6) 创建 SQL 映射的 XML 文件。

在 com.mybatis.pojo 包中创建 SQL 映射的 XML 文件 usersMapper.xml，代码如下：

```
<!DOCTYPE mapper
PUBLIC "-//mybatis.org//DTD Mapper 3.0//EN"
"http://mybatis.org/dtd/mybatis-3-mapper.dtd">
<mapper namespace="com.mybatis.pojo.usersMapper">
    <!-- 数据表users的CRUD操作 -->
    <insert id="addUser" parameterType="Users">
        insert into   users(loginName,loginPwd)
values(#{loginName},#{loginPwd})
    </insert>
    <delete id="deleteUser" parameterType="int">
        delete from users where   id=#{id}
    </delete>
    <update id="updateUser" parameterType="Users">
        update users set loginName=#{loginName}, loginPwd=#{loginPwd} where
id=#{id}
    </update>
    <select id="getUserById" parameterType="int" resultType="Users">
        select * from users where id = #{id}
    </select>
    <select id="getAllUsers" resultType="Users">
        select * from users
    </select>
</mapper>
```

在上述 SQL 映射文件中，<insert>元素用于映射插入语句；<delete>元素用于映射删除语句；<update>元素用于映射更新语句；<select>元素用于映射查询语句。

在这些元素中，id 属性设置在命名空间中唯一的标识符，用于引用这条语句。parameterType 属性指定传入这条语句的参数类的完全限定名或别名。resultType 属性指定从这条语句中返回的期望类型的类的完全限定名或别名，若查询结果是集合，则 resultType 的值应该是集合所包含的元素类型，而不能是集合本身。

(7) 创建属性文件 db.properties。

在 src 目录下创建属性文件 db.properties，保存数据库的连接信息，内容如下：

```
jdbc.driver=com.mysql.jdbc.Driver
jdbc.url=jdbc:mysql://localhost:3306/restrant
jdbc.username=root
jdbc.password=123456
```

(8) 创建 MyBatis 的 XML 配置文件。

在 src 目录下创建 MyBatis 的 XML 配置文件 mybatis-config.xml，代码如下：

```
<?xml version="1.0" encoding="UTF-8" ?>
```

```xml
<!DOCTYPE configuration
PUBLIC "-//mybatis.org//DTD Config 3.0//EN"
"http://mybatis.org/dtd/mybatis-3-config.dtd">
<configuration>
    <!-- 加载属性文件 -->
    <properties resource="db.properties"></properties>
    <!-- 给包中的类注册别名,注册后可以直接使用类名,而不用使用全限定的类名(就是不用包含包
名)。 -->
    <typeAliases>
        <package name="com.mybatis.pojo" />
    </typeAliases>
    <environments default="development">
        <environment id="development">
            <transactionManager type="JDBC" />
            <dataSource type="POOLED">
                <property name="driver" value="${jdbc.driver}" />
                <property name="url" value="${jdbc.url}" />
                <property name="username" value="${jdbc.username}" />
                <property name="password" value="${jdbc.password}" />
            </dataSource>
        </environment>
    </environments>
    <!-- 引用SQL映射的XML文件 -->
    <mappers>
        <mapper resource="com/mybatis/pojo/usersMapper.xml" />
    </mappers>
</configuration>
```

在上述配置文件中,包含了对 MyBatis 系统的核心设置,包含获取数据库连接实例的数据源等信息。<environment>元素体中包含对事务管理和连接池的环境配置。<mappers>元素是包含所有 mapper(映射器)的列表,这些 mapper 的 XML 文件包含 SQL 代码和映射定义信息。

(9) 创建测试类。

创建 JUnit 测试类 MybatisTest.java,存放在 com.mybatis.test 包中。代码如下:

```java
package com.mybatis.test;
import org.apache.ibatis.io.Resources;
import org.apache.ibatis.session.SqlSession;
import org.junit.Test;
……
public class MybatisTest {
    private SqlSessionFactory sqlSessionFactory;
    private SqlSession sqlSession;
    @Before
    public void init() {
        // mybatis 配置文件
        String resource = "mybatis-config.xml";
        // 得到配置文件流
        InputStream inputStream;
        try {
            inputStream = Resources.getResourceAsStream(resource);
            // 创建会话工厂,传入mybatis的配置文件信息
            sqlSessionFactory = new SqlSessionFactoryBuilder()
                    .build(inputStream);
```

```java
            // 通过工厂得到 SqlSession
            sqlSession = sqlSessionFactory.openSession();
        } catch (IOException e) {
            // TODO Auto-generated catch block
            e.printStackTrace();
        }
    }
    // 添加用户
    @Test
    public void testAddUser() {
        Users u = new Users("mybatis1", "123456");
        sqlSession.insert("addUser", u);
    }
    // 根据 id 查询用户
    @Test
    public void testGetUserById() {
        Users u = sqlSession.selectOne("getUserById", 4);
        System.out.println(u);
    }
    // 查询所有用户
    @Test
    public void testGetAllUsers() {
        List<Users> uList = sqlSession.selectList("getAllUsers");
        System.out.println(uList.size());
    }
    // 修改用户
    @Test
    public void testUpdateUser() {
        // 加载编号 id=4 的用户
        Users u = sqlSession.selectOne("getUserById", 4);
        // 修改用户密码
        u.setLoginPwd("123123");
        // 执行更新操作
        int update = sqlSession.update("updateUser", u);
        System.out.println(u);
    }
    // 删除用户
    @Test
    public void testDeleteUser() {
        int delete = sqlSession.delete("deleteUser", 4);
        System.out.println(delete);
    }
    @After
    public void destroy() {
        // 提交事务
        sqlSession.commit();
        // 关闭 session
        sqlSession.close();
    }
}
```

在测试类 MybatisTest 中，首先添加 init()方法，并在方法前面添加@Before 注解。JUnit4 使用 Java5 中的@Before 注解，用于进行初始化。init()方法对于每一个测试方法都要执行一次，方法中代码根据 myBatis 的配置文件信息配置初始化 SqlSessionFactory，获取

SqlSession。

然后添加 destroy()方法，并在方法前面添加@After 注解。JUnit4 使用 Java5 中的@After 注解，用于释放资源。destroy()方法对于每一个测试方法都要执行一次。方法中代码会执行事务提交，释放 SqlSession 资源。

在测试 MybatisTest 中，每一个用@Test 注解修饰的方法称为测试方法，它们的调用顺序为：@Before→@Test→@After。

在测试方法 testAddUser、testGetUserById、testGetAllUsers、testUpdateUser 和 testDeleteUser 中，通过 SqlSession 对象的 insert、selectOne、selectList、update、delete 等方法执行定义在 SQL 映射的 XML 文件中的 INSERT、SELECT、UPDATE 和 DELETE 语句。其中 insert、selectOne、update 和 delete 方法都使用语句的 ID 属性和参数对象，而 selectList 方法不需要参数对象。selectOne 和 selectList 的不同之处在于 selectOne 必须返回一个对象，如果多于一个，或者没有返回(或返回了 null)，那么就会抛出异常。

(10) 测试结果。

为了能在控制台输出 SQL 语句，可以在项目的 src 目录下创建文件 log4j.xml，文件内容查阅源代码。

① 执行测试类 MybatisTest 中的 testAddUser()方法，数据表 users 中成功插入一条新用户记录，同时控制台会打印一条 insert 语句：

```
insert into users(loginName,loginPwd) values(?,?)
```

② 执行测试类 MybatisTest 中的 testGetUserById()方法，控制台输出结果如下：

```
Preparing: select * from users where id = ?
……
Users [id=4, loginName=mybatis1, loginPwd=123456]
```

③ 执行测试类 MybatisTest 中的 testGetAllUsers()方法，控制台输出结果如下：

```
Preparing: select * from users
……
4
```

④ 执行测试类 MybatisTest 中的 testUpdateUser()方法，数据表 users 中编号 id=4 的用户的密码被修改为 123123，控制台会打印一条 update 语句：

```
Preparing: update users set loginName=?, loginPwd=? where id=?
……
Users [id=4, loginName=mybatis1, loginPwd=123123]
```

⑤ 执行测试类 MybatisTest 中的 testDeleteUser()方法，数据表 users 中编号 id=4 的用户记录被成功删除，同时控制台会打印一条 delete 语句：

```
Preparing: delete from users where id=?
……
1
```

该示例中实体类 Users 中的属性名与数据表 users 中的字段名相同，如果属性名与数据表的字段名不相同，那么就需要修改 SQL 映射的 XML 文件 usersMapper.xml。

将项目 mybatis_1 复制并命名为 mybatis_2，再导入 MyEclipse 开发环境中。

修改实体类 Users，将其属性重新命名，使得属性名与数据表 users 的字段名不同。代码如下：

```java
package com.mybatis.pojo;
public class Users {
    private int uid;
    private String uname;
    private String upass;
// 此处省略属性的getter和setter方法
    // 此处省略有参构造方法和无参构造方法
// 此处省略toString()方法
}
```

修改 SQL 映射的 XML 文件 usersMapper.xml，代码如下：

```xml
<?xml version="1.0" encoding="UTF-8"?>
<!DOCTYPE mapper PUBLIC "-//mybatis.org//DTD Mapper 3.0//EN"
"http://mybatis.org/dtd/mybatis-3-mapper.dtd">
<mapper namespace="com.mybatis.pojo.usersMapper">
    <!-- 数据表users的CRUD操作 -->
    <insert id="addUser" parameterType="Users">
        insert into
        users(loginName,loginPwd) values(#{uname},#{upass})
    </insert>
    <delete id="deleteUser" parameterType="int">
        delete from users where
        id=#{uid}
    </delete>
    <update id="updateUser" parameterType="Users">
        update users set
        loginName=#{uname}, loginPwd=#{upass} where id=#{uid}
    </update>
    <select id="getUserById" parameterType="int" resultMap="getUsersMap">
        select *
        from users where id = #{uid}
    </select>
    <select id="getAllUsers" resultMap="getUsersMap">
        select * from users
    </select>
    <resultMap type="Users" id="getUsersMap">
        <id property="uid" column="id" />
        <result property="uname" column="loginName" />
        <result property="upass" column="loginPwd" />
    </resultMap>
</mapper>
```

在<insert>、<delete>、<update>和<select>元素中，SQL 语句参数使用的占位符#{uname}、#{upass}和#{uid}需要与实体类 Users 中的属性一致。

在<select>元素中，resultMap 属性指定了 id 为 getUsersMap 的<resultMap>元素，用来完成查询结果的映射。在<resultMap>元素中，type 属性指定映射结果的类型；<id>子元素和<result>用来映射数据表的列到实体对象的属性，不同的是 id 用来映射标识属性。

修改测试类 MybatisTest 中的错误后，依次执行各个测试方法，结果与前面相同。

16.3 MyBatis 的关联映射

MyBatis 不仅支持单张表的增删改查，还支持多张表之间的关联，包括一对一、一对多和多对多。MyBatis 关联映射可以极大地简化持久层数据的访问。

16.3.1 一对一关联映射

在数据库 restrant 中，新建数据表 admin_detail，用于存储管理员的详细信息，如图 16-2 所示，并在表中任意添加一条测试数据。在数据表 admin 中添加一个整型字段 Adid，并设置其与数据表 admin_detail 的字段 Id 的关联，如图 16-3 所示，并将数据表 admin 中第一条记录的 Adid 字段设置为 1，以引用数据表 admin_detail 中新添加的记录。下面以数据表 admin 和 admin_detail 一对一关联关系为例，介绍如何使用 MyBatis 配置一对一关联映射。

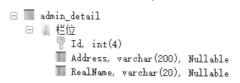

图 16-2 数据表 admin_detail 字段

图 16-3 设置 admin 和 admin_detail 的一对一关联

(1) 将项目 mybatis_1 复制并命名为 mybatis_3，再导入 MyEclipse 开发环境中。
(2) 创建实体类。
在 com.mybatis.pojo 包中创建实体类 AdminDetail.java 和 Admin.java，实体类 AdminDetail 如下：

```
package com.mybatis.pojo;
public class AdminDetail {
    private int id;
    private String address;
    private String realName;
// 此处省略属性的 getter 方法和 setter 方法
    // 此处省略有参构造方法和无参构造方法
```

```
// 此处省略 toString()方法
}
```

实体类 Admin 如下：

```
package com.mybatis.pojo;
public class Admin {
    private int id;
    private String loginName;
    private String loginPwd;
    // 关联属性
    private AdminDetail ad;
// 此处省略属性的 getter 方法和 setter 方法
    // 此处省略有参构造方法和无参构造方法
// 此处省略 toString()方法
}
```

(3) 创建 SQL 映射的 XML 文件。

在项目中创建 com.mybatis.mapper 包，在包中创建 SQL 映射的 XML 文件 adminDetailMapper.xml，添加一个 id="selectAdminDetailById"的<select>元素，根据 id 从数据表 admin_detail 中查询 AdminDetail，返回 AdminDetail 对象。代码如下：

```xml
<?xml version="1.0" encoding="UTF-8" ?>
<!DOCTYPE mapper PUBLIC "-//mybatis.org//DTD Mapper 3.0//EN"
"http://mybatis.org/dtd/mybatis-3-mapper.dtd">
<mapper namespace="com.mybatis.mapper.AdminDetailMapper">
    <!-- 根据 id 查询 AdminDetail，返回 AdminDetail 对象 -->
    <select id="selectAdminDetailById" parameterType="int" resultType="AdminDetail">
        select * from admin_detail where Id=#{id}
    </select>
</mapper>
```

创建 SQL 映射的 XML 文件 adminMapper.xml，代码如下：

```xml
<?xml version="1.0" encoding="UTF-8" ?>
<!DOCTYPE mapper PUBLIC "-//mybatis.org//DTD Mapper 3.0//EN"
"http://mybatis.org/dtd/mybatis-3-mapper.dtd">
<mapper namespace="com.mybatis.mapper.AdminMapper">
    <!-- 根据 id 查询 Admin，返回 resultMap -->
    <select id="getAdminById" parameterType="int" resultMap="getAdminMap">
        select * from admin where Id = #{id}
    </select>
    <!-- 查询语句查询结果映射 -->
    <resultMap type="Admin" id="getAdminMap">
        <id property="id" column="id" />
        <result property="loginName" column="LoginName" />
        <result property="loginPwd" column="LoginPwd" />
        <!-- 一对一关联映射 -->
        <association property="ad" column="Adid"
select="com.mybatis.mapper.AdminDetailMapper.selectAdminDetailById"
javaType="AdminDetail" />
    </resultMap>
</mapper>
```

在 adminMapper.xml 中，添加了一个 id="getAdminById"的<select>元素，由于 Admin 类除了简单的属性 id、loginName 和 loginPwd 之外，还有一个关联的对象 ad，所以返回的是一个 id 为 getAdminMap 的 resultMap。

getAdminMap 中使用了<association>元素来映射一对一的关联关系，select 属性表示会找到 com.mybatis.mapper.AdminDetailMapper 命名空间下 id="selectAdminDetailById"的元素，执行该元素中的 SQL 语句，参数来自 column 属性的 Adid 值，查询出的结果被封装到 property 属性表示的 ad 对象中。

在 com.mybatis.mapper 包中创建接口 AdminMapper.java，声明如下方法：

```
package com.mybatis.mapper;
import com.mybatis.pojo.Admin;
public interface AdminMapper {
    Admin getAdminById(int id);
}
```

(4) 注册 SQL 映射的 XML 文件。

在 MyBatis 的配置文件 mybatis-config.xml 中注册 adminDetailMapper.xml 和 adminMapper.xml，代码如下：

```
<mappers>
    ……
    <mapper resource="com/mybatis/mapper/adminMapper.xml" />
    <mapper resource="com/mybatis/mapper/adminDetailMapper.xml" /></mappers>
```

(5) 测试一对一关联映射。

在测试类 MybatisTest 中添加测试方法 testGetAdminById()，使用@Test 注解修饰，从数据表 admin 获取记录的同时，获取关联的数据表 admin_detail 中的记录。代码如下：

```
@Test
public void testGetAdminById() {
    // 获得 AdminMapper 接口的代理对象
    AdminMapper am = sqlSession.getMapper(AdminMapper.class);
    // 直接调用接口的方法，查询 id=1 的 Admin 对象
    Admin admin = am.getAdminById(1);
    System.out.println(admin);
}
```

执行 testGetAdminById()方法，控制台输出结果如下：

```
Preparing: select * from admin where Id = ?
……
Admin [id=1, loginName=admin, loginPwd=123456, ad=AdminDetail [id=1, address=江苏南京, realName=管理员]]
```

可以看到，查询 Admin 对象时关联的 AdminDetail 对象也查询出来了。

16.3.2 一对多关联映射

在数据库 restrant 中，数据表 mealseries 和 meal 之间存在一对多关联关系。数据表 meal

中的 MealSeriesId 字段引用数据表 mealseries 的 SeriesId 字段。下面以数据表 mealseries 和 meal 之间一对多关联关系为例，介绍如何使用 Mybatis 配置一对多关联映射。

（1）将项目 mybatis-1 复制并命名为 mybatis_4，再导入 MyEclipse 开发环境中。

（2）创建实体类。

在 com.mybatis.pojo 包中创建实体类 Mealseries.java 和 Meal.java，实体类 Mealseries 代码如下：

```java
package com.mybatis.pojo;
import java.util.List;
public class Mealseries {
    private int seriesId;
    private String seriesName;
    // 关联的集合属性
    private List<Meal> meals;
// 此处省略属性的getter方法和setter方法
    // 此处省略有参构造方法和无参构造方法
// 此处省略toString()方法
}
```

实体类 Meal 代码如下：

```java
package com.mybatis.pojo;
public class Meal {
    private int mealId;
    private String mealName;
    private Double mealPrice;
    // 关联属性
    private Mealseries mealseries;
// 此处省略属性的getter方法和setter方法
    // 此处省略有参构造方法和无参构造方法
// 此处省略toString()方法
}
```

（3）创建 SQL 映射的 XML 文件。

在项目中创建 com.mybatis.mapper 包，在包中创建 SQL 映射的 XML 文件 mealseriesMapper.xml 和 mealMapper.xml。mealseriesMapper.xml 代码如下：

```xml
<?xml version="1.0" encoding="UTF-8" ?>
<!DOCTYPE mapper PUBLIC "-//mybatis.org//DTD Mapper 3.0//EN"
"http://mybatis.org/dtd/mybatis-3-mapper.dtd">
<mapper namespace="com.mybatis.mapper.MealseriesMapper">
    <!-- 根据菜系id查询菜系信息，返回resultMap -->
    <select id="selectMealseriesById" parameterType="int"
        resultMap="getMealseriesMap">
        select * from mealseries where SeriesId = #{id}
    </select>
    <!-- 查询语句查询结果映射 -->
    <resultMap type="Mealseries" id="getMealseriesMap">
        <id property="seriesId" column="SeriesId" />
        <result property="seriesName" column="SeriesName" />
        <!-- 一对多关联映射 -->
        <collection property="meals" javaType="ArrayList" column="seriesId"
```

```xml
        ofType="Meal" select=
"com.mybatis.mapper.MealMapper.selectMealBySeriesId">
            <id property="mealId" column="MealId" />
            <result property="mealName" column="MealName" />
            <result property="mealPrice" column="MealPrice" />
        </collection>
    </resultMap>
</mapper>
```

在 mealseriesMapper.xml 文件中，添加了一个 id="selectMealseriesById"的<select>元素，根据菜系编号查询菜系信息。由于 Mealseries 类除了简单的属性 seriesId 和 seriesName 之外，还有一个关联的对象属性 meals，所以返回的是一个<resultMap>元素。由于 meals 是一个 List 集合，所以 id="getMealseriesMap"的<resultMap>元素中使用了<collection>元素来映射一对多关联关系。<collection>元素的 select 属性表示会找到 com.mybatis.mapper.MealMapper 命名空间下 id="selectMealBySeriesId"的元素，执行该元素中的 SQL 语句，参数来自 column 属性的 seriesId 值，查询出的结果被封装到 property 属性表示的 meals 对象中。

mealMapper.xml 代码如下：

```xml
<?xml version="1.0" encoding="UTF-8" ?>
<!DOCTYPE mapper PUBLIC "-//mybatis.org//DTD Mapper 3.0//EN"
"http://mybatis.org/dtd/mybatis-3-mapper.dtd">
<mapper namespace="com.mybatis.mapper.MealMapper">
    <!-- 根据餐品 id 查询餐品信息，返回 resultMap -->
    <select id="selectMealById" parameterType="int" resultMap="getMealMap">
        select * from meal m, mealseries ms where m.MealSeriesId = ms.SeriesId and m.MealId = #{id}
    </select>
    <!-- 根据菜系查询餐品信息，返回 resultMap -->
    <select id="selectMealBySeriesId" parameterType="int" resultMap="getMealMap">
        select * from meal where MealSeriesId = #{id}
    </select>
    <!-- 查询语句查询结果映射 -->
    <resultMap type="Meal" id="getMealMap">
        <id property="mealId" column="MealId" />
        <result property="mealName" column="MealName" />
        <result property="mealPrice" column="MealPrice" />
        <!-- 多对一关联映射 -->
        <association property="mealseries" javaType="Mealseries">
            <id property="seriesId" column="SeriesId" />
            <result property="seriesName" column="SeriesName" />
        </association>
    </resultMap>
</mapper>
```

在 mealMapper.xml 文件中，添加了一个 id="selectMealById"的<select>元素，根据餐品编号查询餐品信息。由于 Meal 类除了简单的属性 mealId、mealName 和 mealPrice 之外，还有一个关联的对象 mealseries，所以返回的是一个 id="getMealMap"的<resultMap>元素。在 getMealMap 中使用<association>元素映射多对一关联关系，由于 id="selectMealById"的<select>元素的 SQL 语句是两个表的连接，关联 mealseries 表的同时查询了菜系信息，因此

<association>元素只简单地封装数据。

在 mealMapper.xml 文件中，还添加了一个 id="selectMealBySeriesId"的<select>元素，根据菜系编号查询所有餐品信息。该查询用于 mealseriesMapper.xml 中一对多关联映射时的关联查询。

接下来，在 com.mybatis.mapper 包中创建接口 MealseriesMapper.java，声明如下方法：

```java
package com.mybatis.mapper;
import com.mybatis.pojo.Mealseries;
public interface MealseriesMapper {
    // 根据id查询菜系信息
    Mealseries selectMealseriesById(int id);
}
```

创建接口 MealMapper.java，声明如下方法：

```java
package com.mybatis.mapper;
import com.mybatis.pojo.Meal;
public interface MealMapper {
    // 根据id查询餐品信息
    Meal selectMealById(int id);
}
```

(4) 注册 SQL 映射的 XML 文件。

在 XML 配置文件 mybatis-config.xml 中注册 mealMapper.xml 和 mealseriesMapper.xml，代码如下：

```xml
<mappers>
    ……
<mapper resource="com/mybatis/mapper/mealMapper.xml" />
    <mapper resource="com/mybatis/mapper/mealseriesMapper.xml" />
</mappers>
```

(5) 测试一对多关联映射。

在测试类 MybatisTest 中添加测试方法 testSelectMealseriesById()，使用@Test 注解修饰，从数据表 mealseries 获取记录的同时，获取与其关联的 meal 表中的记录。代码如下：

```java
@Test
public void testSelectMealseriesById() {
    // 获得MealseriesMapper接口的代理对象
    MealseriesMapper mealseriesMapper = sqlSession
            .getMapper(MealseriesMapper.class);
    // 直接调用接口的方法，查询id=1的Mealseries对象
    Mealseries mealseries = mealseriesMapper.selectMealseriesById(1);
    System.out.println(mealseries);
    // 查看Mealseries对象关联的餐品信息
    List<Meal> meals = mealseries.getMeals();
    for (Meal meal : meals) {
        System.out.println(meal);
    }
}
```

执行 testSelectMealseriesById()方法，控制台输出结果如下：

```
Preparing: select * from mealseries where SeriesId = ?
Parameters: 1(Integer)
Preparing: select * from meal where MealSeriesId = ?
Parameters: 1(Integer)
Total: 9
Total: 1
Mealseries [seriesId=1, seriesName=鲁菜]
Meal [mealId=1, mealName=雪梨肉肘棒, mealPrice=10.0]
Meal [mealId=2, mealName=素锅烤鸭肉, mealPrice=20.0]
……
```

可以看到，MyBatis 执行了查询菜系的 SQL 语句之后，又执行了根据 MealSeriesId 查询餐品信息的 SQL 语句。

(6) 测试多对一关联映射。

在测试类 MybatisTest 中添加测试方法 testSelectMealById()，使用@Test 注解修饰，从数据表 meal 获取记录的同时，获取与其关联的 mealseries 表中的记录。代码如下：

```
@Test
public void testSelectMealById() {
    // 获得 MealMapper 接口的代理对象
    MealMapper mealMapper = sqlSession.getMapper(MealMapper.class);
    // 直接调用接口的方法，查询 id=1 的 Meal 对象
    Meal meal = mealMapper.selectMealById(1);
    System.out.println(meal);
    // 查看 Meal 对象关联的菜系信息
    System.out.println(meal.getMealseries());
}
```

执行 testSelectMealById()方法，控制台输出结果如下：

```
Preparing: select * from meal m, mealseries ms where m.MealSeriesId = ms.SeriesId and m.MealId = ?
Parameters: 1(Integer)
Total: 1
Meal [mealId=1, mealName=雪梨肉肘棒, mealPrice=10.0]
Mealseries [seriesId=1, seriesName=鲁菜]
```

可以看到，MyBatis 执行了一个两张表的关联查询语句，并将查询到的菜系信息封装到了餐品对象的关联属性中。

16.3.3　多对多关联映射

在数据库 restrant 中，数据表 admin 和 functions 之间存在多对多关联关系，它们之间通过中间表 powers 来关联，这个中间表分别与 admin 和 functions 构成多对一关联。中间表 powers 以 aid 和 fid 作为联合主键，其中，aid 字段作为外键参照 admin 表的 id 字段，fid 字段作为外键参照 functions 表的 id 字段。使用 MyBatis 实现数据表 admin 和 functions 多对多关联映射的步骤如下。

(1) 将项目 mybatis_1 复制并命名为 mybatis_5，再导入 MyEclipse 开发环境中。

(2) 创建实体类。

在项目 mybatis_5 的 com.mybatis.pojo 包中创建实体类 Admin.java 和 Functions.java，实体

类 Admin 代码如下：

```java
package com.mybatis.pojo;
import java.util.HashSet;
import java.util.Set;
public class Admin {
    private int id;
    private String loginName;
    private Set fs = new HashSet();// 关联的属性
// 此处省略属性的 getter 方法和 setter 方法
    // 此处省略有参构造方法和无参构造方法
// 此处省略 toString()方法
}
```

实体类 Functions 的代码如下：

```java
package com.mybatis.pojo;
import java.util.HashSet;
import java.util.Set;
public class Functions {
    private int id;
    private String name;
    private Set ais = new HashSet(); // 关联的属性
// 此处省略属性的 getter 方法和 setter 方法
    // 此处省略有参构造方法和无参构造方法
// 此处省略 toString()方法
}
```

(3) 创建 SQL 映射的 XML 文件。

在项目中创建 com.mybatis.mapper 包，在包中创建 SQL 映射的 XML 文件 adminMapper.xml 和 functionsMapper.xml。adminMapper.xml 的代码如下：

```xml
<?xml version="1.0" encoding="UTF-8" ?>
<!DOCTYPE mapper PUBLIC "-//mybatis.org//DTD Mapper 3.0//EN"
"http://mybatis.org/dtd/mybatis-3-mapper.dtd">
<mapper namespace="com.mybatis.mapper.AdminMapper">
    <!-- 根据 id 获取管理员 -->
    <select id="selectAdminById" parameterType="int"
resultMap="getAdminMap">
        select * from admin where id = #{id}
    </select>
    <resultMap type="Admin" id="getAdminMap">
        <id property="id" column="id" />
        <result property="loginName" column="LoginName" />
        <!-- 多对多关联映射 -->
        <collection property="fs" ofType="Functions" column="id"
            select=
"com.mybatis.mapper.FunctionsMapper.selectFunctionsByAdminId">
            <id property="id" column="id" />
            <result property="name" column="name" />
        </collection>
    </resultMap>
</mapper>
```

在 adminMapper.xml 文件中，添加了一个 id="selectAdminById"的<select>元素，根据管

理员 id 获取管理员信息。由于 Admin 类除了简单的属性 id 和 loginName 之外，还有一个关联的集合属性 fs，所以返回的是一个<resultMap>元素。由于 fs 是一个 List 集合，所以 id="getAdminMap"的<resultMap>元素中使用了<collection>元素来映射多对多关联关系。<collection>元素的 select 属性表示会找到 com.mybatis.mapper.FunctionsMapper 命名空间下 id="selectFunctionsByAdminId"的元素，执行该元素中的 SQL 语句，参数来自 column 属性的 id 值，查询出的结果被封装到 property 属性表示的 fs 对象中。

functionsMapper.xml 代码如下：

```xml
<?xml version="1.0" encoding="UTF-8" ?>
<!DOCTYPE mapper PUBLIC "-//mybatis.org//DTD Mapper 3.0//EN"
"http://mybatis.org/dtd/mybatis-3-mapper.dtd">
<mapper namespace="com.mybatis.mapper.FunctionsMapper">
    <!-- 根据管理员 id 获取其功能列表 -->
    <select id="selectFunctionsByAdminId" parameterType="int"
      resultType="Functions">
      select * from functions f where id in (
        select fid from powers where aid = #{id}
      )
    </select>
</mapper>
```

在 functionsMapper.xml 文件中，添加了一个 id="selectFunctionsByAdminId"的<select>元素，根据管理员 id 获取其功能列表。由于管理员和系统功能是多对多的关系，数据库中使用了一个中间表 powers 维护多对多关联关系，此处使用了一个子查询，首先根据管理员 id 到中间表中查询出所有的功能权限 id，然后根据功能权限的 id 到 functions 表中查询出所有的功能权限信息，并将这些信息封装到 Functions 对象中。

接下来，在 com.mybatis.mapper 包中创建接口 AdminMapper.java，声明如下方法：

```java
Admin selectAdminById(int id);
```

创建接口 FunctionsMapper.java，声明如下方法：

```java
Functions selectFunctionsByAdminId(int id);
```

(4) 注册 SQL 映射的 XML 文件。

在配置文件 mybatis-config.xml 中注册 adminMapper.xml 和 functionsMapper.xml，代码如下：

```xml
<mappers>
    ……
    <mapper resource="com/mybatis/mapper/adminMapper.xml" />
    <mapper resource="com/mybatis/mapper/functionsMapper.xml" /></mappers>
```

(5) 测试多对多关联映射。

在测试类 MybatisTest 中添加测试方法 testGetAdminById()，并使用@Test 注解修饰，查看管理员及其功能权限。代码如下：

```java
@Test
public void testGetAdminById() {
    // 获得 AdminMapper 接口的代理对象
```

```
        AdminMapper am = sqlSession.getMapper(AdminMapper.class);
        // 直接调用接口的方法，查询 id=1 的 Admin 对象
        Admin admin = am.selectAdminById(1);
        System.out.println(admin);
        // 查看 Admin 对象关联的功能信息
        Object[] fs = admin.getFs().toArray();
        for (Object f : fs) {
           System.out.println((Functions) f);
        }
}
```

执行 testGetAdminById()方法，控制台输出结果如下：

```
Preparing: select * from admin where id = ?
Parameters: 1(Integer)
Preparing: select * from functions f where id in ( select fid from powers
where aid = ? )
Parameters: 1(Integer)
Total: 12
Total: 1
Admin [id=1, loginName=admin]
Functions [id=6, name=查询订单]
Functions [id=11, name=管理员列表]
Functions [id=1, name=订餐系统管理后台]
……
```

可以看到，MyBatis 执行了 functionsMapper.xml 中定义的子查询，查询出了管理员所关联的所有功能权限信息。

16.4 动态 SQL

在 SQL 语句 where 条件子句中，需要进行一些判断。例如，按名称模糊查询，如果传入的参数是空的，此时查询出的结果很可能是空的，当参数为空时，希望查出全部的信息。此时可以使用动态 SQL，增加一个判断，当参数不符合时，就不去判断此查询条件。

MyBatis 的动态 SQL 是基于 OGNL 表达式的，MyBatis 中用于实现动态 SQL 的元素主要包括 if、choose(when, otherwise)、trim、where、set 、foreach 等。

16.4.1 if 元素

if 元素是简单的条件判断，可用来实现某些简单的条件选择。如果想从数据库 restrant 的数据表 users 中按用户名模糊查询，查询用户名包含 z 的用户列表，使用 if 元素的实现过程如下。

(1) 将项目 mybatis_1 复制并命名为 mybatis_6，再导入 MyEclipse 开发环境中。

(2) 在映射文件 usersMapper.xml 中，添加 id="getUsersByLoginName"的<select>元素，代码如下：

```
<select id="getUsersByLoginName" parameterType="Users"
   resultType="Users">
```

```xml
select * from users u
<if test="loginName != null and loginName != ''">
    where u.loginName LIKE CONCAT(CONCAT('%',#{loginName}),'%')
</if>
</select>
```

如果 loginName 是 null 或空字符串，此语句很可能报错或查询结果为空。通过使用<if>元素的动态 SQL 语句先进行判断，如果值为 null 或等于空字符串，就不进行此条件的判断。

(3) 在测试类 MybatisTest 中添加测试方法 testGetUsersByLoginName()，并用@Test 注解修改，代码如下：

```java
@Test
public void testGetUsersByLoginName() {
    Users cond = new Users();
    cond.setLoginName("z");
    List<Users> users = sqlSession.selectList("getUsersByLoginName", cond);
    for (Users u : users) {
        System.out.println(u);
    }
}
```

(4) 执行 testGetUsersByLoginName()方法，控制台输出结果如下：

```
Preparing: select * from users u where u.loginName LIKE
CONCAT(CONCAT('%',?),'%')
Parameters: z(String)
Total: 2
Users [id=1, loginName=zhangsan, loginPwd=123456]
Users [id=2, loginName=zs, loginPwd=zs]
```

如果在 testGetUsersByLoginName()方法中不设置 loginName 的值，生成的 SQL 语句中就不会包含 where 子句，此时查询结果为所有的用户信息，具体如下：

```
Preparing: select * from users u
Parameters:
Total: 3
Users [id=1, loginName=zhangsan, loginPwd=123456]
Users [id=2, loginName=zs, loginPwd=zs]
Users [id=3, loginName=lisi, loginPwd=lisi]
```

16.4.2　if-where 元素

当 if 元素较多时，SQL 语句会组合成 where and 之类的关键字多余的错误 SQL。此时，可以使用 where 元素来解决。where 元素会判断如果它包含的标签中有返回值的话，它就插入一个 where。如果标签返回的内容是以 and 或 or 开头的，就会将其剔除。如果想从数据表 users 中按用户名模糊查询，同时查询指定状态的用户列表。使用 if-where 元素实现的过程如下。

(1) 在实体类 Users 中添加一个整型属性 status，并为其添加 get 和 set 方法。

(2) 在映射文件 usersMapper.xml 中，添加 id="getUsersByLoginNameAndStatus"的<select>元素，代码如下：

```xml
<select id="getUsersByLoginNameAndStatus" parameterType="Users"
    resultType="Users">
    select * from users u
    <where>
        <if test="loginName!=null and loginName!=''">
            u.loginName LIKE CONCAT(CONCAT('%', #{loginName}),'%')
        </if>
        <if test="status>-1">
            and u.status = #{status}
        </if>
    </where>
</select>
```

（3）在测试类 MybatisTest 中添加测试方法 testGetUsersByLoginNameAndStatus()，并用 @Test 注解修改，代码如下：

```java
@Test
public void testGetUsersByLoginNameAndStatus() {
    Users cond = new Users();
    cond.setLoginName("z");
    cond.setStatus(1);
    List<Users> users = sqlSession.selectList(
            "getUsersByLoginNameAndStatus", cond);
    for (Users u : users) {
        System.out.println(u);
    }
}
```

（4）执行该测试方法，控制台输出结果如下：

```
Preparing: select * from users u WHERE u.loginName LIKE
CONCAT(CONCAT('%', ?),'%') and u.status = ?
Users [id=1, loginName=zhangsan, loginPwd=123456]
Users [id=2, loginName=zs, loginPwd=zs]
```

如果将 cond 对象中 status 设置为-1，则会执行 SQL 语句：

```
Preparing: select * from users u WHERE u.loginName LIKE
CONCAT(CONCAT('%', ?),'%')
```

可以看出，当属性 status 的值不满足测试条件时，where 元素会将 status 条件前多余的 and 关键字剔除。

16.4.3 set-if 元素

当在 update 语句中使用 if 元素时，如果前面的 if 没有执行，生成的 SQL 语句中会包含多余的逗号，从而造成错误。set 元素可给 SQL 语句动态产生 set 关键字，还可将添加到条件结尾处的多余的逗号剔除。如果想更新数据表 users 中某个用户的用户名和密码，使用 set-if 元素实现的过程如下。

（1）在映射文件 usersMapper.xml 中，添加 id="updateUser2"的<update>元素，代码如下：

```xml
<update id="updateUser2" parameterType="Users">
    update users u
```

```xml
    <set>
        <if test="loginName!=null and loginName!=''">
            u.loginName=#{loginName},
        </if>
        <if test="loginPwd!=null and loginPwd!=''">
            u.loginPwd=#{loginPwd}
        </if>
    </set>
    where u.id=#{id}
</update>
```

(2) 在测试类 MybatisTest 中添加测试方法 testUpdateUser2()，并用@Test 注解修改，代码如下：

```java
@Test
public void testUpdateUser2() {
    Users cond = new Users();
    cond.setId(3);
    cond.setLoginName("miaoyong");
    cond.setLoginPwd("123123");
    sqlSession.update("updateUser2", cond);
}
```

(3) 执行该测试方法，控制台输出结果如下：

```
Preparing: update users u SET u.loginName=?, u.loginPwd=? where u.id=?
Parameters: miaoyong(String), 123123(String), 3(Integer)
```

如果在测试方法中没有给 loginPwd 属性指定值，set 元素会将自己包含内容结尾处多余的逗号剔除，控制台输出结果如下：

```
Preparing: update users u SET u.loginName=? where u.id=?
```

16.4.4 trim 元素

trim 元素通过属性 prefix 在自己包含的内容前加上前缀，通过 suffix 属性在内容之后加上后缀，通过 prefixOverrides 属性把包含内容的首部某些内容覆盖(忽略)，通过 suffixOverrides 属性把尾部的某些内容覆盖。因此，trim 元素可用来替代 where 元素和 set 元素实现同样的功能。

使用 trim 元素替代 where 元素，从数据表 users 中按用户名模糊查询，同时查询指定状态的用户列表，实现过程如下。

(1) 在 usersMapper.xml 中，添加 id="getUsersByLoginNameAndStatus_Trim"的<select>元素，代码如下：

```xml
<select id="getUsersByLoginNameAndStatus_Trim" parameterType="Users"
    resultType="Users">
    select * from users u
    <trim prefix="where" prefixOverrides="and|or">
        <if test="loginName!=null and loginName!=''">
            u.loginName LIKE CONCAT(CONCAT('%', #{loginName}),'%')
        </if>
```

```
        <if test="status>-1">
            and u.status = #{status}
        </if>
    </trim>
</select>
```

在 trim 元素中,prefix 属性值为 where,会在自己包含的内容前加上 where 关键字。prefixOverrides 设置为 and|or,会将自己包含内容的首部 and 或 or 关键字剔除,避免出现多余的 and 或 or。

(2) 在测试类 MybatisTest 中添加 testGetUsersByLoginNameAndStatus_Trim()方法,并用 @Test 注解修改,代码如下:

```
@Test
public void testGetUsersByLoginNameAndStatus_Trim() {
    Users cond = new Users();
    cond.setLoginName("z");
    cond.setStatus(1);
    List<Users> users = sqlSession.selectList(
            "getUsersByLoginNameAndStatus_Trim", cond);
    for (Users u : users) {
        System.out.println(u);
    }
}
```

(3) 执行该测试方法,控制台输出结果如下:

```
Preparing: select * from users u where u.loginName LIKE
CONCAT(CONCAT('%', ?),'%') and u.status = ?
Parameters: z(String), 1(Integer)
Users [id=1, loginName=zhangsan, loginPwd=123456]
Users [id=2, loginName=zs, loginPwd=zs]
```

如果在该方法中没有设置 loginName 的值,trim 元素会将 status 条件前多余的 and 关键字剔除,控制台输出的 SQL 语句如下:

```
Preparing: select * from users u where u.status = ?
```

使用 trim 元素替代 set 元素,更新数据表 users 中某个用户的用户名和密码,实现过程如下。

(1) 在映射文件 usersMapper.xml 中,添加一个 id="updateUser2_trim"的<update>元素,代码如下:

```
<update id="updateUser2_trim" parameterType="Users">
    update users u
    <trim prefix="set" suffixOverrides=",">
        <if test="loginName!=null and loginName!=''">
            u.loginName=#{loginName},
        </if>
        <if test="loginPwd!=null and loginPwd!=''">
            u.loginPwd=#{loginPwd}
        </if>
    </trim>
    where u.id=#{id}
```

```
</update>
```

在 trim 元素中，prefix 属性值为 set，会在自己包含的内容前加上 set 关键字。suffixOverrides 设置为 "，"，会将自己包含内容尾部的 "，" 剔除，避免出现多余的逗号。

（2）在测试类 MybatisTest 中添加测试方法 testUpdateUserInfo2_trim，并用@Test 注解修改，代码如下：

```
@Test
public void testUpdateUserInfo2_trim() {
    Users cond = new Users();
    cond.setId(3);
    cond.setLoginName("mmm");
    cond.setLoginPwd("321321");
    sqlSession.update("updateUser2_trim", cond);
}
```

（3）执行该测试方法，控制台输出结果如下：

```
Preparing: update users u set u.loginName=?, u.loginPwd=? where u.id=?
Parameters: mmm(String), 321321(String), 3(Integer)
Updates: 1
```

如果在该测试方法中没有给 loginPwd 属性指定值，trim 元素会将自己所包含内容结尾处多余的逗号剔除，控制台输出结果如下：

```
Preparing: update users u set u.loginName=? where u.id=?
Parameters: mmm(String), 3(Integer)
Updates: 1
```

16.4.5 choose、when、otherwise 元素

如果在查询中不想使用所有的条件，而只是想从多个选项中选择一个。MyBatis 提供了 choose 元素，按顺序判断 when 中的条件是否成立，如果有一个成立，则 choose 结束。当 choose 中所有 when 的条件都不满足时，则执行 otherwise 中的 SQL 语句。如果想从数据表 users 中根据 loginName 或 status 进行查询，当 loginName 不为空时则只按照 loginName 查询，其他条件忽略；否则当 status 大于-1 时则只按照 status 查询；当 loginName 和 status 都为空时，则查询所有用户记录，使用 choose、when、otherwise 元素的实现过程如下。

（1）在映射文件 usersMapper.xml 中，添加一个 id="getUsers_Choose"的<select>元素，代码如下：

```
<select id="getUsers_Choose" parameterType="Users" resultType="Users">
    select * from users u
    <where>
        <choose>
            <when test="loginName!=null and loginName!=''">
                u.loginName LIKE
                CONCAT(CONCAT('%',#{loginName}),'%')
            </when>
            <when test="status>-1">
                and u.status = #{status}
            </when>
```

```
            <otherwise>
            </otherwise>
        </choose>
    </where>
</select>
```

(2) 在测试类 MybatisTest 中添加测试方法 testGetUsers_Choose()，并用@Test 注解修改，代码如下：

```java
@Test
public void testGetUsers_Choose() {
    Users cond = new Users();
    cond.setLoginName("z");
    cond.setStatus(1);
    List<Users> users = sqlSession.selectList("getUsers_Choose", cond);
    for (Users u : users) {
        System.out.println(u);
    }
}
```

(3) 执行该测试方法，控制台输出结果如下：

```
Preparing: select * from users u WHERE u.loginName LIKE
CONCAT(CONCAT('%',?),'%')
Parameters: z(String)
Users [id=1, loginName=zhangsan, loginPwd=123456]
Users [id=2, loginName=zs, loginPwd=zs]
```

当条件 loginName 不为空且 status 大于-1 时，只按照 loginName 查询，而忽略条件 status。

16.4.6 foreach 元素

foreach 元素主是要迭代一个集合，通常是用于 in 条件。例如，SQL 中的条件为 where id in 一大串的 id，这时可使用 foreach 元素，而不必去拼接 id 字符串。

foreach 元素可以向 SQL 语句传递数组、List<E>等。List<E>实例使用 list 作为键，数组实例使用 array 作为键。

如果想从数据表 users 中查询 id 为 1 和 3 的用户记录，使用 foreach 元素的 List<E>实例的实现过程如下。

(1) 在映射文件 usersMapper.xml 中，添加 id="getUsersByIds"的<select>元素，代码如下：

```xml
<select id="getUsersByIds" resultType="Users">
    select * from users u where u.id in
    <foreach collection="list" item="ids" open="(" separator=","
        close=")">
        #{ids}
    </foreach>
</select>
```

foreach 元素的属性主要包括 item、index、collection、open、separator、close 等。item 属

性表示集合中每个元素迭代时的别名。index 属性指定一个变量名称，表示每次迭代到的位置。open 表示该语句的开始符号。separator 属性表示每次迭代之间的分隔符号。close 属性表示该语句的结束符号。collection 属性需要根据具体情况进行设置，常用有以下两种情况。

- 如果向 SQL 语句传递的是单参数且参数类型为 List<E>时，collection 属性值为 list。
- 如果向 SQL 语句传递的是单参数且参数类型为 array 数组时，collection 属性值为 array。

(2) 在测试类 MybatisTest 中添加测试方法 testGetUsersByIds()，并用@Test 注解修改，代码如下：

```java
@Test
public void testGetUsersByIds() {
    List<Integer> ids = new ArrayList<Integer>();
    ids.add(1);
    ids.add(3);
    List<Users> users = sqlSession.selectList("getUsersByIds", ids);
    for (Users u : users) {
        System.out.println(u);
    }
}
```

(3) 执行该测试方法，控制台输出如下：

```
Preparing: select * from users u where u.id in ( ? , ? )
Parameters: 1(Integer), 3(Integer)
Total: 2
Users [id=1, loginName=zhangsan, loginPwd=123456]
Users [id=3, loginName=mmm, loginPwd=321321]
```

使用 foreach 元素的 array 实例，实现从数据表 user 中查询 id 为 1 和 2 的用户记录，过程如下。

(1) 在映射文件 usersMapper.xml 中，添加 id="getUsersByIds2"的<select>元素，代码如下：

```xml
<select id="getUsersByIds2" resultType="Users">
    select * from users u where u.id in
    <foreach collection="array" item="ids" open="(" separator=","
        close=")">
        #{ids}
    </foreach>
</select>
```

(2) 在测试类 MybatisTest 中添加测试方法 testGetUsersByIds2()，并用@Test 注解修改，代码如下：

```java
@Test
public void testGetUsersByIds2() {
    int[] ids = new int[2];
    ids[0] = 1;
    ids[1] = 2;
    List<Users> users = sqlSession.selectList("getUsersByIds2", ids);
    for (Users u : users) {
        System.out.println(u);
```

    }
}
```

(3) 执行该测试方法，控制台输出结果如下：

```
Preparing: select * from users u where u.id in ( ? , ? )
Parameters: 1(Integer), 2(Integer)
Total: 2
Users [id=1, loginName=zhangsan, loginPwd=123456]
Users [id=2, loginName=zs, loginPwd=zs]
```

16.5 MyBatis 的注解配置

MyBatis 是一个 XML 驱动的框架。前面介绍的 MyBatis 的增删改查、关联映射、动态 SQL 等知识，其所有的配置都是通过 XML 完成的。编写大量的 XML 配置比较烦琐。到了 MyBatis 3，可以使用 MyBatis 提供的基于注解的配置方式。

16.5.1 基于注解的增删改查

MyBatis 提供@Select、@Insert、@Update 和@Delete 注解完成常见的增删改查 SQL 语句映射。下面以数据表 users 为例，基于注解实现该表的增删改查操作。

(1) 将项目 mybatis_1 复制并命名为 mybatis_7，再导入 MyEclipse 开发环境中。

(2) 将项目 mybatis_7 中 com.mybatis.pojo 包中的 usersMapper.xml 文件删除，新建接口 UsersMapper.java，存放在 com.mybatis.mapper 包中。代码如下：

```java
package com.mybatis.mapper;
import java.util.List;
import org.apache.ibatis.annotations.Delete;
import org.apache.ibatis.annotations.Insert;
import org.apache.ibatis.annotations.Select;
import org.apache.ibatis.annotations.Update;
import com.mybatis.pojo.Users;
public interface UsersMapper {
    @Insert("insert into users(loginName,loginPwd) values(#{loginName},#{loginPwd})")
    public int addUser(Users u);
    @Delete("delete from users where id=#{id}")
    public int deleteUserById(int id);
    @Update("update users set loginName=#{loginName},loginPwd=#{loginPwd} where id=#{id}")
    public int updateUser(Users u);
    @Select("select * from users where id=#{id}")
    public Users getUserById(int id);
    @Select("select * from users")
    public List<Users> getAllUsers();
}
```

在接口 UsersMapper.java 中，声明了 addUser、deleteUserById、updateUser、getUserById 和 getAllUsers 等方法，分别对应数据表 users 的插入、删除、更新、根据 id 查询用户和查询
```

所有用户这 5 个操作，并分别使用@Insert、@Delete、@Update 和@Select 注解替代了之前的 XML 配置，这些注解中的每一个代表了执行的真实 SQL 语句。

(3) 修改 MyBatis 的配置文件 mybatis-config.xml。

在 mybatis-config.xml 文件中，先将原先注册的 usersMapper.xml 文件删除或注释掉，再添加对接口 UsersMapper 的注册，代码如下：

```xml
<mappers>
    <!-- <mapper resource="com/mybatis/pojo/usersMapper.xml" /> -->
    <mapper class="com.mybatis.mapper.UsersMapper" />
</mappers>
```

(4) 修改测试类 MybatisTest 中的测试方法。

其中，testAddUser()方法修改如下：

```java
@Test
public void testAddUser() {
    Users u = new Users("mybatis1", "123456");
    UsersMapper um = sqlSession.getMapper(UsersMapper.class);
    int insert = um.addUser(u);
    System.out.println(insert);
}
```

testGetUserById 方法修改如下：

```java
@Test
public void testGetUserById() {
    UsersMapper um = sqlSession.getMapper(UsersMapper.class);
    Users u = um.getUserById(4);
    System.out.println(u);
}
```

testGetAllUsers ()方法修改如下：

```java
@Test
public void testGetAllUsers() {
    UsersMapper um = sqlSession.getMapper(UsersMapper.class);
    List<Users> uList = um.getAllUsers();
    System.out.println(uList.size());
}
```

testUpdateUser ()方法修改如下：

```java
@Test
public void testUpdateUser() {
    UsersMapper um = sqlSession.getMapper(UsersMapper.class);
    // 加载编号 id=4 的用户
    Users u = um.getUserById(4);
    // 修改用户密码
    u.setLoginPwd("123123");
    // 执行更新操作
    int update = um.updateUser(u);
    System.out.println(u);
}
```

testDeleteUser ()方法修改如下：

```java
@Test
public void testDeleteUser() {
    UsersMapper um = sqlSession.getMapper(UsersMapper.class);
    int delete = um.deleteUserById(4);
    System.out.println(delete);
}
```

执行这些测试方法，结果与基于 XML 映射文件时相同。

16.5.2　基于注解的一对一关联映射

以 16.3.1 小节使用的数据表 admin 和 admin_detail 为例，基于注解实现这两张表之间的一对一关联映射。

(1) 将项目 mybatis_3 中与数据表 admin 和 admin_detail 对应的实体类 Admin.java 和 AdminDetail.java 复制到项目 mybatis_7 的 com.mybatis.pojo 包中。

(2) 在 com.mybatis.mapper 包中创建接口 AdminMapper.java 和 AdminDetailMapper.java。AdminDetailMapper.java 代码如下：

```java
package com.mybatis.mapper;
import org.apache.ibatis.annotations.Select;
import com.mybatis.pojo.AdminDetail;
public interface AdminDetailMapper {
    @Select("select * from admin_detail where id = #{id} ")
    AdminDetail selectAdminDetailById(int id);
}
```

AdminMapper.java 代码如下：

```java
package com.mybatis.mapper;
import org.apache.ibatis.annotations.One;
import org.apache.ibatis.annotations.Result;
import org.apache.ibatis.annotations.Results;
import org.apache.ibatis.annotations.Select;
import com.mybatis.pojo.Admin;
public interface AdminMapper {
    @Select("select * from admin where id = #{id} ")
    @Results({
        @Result(id = true, column = "id", property = "id"),
        @Result(column = "LoginName", property = "loginName"),
        @Result(column = "LoginPwd", property = "loginPwd"),
        @Result(column = "Adid", property = "ad",
            one = @One(select =
"com.mybatis.mapper.AdminDetailMapper.selectAdminDetailById")) })
    Admin selectAdminById(int id);
}
```

selectAdminById 方法中使用了@Select 注解，根据 id 查询 Admin 对象。由于需要将 Admin 关联的 AdminDetail 对象也查出来，因此 Admin 对象的 ad 属性使用了一个@Result 注解来映射结果。column="Adid", property="ad"表示 Admin 的 ad 属性对应数据表 admin 的 Adid

字段，属性 one 表示这是一个一对一关联关系，@One 注解的 select 属性表示会找到并执行 com.mybatis.mapper 包中 AdminDetailMapper 接口里定义的方法 selectAdminDetailById。

(3) 修改 MyBatis 配置文件 mybatis-config.xml。

在 mybatis-config.xml 文件中，添加对接口 AdminMapper 和 AdminDetailMapper 的注册，代码如下：

```xml
<mappers>
    ……
    <mapper class="com.mybatis.mapper.AdminMapper" />
    <mapper class="com.mybatis.mapper.AdminDetailMapper" />
</mappers>
```

(4) 测试一对一关联映射。

在测试类 MybatisTest 中添加测试方法 testOne2One()，使用@Test 注解修饰，从数据表 admin 获取记录的同时，获取关联的数据表 admin_detail 中的记录。代码如下：

```java
@Test
public void testOne2One() {
    // 获取 AdminMapper 实例
    AdminMapper adminMapper = sqlSession.getMapper(AdminMapper.class);
    // 根据 id 查询 Admin 对象，同时获取关联的 AdminDetail 对象
    Admin admin= adminMapper.selectAdminById(1);
    // 查看 Admin 对象
    System.out.println(admin);
    // 查看 Admin 关联的对象 AdminDetail
    System.out.println(admin.getAd());
}
```

执行 testOne2One()方法，控制台输出结果如下：

```
Preparing: select * from admin where id = ?
Parameters: 1(Integer)
Preparing: select * from admin_detail where id = ?
Parameters: 1(Integer)
Total: 1
Total: 1
Admin [id=1, loginName=admin, loginPwd=123456, ad=AdminDetail [id=1, address=江苏南京, realName=管理员]]
AdminDetail [id=1, address=江苏南京, realName=管理员]
```

可以看到，查询 Admin 对象时关联的 AdminDetail 对象也查询出来了。

16.5.3 基于注解的一对多关联映射

以 16.3.2 小节使用的数据表 mealseries 和 meal 为例，基于注解实现这两张表之间的一对多关联映射。

(1) 将项目 mybatis_4 中与数据表 mealseries 和 meal 对应的实体类 Admin.java 和 AdminDetail.java 复制到项目 mybatis_7 的 com.mybatis.pojo 包中。

(2) 在 com.mybatis.mapper 包中创建接口 MealMapper.java 和 MealseriesMapper.java。MealMapper.java 代码如下：

```java
import java.util.List;
import org.apache.ibatis.annotations.One;
import org.apache.ibatis.annotations.Result;
import org.apache.ibatis.annotations.Results;
import org.apache.ibatis.annotations.Select;
import com.mybatis.pojo.Meal;
public interface MealMapper {
    // 根据菜系编号查询所有餐品
    @Select("select * from meal where MealSeriesId = #{id} ")
    @Results({ @Result(id = true, column = "MealId", property = "mealId"),
@Result(column = "MealName", property = "mealName"),
@Result(column = "MealPrice", property = "mealPrice") })
    List<Meal> selectMealByMealSeriesId(int mealSeriesId);
    // 根据餐品编号获取餐品信息
    @Select("select * from meal where MealId = #{id} ")
    @Results({
        @Result(id = true, column = "MealId", property = "mealId"),
        @Result(column = "MealName", property = "mealName"),
        @Result(column = "MealPrice", property = "mealPrice"),
        @Result(column = "MealSeriesId", property = "mealseries",
        one = @One(select =
"com.mybatis.mapper.MealseriesMapper.selectMealseriesById")) })
    Meal selectMealById(int mealId);
}
```

selectMealById 方法使用@Select 注解，根据餐品编号查询餐品信息。由于需要将 Meal 对应的 Mealseries 数据查询出来，因此 Meal 关联的属性 mealseries 使用了一个@Result 结果映射。column = "MealSeriesId", property = "mealseries"表示 Meal 的 mealseries 属性对应数据表 meal 的 MealSeriesId 字段，属性 one 表示这是一个一对一关联关系，@One 注解的 select 属性表示会找到并执行 com.mybatis.mapper 包中 MealseriesMapper 接口里定义的方法 selectMealseriesById。

MealseriesMapper.java 代码如下：

```java
package com.mybatis.mapper;
import org.apache.ibatis.annotations.Many;
……
public interface MealseriesMapper {
    // 根据菜系编号获取菜系信息
    @Select("select * from mealseries where SeriesId = #{id} ")
    @Results({  @Result(id = true, column = "SeriesId", property =
"seriesId"),@Result(column = "SeriesName", property = "seriesName"),
    @Result(column = "SeriesId", property = "meals", many = @Many(select =
"com.mybatis.mapper.MealMapper.selectMealByMealSeriesId")) })
    Mealseries selectMealseriesById(int id);
}
```

selectMealseriesById 方法中使用了@Select 注解，根据菜系编号获取菜系信息。由于需要将 Mealseries 关联的 Meal 对象也要查出来，因此 Mealseries 对象的 meals 属性使用了一个 @Result 注解来映射结果。column="SeriesId"表示要使用 SeriesId 作为查询条件，属性 many 表示这是一个一对多关联关系，@Many 注解的 select 属性表示会找到并执行

com.mybatis.mapper 包中 MealMapper 接口里定义的方法 selectMealByMealSeriesId。

(3) 修改 MyBatis 配置文件 mybatis-config.xml。

在 mybatis-config.xml 文件中，添加对接口 MealMapper 和 MealseriesMapper 的注册，代码如下：

```xml
<mappers>
   ……
   <mapper class="com.mybatis.mapper.MealMapper" />
   <mapper class="com.mybatis.mapper.MealseriesMapper" />
</mappers>
```

(4) 测试一对多关联映射。

在测试类 MybatisTest 中添加测试方法 testOne2Many()，使用@Test 注解修饰，从数据表 mealseries 获取记录的同时，获取关联的数据表 meal 中的记录。代码如下：

```java
@Test
public void testOne2Many() {
    // 获得 MealseriesMapper 接口的代理对象
    MealseriesMapper mealseriesMapper =
sqlSession.getMapper(MealseriesMapper.class);
    // 直接调用接口的方法，查询 id=1 的 Mealseries 对象
    Mealseries mealseries = mealseriesMapper.selectMealseriesById(1);
    System.out.println(mealseries);
    // 查看 Mealseries 对象关联的餐品信息
    List<Meal> meals = mealseries.getMeals();
    for (Meal meal : meals) {
        System.out.println(meal);
    }
}
```

执行 testOne2Many()方法，控制台输出结果如下：

```
Preparing: select * from mealseries where SeriesId = ?
Preparing: select * from meal where MealSeriesId = ?
Mealseries [seriesId=1, seriesName=鲁菜]
Meal [mealId=1, mealName=雪梨肉肘棒, mealPrice=10.0]
Meal [mealId=2, mealName=素锅烤鸭肉, mealPrice=20.0]
……
```

可以看到，查询 Mealseries 对象时关联的 Meal 对象也查询出来了。

(5) 测试多对一关联映射。

在测试类 MybatisTest 中添加测试方法 testMany2One()，使用@Test 注解修饰，从数据表 meal 获取记录的同时，获取关联的数据表 mealseries 中的记录。代码如下：

```java
@Test
public void testMany2One() {
    // 获得 MealMapper 接口的代理对象
    MealMapper mealMapper = sqlSession.getMapper(MealMapper.class);
    // 直接调用接口的方法，查询 id=1 的 Meal 对象
    Meal meal = mealMapper.selectMealById(1);
    System.out.println(meal);
    // 查看 Meal 对象关联的菜系信息
```

```
        System.out.println(meal.getMealseries());
}
```

执行 testMany2One()方法,控制台输出结果如下:

```
Preparing: select * from meal where MealId = ?
Preparing: select * from mealseries where SeriesId = ?
Meal [mealId=1, mealName=雪梨肉肘棒, mealPrice=10.0]
Mealseries [seriesId=1, seriesName=鲁菜]
```

可以看到,查询 Meal 对象时关联的 Mealseries 对象也查询出来了。

16.5.4　基于注解的多对多关联映射

以 16.3.3 小节使用的数据表 admin 和 functions 为例,基于注解实现这两张表之间的多对多关联映射。

(1) 将项目 mybatis_1 复制并命名为 mybatis_8,再导入 MyEclipse 开发环境中。

(2) 将项目 mybatis_5 中与数据表 admin 和 functions 对应的实体类 Admin.java 和 Functions.java 复制到项目 mybatis_8 的 com.mybatis.pojo 包中。

(3) 创建接口 AdminMapper.java 和 FunctionsMapper.java,存放在 com.mybatis.mapper 包中,FunctionsMapper.java 代码如下:

```
package com.mybatis.mapper;
import java.util.List;
import org.apache.ibatis.annotations.Select;
import com.mybatis.pojo.Functions;
public interface FunctionsMapper {
    // 根据管理员 id 获取关联的功能权限信息
    @Select("select * from functions where id in (select fid from powers where aid = #{id} )")
    List<Functions> selectByAdminId(int adminId);
}
```

AdminMapper.java 代码如下:

```
package com.mybatis.mapper;
import org.apache.ibatis.annotations.Many;
import org.apache.ibatis.annotations.Result;
import org.apache.ibatis.annotations.Results;
import org.apache.ibatis.annotations.Select;
import com.mybatis.pojo.Admin;
public interface AdminMapper {
    @Select("select * from admin where id = #{id} ")
    @Results({
            @Result(id = true, column = "id", property = "id"),
            @Result(column = "LoginName", property = "loginName"),
            @Result(column = "id", property = "fs", many = @Many(select = "com.mybatis.mapper.FunctionsMapper.selectByAdminId")) })
    Admin selectById(int id);
}
```

selectById 方法中使用了@Select 注解,根据 id 查询 Admin 对象。由于需要将 Admin 关联的 Functions 对象也要查出来，因此 Admin 对象的 fs 属性使用了一个@Result 注解来映射结果。column="id"表示要使用 id 作为查询条件，属性 many 表示这是一个一对多关联关系，@Many 注解的 select 属性表示会找到并执行 com.mybatis.mapper 包中 FunctionsMapper 接口里定义的方法 selectByAdminId。

(4) 修改 MyBatis 配置文件 mybatis-config.xml。

在 mybatis-config.xml 文件中，添加对接口 AdminMapper.java 和 FunctionsMapper.java 的注册，代码如下：

```xml
<mappers>
    ……
    <mapper class="com.mybatis.mapper.AdminMapper" />
    <mapper class="com.mybatis.mapper.FunctionsMapper" />
</mappers>
```

(5) 测试多对多关联映射。

在测试类 MybatisTest 中添加测试方法 testM2M()，使用@Test 注解修饰，查看管理员及其功能权限。代码如下：

```java
@Test
public void testM2M() {
    // 获得 AdminMapper 接口的代理对象
    AdminMapper am = sqlSession.getMapper(AdminMapper.class);
    // 直接调用接口的方法，查询 id=1 的 Admin 对象
    Admin admin = am.selectById(1);
    System.out.println(admin);
    // 查看 Admin 对象关联的功能信息
    Object[] fs = admin.getFs().toArray();
    for (Object f : fs) {
        System.out.println((Functions) f);
    }
}
```

执行 testM2M()方法，控制台输出结果如下：

```
Preparing: select * from admin where id = ?
Preparing: select * from functions where id in (select fid from powers where aid = ? )
Admin [id=1, loginName=admin]
Functions [id=2, name=餐品管理]
Functions [id=10, name=用户列表]
Functions [id=3, name=餐品列表]
……
```

可以看到，查询 Admin 对象时关联的 Functions 对象也查询出来了。

16.5.5 基于注解的动态 SQL

MyBatis 提供了@SelectProvider、@InsertProvider、@UpdateProvider 和@DeleteProvider 等注解来构建动态 SQL 语句，然后再执行这些 SQL 语句。动态 SQL Provider 方法可以无参，也可用 Java 对象或 Map 对象作为参数。

下面以数据库 restrant 中 users 表的增删改查为例，介绍基于注解的动态 SQL。

1．使用@SelectProvider 注解动态查询数据

(1) 将项目 mybatis_1 复制并命名为 mybatis_9，再导入 MyEclipse 开发环境中。

(2) 将项目 mybatis_9 的 com.mybatis.pojo 包中 usersMapper.xml 文件删除，新建接口 UsersMapper.java，存放在 com.mybatis.mapper 包中。在 UsersMapper 接口中添加如下方法：

```java
package com.mybatis.mapper;
import java.util.List;
import java.util.Map;
import org.apache.ibatis.annotations.SelectProvider;
import com.mybatis.pojo.Users;
public interface UsersMapper {
    @SelectProvider(type=UsersDynaSqlProvider.class,
method="selectWithParam")
    List<Users> selectUsersByCond(Map<String, Object> param);
}
```

selectUsersByCond 方法中使用了@SelectProvider 注解，指定使用 UsersDynaSqlProvider 类中定义的方法 selectWithParam，该方法提供需要执行的 SQL 语句。

(3) 在 com.mybatis.mapper 包中，创建 UsersDynaSqlProvider 类，添加 selectWithParam 方法，代码如下：

```java
package com.mybatis.mapper;
import java.util.Map;
import org.apache.ibatis.jdbc.SQL;
public class UsersDynaSqlProvider {
    public String selectWithParam(Map<String, Object> param){
        return new SQL(){
            {
                SELECT("*");
                FROM("users");
                if(param.get("id")!=null){
                    WHERE("id = #{id} ");
                }
                if(param.get("loginName")!=null){
                    WHERE("loginName = #{loginName} ");
                }
                if(param.get("loginPwd")!=null){
                    WHERE("loginPwd = #{loginPwd} ");
                }
            }
        }.toString();
    }
}
```

selectWithParam 方法根据参数 Map 中的内容构建动态 SELECT 语句。

(4) 修改 MyBatis 的配置文件 mybatis-config.xml。

在 mybatis-config.xml 文件中，先将原先注册的 usersMapper.xml 文件删除或注释掉，再添加对接口 UsersMapper 的注册，代码如下：

```xml
<mappers>
    <!-- <mapper resource="com/mybatis/pojo/usersMapper.xml" /> -->
    <mapper class="com.mybatis.mapper.UsersMapper" />
</mappers>
```

(5) 在测试类 MybatisTest 中，添加测试方法 testSelectUsersByCond()，使用@Test 注解修饰。代码如下：

```java
@Test
public void testSelectUsersByCond() {
    UsersMapper um = sqlSession.getMapper(UsersMapper.class);
    Map<String, Object> param = new HashMap<String, Object>();
    param.put("loginName", "zhangsan");
    param.put("loginPwd", "123456");
    List<Users> users = um.selectUsersByCond(param);
    System.out.println(users);
}
```

执行 testSelectUsersByCond()方法，控制台输出结果如下：

```
Preparing: SELECT * FROM users WHERE (loginName = ? AND loginPwd = ?
Parameters: zhangsan(String), 123456(String)
Total: 1
[Users [id=1, loginName=zhangsan, loginPwd=123456]]
```

可以看出，因为 Map 中设置了 loginName 和 loginPwd 两个参数，所有执行的 SQL 语句中包含这两个条件。如果 Map 中只设置 loginName 这个参数，控制台输出的 SQL 语句中只包含 loginName 这个条件，代码如下：

```
Preparing: SELECT * FROM users WHERE (loginName = ? )
Parameters: zhangsan(String)
```

当然，UsersDynaSqlProvider 类中的 selectWithParam 方法也可以传递 Users 对象作为参数。

2. 使用@InsertProvider 注解动态插入数据

(1) 在接口 UsersMapper.java 中添加如下方法：

```java
@InsertProvider(type = UsersDynaSqlProvider.class, method = "insertUsers")
@Options(useGeneratedKeys = true, keyProperty = "id")
int insertUsers(Users u);
```

insertUsers 方法中使用了@Options 注解，表示在向数据表 users 插入数据时，自动将主键字段 id 的自增值赋值给对象 u 的属性 id。

(2) 在 UsersDynaSqlProvider 类中添加 insertUsers 方法，代码如下：

```java
public String insertUsers(Users u) {
    return new SQL() {
        {
            INSERT_INTO("users");
            if (u.getLoginName() != null) {
                VALUES("LoginName", "#{loginName}");
            }
```

```
            if (u.getLoginPwd() != null) {
                VALUES("LoginPwd", "#{loginPwd}");
            }
        }
    }.toString();
}
```

(3) 在测试类 MybatisTest 中，添加测试方法 testInsertUsers()，使用@Test 注解修饰。代码如下：

```
@Test
public void testInsertUsers() {
    UsersMapper um = sqlSession.getMapper(UsersMapper.class);
    Users u=new Users();
    u.setLoginName("mybatis2");
    u.setLoginPwd("123456");
    um.insertUsers(u);
    System.out.println("插入的用户编号："+u.getId());
}
```

执行 testInsertUsers()方法，控制台输出结果如下：

```
Preparing: INSERT INTO users (LoginName, LoginPwd) VALUES (?, ?)
Parameters: mybatis2(String), 123456(String)
Updates: 1
插入的用户编号：6
```

读者可以通过只设置 loginName 或 loginPwd 参数，来观察控制台输出的 SQL 语句。

3. 使用@UpdateProvider 注解动态更新数据

(1) 在接口 UsersMapper.java 中添加下面两个方法：

```
// 根据id查询用户
@SelectProvider(type = UsersDynaSqlProvider.class, method = "selectWithParam")
Users selectUsersById(Map<String, Object> param);
// 动态更新
@UpdateProvider(type = UsersDynaSqlProvider.class, method = "updateUsers")
int updateUsers(Users u);
```

selectUsersById 方法根据用户 id 查询用户信息；updateUsers 方法动态更新用户信息。

(2) 在 UsersDynaSqlProvider 类中添加 updateUsers 方法，代码如下：

```
public String updateUsers(Users u) {
    return new SQL() {
        {
            UPDATE("users");
            if (u.getLoginName() != null) {
                SET("LoginName = #{loginName}");
            }
            if (u.getLoginPwd() != null) {
                SET("LoginPwd = #{loginPwd}");
            }
            WHERE ("id = #{id} ");
        }
```

```
    }.toString();
}
```

(3) 在测试类 MybatisTest 中，添加测试方法 testUpdateUsers()，使用@Test 注解修饰。代码如下：

```
@Test
public void testUpdateUsers() {
    UsersMapper um = sqlSession.getMapper(UsersMapper.class);
    // 使用 Map 封装查询条件
    Map<String, Object> param = new HashMap<String, Object>();
    param.put("id", "2");
    // 查询 id=2 的用户
    Users u = um.selectUsersById(param);
    // 修改该用户的密码
    u.setLoginPwd("666666");
    // 动态更新
    um.updateUsers(u);
}
```

执行 testUpdateUsers()方法，控制台输出结果如下：

```
Preparing: SELECT * FROM users WHERE (id = ?)
Preparing: UPDATE users SET LoginName = ?, LoginPwd = ? WHERE (id = ?)
Parameters: zs(String), 666666(String), 2(Integer)
```

查看数据表 users，可见 id=2 的用户 zs 的密码被修改为 666666。

4．使用@DeleteProvider 注解动态删除数据

(1) 在接口 UsersMapper.java 中添加如下方法：

```
@DeleteProvider(type = UsersDynaSqlProvider.class, method = "deleteUsers")
void deleteUsers(Map<String, Object> param);
```

(2) 在 UsersDynaSqlProvider 类中添加 deleteUsers 方法，代码如下：

```
public String deleteUsers(Map<String, Object> param) {
    return new SQL() {
        {
            DELETE_FROM("users");
            if (param.get("id") != null) {
                WHERE("id = #{id} ");
            }
            if (param.get("loginName") != null) {
                WHERE("loginName = #{loginName} ");
            }
            if (param.get("loginPwd") != null) {
                WHERE("loginPwd = #{loginPwd} ");
            }
        }
    }.toString();
}
```

(3) 在测试类 MybatisTest 中，添加测试方法 testDeleteUsers()，使用@Test 注解修饰。代码如下：

```
@Test
public void testDeleteUsers() {
    UsersMapper um = sqlSession.getMapper(UsersMapper.class);
    // 使用 Map 封装查询条件
    Map<String, Object> param = new HashMap<String, Object>();
    param.put("loginName", "mybatis1");
    param.put("loginPwd", "123456");
    // 动态删除
    um.deleteUsers(param);
}
```

执行 testDeleteUsers()方法，控制台输出结果如下：

```
Preparing: DELETE FROM users WHERE (loginName = ? AND loginPwd = ? )
Parameters: mybatis1(String), 123456(String)
```

读者可以向 Map 中添加不同的参数组合，来观察控制台输出的 SQL 语句。

16.6 MyBatis 的缓存

MyBatis 提供的查询缓存分为一级缓存和二级缓存，可以有效地提高数据库查询的性能。

16.6.1 一级缓存

MyBatis 的一级缓存是 SqlSession 级别的缓存，当在同一个 SqlSession 中执行两次相同的 SQL 语句时，会将第一次执行查询的数据存入一级缓存中，第二次查询时会从缓存中获取数据，而不用再去数据库查询，从而提高查询性能。但如果 SqlSession 执行 insert、delete 和 update 操作，并提交到数据库，或者 SqlSession 结束后，这个 SqlSession 中的一级缓存就不存在了。

下面通过示例测试 MyBatis 的一级缓存。

(1) 将项目 mybatis_1 复制并命名为 mybatis_10，再导入 MyEclipse 开发环境中。

(2) 新建接口 UsersMapper.java，存放在 com.mybatis.mapper 包中。在接口 UsersMapper 中添加如下方法：

```
package com.mybatis.mapper;
import com.mybatis.pojo.Users;
public interface UsersMapper {
    // 根据 id 查询用户
    Users getUserById(int id);
}
```

(3) 修改 SQL 的映射文件 usersMapper.xml 的命名空间，代码如下：

```
<mapper namespace="com.mybatis.mapper.UsersMapper">
```

(4) 在测试类 MybatisTest 中，添加测试方法 testFirstLevelCache()，使用@Test 注解修饰。代码如下：

```
@Test
```

```java
public void testFirstLevelCache() {
    // 获得UsersMapper对象
    UsersMapper um = sqlSession.getMapper(UsersMapper.class);
    // 查询id=1的Users对象
    Users u1 = um.getUserById(1);
    System.out.println(u1);
    // 再次查询id=1的Users对象
    Users u2 = um.getUserById(1);
    System.out.println(u2);
}
```

执行测试方法 testFirstLevelCache()，控制台输出结果如下：

```
Preparing: select * from users where id = ?
Parameters: 1(Integer)
Total: 1
Users [id=1, loginName=zhangsan, loginPwd=123456]
Users [id=1, loginName=zhangsan, loginPwd=123456]
```

可以看出，第一次查询 id=1 的 Users 对象时发出了一条 SQL 语句，由于 MyBatis 默认开启了一级缓存，因此一级缓存 SqlSession 中缓存了 id=1 的 Users 对象。第二次再查询 id=1 的 Users 对象时，直接从一级缓存中获取数据，而不用查询数据库，所以第二次没有发出 SQL 语句。

在测试类 MybatisTest 中，添加测试方法 testFirstLevelCache_1()，并使用@Test 注解修饰。在第一次查询后关闭 SqlSession，然后开启一个新的 SqlSession，再次执行查询。代码如下：

```java
@Test
public void testFirstLevelCache_1() {
    // 获得UsersMapper对象
    UsersMapper um = sqlSession.getMapper(UsersMapper.class);
    // 查询id=1的Users对象
    Users u1 = um.getUserById(1);
    System.out.println(u1);
    sqlSession.commit();
    // 关闭sqlSession，即清空一级缓存
    sqlSession.close();
    // 开始一个新的sqlSession
    sqlSession = sqlSessionFactory.openSession();
    // 再次获得UsersMapper对象
    um = sqlSession.getMapper(UsersMapper.class);
    // 再次查询id=1的Users对象
    Users u2 = um.getUserById(1);
    System.out.println(u2);
}
```

执行测试方法 testFirstLevelCache_1()，控制台输出结果如下：

```
DEBUG 06-22 16:47:12 ==> Preparing: select * from users where id = ?
DEBUG 06-22 16:47:12 ==> Parameters: 1(Integer)
DEBUG 06-22 16:47:12 <==      Total: 1
Users [id=1, loginName=zhangsan, loginPwd=123456]
DEBUG 06-22 16:47:12 ==> Preparing: select * from users where id = ?
```

```
DEBUG 06-22 16:47:12 ==> Parameters: 1(Integer)
DEBUG 06-22 16:47:12 <==        Total: 1
Users [id=1, loginName=zhangsan, loginPwd=123456]
```

可以看出，关闭 SqlSession 后一级缓存会被清空，所以第二次查询时，一级缓存中查询不到数据，会发出第二条 SQL 语句查询数据库。

16.6.2 二级缓存

MyBatis 的二级缓存是 mapper 级别的缓存，多个 SqlSession 共用二级缓存，它们使用同一个 Mapper 的 SQL 语句去操作数据库，获得的数据会存放在二级缓存中。

MyBatis 默认没有开启二级缓存，需要在 MyBatis 的配置文件 mybatis-config.xml 中开启二级缓存，配置如下：

```xml
<settings>
    <setting name="cacheEnabled" value="true" />
</settings>
```

需要注意的是，settings 元素要放在 properties 元素之后、typeAliases 元素之前，否则配置文件会报错。

在 usersMapper.xml 映射文件中开启二缓存，usersMapper.xml 下的 SQL 语句执行结束后，会将结果存储到它的二级缓存中。代码如下：

```xml
<cache eviction="LRU" flushInterval="30000" size="512" readOnly="true" />
```

上述配置开启当前 mapper 的 namespace 下的二级缓存，该元素的属性含义如下。

(1) flushInterval 属性。表示刷新间隔，可以被设置为任意的正整数，单位是毫秒。默认不设置，表示没有刷新间隔，仅仅调用语句时刷新缓存。

(2) Size 属性。表示引用数目，可以被设置为任意正整数，默认值是 1024。readOnly 属性设置是否只读，true 表示只读，false 表示可读写。只读的缓存会给所有调用者返回缓存对象的相同实例，因此这些对象不能被修改，这提供了重要的性能优势。可读写的缓存会返回缓存对象的复制(通过发序列化)。这会慢一些，但是安全，因此默认是 false。

(3) eviction 属性。表示收回策略，有 LRU、FIFO、SOFT、WEAK 等策略，默认为 LRU。LRU 表示最近最少使用策略，即移除最近最少使用的对象。FIFO 表示先进先出策略，按对象进入缓存的顺序来移除它们。SOFT 表示软引用策略，移除基于垃圾回收器状态和软引用规则的对象。WEAK 表示弱引用策略，更积极地移除基于垃圾收集器状态和弱引用规则的对象。

再执行测试类 MybatisTest 中的测试方法 testFirstLevelCache_1()，控制台输出结果如下：

```
Cache Hit Ratio [com.mybatis.mapper.UsersMapper]: 0.0
DEBUG 06-22 20:51:33 ==> Preparing: select * from users where id = ?
DEBUG 06-22 20:51:33 ==> Parameters: 1(Integer)
DEBUG 06-22 20:51:33 <==        Total: 1    (BaseJdbcLogger.java:142)
Users [id=1, loginName=zhangsan, loginPwd=123456]
DEBUG 06-22 20:51:33 Cache Hit Ratio [com.mybatis.mapper.UsersMapper]: 0.5
Users [id=1, loginName=zhangsan, loginPwd=123456]
```

可以看出，第一次查询 id=1 的 Users 对象时执行了一条 select 语句，然后关闭 SqlSession，一级缓存被清空。第二次查询 id=1 的 Users 对象时，先查找一级缓存，没有找到 id=1 的 Users 对象，再去查找二级缓存，找到了 id=1 的 Users 对象，所以不会再次执行 select 语句。

在映射文件 usersMapper.xml 中，如果给 id="getUserById"的 select 元素添加 useCache="false" 属性值，则表示禁用当前 select 语句的二级缓存，即每次查询都会发出 SQL 语句，默认为 true，表示该 SQL 语句使用二级缓存。

再次执行测试方法 testFirstLevelCache_1()，观察控制台输出。由于该 select 语句禁用了二级缓存，因此第二次查询时会再次发出 select 语句。

16.7 小　　结

本章介绍了另一个优秀的持久层框架——MyBatis，基于 XML 配置文件和注解介绍了单张表的增删改查、多表的关联映射、动态 SQL 等内容，以及 MyBatis 的缓存。

对比前面学习的持久层框架——Hibernate，这两个框架都是 ORM 对象关系映射框架，都是用于将数据持久化的框架技术。Hibernate 较深度地封装了 JDBC，对开发者编写 SQL 的能力要求不高，只要通过 HQL 语句操作对象即可完成对数据持久化的操作了。另外，Hibernate 可移植性好，如：一个项目开始使用的是 MySQL 数据库，现在决定使用 Oracle 数据库，由于不同的数据库 SQL 标准还是有差距的，手动修改会存在很大的困难，使用 Hibernate 只需要改变数据库方言即可。使用 Hibernate 框架，数据库的移植变得非常方便。但是 Hibernate 也存在着诸多不足，比如在实际开发过程中会生成很多不必要的 SQL 语句耗费程序资源，优化起来也不是很方便，且对存储过程的支持也不够强大。Mybatis 也是对 JDBC 的封装，但是封装得没有 Hibernate 那么深，通过在配置文件中编写 SQL 语句，可以根据需求定制 SQL 语句，数据优化起来较 Hibernate 容易得多。Mybatis 要求程序员编写 SQL 的能力要比 Hibernate 高得多，且可移植性也不是很好。涉及大数据的系统使用 Mybatis 比较好，因为优化方便。涉及的数据量不大且对优化要求不高的系统，可以使用 Hibernate。

第 17 章
Spring 的基本应用

前面学习了进行持久层开发的两个流行的框架 Hibernate 和 MyBatis。本章介绍进行逻辑层开发的流行框架 Spring。Spring 框架从某种程度上来看，充当了黏合剂和润滑剂的角色，它对 Hibernate 和 Struts 2 等框架提供了良好的支持，能够将相应的 Java Web 系统柔顺地整合起来，并让它们更易使用。同时，Spring 本身还提供了声明式事务等企业级开发不可或缺的功能。

17.1 认识 Spring 框架

Spring 作为实现 J2EE 的一个全方位应用程序框架，为开发企业级应用提供了一个健壮、高效的解决方案。Spring 框架具有以下几个特点。

(1) 非侵入式。所谓非侵入式，是指 Spring 框架的 API 不会在业务逻辑上出现，也就是说业务逻辑应该是纯净的，不能出现与业务逻辑无关的代码。针对应用而言，这样才能将业务逻辑从当前应用中剥离出来，从而在其他应用中实现复用；针对框架而言，由于业务逻辑中没有 Spring 的 API，所以业务逻辑也可以从 Spring 框架快速地移植到其他框架。

(2) 容器。Spring 提供容器功能，容器可以管理对象的生命周期，对象与对象之间的依赖关系。可以写一个配置文件(通常是 xml 文件)，在上面定义对象的名字，是否是单例，以及设置与其他对象的依赖关系。那么在容器启动之后，这些对象就被实例化好了，直接用就好了，而且依赖关系也建立好了。

(3) IOC。控制反转，即依赖关系的转移，如果以前都是依赖于实现，那么现在反转为依赖于抽象，其核心思想就是要面向接口编程。

(4) 依赖注入。对象与对象之间依赖关系的实现，包括接口注入、构造注入、set 方法注入，在 Spring 中只支持后两种。

(5) AOP。面向切面编程，将日志、安全、事务管理等服务(或功能)理解成一个"切面"，以前这些服务通常是直接写在业务逻辑的代码当中的，有两个缺点：首先是业务逻辑不纯净；其次是这些服务被很多业务逻辑反复使用，不能做到复用。AOP 解决了上述问题，可以把这些服务剥离出来形成一个"切面"，可以实现复用；然后将"切面"动态地插入业务逻辑中，让业务逻辑能够方便地使用"切面"提供的服务。

其他还有一些特点但不是 Spring 的核心，如：对 JDBC 的封装与简化，提供事务管理功能，对 O/R mapping 工具(Hibernate、MyBatis)的整合；提供 MVC 解决方案，也可以与其他 Web 框架(Struts、JSF)进行整合；还有对 JNDI、mail 等服务进行封装。

Spring 框架(Spring Framework)不断在发展和完善，但基本与核心的部分已经相当稳定，包括 Spring 的依赖注入容器、AOP 实现和对持久层的支持。Spring Framework 包含的内容如图 17-1 所示。

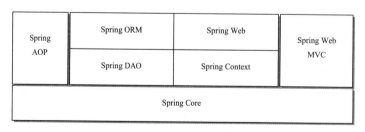

图 17-1　Spring Framework 包含的内容

其中，Spring Core 是最基础的，作为 Spring 依赖注入容器的部分。Spring AOP 是基于 Spring Core 的，典型的应用之一就是声明式事务。Spring Core 对 JDBC 提供了支持，简化了

JDBC 编码，同时使代码更加健壮。Spring ORM 对 Hibernate 等持久层框架提供了支持。Spring 可以在 Java SE 中使用，也可以在 Java EE 中使用。Spring Context 为企业级开发提供了便捷和集成的工具。Spring Web 为 Web 应用程序的开发提供了支持。

下面通过示例演示 Spring 框架的简单应用，其中只用到了 Spring 框架而没有使用其他技术，这样能使初学者更加容易理解。实现步骤如下。

(1) 创建一个名为 spring_1 的 Java 项目，在项目中新建文件夹 lib，用于存放项目所需的 jar 包。

(2) 从 Spring 官方网站下载 Spring，以 spring-framework-4.3.5.RELEASE-dist.zip 为例，解压后将其 libs 目录下的 spring-beans-4.3.5.RELEASE.jar、spring-context-4.3.5.RELEASE.jar、spring-core-4.3.5.RELEASE.jar 和 spring-expression-4.3.5.RELEASE.jar 这 4 个文件复制到项目 spring-1 的 lib 目录中，即完成了 Spring 的安装。

(3) 将 Spring 依赖的日志包 commons-logging-1.1.3.jar 也复制到 lib 目录中。

(4) 选中该项目 lib 目录下的所有 jar 包，右击并选择 Build Path→Add to Build Path 命令，将这些 jar 包添加到项目的构建路径中。

(5) 在 spring_1 项目中创建 com.shw 包，在包中新建一个名为 HelloWorld 的类。代码如下：

```java
package com.shw;
public class HelloWorld {
    private String userName;
    public void setUserName(String userName) {
        this.userName = userName;
    }
    public void show() {
        System.out.println(userName + "：欢迎您学习Spring框架");
    }
}
```

(6) 在项目 src 目录下创建 applicationContext.xml 文件，内容如下：

```xml
<?xml version="1.0" encoding="UTF-8"?>
<beans xmlns="http://www.springframework.org/schema/beans"
    xmlns:xsi="http://www.w3.org/2001/XMLSchema-instance"
xmlns:p="http://www.springframework.org/schema/p"
    xsi:schemaLocation="http://www.springframework.org/schema/beans
    http://www.springframework.org/schema/beans/spring-beans-4.3.xsd">
    <!-- 配置一个 bean -->
    <bean id="helloWorld" class="com.shw.HelloWorld">
        <!-- 为属性赋值 -->
        <property name="userName" value="zhangsan"></property>
    </bean>
</beans>
```

在 applicationContext.xml 文件中，通过<bean>元素来实例化 HelloWorld 类，id 属性用来标识实例名 helloWorld，class 属性指定待实例化的全路径类名 com.shw.HelloWorld。子元素<property>用来为类中的属性赋值，name 属性指定 HelloWorld 类中的属性 userName，value 属性给 userName 指定了值 zhangsan。

(7) 在 com.shw 包中创建测试类 TestHelloWorld，加载 applicationContext.xml 配置文件，获取 HelloWorld 类的实例，然后调用类中的 show()方法在控制台输出信息。代码如下：

```java
package com.shw;
import org.springframework.context.ApplicationContext;
import org.springframework.context.support.ClassPathXmlApplicationContext;
public class TestHelloWorld {
    public static void main(String[] args) {
        // 加载 applicationContext.xml 配置
        ApplicationContext ctx = new ClassPathXmlApplicationContext(
            "applicationContext.xml");
        // 获取配置中的实例
        HelloWorld helloWorld = (HelloWorld) ctx.getBean("helloWorld");
        // 调用方法
        helloWorld.show();
    }
}
```

在测试类 TestHelloWorld 中，首先通过 ClassPathXmlApplicationContext 类加载 applicationContext.xml 配置文件以初始化一个 Spring 容器，返回的 ApplicationContext 是一个接口，具有维护不同 Bean 以及它们依赖项的高级工厂能力，然后使用 getBean 方法来检索 HelloWorld 类的实例，最后通过实例 helloWorld 调用 HelloWorld 类的 show()方法。

执行测试类 TestHelloWorld，控制台输出结果如下：

```
zhangsan: 欢迎您学习 Spring 框架
```

17.2 了解 Spring 的核心机制：依赖注入/控制反转

Spring 的核心机制就是 IoC(控制反转)容器，IoC 的另外一个称呼是依赖注入(DI)。通过依赖注入，Java EE 应用中的各种组件不需要以硬编码的方法进行耦合，当一个 Java 实例需要其他 Java 实例时，系统自动提供需要的实例，无须程序显式获取。因此，依赖注入实现了组件之间的解耦。

依赖注入和控制反转含义相同，当某个 Java 实例需要另一个 Java 实例时，传统的方法是由调用者来创建被调用者的实例(如：使用 new 关键字获得被调用者实例)。

采用依赖注入方式时，被调用者的实例不再需要由调用者来创建，称为控制反转。被调用者的实例通常是由 Spring 容器来完成的，然后注入调用者，调用者便获得了被调用者的实例，称为依赖注入。

Spring 提倡面向接口的编程，依赖注入的基本思想是：明确地定义组件接口，独立开发各个组件，然后根据组件的依赖关系组装运行。下面以一个简单的登录验证为例，介绍 Spring 依赖注入的运用。

(1) 将项目 spring_1 复制并命名为 spring_2，再导入 MyEclipse 开发环境中。

(2) 编写 DAO 层。

在项目 spring_2 的 src 目录下，新建包 com.shw.dao，在包中新建一个接口 UsersDAO.java，在接口中添加方法 login()，代码如下：

```java
package com.shw.dao;
public interface UsersDAO {
    public boolean login(String loginName,String loginPwd);
}
```

创建接口 UsersDAO 的实现类 UsersDAOImpl，存放在 com.shw.dao.impl 包中，实现 login()方法。代码如下：

```java
package com.shw.dao.impl;
import com.shw.dao.UsersDAO;
public class UsersDAOImpl implements UsersDAO {
    @Override
    public boolean login(String loginName, String loginPwd) {
        if (loginName.equals("admin") && loginPwd.equals("123456")) {
            return true;
        }
        return false;
    }
}
```

在登录验证时为了简化 DAO 层代码，暂时没有用到数据库。如果用户名为 admin，密码为 123456，则登录成功。

(3) 编写 Service 层。

在 src 目录下新建包 com.shw.service，在包中新建一个接口 UsersService.java，在接口中添加方法 login()，代码如下：

```java
package com.shw.service;
public interface UsersService {
    public boolean login(String loginName,String loginPwd);
}
```

创建接口 UsersService 的实现类 UsersServiceImpl.java，存放在 com.shw.service.impl 包中，实现 login()方法。代码如下：

```java
package com.shw.service.impl;
import com.shw.dao.UsersDAO;
import com.shw.service.UsersService;
public class UsersServiceImpl implements UsersService {
    // 使用接口UsersDAO声明对象，添加set()方法，用于依赖注入
    UsersDAO usersDAO;
    public void setUsersDAO(UsersDAO usersDAO) {
        this.usersDAO = usersDAO;
    }
    @Override
    public boolean login(String loginName, String loginPwd) {
        return usersDAO.login(loginName, loginPwd);
    }
}
```

在上述代码中，没有采用传统的 new UsersDAOImpl()方式获取数据访问层 UsersDAOImpl 类的实例，只用 UsersDAO 接口声明了对象 usersDAO，并为其添加 set()方法，用于依赖注入。

UsersDAOImpl 类的实例化和对象 usersDAO 的注入将在 applicationContext.xml 配置文件中完成。

(4) 配置 applicationContext.xml 文件。

为了创建 UsersDAOImpl 类和 UsersServiceImpl 类的实例，需要添加<bean>标记，并配置其相关属性。代码如下：

```xml
<!-- 配置创建 UsersDAOImpl 的实例 -->
<bean id="usersDAO" class="com.shw.dao.impl.UsersDAOImpl">
</bean>
<!-- 配置创建 UsersServiceImpl 的实例 -->
<bean id="usersService" class="com.shw.service.impl.UsersServiceImpl">
    <!-- 依赖注入数据访问层组件 -->
    <property name="usersDAO" ref="usersDAO" />
</bean>
```

<bean>元素用来定义 Bean 的实例化信息，class 属性指定类全名(包名+类名)，id 属性指定生成的 Bean 实例名称。在上述配置中，首先通过一个<bean>元素创建了 UsersDAOImpl 类的实例，在使用另一个<bean>元素创建 UsersServiceImpl 类的实例时，使用了<property>元素，该元素是<bean>元素的子元素，用于调用 Bean 实例中的相关 Set()方法完成属性值的赋值，从而实现依赖关系的注入。<property>元素中的 name 属性指定 Bean 实例中的相应属性的名称，这里 name 属性设置为 usersDAO，代表 UsersServiceImpl 类中的 usersDAO 属性需要注入值。name 属性的值可以通过 ref 属性或者 value 属性指定。当使用 ref 属性时，表示对 Spring IoC 容器中某个 Bean 的实例的引用。这里引用了前一个<bean>元素中创建的 UsersDAOImpl 类的实例 usersDAO，并将该实例赋值给 UsersServiceImpl 类中的 usersDAO 属性，从而实现了依赖关系的注入。UsersServiceImpl 类的 usersDAO 属性值是通过调用 setUsersDAO()方法完成注入的，这种注入方式称为设值注入。设值注入方式是 Spring 推荐使用的。

(5) 编写测试类。

在 com.shw 包中创建测试类 TestSpringDI，代码如下：

```java
package com.shw;
import org.springframework.context.ApplicationContext;
import org.springframework.context.support.ClassPathXmlApplicationContext;
import com.shw.service.UsersService;
public class TestSpringDI {
    public static void main(String[] args) {
        // 加载 applicationContext.xml 配置
        ApplicationContext ctx = new ClassPathXmlApplicationContext(
                "applicationContext.xml");
        // 获取配置中的 UsersServiceImpl 实例
        UsersService usersService = (UsersService) ctx.getBean("usersService");
        boolean flag = usersService.login("admin", "123456");
        if (flag) {
            System.out.println("登录成功");
        } else {
            System.out.println("登录失败");
        }
    }
```

```
    }
}
```

在测试类 TestSpringDI 中，首先通过 ClassPathXmlApplicationContext 类加载 Spring 配置文件 applicationContext.xml，然后从配置文件中获取 UsersServiceImpl 类的实例，最后调用 login()方法。运行测试类，当用户名为 admin，密码为 123456 时，控制台输出"登录成功"，否则输出"登录失败"。

17.3 小　　结

本章主要介绍了使用 Spring 框架进行逻辑层开发的基本知识，先通过一个简单的 Hello World 示例演示 Spring 框架的简单应用，然后以一个简单的登录验证为例，讲述了 Spring 的核心机制：依赖注入/控制反转。

第 18 章
Spring Bean 的装配模式

作为 Spring 核心机制的依赖注入，改变了传统编程习惯，对组件的实例化不再由应用程序完成，转而交由 Spring 容器完成，需要时注入应用程序中，从而将组件之间的依赖关系进行了解耦。这一切都离不开 Spring 配置文件中使用的<bean>元素。下面深入学习 Spring 中的 Bean。

18.1 Bean 工厂 ApplicationContext

Spring IoC 设计的核心是 Bean 容器，BeanFactory 采用了 Java 经典的工厂模式，通过从 XML 配置文件中读取 JavaBean 的定义，来实现 JavaBean 的创建、配置和管理。所以 BeanFactory 可以称为 "IoC 容器"。而 ApplicationContext 扩展了 BeanFactory 容器并添加了对 I18N(国际化)、资源访问、事件传播等方面的良好支持，使之成为 Java EE 应用中首选的 IoC 容器，可应用在 Java APP 和 Java Web 中。

ApplicationContext 的中文含义是"应用上下文"，它继承自 BeanFactory 接口。ApplicationContext 接口有三个常用的实现类，具体如下。

(1) ClassPathXmlApplicationContext。

ClassPathXmlApplicationContext 类从类路径 ClassPath 中寻找指定的 XML 配置文件，找到并装载 ApplicationContext 的实例化工作。例如：

```
ApplicationContext context=new ClassPathXmlApplicationContext(String configLocation);
```

configLocation 参数指定 Spring 配置文件的名称和位置，如"applicationContext.xml"。

(2) FileSystemXmlApplicationContext。

FileSystemXmlApplicationContext 类从指定的文件系统路径中寻找指定的 XML 配置文件，找到并装载 ApplicationContext 的实例化工作。例如：

```
ApplicationContext context=new FileSystemXmlApplicationContext(String configLocation);
```

与 ClassPathXmlApplicationContext 的区别在于读取 Spring 配置文件的方式，FileSystemXmlApplicationContext 不再从类路径中读取配置文件，而是通过参数指定配置文件的位置，可以获取类路径之外的资源。

(3) XmlWebApplicationContext。

XmlWebApplicationContext 类从 Web 应用中寻找指定的 XML 配置文件，找到并装载完成 ApplicationContext 的实例化工作。

可以通过实例化其中的任何一个类来创建 Spring 的 ApplicationContext 容器。这些实现类的主要区别在于装载 Spring 配置文件实例化 ApplicationContext 容器的方式不同。在实例化 ApplicationContext 后，同样通过 getBean 方法从 ApplicationContext 容器中获取装配好的 Bean 实例以供使用。

在 Java 项目中通过 ClassPathXmlApplicationContext 类手工实例化 ApplicationContext 容器通常是不二之选。但对于 Web 项目就不行了，Web 项目的启动是由相应的 Web 服务器负责的。因此，在 Web 项目中 ApplicationContext 容器的实例化工作最好交由 Web 服务器来完成。Spring 为此提供了如下两种方式。

(4) 基于 ContextLoaderListener 实现。

这种方式只适用于 Servlet 2.4 及以上规范的 Servlet，需要在 web.xml 中添加如下代码：

```xml
<!-- 指定 Spring 配置文件的位置，多个配置文件以逗号分隔 -->
<context-param>
    <param-name>contextConfigLocation</param-name>
    <param-value>classpath:applicationContext.xml</param-value>
</context-param>
<!-- 指定以 Listener 方式启动 Spring 容器 -->
<listener>
   <listener-class>org.springframework.web.context.ContextLoaderListener</listener-class>
</listener>
```

(5) 基于 ContextLoaderServlet 实现。

该方式需要在 web.xml 中添加如下代码：

```xml
<!-- 指定 Spring 配置文件的位置，多个配置文件以逗号分隔 -->
<context-param>
    <param-name>contextConfigLocation</param-name>
    <param-value>classpath:applicationContext.xml</param-value>
</context-param>
<!-- 指定以 Servlet 方式启动 Spring 容器 -->
<servlet>
  <servlet-name>context</servlet-name>
  <servlet-class>org.springframework.web.context.ContextLoaderServlet</servlets-class>
  <load-on-startup>1</load-on-startup>
</ servlet >
```

在后面章节中讲解 Spring 与 Struts 整合开发时，将采用基于 ContextLoaderListener 的方式来实现由 Web 服务器实例化 ApplicationContext 容器。

18.2　Bean 的作用域

容器最重要的任务是创建并管理 JavaBean 的生命周期，创建 Bean 之后，需要了解 Bean 在容器中是如何在不同作用域下工作的。

Bean 的作用域就是 Bean 实例的生存空间或有效范围。Spring 为 Bean 实例定义了 5 种作用域来满足不同情况下的应用需求，具体介绍如下。

- singleton：在每个 Spring IoC 容器中，一个 bean 定义对应一个对象实例。
- prototype：一个 bean 定义对应多个对象实例。
- request：在一次 Http 请求中，容器会返回该 Bean 的同一个实例，而对于不同的用户请求，会返回不同的实例。该作用域仅在基于 Web 的 Spring ApplicationContext 情形下有效。
- session：在一次 HTTP Session 中，容器会返回该 Bean 的同一个实例。而对于不同的 HTTP Session 请求，会返回不同的实例。该作用域仅在基于 Web 的 Spring ApplicationContext 情形下有效。
- global session：在一个全局的 HTTP Session 中，容器会返回该 Bean 的同一个实例。在典型情况下，仅在使用 portlet context 时有效。该作用域仅在基于 Web 的 Spring

ApplicationContext 情形下有效。

1) singleton(单实例)作用域

这是 Spring 容器默认的作用域。当一个 bean 的作用域为 singleton，Spring IoC 容器中只会存在一个共享的 bean 实例，并且所有对 bean 的请求，只要 id 与该 bean 定义相匹配，则只会返回 bean 的同一实例。换言之，当把一个 bean 定义设置为 singleton 作用域时，Spring IoC 容器只会创建该 bean 定义的唯一实例。这个单一实例会被存储到单例缓存(singleton cache)中，并且所有针对该 bean 的后续请求和引用都将返回被缓存的对象实例。单例模式对无会话状态的 Bean(如 DAO 组件、业务逻辑组件)来说是最理想的选择。

要在 Spring 配置文件 applicationContext.xml 中将 bean 定义成 singleton，可以这样配置：

```xml
<bean id="helloWorld" class="com.shw.HelloWorld" scope="singleton">
    <property name="userName" value="zhangsan"></property>
</bean>
```

将项目 spring_1 复制并命名为 spring_3，再导入 MyEclipse 开发环境中。在项目 spring_3 的 com.shw 包中创建测试类 TestBeanScope，在 main()方法中测试 singleton 作用域，代码如下：

```java
package com.shw;
import org.springframework.context.ApplicationContext;
import org.springframework.context.support.ClassPathXmlApplicationContext;
public class TestBeanScope {
    public static void main(String[] args) {
        // 加载applicationContext.xml 配置
        ApplicationContext context = new ClassPathXmlApplicationContext(
            "applicationContext.xml");
        // 获取配置中的实例
        HelloWorld hw1 = (HelloWorld) context.getBean("helloWorld");
        HelloWorld hw2 = (HelloWorld) context.getBean("helloWorld");
        System.out.println(hw1 == hw2);
    }
}
```

运行测试类 TestBeanScope，控制台输出结果为 true，说明只创建了一个 HelloWorld 类的实例。

2) prototype(原型模式)作用域

prototype 作用域的 bean 会导致在每次对该 bean 请求时都会创建一个新的 bean 实例。对需要保持会话状态的 Bean(如 Struts 2 中充当控制器的 Action 类)应该使用 prototype 作用域。Spring 不能对一个原型模式 Bean 的整个生命周期负责，容器在初始化、装配好一个原型模式实例后，将它交由客户端，就不再过问了。因此，客户端要负责原型模式实例的生命周期管理。

在 Spring 配置文件中将 bean 定义成 prototype，可以这样配置：

```xml
<bean id="helloWorld" class="com.shw.HelloWorld" scope="prototype">
    <property name="userName" value="zhangsan"></property>
</bean>
```

再次运行测试类 TestBeanScope，控制台输出结果为 false，说明创建了两个 HelloWorld 类的实例。其他作用域，如 request、session 以及 global session 仅在基于 Web 的应用中使用。

18.3 基于 Annotation 的 Bean 装配

在 Spring 中尽管使用 XML 配置文件可以实现 Bean 的装配工作，但如果应用中 Bean 的数量较多，会导致 XML 配置文件过于臃肿，从而给维护和升级带来一定的困难。从 JDK 5 开始提供了名为 Annotation(注解)的功能，Spring 正是利用这一特性。Spring 逐步完善对 Annotation 注解技术的全面支持，使 XML 配置文件不再臃肿，向"零配置"迈进。

Spring 中定义了一系列 Annotation 注解，具体介绍如下。

(1) @Component 注解。

@Component 是一个泛化的概念，仅仅表示一个组件(Bean)，可以作用在任何层次。

(2) @Repository 注解。

@Repository 注解用于将数据访问层(DAO 层)的类标识为 Spring 的 Bean。

(3) @Service 注解。

@Service 通常作用在业务层，但是目前该功能与@Component 相同。

(4) @Controller 注解。

@Controller 标识表示层组件，但是目前该功能与@Component 相同。

通过在类上使用@Component、@Repository、@Service 和@Controller 注解，Spring 会自动创建相应的 BeanDefinition 对象，并注册到 ApplicationContext 中。这些类就成了 Spring 受管组件。

(5) @Autowired 注解。

用于对 Bean 的属性变量、属性的 set 方法及构造函数进行标注，配合对应的注解处理器完成 Bean 的自动配置工作。@Autowired 注解默认按照 Bean 类型进行装配。@Autowired 注解加上@Qualifier 注解，可直接指定一个 Bean 实例名称来进行装配。

(6) @Resource 注解。

作用相当于@Autowired，配置对应的注解处理器完成 Bean 的自动配置工作。区别在于：@Autowired 默认按照 Bean 类型进行装配，@Resource 默认按照 Bean 实例名称进行装配。@Resource 包括 name 和 type 两个重要属性。Spring 将 name 属性解析为 Bean 实例的名称，type 属性解析为 Bean 实例的类型。如果指定 name 属性，则按照实例名称进行装配；如果指定 type，则按照 Bean 类型进行装配。如果都不指定，则先按照 Bean 实例名称装配，如果不能匹配，再按照 Bean 类型进行装配，如果都无法匹配，则抛出 NoSuchBeanDefinitionException 异常。

(7) @Qualifier 注解。

与@Autowired 注解配合，将默认按 Bean 类型进行装配修改为按 Bean 实例名称进行装配，Bean 的实例名称由@Qualifier 注解的参数指定。

在 17.2 小节以登录验证为例讲述依赖注入时，使用了基于 XML 的 Bean 装配。下面将该示例的依赖关系通过@Resource 注解进行装配，实现过程如下。

(1) 将项目 spring_2 复制并命名为 spring_4，再导入 MyEclipse 开发环境中。

(2) 将 spring-aop-4.3.5.RELEASE.jar 文件添加到项目 spring_4 的 lib 目录中，再将该 jar 包添加到项目的构建路径中。

(3) 修改 UsersDAO 接口的实现类 UsersDAOImpl，代码如下：

```java
package com.shw.dao.impl;
import org.springframework.stereotype.Repository;
import com.shw.dao.UsersDAO;
@Repository("usersDAO")
public class UsersDAOImpl implements UsersDAO {
    @Override
    public boolean login(String loginName, String loginPwd) {
        if (loginName.equals("admin") && loginPwd.equals("123456")) {
            return true;
        }
        return false;
    }
}
```

在 UsersDAOImpl 类上使用了@Repository 注解，将数据访问层的类 UsersDAOImpl 标识为 Spring Bean，通过 value 属性值标识该 Bean 名称为 usersDAO(value 可以默认)。

(4) 修改 UsersService 接口的实现类 UsersServiceImpl，代码如下：

```java
package com.shw.service.impl;
import org.springframework.beans.factory.annotation.Autowired;
import org.springframework.stereotype.Service;
import com.shw.dao.UsersDAO;
import com.shw.service.UsersService;
@Service("usersService")
public class UsersServiceImpl implements UsersService {
    @Autowired
    UsersDAO usersDAO;
    @Override
    public boolean login(String loginName, String loginPwd) {
        return usersDAO.login(loginName, loginPwd);
    }
}
```

在 UsersServiceImpl 类上使用了@Service 注解，将业务逻辑层的类 UsersServiceImpl 标识为 Spring Bean，该 Bean 的名称为 usersService。在 UsersDAO 类型的属性 usersDAO 上使用了@Autowired 注解，可将步骤 3 中由 Spring 容器实例化的名称为 usersDAO 的 Bean 装配到属性 usersDAO。@Autowired 注解自动装配具有兼容类型的单个 Bean 属性，可以加在构造器、普通字段、一切具有参数的方法上。

(5) 修改 Spring 配置文件。

为了让 Spring 能够扫描类路径中的类，并识别出@Component、@Repository、@Service 和@Controller 注解，需要在 Spring 配置文件中启用 Bean 的自动扫描功能，可以通过 <context:component-scan /> 元素来实现。代码如下：

```xml
<?xml version="1.0" encoding="UTF-8"?>
<beans xmlns="http://www.springframework.org/schema/beans"
    xmlns:xsi="http://www.w3.org/2001/XMLSchema-instance"
```

```
xmlns:p="http://www.springframework.org/schema/p"
    xmlns:context="http://www.springframework.org/schema/context"
    xsi:schemaLocation="http://www.springframework.org/schema/beans
http://www.springframework.org/schema/beans/spring-beans-4.3.xsd
        http://www.springframework.org/schema/context
http://www.springframework.org/schema/context/spring-context-4.3.xsd">
    <!-- 配置自动扫描的基包 -->
    <context:component-scan base-package="com.shw" />
</beans>
```

为了能正常使用<context>元素，需要引入 context 命名空间。base-package 属性指定需要扫描的基包，Spring 容器将会扫描这个基包中及其子包中的所有类，当需要扫描多个包时，可以使用逗号分隔。对于扫描到的组件，Spring 有默认的命名策略，使用非限定类名，第一个字母小写，也可以在注解中通过 value 属性值标识组件的名称。<context:component-scan /> 元素还会自动注册<AutowiredAnnotationBeanPostProcessor 实例，该实例可以自动装配具有@Autowired、@Resource 和@Inject 注解的属性。

运行测试类 TestSpringDI，运行效果同项目 spring_2，控制台输出结果为"登录成功"。

18.4 小　　结

本章主要介绍了 Bean 工厂 ApplicationContext、Bean 的作用域和基于 Annotation 注解的 Bean 装配。使用 Annotation 注解技术，使 XML 配置文件不再臃肿，从而向"零配置"迈进。

第 19 章
面向切面编程 (Spring AOP)

　　AOP(Aspect Oriented Programming，面向切面编程)是一种编程范式，旨在通过允许横切关注点的分离，提高模块化。AOP 提供切面来将跨越对象关注点模块化。目前有许多 AOP 框架，其中最流行的两个框架为 Spring AOP 和 AspectJ。

19.1 AOP 简介

在传统的业务处理代码中，通常会进行日志记录、参数合法性验证、异常处理、事务控制等操作。甚至常常要关心这些操作的代码是否处理正确。例如：哪里的业务日志忘记做了，哪里的事务是否在异常时忘记添加事务回滚的代码，更为担心的是，如果需要修改系统日志的格式或者安全验证的策略等，会有多少地方的代码要修改等。

日志、事务、安全验证等这些"通用的"、散布在系统各处的需要在实现业务逻辑时关注的事情称为"切面"，也可称为"关注点"。如果能将这些"切面"集中处理，然后在具体运行时，再由容器动态织入这些"切面"。这样至少有以下两个好处。

(1) 减少"切面"代码里的错误，处理策略改变时还能做到统一修改。
(2) 在编写业务逻辑时可以专心于核心业务。

因此，AOP 要做的事件就是从系统中分离出"切面"，然后集中实现，从而独立地编写业务代码和方面代码，在系统运行时，再将方面"织入"到系统中。

在使用 AOP 时，会涉及切面、通知、切入点、目标对象、代理对象、织入等概念。下面对这些概念做简要介绍。

(1) 切面。是共有功能的实现，如日志切面、事务切面、权限切面等等。在实际应用中通常是存放共有功能实现的普通 Java 类，要被 AOP 容器识别为切面，需要在配置中通过 <bean>标记指定。

(2) 通知。是切面的具体实现。以目标方法为参照点。根据放置的位置不同，可以分为前置通知、后置通知、异常通知、环绕通知和最终通知 5 种。切面类中的某个方法具体属于哪类通知，需要在配置中指定。

(3) 切入点。用于定义通知应该织入哪些连接点上。

(4) 目标对象。指将要织入切面的对象，即那些被通知的对象。这些对象中只包含核心业务逻辑代码，所有日志、事务、安全验证等方面的功能等待 AOP 容器的织入。

(5) 代理对象。将通知应用到目标对象之后，被动态创建的对象，代理对象的功能相当于目标对象中实现的核心业务逻辑功能加上方面(日志、事务、安全验证)代码实现的功能。

(6) 织入。将切面应用到目标对象，从而创建一个新的代理对象的过程。

19.2 基于 XML 配置文件的 AOP 实现

Spring AOP 通知包括前置通知、返回通知、正常返回通知、异常通知和环绕通知。本节将基于 XML 配置文件的方式实现前置通知、返回通知、异常通知和环绕通知。

19.2.1 前置通知

前置通知在连接点(所织入的业务方法)前面执行，不会影响连接点的执行，除非此处抛出异常。下面通过示例演示如何实现前置通知，其过程如下。

(1) 将项目 spring_1 复制并命名为 spring_5，再导入 MyEclipse 开发环境中。

(2) 将 spring-aop-4.3.5.RELEASE.jar、spring-aspects-4.3.5.RELEASE.jar、aopalliance-1.0 和 aspectjweaver-1.8.6.jar 文件添加到项目 spring_5 的 lib 目录中，再将该 jar 包添加到项目的构建路径中。

(3) 创建包 com.shw.service，在包中创建接口 MealService，添加方法 browse()，模拟用户浏览餐品的业务。代码如下：

```java
package com.shw.service;
public interface MealService {
    public void browse(String loginName,String mealName);
}
```

(4) 创建接口 MealService 的实现类 MealServiceImpl，存放在 com.shw.service.impl 包中，实现 browse()方法。代码如下：

```java
package com.shw.service.impl;
import com.shw.service.MealService;
public class MealServiceImpl implements MealService {
    @Override
    public void browse(String loginName, String mealName) {
        System.out.println("执行业务方法browse");
    }
}
```

(5) 创建包 com.shw.aop，在包中创建日志通知类 LogAdvice，在类中编写用于生成日志记录的方法 myBeforeAdvice，代码如下：

```java
package com.shw.aop;
import java.text.SimpleDateFormat;
import java.util.Arrays;
import java.util.Date;
import java.util.List;
import org.aspectj.lang.JoinPoint;
public class LogAdvice {
    // 此方法将作为前置通知
    public void myBeforeAdvice(JoinPoint joinpoint) {
        // 获取业务方法参数
        List<Object> args = Arrays.asList(joinpoint.getArgs());
        // 日志格式字符串
        String logInfoText = "前置通知: "
                + new SimpleDateFormat("yyyy-MM-dd HH:mm:ss")
                    .format(new Date()) + " " + args.get(0).toString()
                + " 浏览餐品 " + args.get(1).toString();
        // 将日志信息输出到控制台
        System.out.println(logInfoText);
    }
}
```

myBeforeAdvice 方法中使用了 JoinPoint 接口类型的参数 joinpoint，通过 joinpoint 的 getArgs()方法，myBeforeAdvice 就能获得业务方法 browse 的参数 loginName 和 mealName。

(6) 编辑 Spring 配置文件。

在 Spring 配置文件 applicationContext.xml 中，采用 AOP 配置方式将日志类 LogAdvice 与业务组件 MealService 原本是两个互不相关的类和接口通过 AOP 元素进行装配，实现将日志通知类 LogAdvice 中的日志通知织入 MealService 中，以实现预期的日志记录。applicationContext.xml 配置文件内容如下：

```xml
<?xml version="1.0" encoding="UTF-8"?>
<beans xmlns="http://www.springframework.org/schema/beans"
    xmlns:xsi="http://www.w3.org/2001/XMLSchema-instance"
    xmlns:p="http://www.springframework.org/schema/p"
    xmlns:aop="http://www.springframework.org/schema/aop"
    xsi:schemaLocation="http://www.springframework.org/schema/aop
http://www.springframework.org/schema/aop/spring-aop-4.3.xsd
        http://www.springframework.org/schema/beans
http://www.springframework.org/schema/beans/spring-beans-4.3.xsd">
    <!-- 实例化业务类的 Bean -->
    <bean id="mealService" class="com.shw.service.impl.MealServiceImpl">
    </bean>
    <!--实例化日志通知类(切面)的 Bean -->
    <bean id="logAdvice" class="com.shw.aop.LogAdvice"></bean>
    <!-- 配置 aop -->
    <aop:config>
        <!-- 配置日志切面 -->
        <aop:aspect id="logaop" ref="logAdvice">
            <!-- 定义切入点,切入点采用正则表达式，含义是对
com.shw.service.ProductInfoService 中的所有方法，都进行拦截 -->
            <aop:pointcut id="logpointcut"
                expression="execution(* com.shw.service.MealService.*(..))" />
            <!-- 将日志通知类中的 myBeforeAdvice 方法指定为前置通知 -->
            <aop:before method="myBeforeAdvice" pointcut-ref="logpointcut" />
        </aop:aspect>
    </aop:config>
</beans>
```

由于 Spring 的 AOP 配置标签是放置于 aop 命名空间之下的，需要在 Spring 配置文件的 <beans>标记中导入 AOP 命名空间及其配套的 schemaLocation。在配置文件中，首先实例化业务类 MealServiceImpl 的 Bean，然后实例化日志通知类(切面) LogAdvice 的 Bean，最后通过<aop:config>元素进行 AOP 的配置。在配置 AOP 时，通过<aop:aspect>子元素配置日志切面；在配置日志切面时，先通过<aop:pointcut>子元素定义切入点，切入点采用正则表达式 execution(* com.shw.service.MealService.*(..))，含义是对 com.shw.service.MealService 包中的所有方法，都进行拦截。再通过<aop:before>子元素将日志通知类中的 myBeforeAdvice 方法指定为前置通知。

(7) 在 com.shw 包中创建测试类 TestAOP.java，代码如下：

```java
package com.shw;
import org.springframework.context.ApplicationContext;
import org.springframework.context.support.ClassPathXmlApplicationContext;
import com.shw.service.MealService;
public class TestAOP {
    public static void main(String[] args) {
```

```
        // 加载 applicationContext.xml 配置
        ApplicationContext ctx = new ClassPathXmlApplicationContext(
                "applicationContext.xml");
        // 获取配置中的实例
        MealService mealService = (MealService) ctx
                .getBean("mealService");
        // 调用业务方法
        mealService.browse("zhangsan", "素锅烤鸭肉");
    }
}
```

执行测试类 TestAOP，控制台输出结果如下：

```
前置通知：2017-06-10 21:49:23 zhangsan 浏览餐品 素锅烤鸭肉
执行业务方法browse
```

从控制台输出可以看出，在业务方法 browse 执行前，先输出了日志通知类 LogAdvice 中 myBeforeAdvice 方法产生的日志记录。

19.2.2 返回通知

返回通知在连接点执行完成后执行，不管是正常执行完成，还是抛出异常，都会执行返回通知中的内容。下面通过示例演示如何实现返回通知，其过程如下。

(1) 在日志通知类 LogAdvice 中添加方法 myAfterReturnAdvice，作为返回通知。代码如下：

```
public void myAfterReturnAdvice(JoinPoint jionpoint) {
    // 获取方法参数
    List<Object> args = Arrays.asList(joinpoint.getArgs());
    // 日志格式字符串
    String logInfoText = "返回通知: "
            + new SimpleDateFormat("yyyy-MM-dd HH:mm:ss")
                    .format(new Date()) + " " + args.get(0).toString()
            + " 浏览餐品 " + args.get(1).toString();
    // 将日志信息输出到控制台
    System.out.println(logInfoText);
}
```

(2) 在 Spring 配置文件 applicationContext.xml 中的<aop:aspect>元素内添加 <aop:after-returning>元素，将 LogAdvice 日志通知类中的 myAfterReturnAdvice 方法指定为返回通知。代码如下：

```
<aop:after-returning method="myAfterReturnAdvice"          pointcut-ref="logpointcut" />
```

将 applicationContext.xml 中前置通知的<aop:before>配置加以注释，再执行测试类 TestAOP，控制台输出结果如下：

```
执行业务方法browse
返回通知：2017-06-11 20:22:21 zhangsan 浏览餐品 素锅烤鸭肉
```

从控制台输出可以看出，在业务方法 browse 执行后才输出日志通知类 LogAdvice 中

myAfterReturnAdvice 方法产生的日志记录。

19.2.3 异常通知

异常通知在连接点抛出异常后执行，下面通过示例演示如何实现异常通知，其过程如下。

(1) 在日志通知类 LogAdvice 中添加方法 myThrowingAdvice，作为异常通知。代码如下：

```java
public void myThrowingAdvice(JoinPoint jionpoint, Exception e) {
    // 获取被调用的类名
    String targetClassName = joinpoint.getTarget().getClass().getName();
    // 获取被调用的方法名
    String targetMethodName = joinpoint.getSignature().getName();
    // 日志格式字符串
    String logInfoText = "异常通知：执行" + targetClassName + "类的"
        + targetMethodName + "方法时发生异常";
    // 将日志信息输出到控制台
    System.out.println(logInfoText);
}
```

(2) 修改 MealServiceImpl 类中 show 方法，人为抛出一个异常。代码如下：

```java
package com.shw.service.impl;
import com.shw.service.MealService;
public class MealServiceImpl implements MealService {
    @Override
    public void browse(String loginName, String mealName) {
        System.out.println("执行业务方法 browse");
        // 演示异常通知时，人为抛出该异常
        throw new RuntimeException("这是特意抛出的异常信息！");
    }
}
```

(3) 在 Spring 配置文件 applicationContext.xml 中的<aop:aspect>元素内添加<aop:after-throwing>元素，将 LogAdvice 日志通知类中的 myThrowingAdvice 方法指定为异常通知。代码如下：

```xml
<aop:after-throwing method="myThrowingAdvice" pointcut-ref="logpointcut" throwing="e" />
```

将 applicationContext.xml 中前置通知的<aop:before>和返回通知的<aop:after-returning>配置加以注释，再执行测试类 TestAOP，控制台输出结果如下：

```
执行业务方法 browse
异常通知：执行 com.shw.service.impl.MealServiceImpl 类的 browse 方法时发生异常
Exception in thread "main" java.lang.RuntimeException: 这是特意抛出的异常信息！
```

从控制台输出可以看出，在执行业务方法 browse 时，输出日志通知类 LogAdvice 中 myThrowingAdvice 方法产生的日志记录。

19.2.4 环绕通知

环绕通知围绕在连接点前后,比如一个方法调用的前后,这是最强大的通知类型,能在方法调用前后自定义一些操作。环绕通知还需要负责决定是继续处理 joinpoint(调用 ProceedingJoinPoint 的 proceed 方法)还是中断执行。下面通过示例演示如何实现环绕通知,其过程如下。

(1) 在日志通知类 LogAdvice 中添加方法 myAroundAdvice,作为环绕通知。代码如下:

```java
public void myAroundAdvice(ProceedingJoinPoint joinpoint) throws Throwable
{
    long beginTime = System.currentTimeMillis();
    joinpoint.proceed();
    long endTime = System.currentTimeMillis();
    // 获取被调用的方法名
    String targetMethodName = joinpoint.getSignature().getName();
    // 日志格式字符串
    String logInfoText = "环绕通知: " + targetMethodName + "方法调用前时间" + beginTime + "毫秒," + "调用后时间" + endTime + "毫秒";
    // 将日志信息输出到控制台
    System.out.println(logInfoText);
}
```

(2) 修改 MealServiceImpl 类中 browse 方法,通过 while 循环延长方法的执行时间。代码如下:

```java
public void browse(String loginName, String mealName) {
    System.out.println("执行业务方法browse");
    int i = 100000000;
    while (i > 0) {
        i--;
    }
}
```

(3) 在 Spring 配置文件 applicationContext.xml 中的<aop:aspect>元素内添加<aop:around>元素,将 LogAdvice 日志通知类中的 myAroundAdvice 方法指定为环绕通知。代码如下:

```xml
<aop:around method="myAroundAdvice" pointcut-ref="logpointcut" />
```

将 applicationContext.xml 中前置通知的<aop:before>、返回通知的<aop:after-returning>和异常通知的<aop:after-throwing>元素配置加以注释,再执行测试类 TestAOP,控制台输出结果如下:

```
执行业务方法browse
环绕通知: browse方法调用前时间1497236740026毫秒,调用后时间1497236740028毫秒
```

从控制台输出结果可以看出,通过环绕通知可以记录业务方法 browse 执行前后的时间。

19.3 基于@AspectJ 注解的 AOP 实现

基于 XML 配置文件的 AOP 实现免不了在 Spring 配置文件中配置大量信息,不仅配置麻烦,而且造成配置文件的臃肿。Annotation 注解技术具有为配置文件瘦身的本领。Spring 为 AOP 的实现提供了一套 Annotation 注解,用以取代 Spring 配置文件中为实现 AOP 功能所配置的臃肿代码。这些 Annotation 注解的说明如下。

- @AspectJ:用于定义一个切面。
- @Pointcut:用于定义一个切入点,切入点的名称由一个方面名称定义。
- @Before:用于定义一个前置通知。
- @AfterReturning:用于定义一个后置通知。
- @AfterThrowing:用于定义一个异常通知。
- @Around:用于定义一个环绕通知。

使用@AspectJ 注解重新实现 19.2 小节中 LogAdvice 日志类功能,具体步骤如下。

(1) 将项目 spring_5 复制并命名为 spring_6,再导入 MyEclipse 开发环境中。

(2) 在项目 spring_6 中,修改 MealService 接口的实现类 MealServiceImpl,在类上添加@Component("mealService")注解,在 Spring 容器中自动创建 MealServiceImpl 类的 Bean 实例。代码如下:

```
@Component("mealService")
public class MealServiceImpl implements MealService {
    ……
}
```

(3) 修改日志通知类 LogAdvice,使用注解定义 Bean、切面、切点和四种类型通知。代码如下:

```
package com.shw.aop;
……
import org.springframework.stereotype.Component;
@Aspect
@Component
public class LogAdvice {
    @Pointcut("execution(* com.shw.service.MealService.*(..))")
    private void allMethod() {
    } // 定义切入点名字
    @Before("allMethod()")
    // 此方法将作为前置通知
    public void myBeforeAdvice(JoinPoint joinpoint) {
        // 获取业务方法参数
        List<Object> args = Arrays.asList(joinpoint.getArgs());
        // 日志格式字符串
        String logInfoText = "前置通知: "
                + new SimpleDateFormat("yyyy-MM-dd HH:mm:ss")
                    .format(new Date()) + " " + args.get(0).toString() + "浏览餐品 " + args.get(1).toString();
        // 将日志信息输出到控制台
        System.out.println(logInfoText);
```

```java
    }
    @AfterReturning("allMethod()")
    // 返回通知
    public void myAfterReturnAdvice(JoinPoint joinpoint) {
        // 获取方法参数
        List<Object> args = Arrays.asList(joinpoint.getArgs());
        // 日志格式字符串
        String logInfoText = "返回通知："
                + new SimpleDateFormat("yyyy-MM-dd HH:mm:ss")
                        .format(new Date()) + " " + args.get(0).toString() + " 浏览餐品 " + args.get(1).toString();
        // 将日志信息输出到控制台
        System.out.println(logInfoText);
    }
    @AfterThrowing(pointcut="allMethod()",throwing="e")
    // 异常通知
    public void myThrowingAdvice(JoinPoint joinpoint, Exception e) {
        // 获取被调用的类名
        String targetClassName = joinpoint.getTarget().getClass().getName();
        // 获取被调用的方法名
        String targetMethodName = joinpoint.getSignature().getName();
        // 日志格式字符串
        String logInfoText = "异常通知：执行" + targetClassName + "类的" + targetMethodName + "方法时发生异常";
        // 将日志信息输出到控制台
        System.out.println(logInfoText);
    }
    @Around("allMethod()")
    // 环绕通知
    public void myAroundAdvice(ProceedingJoinPoint joinpoint) throws Throwable {
        long beginTime = System.currentTimeMillis();
        joinpoint.proceed();
        long endTime = System.currentTimeMillis();
        // 获取被调用的方法名
        String targetMethodName = joinpoint.getSignature().getName();
        // 日志格式字符串
        String logInfoText = "环绕通知：" + targetMethodName + "方法调用前时间" + beginTime + "毫秒," + "调用后时间" + endTime + "毫秒";
        // 将日志信息输出到控制台
        System.out.println(logInfoText);
    }
}
```

在 LogAdvice 类上，首先使用@Component 注解在 Spring 容器中自动创建 LogAdvice 类的 Bean 实例，使用@Aspect 注解定义一个切面；然后使用@Pointcut 注解定义一个切入点，切入点的名字为 allMethod()，切入点的正则表达式 execution(* com.shw.service.MealService.*(..))，含义是对 com.shw.service.MealService 接口中的所有方法都进行拦截；再分别使用@Before、@AfterReturning、@AfterThrowing 和@Around 注解定义前置通知、返回通知、异常通知和环绕通知，这些通知中的代码含义与前一小节相同，此处不再赘述。

（4）修改 Spring 配置文件，配置自动扫描的包，并开启基于@AspectJ 切面的注解处理器。代码如下：

```xml
<?xml version="1.0" encoding="UTF-8"?>
<beans xmlns="http://www.springframework.org/schema/beans"
    xmlns:xsi="http://www.w3.org/2001/XMLSchema-instance"
xmlns:p="http://www.springframework.org/schema/p"
    xmlns:aop="http://www.springframework.org/schema/aop"
    xmlns:context="http://www.springframework.org/schema/context"
    xsi:schemaLocation="http://www.springframework.org/schema/aop
http://www.springframework.org/schema/aop/spring-aop-4.3.xsd
        http://www.springframework.org/schema/beans
http://www.springframework.org/schema/beans/spring-beans-4.3.xsd
        http://www.springframework.org/schema/context
http://www.springframework.org/schema/context/spring-context-4.3.xsd">
    <!-- 配置自动扫描的包 -->
    <context:component-scan base-package="com.shw" />
    <!-- 开启基于@AspectJ 切面的注解处理器 -->
    <aop:aspectj-autoproxy />
</beans>
```

由于 Spring 配置文件中使用了<context>和<aop>元素，因此需要引入 context 和 aop 命名空间及其配套的 schemaLocation。

在日志通知类 LogAdvice 中，依次启用一个通知方法进行测试，将其他通知方法加以注释，并根据所测试通知类型的需要，修改 MealServiceImpl 类中 browse 方法代码(参考 19.2 节)，再运行测试类 TestAOP，效果与 19.2 节相同。

基于@AspectJ 注解的 AOP 实现效果与基于 XML 配置文件的 AOP 实现效果相同，但 Spring 配置文件变得更为简洁。

19.4 小　　结

本章主要介绍了 Spring AOP 的相关概念，并以日志通知为例先后讲解了基于 XML 配置文件的 AOP 实现和基于@AspectJ 注解的 AOP 实现。通过对比分析可知，使用 Spring 为 AOP 的实现提供的一组注解，极大地简化了 Spring 的配置。

第 III 篇

整合和实例部分

- 第 20 章　Spring 整合 Struts 2 与 Hibernate
- 第 21 章　Spring MVC
- 第 22 章　Spring 整合 Struts 2 与 Hibernate 实现网上订餐系统前台
- 第 23 章　Spring 整合 Spring MVC 与 Hibernate 实现网上订餐系统后台
- 第 24 章　Spring 整合 Spring MVC 与 MyBatis 实现新闻发布系统

第 20 章 Spring 整合 Struts 2 与 Hibernate

Spring 的依赖注入和 AOP，大家感觉到 Spring 的魅力所在，然而 Spring 的强大之处还在于：对 Hibernate 提供的支持，并集成 Struts 2 框架，让 S2SH (Struts 2+Spring+Hibernate)成为 Java EE 应用开发的经典框架组合之一。Spring 整合 Struts 2 与 Hibernate 可以采用基于 XML 配置文件和基于 Annotation 注解两种方式。本章以用户登录功能为例，介绍这两种整合方式。

20.1 基于 XML 配置的 S2SH 整合

采用 XML 配置文件方式整合 S2SH 实现用户登录的步骤包括环境搭建、创建实体类及映射文件、Spring 整合 Hibernate、DAO 层开发、Service 层开发、Action 开发、Spring 整合 Struts 2 和创建页面。

20.1.1 环境搭建

在 MyEclipse 中创建一个名为 s2sh 的 Web 项目，选择 Java EE version 为 JavaEE 7，Java version 为 1.8，Target runtime 为 Apache Tomcat v8.0。然后依次添加 Hibernate、Spring 和 Struts 2 框架的 jar 包，以及相关 jar 包。

1. 添加 Hibernate 框架及相关的 jar 包

将 hibernate-release-5.2.6.Final.zip 解压后的 lib\required 目录下如图 20-1 所示的 10 个 jar 包，以及 MySQL 数据库驱动包 mysql-connector-java-5.1.18-bin.jar、Hibernate 事务管理包 jboss-transaction-api_1.1_spec-1.0.1.Final.jar、连接池核心包 c3p0-0.9.5.2.jar、c3p0 连接池的依赖包 mchange-commons-java-0.2.11.jar 这 4 个包复制到项目 s2sh 的 WebRoot\WEB-INF\lib 目录中。

2. 添加 Spring 框架及相关的 jar 包

将 spring-framework-4.3.5.RELEASE-dist.zip 解压后的 libs 目录下的如图 20-2 所示的 12 个 jar 包，以及相关的 aopalliance-1.0.jar、aspectjweaver-1.8.6.jar 和 cglib-3.2.0.jar 这 3 个 jar 包复制到项目 s2sh 的 WebRoot\WEB-INF\lib 目录中。

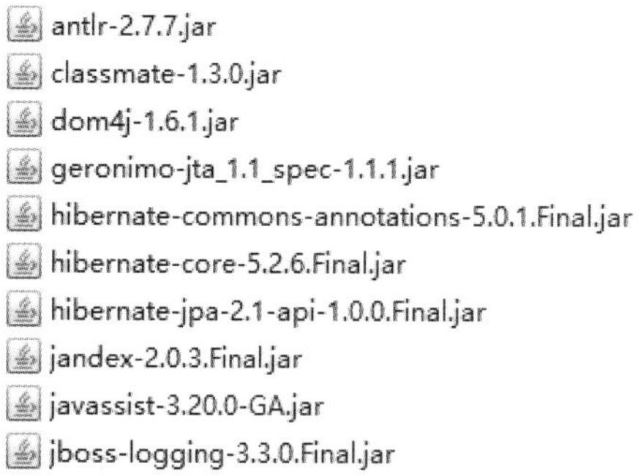

图 20-1　Hibernate 所需的 jar 包

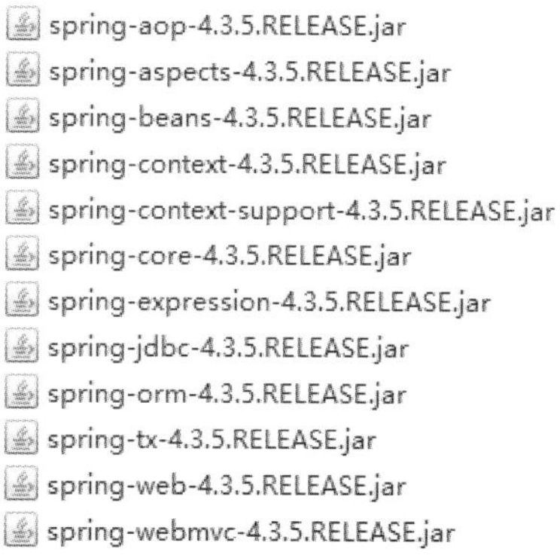

图 20-2　Spring 所需的 jar 包

3. 添加 Struts 2 框架的 jar 包

将 struts-2.5.8-lib.zip 解压后的 lib 目录下如图 20-3 所示的 14 个 jar 包复制到项目 s2sh 的 WebRoot\WEB-INF\lib 目录中。

项目最终的目录结构如图 20-4 所示。

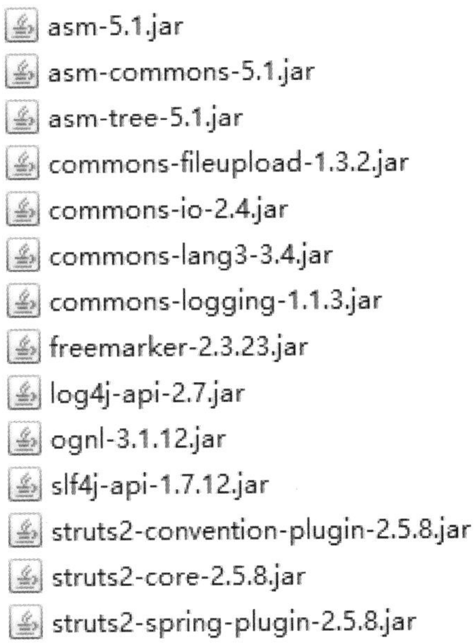

图 20-3　Struts 2 所需的 jar 包 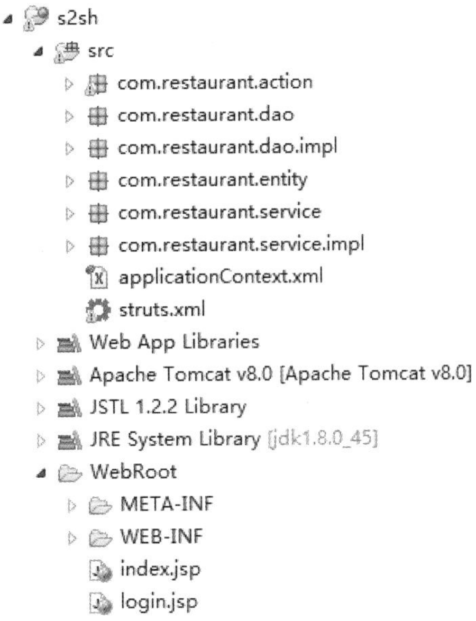　　　　图 20-4　项目 s2sh 的目录结构

20.1.2　创建实体类及映射文件

1．创建实体类

在 src 目录下创建包 com.restaurant.entity，在包中创建实体类 Users，与数据库 restrant 中的数据表 users 对应。代码如下：

```
package com.digital.entity;
import java.util.Date;
public class UserInfo {
    private int id;
    private String loginName;
    private String loginPwd;
    //此处省略了上述属性的 get 方法和 set 方法
}
```

2．创建映射文件

在 com.restaurant.entity 包中创建与实体类 Users 对应的映射文件 Users.hbm.xml。代码如下：

```
<?xml version="1.0" encoding="UTF-8"?>
<!DOCTYPE hibernate-mapping PUBLIC "-//Hibernate/Hibernate Mapping DTD
3.0//EN" "http://www.hibernate.org/dtd/hibernate-mapping-3.0.dtd">
<hibernate-mapping package="com.restaurant.entity">
    <class name="Users" table="users" catalog="restrant">
        <id name="id" type="java.lang.Integer">
            <column name="id" />
```

```xml
            <generator class="native"></generator>
        </id>
        <property name="loginName" type="java.lang.String">
            <column name="LoginName" length="20" not-null="true" />
        </property>
        <property name="loginPwd" type="java.lang.String">
            <column name="LoginPwd" length="20" not-null="true" />
        </property>
    </class>
</hibernate-mapping>
```

20.1.3 Spring 整合 Hibernate

Spring 整合 Hibernate 的目的在于由 Spring 的 IoC 容器来管理 Hibernate 的 SessionFactory，同时让 Hibernate 使用 Spring 的声明式事务，这些目的的实现是在 Spring 配置文件中完成的。在 src 目录下创建 Spring 配置文件 applicationContext.xml，基于 XML 配置数据源 dataSource、配置 Hibernate 的 sessionFactory 实例、声明 Hibernate 事务管理器、定义事务通知、定义切面，并将事务通知和切面组合。代码如下：

```xml
<?xml version="1.0" encoding="UTF-8"?>
<beans xmlns="http://www.springframework.org/schema/beans"
    xmlns:xsi="http://www.w3.org/2001/XMLSchema-instance"
    xmlns:p="http://www.springframework.org/schema/p"
    xmlns:tx="http://www.springframework.org/schema/tx"
    xmlns:aop="http://www.springframework.org/schema/aop"
    xsi:schemaLocation="http://www.springframework.org/schema/aop
http://www.springframework.org/schema/aop/spring-aop-4.3.xsd
        http://www.springframework.org/schema/beans
http://www.springframework.org/schema/beans/spring-beans-4.3.xsd
        http://www.springframework.org/schema/tx
http://www.springframework.org/schema/tx/spring-tx-4.3.xsd">
    <!-- 配置数据源 -->
    <bean id="dataSource"
        class="com.mchange.v2.c3p0.ComboPooledDataSource">
        <property name="driverClass" value="com.mysql.jdbc.Driver" />
        <property name="jdbcUrl" value="jdbc:mysql:///restrant" />
        <property name="user" value="root" />
        <property name="password" value="123456" />
        <property name="minPoolSize" value="5" />
        <property name="maxPoolSize" value="10" />
    </bean>
    <!-- 配置 Hibernate 的 sessionFactory 实例 -->
    <bean id="sessionFactory"
        class="org.springframework.orm.hibernate5.LocalSessionFactoryBean">
        <!-- 配置数据源属性 -->
        <property name="dataSource">
            <ref bean="dataSource" />
        </property>
        <!-- 配置 Hibernate 的基本属性-->
        <property name="hibernateProperties">
```

```xml
            <props>
                <prop key="hibernate.dialect">
                    org.hibernate.dialect.MySQLDialect
                </prop>
            </props>
        </property>
        <!-- 配置 Hibernate 映射文件的位置及名称-->
        <property name="mappingResources">
            <list>
                <value>com/restaurant/entity/Users.hbm.xml</value>
            </list>
        </property>
    </bean>
    <!-- 声明 Hibernate 事务管理器 -->
    <bean id="transactionManager"
        class="org.springframework.orm.hibernate5.HibernateTransactionManager">
        <property name="sessionFactory" ref="sessionFactory" />
    </bean>
    <!-- 定义事务通知,需要事务管理器 -->
    <tx:advice id="txAdvice" transaction-manager="transactionManager">
        <!-- 指定事务传播规则 -->
        <tx:attributes>
            <!-- 对所有方法应用 REQUIRED 事务规则 -->
            <tx:method name="*" propagation="REQUIRED" />
        </tx:attributes>
    </tx:advice>
    <!--定义切面,并将事务通知和切面组合(定义哪些方法应用事务规则) -->
    <aop:config>
    <!-- 对com.restaurant.service 包下的所有类的所有方法都应用事务规则 -->
        <aop:pointcut id="serviceMethods" expression="execution(* com.restaurant.service.*.*(..))" />
        <!-- 将事务通知和切面组合 -->
        <aop:advisor advice-ref="txAdvice" pointcut-ref="serviceMethods" />
    </aop:config>
</beans>
```

在配置文件 applicationContext.xml 中使用了<bean>、<tx>和<aop>元素，首先需要引入 bean、tx 和 aop 命名空间及其配套的 schemaLocation。然后依次实例化连接池核心包 c3p0-0.9.5.2.jar 中的 ComboPooledDataSource 类，在 Spring IoC 容器中配置了数据源，数据源实例名为 dataSource。使用 LocalSessionFactoryBean 类配置 Hibernate 的 sessionFactory 实例，实例名为 sessionFactory，在配置 sessionFactory 实例时，需要配置数据源属性、Hibernate 的基本属性、Hibernate 映射文件的位置及名称。使用 HibernateTransactionManager 类声明 Hibernate 事务管理器，该实例名为 transactionManager。使用<tx:advice>元素定义事务通知(需要事务管理器)，<tx:advice>的子元素<tx:attributes>用于指定事务传播规则，在<tx:attributes>的子元素<tx:method>中配置了对所有的方法应用 REQUIRED 事务规则，表示当前方法必须运行在一个事务环境中，如果一个现有事务正在运行中，该方法将运行在这个事务中，否则，就要开始一个新的事务。如果不配置该规则，在持久层开发时就会无法调用 sessionFactory 的

getCurrentSession 方法获取 session 对象。Spring 的事务规则也就是事务传播行为，常见的事务传播行为如表 20-1 所示。

表 20-1 常见的事务传播行为

名称	说明
REQUIRED	表示当前方法必须运行在一个事务环境中，如果一个现有事务正在运行中，该方法将运行在这个事务中，否则，就要开始一个新的事务
REQUIRESNEW	表示当前方法必须运行在自己的事务里
SUPPORTS	表示当前方法不需要在事务处理环境中，但如果有一个事务正在运行的话，则这个方法也可以运行在这个事务中
MANDATORY	表示当前方法必须运行在一个事务上下文中，否则就抛出异常
NEVER	表示当前方法不应该运行在一个事务上下文中，否则就抛出异常

事务管理的主要任务是事务的创建、事务的回滚与事务的提交，是否需要创建事务及如何创建事务是由事务传播行为控制的，通常数据的读取可以不需要事务管理，或者可以指定为只读事务，而对于数据的增加、删除和修改操作，则有必要进行事务管理。如果没有指定事务的传播行为，Spring 默认将采用 REQUIRED。

使用<aop:config>元素定义切面，并将事务通知和切面组合，即定义哪些方法应用事务规则。在<aop:config>的子元素<aop:pointcut>中配置了对 com.digital.service 包下的所有类的所有方法都应用事务规则，在<aop:config>的子元素<aop:advisor>中将事务通知和切面组合。

至此就完成了 Spring 对 Hibernate 的整合。

20.1.4 DAO 层开发

在项目 s2sh 中创建包 com.restaurant.dao，在包中创建接口 UsersDAO，在 UsersDAO 接口中声明方法 search，用于登录验证，代码如下：

```
package com.restaurant.dao;
import java.util.List;
import com.restaurant.entity.*;
public interface UsersDAO {
    public List<Users> search(Users cond);
}
```

创建 UsersDAO 接口的实现类 UsersDAOImpl，存放在 com.restaurant.dao.impl 包中，实现 search 方法。代码如下：

```
package com.restaurant.dao.impl;
import java.util.List;
import org.hibernate.query.Query;
……
public class UsersDAOImpl implements UsersDAO {
    SessionFactory sessionFactory;
    public void setSessionFactory(SessionFactory sessionFactory) {
        this.sessionFactory = sessionFactory;
```

```java
    }
    @Override
    @SuppressWarnings("unchecked")
    public List<Users> search(Users cond) {
        List<Users> uList = null;
        // 获得 session
        Session session = sessionFactory.getCurrentSession();
        // 创建 HQL 语句
        String hql = "from Users u where u.loginName = ? and u.loginPwd = ?";
        // 执行查询
        Query<Users> query = session.createQuery(hql);
        query.setParameter(0, cond.getLoginName());
        query.setParameter(1, cond.getLoginPwd());
        uList = query.getResultList();
        // 返回结果
        return uList;
    }
}
```

在实现类 UsersDAOImpl 中，声明了 SessionFactory 类型的属性 sessionFactory，并给该属性添加 set 方法，用于接收 Spring 配置文件中 LocalSessionFactoryBean 类的 Bean 实例注入，这样就可以调用 sessionFactory 的 getCurrentSession()方法获得 session 对象进行持久化操作了。

在 Spring 配置文件的<beans>元素中使用子元素<bean>配置 UsersDAOImpl 类的 Bean 实例，并为其属性 sessionFactory 注入值。代码如下：

```xml
<!-- 实例化 UsersDAOImpl 类 -->
<bean id="usersDAO" class="com.restaurant.dao.impl.UsersDAOImpl">
    <property name="sessionFactory" ref="sessionFactory" />
</bean>
```

20.1.5　Service 层开发

在项目 s2sh 中创建包 com.restaurant.service，在包中创建接口 UsersService，在 UsersService 接口中声明方法 login，用于登录验证，代码如下：

```java
package com.restaurant.service;
import java.util.List;
import com.restaurant.entity.Users;
public interface UsersService {
    public List<Users> login(Users cond);
}
```

创建 UsersService 接口的实现类 UsersServiceImpl，存放在 com.restaurant.service.impl 包中，实现 login 方法。代码如下：

```java
package com.restaurant.service.impl;
import java.util.List;
……
public class UsersServiceImpl implements UsersService {
    UsersDAO usersDAO;
    public void setUsersDAO(UsersDAO usersDAO) {
```

```
        this.usersDAO = usersDAO;
    }
    @Override
    public List<Users> login(Users cond) {
        return usersDAO.search(cond);
    }
}
```

在 UsersServiceImpl 类中使用 UsersDAO 接口声明了属性 usersDAO，并为该属性添加了 set 方法，用于接收 Spring 配置文件中的 UsersDAOImpl 类的 Bean 实例 usersDAO 注入。在 Spring 配置文件的<beans>元素中使用子元素<bean>配置 UsersServiceImpl 类的 Bean 实例，并为其属性 usersDAO 注入值。代码如下：

```
<!-- 实例化 UsersServiceImpl 类 -->
<bean id="usersService"
class="com.restaurant.service.impl.UsersServiceImpl">
    <property name="usersDAO" ref="usersDAO" />
</bean>
```

20.1.6 Action 开发

在项目 src 目录下创建包 com.restaurant.action，用于存放 Action。在包中创建类 UsersAction.java，代码如下：

```
package com.restaurant.action;
import java.util.List;
import com.opensymphony.xwork2.ActionSupport;
import com.restaurant.entity.Users;
import com.restaurant.service.UsersService;
public class UsersAction extends ActionSupport {
    Users u;
    public Users getU() {
        return u;
    }
    public void setU(Users u) {
        this.u = u;
    }
    UsersService usersService;
    public void setUsersService(UsersService usersService) {
        this.usersService = usersService;
    }
    public String doLogin() throws Exception {
        List<Users> uList = usersService.login(u);
        if (uList.size() > 0) {
            // 登录成功，转发到 index.jsp
            return "index";
        } else {
            // 登录失败，重定向到 login.jsp
            return "login";
        }
    }
}
```

在 UsersAction 类中使用 UsersService 接口声明了属性 usersService，并为该属性添加了 set 方法，用于接收 Spring 配置文件中的 UsersServiceImpl 类的 Bean 实例 usersService 注入。

20.1.7　Spring 整合 Struts 2

Spring 整合 Struts 2 的目的在于使用 Spring IoC 容器来管理 Struts 2 的 Action，整合的步骤如下：

(1) 在 web.xml 配置文件中指定以 Listener 方式启动 Spring，并配置 Struts 2 的 StrutsPrepareAndExecuteFilter。代码如下：

```xml
<?xml version="1.0" encoding="UTF-8"?>
<web-app xmlns:xsi="http://www.w3.org/2001/XMLSchema-instance"
    xmlns="http://java.sun.com/xml/ns/javaee"
    xsi:schemaLocation="http://java.sun.com/xml/ns/javaee
http://java.sun.com/xml/ns/javaee/web-app_3_0.xsd"
    id="WebApp_ID" version="3.0">
    <!-- 指定以 Listener 方式启动 Spring -->
    <listener>
        <listener-class>org.springframework.web.context.ContextLoaderListener
        </listener-class>
    </listener>
    <!-- 指定 Spring 配置文件的位置 -->
    <context-param>
        <param-name>contextConfigLocation</param-name>
        <param-value>classpath:applicationContext.xml</param-value>
    </context-param>
<!-- 配置 Struts2 的核心控制器 -->
    <filter>
        <filter-name>struts2</filter-name>            <filter-class>org.apache.struts2.dispatcher.filter.
StrutsPrepareAndExecuteFilter
        </filter-class>
    </filter>
    <filter-mapping>
        <filter-name>struts2</filter-name>
        <url-pattern>/*</url-pattern>
    </filter-mapping>
</web-app>
```

(2) 在 Spring 配置文件中配置 UsersAction 类，代码如下：

```xml
<!-- 实例化 UsersAction 类，并为其中属性 usersService 注入值 -->
<bean name="usersAction" class="com.restaurant.action.UsersAction"
scope="prototype">
    <property name="usersService" ref="usersService" />
</bean>
```

为了保证对每个用户的请求都会创建一个新的 Bean 实例，在配置 UsersAction 的实例时，需要将<bean>元素的 scope 属性设置为 prototype(原型模式)。

(3) 在项目 src 目录下创建 Struts 2 的配置文件 struts.xml，代码如下：

```xml
<?xml version="1.0" encoding="UTF-8"?>
<!DOCTYPE struts PUBLIC "-//Apache Software Foundation//DTD Struts
Configuration 2.5//EN" http://struts.apache.org/dtds/struts-2.5.dtd">
<struts>
    <constant name="struts.i18n.encoding" value="utf-8"></constant>
<!--定义名称为restaurant的包,继承struts 2的默认包,指定命名空间为"/" -->
    <package name="restaurant" namespace="/" extends="struts-default">
        <!-- 为类中的方法配置映射 -->
        <action name="doLogin" class="usersAction" method="doLogin">
            <result name="index" type="dispatcher">index.jsp</result>
            <result name="login" type="redirect">login.jsp</result>
        </action>
    </package>
</struts>
```

Spring 整合 Struts 2 后,<action>元素中 class 属性不再使用 UsersAction 的全类名,而是引用 Spring 配置文件中 UsersAction 类的 Bean 实例名 usersAction。

20.1.8 创建页面

创建登录页 login.jsp,表单部分代码如下:

```
<s:form action="doLogin">
    <table>
        <tr>
            <s:textfield name="u.loginName" label="用户名" />
        </tr>
        <tr>
            <s:textfield name="u.loginPwd" label="密码" />
        </tr>
        <tr>
            <s:submit value="登录" />
        </tr>
    </table>
</s:form>
编辑登录成功页 index.jsp
<body>
    欢迎您,登录成功!
</body>
```

部署项目,启动 Tomcat,在浏览器中输入 http://localhost:8080/s2sh/login.jsp,打开如图 20-5 所示的登录页面。输入数据表 users 中正确的用户名和密码,单击"确定"按钮,页面转到 index.jsp,如图 20-6 所示。

图 20-5　登录页面

图 20-6　登录成功页面

20.2　基于 Annotation 注解的 S2SH 整合

Annotation 注解的好处在于，将配置信息直接写在程序中，将配置与程序进行完美结合，使传统的 XML 配置文件得以简化，同时也提高了程序的可读性和可维护性。下面使用 Annotation 注解技术实现 20.1 节中的用户登录功能。

采用 Annotation 注解方式整合 S2SH 实现用户登录的步骤如下。

(1) 将项目 s2sh 复制并命名为 s2sh_annotation，再导入 MyEclipse 开发环境中。

(2) 在项目 s2sh_annotation 的实体类 Users 中采用注解方式实现类的属性与数据表 users 的字段之间的映射关系，修改后的代码如下：

```java
package com.restaurant.entity;
import javax.persistence.*;
@Entity
@Table(name = "users", catalog = "restrant")
public class Users {
    private int id;
    private String loginName;
    private String loginPwd;
    @Id
    @GeneratedValue(strategy = GenerationType.IDENTITY)
    @Column(name = "id", unique = true, nullable = false)
    public int getId() {
        return id;
    }
    ……
    @Column(name = "loginName", length = 20)
    public String getLoginName() {
        return loginName;
    }
    ……
    @Column(name = "loginPwd", length = 20)
    public String getLoginPwd() {
        return loginPwd;
    }
    ……
}
```

采用注解方式实现实体类 Users 后，就不再需要原先的映射文件 Users.hbm.xml 了，因此将其删除。

(3) 使用@Repository 和@Autowired 注解修改 UsersDAOImpl 类，代码如下：

```java
package com.restaurant.dao.impl;
……
//使用@Repository注解在Spring容器中注册实例名为usersDAO的UsersDAOImpl实例
@Repository("usersDAO")
public class UsersDAOImpl implements UsersDAO {
    // 通过@Autowired注解注入Spring容器中的SessionFactory实例
    @Autowired
    SessionFactory sessionFactory;
```

```
    @Override
    @SuppressWarnings("unchecked")
    public List<Users> search(Users cond) {
        ……
    }
}
```

(4) 使用@Service、@Autowired 和@Transactional 注解修改 UsersServiceImpl 类，代码如下：

```
package com.restaurant.service.impl;
……
// 使用@Service 注解在 Spring 容器中注册名为 userInfoService 的 UsersServiceImpl 实例
@Service("userInfoService")
// 使用@Transactional 注解实现事务管理
@Transactional
public class UsersServiceImpl implements UsersService {
    // 使用@Autowired 注解注入 UserInfoDAOImpl 实例
    @Autowired
    UsersDAO usersDAO;
    @Override
    public List<Users> login(Users cond) {
        return usersDAO.search(cond);
    }
}
```

在 UsersServiceImpl 类中使用 Spring 为事务管理提供的@Transactional 注解，通过为@Transactional 指定不同的参数，可以满足不同的事务要求。@Transactional 常见的参数如表 20-2 所示。

表 20-2 @Transactional 参数表

参 数 名	说 明
propagation	设置事务的传播规则，常用的事务规则可参见表 20-1。 格式如：@Transactional(propagation=Propagation.REQUIRED)
rollbackFor	需要回滚的异常类，当方法中抛出异常时，则进行事务回滚。 单一异常类格式：@Transactional(rollbackFor=RuntimeException.class) 多个异常类格式： @Transactional(rollbackFor={RuntimeException.class,Exception.class})
rollbackForClassName	需要回滚的异常类名，当方法抛出指定异常名称时，则进行回滚。 单一异常类名称格式： @Transactional(rollbackForClassName="RuntimeException") 多个异常类名称格式： @Transactional(rollbackForClassName={"RuntimeException","Exception"})
isolation	事务隔离级别，用于处理多个事务并发，基本不需要设置
timeout	设置事务的超时秒数
readOnly	事务是否只读，设置为 true 表示只读

(5) 使用@Controller 与@Scope、@Action 与@Result 和@Autowired 注解修改 UsersAction 类，代码如下：

```java
package com.restaurant.action;
……
//使用@Controller 注解在 Spring 容器中注册 UsersAction 实例
@Controller
//使用@Scope("prototype")指定原型模式
@Scope("prototype")
public class UsersAction extends ActionSupport {
    Users u;
    public Users getU() {
        return u;
    }
    public void setU(Users u) {
        this.u = u;
    }
    // 使用@Autowired 注解注入 UsersServiceImpl 实例
    @Autowired
    UsersService usersService;
    // 使用@Action 注解与@Result 实现 Action 的 Struts 配置
    @Action(value = "/doLogin", results = {
            @Result(name = "index", type = "dispatcher", location = "/index.jsp"),
            @Result(name = "login", type = "redirect", location = "/login.jsp") })
    public String doLogin() throws Exception {
        List<Users> uList = usersService.login(u);
        if (uList.size() > 0) {
            // 登录成功，转发到 index.jsp
            return "index";
        } else {
            // 登录失败，重定向到 login.jsp
            return "login";
        }
    }
}
```

(6) 修改 Spring 配置文件 applicationContext.xml。代码如下：

```xml
<?xml version="1.0" encoding="UTF-8"?>
<beans xmlns="http://www.springframework.org/schema/beans"
   xmlns:xsi="http://www.w3.org/2001/XMLSchema-instance"
xmlns:p="http://www.springframework.org/schema/p"
   xmlns:tx="http://www.springframework.org/schema/tx"
xmlns:aop="http://www.springframework.org/schema/aop"
   xmlns:context="http://www.springframework.org/schema/context"
   xsi:schemaLocation="http://www.springframework.org/schema/aop
http://www.springframework.org/schema/aop/spring-aop-4.3.xsd
       http://www.springframework.org/schema/beans
http://www.springframework.org/schema/beans/spring-beans-4.3.xsd
       http://www.springframework.org/schema/context
http://www.springframework.org/schema/context/spring-context-4.3.xsd
       http://www.springframework.org/schema/tx
http://www.springframework.org/schema/tx/spring-tx-4.3.xsd">
```

```xml
<!-- 配置数据源 -->
<bean id="dataSource"
    class="com.mchange.v2.c3p0.ComboPooledDataSource">
    <property name="driverClass" value="com.mysql.jdbc.Driver" />
    <property name="jdbcUrl" value="jdbc:mysql:///restrant" />
    <property name="user" value="root" />
    <property name="password" value="123456" />
    <property name="minPoolSize" value="5" />
    <property name="maxPoolSize" value="10" />
</bean>
<!-- 配置Hibernate的sessionFactory实例 -->
<bean id="sessionFactory"
    class="org.springframework.orm.hibernate5.LocalSessionFactoryBean">
    <!-- 配置数据源属性 -->
    <property name="dataSource">
        <ref bean="dataSource" />
    </property>
    <!-- 配置 Hibernate 的基本属性 -->
    <property name="hibernateProperties">
        <props>
            <prop key="hibernate.dialect">
                org.hibernate.dialect.MySQLDialect
            </prop>
        </props>
    </property>
    <!-- 配置 Hibernate 基于注解的实体类的位置及名称 -->
    <property name="annotatedClasses">
        <list>
            <value>com.restaurant.entity.Users</value>
        </list>
    </property>
</bean>
<!-- 声明Hibernate 事务管理器 -->
<bean id="transactionManager"
    class="org.springframework.orm.hibernate5.HibernateTransactionManager">
    <property name="sessionFactory" ref="sessionFactory" />
</bean>
<!-- 开启注解处理器 -->
<context:annotation-config />
<!-- 开启 Spring 的 Bean 自动扫描机制来检查与管理 Bean 实例 -->
<context:component-scan base-package="com.restaurant" />
<!-- 基于@Transactional 注解方式的事务管理 -->
<tx:annotation-driven transaction-manager="transactionManager" />
</beans>
```

由于使用了注解技术，首先需要在<beans>标记中添加与 context 相关的命名空间。由于使用了 Annotation 注解，需要在 Spirng 配置文件中开启注解处理器，代码如下：

```xml
<!-- 开启注解处理器 -->
<context:annotation-config />
```

与基于 AOP 的事务管理配置不同的是，Annotation 方式事务管理不再需要在配置文件中定义事务通知和切面，及将事务通知和切面组合，只需要配置一个基于@Transactional 注解方式的事务管理。因此，可将之前基于 AOP 的事务管理配置删除。基于@Transactional 注解方式的事务管理配置如下：

```xml
<!-- 基于@Transactional 注解方式的事务管理 -->
<tx:annotation-driven transaction-manager="transactionManager" />
```

（7）修改 Struts 2 配置文件 struts.xml。代码如下：

```xml
<?xml version="1.0" encoding="UTF-8" ?>
<!DOCTYPE struts PUBLIC "-//Apache Software Foundation//DTD Struts
Configuration 2.3//EN" "http://struts.apache.org/dtds/struts-2.3.dtd">
<struts>
    <constant name="struts.i18n.encoding" value="utf-8"></constant>
</struts>
```

由于项目 s2sh_annotation 开始是从项目 s2sh 复制而来的，因此项目 s2sh_annotation 的部署名称也为"s2sh"。为了避免重复，需要修改项目 s2sh_annotation 的部署名称。在 MyEclipse 中依次选择 Project→Properties→MyEclipse→Deployment Assembly，将 Web Context Root 修改为"/s2sh_annotation"即可。

部署项目，启动 Tomcat，在浏览器中输入 http://localhost:8080/s2sh_annotation/login.jsp 进行测试，效果与 20.1 节相同。

20.3 小　　结

本章以用户登录功能实现为例，先后介绍了基于 XML 配置文件和基于 Annotation 注解的 Struts 2、Spring 和 Hibernate 整合。通过对比分析，可以看出基于 Annotation 注解的整合方式简化了配置，也提高了程序的可读性和可维护性。

第 21 章
Spring MVC

对 Web 应用来说，表示层是个不可或缺的重要环节。前面介绍的 Struts 2 框架就是一个优秀的 Web 框架，是在 WebWork 的技术基础上开发的全新 MVC 框架。除了 Struts 2 框架，Spring 框架也为表示层提供了一个优秀的 Web 框架，即 Spring MVC。由于 Spring MVC 采用了松耦合可插拔组件结构，比其他 MVC 框架具有更大的扩展性和灵活性。通过注解，Spring MVC 使得 POJO 成为处理用户请求的控制器，无须实现任何接口。

21.1 Spring MVC 概述

Spring MVC 是基于 Model2 实现的技术框架。在 Spring MVC 中，Action 被称为 Controller(控制器)。Spring 的 Web 框架围绕 DispatcherServlet(分发器)设计的，其作用是将用户请求分发到不同的控制器(又称处理器)。Spring Web 框架中默认的控制器接口为 Controller，可以通过实现这个接口来创建自己的控制器，但建议使用 Spring 已经实现的一系列控制器，如 AbstractController、AbstractCommandController 和 SimpleFormController 等。Spring MVC 框架还包括可配置的处理器映射、视图解析、本地化、主题解析，同时支持文件上传。

Spring MVC 是基于 Model2 实现的框架，所以它底层的机制也是 MVC。Spring MVC 的工作原理如图 21-1 所示。

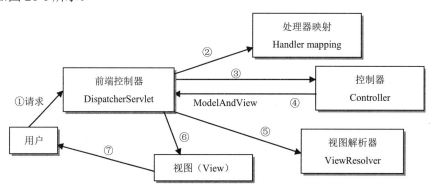

图 21-1　Spring MVC 的工作原理

从图 21-1 可以看出，Spring MVC 框架的各个组件各负其责。

① 客户端发出一个 HTTP 请求，Web 应用服务器接收这个请求，如果与 web.xml 配置文件中指定的 DispatcherServlet 请求映射路径匹配，Web 容器将该请求转交给 DispatcherServlet 处理。

② DispatcherServlet 接收这个请求后，根据请求的信息(URL 或请求参数等)按照某种机制寻找恰当的映射处理器来处理这个请求。

③ DispatcherServlet 根据映射处理器(Handler mapping)来选择并决定将请求派送给哪个控制器。

④ 控制器处理这个请求，并返回一个 ModelAndView 给 DispatcherServlet，ModelAndView 包含了视图逻辑名和模型数据信息。

⑤ 由于 ModelAndView 中包含的是视图逻辑名而非真正的视图对象，因此 DispatcherServlet 需要通过 ViewResolver 完成视图逻辑名到真实视图对象的解析功能。

⑥ 得到真实的视图对象后，DispatcherServlet 就使用 View 对象对 ModelAndView 中的模型数据进行渲染。

⑦ 最终客户端得到返回的响应，可能是一个普通的 HTML 页面，也可能是一个 Excel、PDF 文档等视图形式。

21.2 Spring MVC 常用注解

Spring MVC 常用注解包括@Controller、@RequestMapping、@PathVariable、@RequestParam、@SessionAttributes、@ModelAttribute、@ResponseBody 等。

21.2.1 基于注解的处理器

在使用注解的 Spring MVC 中，处理器是基于@Controller 和@RequestMapping 这两个注解的，@Controller 用于声明一个控制器类，@RequestMapping 用于声明对应请求的映射关系，这样就可以提供一个非常灵活的匹配和处理方式。

下面通过一个简单的 Hello World 示例演示如何使用注解声明控制器与请求映射。具体实现步骤如下。

(1) 创建一个名为 springmvc_1 的 Web 项目，将第 20 章中图 20-2 所示的 Spring 框架 jar 包，以及 aopalliance-1.0.jar、aspectjweaver-1.8.6.jar、cglib-3.2.0.jar 和 commons-logging-1.1.3.jar 这 4 个相关 jar 包添加到项目 springmvc_1 的 WebRoot\WEB-INF\lib 目录中。

(2) 在 web.xml 文件中配置 Spring MVC 的前端控制器 DispatcherServlet。代码如下：

```xml
<!-- 配置DispatcherServlet -->
<servlet>
    <servlet-name>dispatcherServlet</servlet-name>
    <servlet-class>org.springframework.web.servlet.DispatcherServlet
    </servlet-class>
    <!-- 配置Spring MVC配置文件的位置及名称 -->
    <init-param>
        <param-name>contextConfigLocation</param-name>
        <param-value>classpath:springmvc.xml</param-value>
    </init-param>
    <load-on-startup>1</load-on-startup>
</servlet>
<servlet-mapping>
    <servlet-name>dispatcherServlet</servlet-name>
    <url-pattern>/</url-pattern>
</servlet-mapping>
```

上述配置的目的在于，让 Web 容器使用 Spring MVC 的 DispatcherServlet，并通过设置 url-pattern 为"/"，将所有的 URL 请求都映射到这个前端控制器 DispatcherServlet。

(3) 在项目 springmvc_1 的 src 目录下创建 Spring MVC 配置文件 springmvc.xml，代码如下：

```xml
<?xml version="1.0" encoding="UTF-8"?>
<beans xmlns="http://www.springframework.org/schema/beans"
   xmlns:xsi="http://www.w3.org/2001/XMLSchema-instance"
xmlns:aop="http://www.springframework.org/schema/aop"
   xmlns:context="http://www.springframework.org/schema/context"
   xmlns:mvc="http://www.springframework.org/schema/mvc"
   xsi:schemaLocation="http://www.springframework.org/schema/beans
```

```xml
        http://www.springframework.org/schema/beans/spring-beans-4.3.xsd
        http://www.springframework.org/schema/context
        http://www.springframework.org/schema/context/spring-context-4.3.xsd
        http://www.springframework.org/schema/mvc
        http://www.springframework.org/schema/mvc/spring-mvc-4.3.xsd">
    <!-- 配置自动扫描的包 -->
    <context:component-scan base-package="com.springmvc" />
    <!-- 配置视图解析器，将控制器方法返回的逻辑视图解析为物理视图 -->
    <bean class=
"org.springframework.web.servlet.view.InternalResourceViewResolver">
        <property name="prefix" value="/"></property>
        <property name="suffix" value=".jsp"></property>
    </bean>
</beans>
```

在 springmvc.xml 文件中，首先要引入 beans、aop、context 和 mvc 命名空间；然后添加 <context:component-scan> 元素，通过 base-package 属性指定扫描 com.springmvc 包内被 @Repository、@Service 和 @Controller 等注解修饰的类，然后注册到 Spring IoC 容器中；最后通过 <bean> 元素配置视图解析器 InternalResourceViewResolver，将控制器方法返回的逻辑视图解析为物理视图。

(4) 创建处理请求的类，并使用注解标识为处理器。

在项目 src 目录下创建包 com.springmvc.controller，在包中创建类 HelloController，代码如下：

```java
package com.springmvc.controller;
import org.springframework.stereotype.Controller;
import org.springframework.web.bind.annotation.RequestMapping;
@Controller
public class HelloController {
    @RequestMapping("/hello")
    public String sayHello() {
        System.out.println("Hello World");
        return "success";
    }
}
```

在控制器类 HelloController 中，通过 @Controller 注解将 HelloController 声明为一个处理器类，通过 @RequestMapping 注解将用户对 sayHello() 方法的请求映射为 "/helloWorld"，这种映射是根据请求的 URL 进行映射的，是通过 @RequestMapping 注解的 value 属性来实现的(value="/hello")，该属性可默认不写("/hello")。

(5) 编写视图。

编辑 index.jsp 页面，在 <body> 元素中添加一个超链接，代码如下：

```html
<a href="hello">Hello World</a><br><br>
```

在项目 WebRoot 目录下创建一个目标页面 success.jsp，代码如下：

```html
<body>
    欢迎学习 Spring MVC
</body>
```

部署项目，启动 Tomcat，在浏览器中输入 http://localhost:8080/springmvc_1/ index.jsp，单击 index.jsp 页面中的 Hello World 超链接，页面转发到 success.jsp。

@RequestMapping 注解除了可以修饰类中的方法外，还可以用来修饰类。修改控制器类 HelloController，使用@RequestMapping 注解修饰 HelloController 类。

```
package com.springmvc.controller;
import org.springframework.stereotype.Controller;
import org.springframework.web.bind.annotation.RequestMapping;
@RequestMapping("/springmvc")
@Controller
public class HelloController {
    @RequestMapping("/hello")
    public String sayHello() {
        ……
    }
}
```

通过在类上添加@RequestMapping 注解可将请求分路径，此时 index.jsp 页面中的 Hello World 超链接要做如下修改：

```
<a href="springmvc/hello">Hello World</a><br><br>
```

21.2.2 请求映射方式

Spring MVC 可以根据请求方式、Ant 风格的 URL 路径和 REST 风格的 URL 路径进行映射。

1. 根据请求方式进行映射

@RequestMapping 注解除了可以根据请求的 URL 进行映射外，还可以根据请求方式进行映射。如果想根据请求方式进行映射，可通过设置 method 属性来实现。

在 HelloController 类中添加方法 requestMethod，使用@RequestMapping 注解的 method 属性指定该方法的请求方式为 POST。代码如下：

```
@RequestMapping(value = "/requestMethod", method = RequestMethod.POST)
public String requestMethod() {
    return "success";
}
```

在 index.jsp 页面中添加一个 Request Method 超链接，代码如下：

```
<a href="springmvc/requestMethod">Request Method</a>
```

重启 Tomcat，浏览页面 index.jsp，单击 Request Method 超链接，浏览器会显示错误信息 HTTP Status 405 - Request method 'GET' not supported。错误原因在于：requestMethod 方法被设置为处理 POST 方式的请求，而通过超链接发出请求的方式为 GET 方式。

在 index.jsp 页面中添加一个表单，通过表单提交请求，代码如下：

```
<form action="springmvc/requestMethod" method="post">
   <input type="submit" value="Request Method">
</form>
```

浏览页面 index.jsp，单击 Request Method 提交按钮，页面成功转发到 success.jsp。

2. Ant 风格的 URL 路径映射

Ant 风格的 URL 支持"？""*"和"**"三种匹配符，"？"符合匹配文件名中的一个字符，"*"符号匹配文件名中的任意字符，"**"符号匹配多层路径。

在 HelloController 类中添加方法 pathAnt，以"*"符号为例演示 Ant 风格的 URL 路径映射。代码如下：

```
@RequestMapping("/*/pathAnt")
public String pathAnt(){
    System.out.println("Path Ant");
    return "success";
}
```

在 index.jsp 页面中添加一个 Path Ant 超链接，代码如下：

```
<a href="springmvc/my/pathAnt">Path Ant</a><br><br>
```

重启 Tomcat，浏览页面 index.jsp，单击 Path Ant 链接，页面成功转发到 success.jsp。

3. REST 风格的 URL 路径映射

REST(Representational State Transfer,表现层状态转化)是当前流行的一种互联网软件架构，采用 REST 风格可以有效降低开发的复杂性，提高系统的可伸缩性。

在 Web 开发中，REST 使用 HTTP 协议连接器来标识对资源的操作(获取/查询、创建、删除、修改)，用 HTTP Method(请求方法)标识操作类型。HTTP GET 标识获取和查询资源，HTTP POST 标识创建资源，HTTP PUT 标识修改资源，HTTP DELETE 标识删除资源。

这样，URI 加上 HTTP Method 构成了 REST 风格数据处理的核心，URI 确定操作的对象，HTTP Method 确定操作的方式。例如，"/Users/1 HTTP GET"表示获取 id 为 1 的 Users 对象，"/Users/1 HTTP DELETE"表示删除 id 为 1 的 Users 对象，"/Users/1 HTTP PUT"表示更新 id 为 1 的 Users 对象，"/Users HTTP POST"表示新增 Users 对象。

由于 form 表单只支持 GET 和 POST 请求，而不支持 DELETE 和 PUT 等请求方式，Spring 提供了一个过滤器 HiddenHttpMethodFilter，可以将 DELETE 和 PUT 请求转换为标准的 HTTP 方式，即能将 POST 请求转为 DELETE 或 PUT 请求。

下面通过一个简单的示例演示 REST 风格的 URL 路径映射，实现过程如下：

(1) 在 web.xml 文件中配置过滤器 HiddenHttpMethodFilter，代码如下：

```
<!-- 配置 HiddenHttpMethodFilter,可将 POST 请求转为 DELETE 或 PUT 请求 -->
<filter>
    <filter-name>HiddenHttpMethodFilter</filter-name>
    <filter-class>org.springframework.web.filter.HiddenHttpMethodFilter
</filter-class>
</filter>
<filter-mapping>
    <filter-name>HiddenHttpMethodFilter</filter-name>
    <url-pattern>/*</url-pattern>
</filter-mapping>
```

(2) 处理 GET 请求。

① 在 index.jsp 页面中添加 Rest GET 超链接，代码如下：

```
<a href="springmvc/rest/1">Rest GET</a>
```

② 在 HelloController 类中添加方法 restGET，处理 GET 方式请求，代码如下：

```java
@RequestMapping(value = "/rest/{id}", method = RequestMethod.GET)
public String restGET(@PathVariable("id") Integer id) {
    System.out.println("Rest GET:" + id);
    return "success";
}
```

重启 Tomcat，浏览 index.jsp 页面，单击 Rest GET 链接，控制台输出结果如下：

```
Rest GET:1
```

(3) 处理 POST 请求。

① 在 index.jsp 页面中添加一个表单，代码如下：

```
<form action="springmvc/rest" method="post">
    <input type="submit" value="Rest POST">
</form>
```

② 在 HelloController 类中添加方法 restPOST，处理 POST 方式请求，代码如下：

```java
@RequestMapping(value = "/rest", method = RequestMethod.POST)
public String restPOST() {
    System.out.println("Rest POST");
    return "success";
}
```

重启 Tomcat，浏览 index.jsp 页面，单击 Rest POST 按钮，控制台输出结果如下：

```
Rest POST
```

(4) 处理 DELETE 请求。

① 在 index.jsp 页面中添加一个表单，代码如下：

```
<form action="springmvc/rest/1" method="post">
    <input type="hidden" name="_method" value="DELETE">
    <input type="submit" value="Rest DELETE">
</form>
```

表单中使用了一个名称为"_method"的隐藏域，并给其赋值为 DELETE。过滤器 HiddenHttpMethodFilter 正是通过"_method"的值，将 POST 请求转为 DELETE。

② 在 HelloController 类中添加方法 restDELETE，处理 DELETE 方式请求，代码如下：

```java
@RequestMapping(value = "/rest/{id}", method = RequestMethod.DELETE)
public String restDELETE(@PathVariable("id") Integer id) {
    System.out.println("Rest DELETE:" + id);
    return "redirect:/springmvc/doTransfer";
}
```

restDELETE 方法返回值使用了重定向，将请求重定向到 springmvc/doTransfer，该映射对应的处理方法如下：

```
@RequestMapping("/doTransfer")
public String doTransfer() {
    return "success";
}
```

重启 Tomcat，浏览 index.jsp 页面，单击 Rest DELETE 按钮，控制台输出结果如下：

```
Rest DELETE:1
```

从控制台输出结果可以看出，单击 Rest DELETE 按钮发出的请求被成功提交给了 HelloController 类中的 restDELETE 方法来处理。

(5) 处理 PUT 请求。

① 在 index.jsp 页面中添加一个表单，代码如下：

```
<form action="springmvc/rest/1" method="post">
    <input type="hidden" name="_method" value="PUT">
    <input type="submit" value="Rest PUT">
</form>
```

表单中使用了一个名称为"_method"的隐藏域，并将其赋值为 PUT。过滤器 HiddenHttpMethodFilter 正是通过"_method"的值，将 POST 请求转为 PUT。

② 在 HelloController 类中添加方法 restPUT，处理 PUT 方式请求，代码如下：

```
@RequestMapping(value = "/rest/{id}", method = RequestMethod.PUT)
public String restPUT(@PathVariable("id") Integer id) {
    System.out.println("Rest PUT:" + id);
    return "redirect:/springmvc/doTransfer";
}
```

重启 Tomcat，浏览 index.jsp 页面，单击 Rest PUT 按钮，控制台输出结果如下：

```
Rest PUT:1
```

从控制台输出结果可以看出，单击 Rest PUT 按钮发出的请求被成功提交给了 HelloController 类中的 restPUT 方法来处理。

21.2.3 绑定控制器类处理方法入参

Spring MVC 支持将多种途径传递的参数绑定到控制器类的处理方法的输入参数中。

1. 映射 URL 绑定的占位符到方法入参

使用@PathVariable 注解可以将 URL 中的占位符绑定到控制器方法的入参中，在 HelloController 类中添加方法 pathVariable，使用@PathVariable 注解映射 URL 中的占位符到目标方法的参数中。代码如下：

```
@RequestMapping("/pathVariable/{id}")
public String pathVariable(@PathVariable("id") Integer id) {
    System.out.println("Path Variable:" + id);
    return "success";
}
```

在 index.jsp 页面中添加一个 Path Variable 超链接，代码如下：

```
<a href="springmvc/pathVariable/1">Path Variable</a><br><br>
```

URL 中的占位符{id}通过注解@PathVariable("id")绑定到 pathVariable 方法的入参 id 中，重启 Tomcat，浏览页面 index.jsp，单击 Path Variable 链接，控制台输出 Path Variable:1。

2. 绑定请求参数到控制器方法参数

在控制器方法入参处使用@RequestParam 注解可以将请求参数传递给方法，通过@RequestParam 注解的 value 属性指定参数名，required 属性指定参数是否必须，默认为 true，表示请求参数中必须包含对应的参数，如果不存在，则抛出异常。

在 HelloController 类中添加方法 requestParam，使用@RequestParam 注解绑定请求参数到控制器方法参数。代码如下：

```java
@RequestMapping("/requestParam")
public String requestParam(
        @RequestParam(value = "loginName") String loginName,
        @RequestParam(value = "loginPwd") String loginPwd) {
    System.out.println("Request Param:" + loginName + " " + loginPwd);
    return "success";
}
```

在 index.jsp 页面中添加一个 Request Param 超链接，代码如下：

```
<a href="springmvc/requestParam?loginName=my&loginPwd=123456">
Request Param</a> <br><br>
```

重启 Tomcat，浏览页面 index.jsp，单击 Request Param 链接，控制台输出 Request Param:my 123456。

3. 将请求参数绑定到控制器方法的表单对象

Spring MVC 会按照参数名和属性名进行自动匹配，自动为该对象填充属性值，并且支持级联。在项目的 src 目录下创建包 com.springmvc.entity，在包中创建实体类 Users.java 和 Address.java。实体类 Address 代码如下：

```java
package com.springmvc.entity;
public class Address {
    private String province;
    private String city;
    //此处省略属性的get和set方法
    //此处省略构造方法
//此处省略toString()方法
}
```

实体类 Users 代码如下：

```java
package com.springmvc.entity;
public class Users {
    private String loginName;
    private String loginPwd;
    private Address address;
```

```
    //此处省略属性的get和set方法
    //此处省略构造方法
//此处省略toString()方法
}
```

然后在 HelloController 类中添加方法 saveUsers，将请求参数绑定到控制器方法的表单对象 Users 中。代码如下：

```
@RequestMapping("/saveUsers")
public String saveUsers(Users u) {
    System.out.println(u);
    return "success";
}
```

最后在 index.jsp 页面创建一个表单，代码如下：

```
<form action="springmvc/saveUsers" method="post">
    loginName:<input type="text" name="loginName"><br>
    loginPwd:<input type="password" name="loginPwd"><br>
    province:<input type="text" name="address.province"><br>
    city:<input type="text" name="address.city"><br> <input
        type="submit" value="提交">
</form>
```

重启 Tomcat，浏览页面 index.jsp，在表单中输入用户名 my，密码 123456，省份 JiangSu 和城市 NanJing，单击"提交"按钮，控制台输出结果如下：

```
Users [loginName=my, loginPwd=123456, address=Address [province=JiangSu, city=NanJing]]
```

4. 将请求参数绑定到控制器方法的 Map 对象

Spring MVC 注解可以将表单数据传递到控制器方法中的 Map 类型的入参中，在 com.springmvc.entity 包中创建 UsersMap 类，定义 Map<String, Users>类型的属性 uMap，并为该属性提供 get 和 set 方法，代码如下：

```
package com.springmvc.entity;
import java.util.Map;
public class UsersMap {
    private Map<String, Users> uMap;
    public Map<String, Users> getuMap() {
        return uMap;
    }
    public void setuMap(Map<String, Users> uMap) {
        this.uMap = uMap;
    }
}
```

在 HelloController 类中添加方法 getUsers，实现将请求参数绑定到控制器方法的 Map 对象，并遍历 Map，将 Map 中的内容输出到控制台。代码如下：

```
@RequestMapping("/getUsers")
public String getUsers(UsersMap uMap) {
    Set set = uMap.getuMap().keySet();
```

```
    Iterator iterator = set.iterator();
    while (iterator.hasNext()) {
        Object keyName = iterator.next();
        Users u = uMap.getuMap().get(keyName);
        System.out.println(u);
    }
    return "success";
}
```

在 index.jsp 页面中创建一个表单，代码如下：

```
<form action="springmvc/getUsers" method="post">
   loginName1:<input type="text" name="uMap['u1'].loginName"><br>
   loginPwd1:<input type="password" name="uMap['u1'].loginPwd"><br>
   province1:<input type="text" name="uMap['u1'].address.province"><br>
   city1:<input type="text" name="uMap['u1'].address.city"><br>
   loginName2:<input type="text" name="uMap['u2'].loginName"><br>
   loginPwd2:<input type="password" name="uMap['u2'].loginPwd"><br>
   province2:<input type="text" name="uMap['u2'].address.province"><br>
   city2:<input type="text" name="uMap['u2'].address.city"><br>
   <input type="submit" value="提交">
</form>
```

重启 Tomcat，浏览页面 index.jsp，在表单中分别输入两个用户的信息。第一个用户的用户名为 my，密码为 123456，省份为 JiangSu 和城市为 NanJing；第二个用户的用户名为 sj，密码为 123456，省份为 JiangSu 和城市为 YangZhou。单击"提交"按钮，控制台输出结果如下：

```
Users [loginName=my, loginPwd=123456, address=Address [province=JiangSu,
city=NanJing]]
Users [loginName=sj, loginPwd=123456, address=Address [province=JiangSu,
city=YangZhou]]
```

从控制台输出结果可以看出，表单中输入的两个用户的信息成功传递到控制器方法 getUsers 的 UsersMap 类型的参数 uMap 中。

21.2.4　控制器类处理方法的返回值类型

在前面的示例中，当控制器处理完请求时，会以字符串的形式返回逻辑视图名。除了 String 类型外，Spring MVC 返回类型还包括 ModelAndView、Model、ModelMap、Map 等。

如果返回类型是 ModelAndView，则其中可包含视图和模型信息，且 Spring MVC 会将模型信息存放到 request 域中。在 HelloController 类中添加方法 returnModelAndView，返回 ModelAndView 类型。代码如下：

```
@RequestMapping("/returnModelAndView")
public ModelAndView returnModelAndView() {
    String viewName="success";
    ModelAndView mv=new ModelAndView(viewName);
    Users u=new Users("zhangsan", "123456", new Address("jiangsu",
"nanjing"));
    mv.addObject("u", u);
```

```
        return mv;
}
```

在 index.jsp 页面中添加一个 ModelAndView 超链接，代码如下：

```
<a href="springmvc/returnModelAndView">ModelAndView</a><br><br>
```

在 success.jsp 页面中添加用于访问 ModelAndView 对象中保存的 Users 对象的代码如下：

```
ModelAndView:${requestScope.u }
```

重启 Tomcat，浏览页面 index.jsp，单击 ModelAndView 链接，success.jsp 页面显示如下：

```
欢迎学习 Spring MVC, ModelAndView:Users [loginName=zhangsan, loginPwd=123456,
address=Address [province=jiangsu, city=nanjing]]
```

存入 ModelAndView、Model、ModelMap、Map 中的数据对象，可以通过 request 作用域来访问。

21.2.5 保存模型属性到 HttpSession

通过在控制器类上标注@SessionAttributes 注解，可将模型数据保存到 HttpSession 中，以便多个请求之间共用该模型属性。在 HelloController 类中添加方法 sessionAttributes，先将模型属性 users 保存到 ModelMap 中。代码如下：

```
@RequestMapping("/sessionAttributes")
public String sessionAttributes(ModelMap model) {
    Users user = new Users("zhangsan", "123456", new Address("jiangsu",
        "nanjing"));
    model.put("user", user);
    return "success";
}
```

在控制器类 HelloController 上标注@SessionAttributes 注解，将 Users 类型的模型属性 user 存入 HttpSession 中。代码如下：

```
@SessionAttributes(value={"user"})
```

在 index.jsp 页面中添加一个 Session Attributes 超链接，代码如下：

```
<a href="springmvc/sessionAttributes">Session Attributes</a><br><br>
```

在 success.jsp 页面中添加用于访问保存到 HttpSession 中的 Users 对象的代码如下：

```
HttpSession 中保存的 user: ${sessionScope.user }
```

重启 Tomcat，浏览页面 index.jsp，单击 Session Atttributes 链接，success.jsp 页面显示如下：

```
HttpSession 中保存的 user: Users [loginName=zhangsan, loginPwd=123456,
address=Address [province=jiangsu, city=nanjing]]
```

21.2.6 在控制器类方法之前执行的方法

如果想让一个方法在控制器类的所有处理方法之前执行，可以通过在该方法上标注 @ModelAttribute 注解来实现。

在 HelloController 类中添加方法 getUsers，在方法上标注 @ModelAttribute 注解。在 getUsers 方法中实例化 Users 对象 u，保存到 Model 中，getUsers 方法的返回值为对象 u。代码如下：

```java
@ModelAttribute
public Users getUsers(Model model) {
    Users u = new Users("lisi", "123456", new Address("JiangSu", "SuZhou"));
    model.addAttribute("u", u);
    return u;
}
```

然后在 HelloController 类中添加方法 modelAttribute，代码如下：

```java
@RequestMapping("/modelAttribute")
public String modelAttribute(Users u) {
    System.out.println(u);
    return "success";
}
```

在 index.jsp 页面中添加一个 Model Attribute 超链接，代码如下：

```
<a href="springmvc/modelAttribute">Model Attribute</a><br><br>
```

重启 Tomcat，浏览页面 index.jsp，单击 Model Attribute 链接，控制台输出结果如下：

```
信息: Server startup in 8582 ms
Users [loginName=lisi, loginPwd=123456, address=Address [province=JiangSu, city=SuZhou]]
```

单击 Model Attribute 链接时，请求被 HelloController 类中的 modelAttribute 方法处理，方法中没有对对象 u 进行初始化，而控制台输出结果显示对象 u 已经被初始化过，这一初始化过程显然是在使用 @ModelAttribute 注解标注的 getUsers 方法中完成的。也就是说，在执行方法 modelAttribute 前，先调用了 getUsers 方法，实例化对象 u 后将其存入 Model，又因为 getUsers 方法返回了该对象 u，被传递给 modelAttribute 方法的参数 u。

success.jsp 页面显示如下：

```
欢迎学习 Spring MVC, ModelAndView:Users [loginName=lisi, loginPwd=123456, address=Address [province=JiangSu, city=SuZhou]]
HttpSession 中保存的 user:
```

在 getUsers 方法中，对象 u 被存入 Model，访问 request 作用域可以获得对象 u 的值。由于对象 u 没有保存在 HttpSession 中，访问 session 作用域无法获取对象 u 的值。

21.2.7 Spring MVC 返回 JSON 数据

如果想让控制器类的处理方法返回 JSON 数据，可以使用 HttpMessageConverter 类来实现。如果不想显式创建 HttpMessageConverter 的 Bean 实例，可以在 Spring MVC 配置文件

springmvc.xml 中添加<mvc:annotation-driven>元素。代码如下：

```
<mvc:annotation-driven />
```

下面通过示例介绍如何使用 Spring MVC 返回 JSON 数据，实现过程如下。

（1）添加 jar 包。

将 jackson-annotations-2.6.0.jar、jackson-core-2.6.0.jar 和 jackson-databind-2.6.0.jar 这三个 jar 包复制到项目 springmvc_1 的 WebRoot\WEB-INF\lib 目录下。

（2）引入 jQuery 资源文件。

在项目 WebRoot 目录下创建一个文件夹 scripts，将 jQuery 资源文件 jquery.min.js 复制到其中。

（3）处理静态资源文件 jquery.min.js。

在 Spring MVC 配置文件中添加<mvc:default-servlet-handler />元素，代码如下：

```
<mvc:default-servlet-handler />
```

该元素将在 Spring MVC 上下文中定义一个 DefaultServletHttpRequestHandler，它会对进入 DispatcherServlet 的请求进行筛查，如果发现是没有经过映射的请求，就将该请求交由 Web 应用服务器默认的 Servlet 处理。如果不是静态资源的请求，才由 DispatcherServlet 继续处理。如果没有添加<mvc:default-servlet-handler />元素，Web 容器启动时会抛出如下异常：

警告：No mapping found for HTTP request with URI [/springmvc_1/scripts/jquery.min.js] in DispatcherServlet with name 'dispatcherServlet'

（4）在 HelloController 类中添加方法 returnJson，返回 JSON 格式数据。代码如下：

```
@ResponseBody
@RequestMapping("/returnJson")
public Collection<Users> returnJson() {
    Map<Integer, Users> us = new HashMap<Integer, Users>();
    us.put(1, new Users("zhangsan", "123456", new Address("Jiangsu",
            "NanJing")));
    us.put(2, new Users("lisi", "123456",
            new Address("Jiangsu", "YangZhou")));
    us.put(3, new Users("wangwu", "123456",
            new Address("Jiangsu", "SuZhou")));
    return us.values();
}
```

returnJson 方法的返回类型为 Collection<Users>，但在标注@ResponseBody 注解后，返回类型就转变为 JSON 格式了。代码如下：

```
在 index.jsp 页面的<head></head>标签中引入资源文件 jquery.min.js
<script type="text/javascript" src="scripts/jquery.min.js"></script>
```

在 index.jsp 页面中添加一个 Test Json 超链接，代码如下：

```
<a href="javascript:void(0)" id="returnJson" onclick="getUsersJson()">Test Json</a><br><br>
```

单击 Test Json 链接，将执行一个 javascript 脚本函数 getUsersJson ()。在 index.jsp 页面的

<head></head>标签中,创建函数 getUsersJson()。代码如下:

```html
<script type="text/javascript">
    function getUsersJson() {
        var url = "springmvc/returnJson";
        var args = {};
        $.post(url, args, function(data) {
        });
    }
</script>
```

在 getUsersJson 函数中,使用$.post 将请求提交到控制器类 HelloController 中的 returnJson 方法,参数 data 就是 returnJson 方法执行后返回的 JSON 格式的数据。

重启 Tomcat,使用火狐浏览器浏览页面 index.jsp,单击 Test Json 链接,通过其中配置的 Firebug 可以更方便地查看参数 data 的内容,如图 21-2 所示。

图 21-2　查看 JSON 格式的数据

21.3　直接页面转发、自定义视图与页面重定向

1. 直接页面转发

如果想不经过控制器类的处理方法直接转发到页面,可以通过使用<mvc:view-controller>元素来实现。在 Spring MVC 配置文件 springmvc.xml 中,添加<mvc:view-controller>元素,其配置如下:

```xml
<mvc:view-controller path="/success" view-name="success"/>
```

重启 Tomcat,在浏览器中直接输入 http://localhost:8080/springmvc_1/success,页面成功转发到 success.jsp。

需要注意的是,如果在 springmvc.xml 文件中没有添加<mvc:annotation-driven>元素,代码如下:

```xml
<mvc:annotation-driven />
```

浏览 index.jsp 页面中的其他超链接时,会出现问题。使用<mvc:annotation-driven>元素后,会自动注册 RequestMappingHandlerMapping、RequestMappingHandlerAdapter 和 ExceptionHandlerExceptionResolver 这三个 Bean,就可以解决问题了。

2. 定义视图

通过使用 BeanNameViewResolver 类可以实现用户自定义的视图，创建一个类 MyView，实现 View 接口，存放在 com.springmvc.view 包中，实现一个简单的自定义视图。代码如下：

```java
package com.springmvc.view;
import java.util.Map;
import javax.servlet.http.HttpServletRequest;
import javax.servlet.http.HttpServletResponse;
import org.springframework.stereotype.Component;
import org.springframework.web.servlet.View;
@Component
public class MyView implements View {
    @Override
    public String getContentType() {
        return "text/html";
    }
    @Override
    public void render(Map<String, ?> arg0, HttpServletRequest request,
            HttpServletResponse response) throws Exception {
        response.getWriter().println("hello,this is my view");
    }
}
```

在 MyView 类上标注了@Component 注解，Spring 会为该类创建 Bean 实例。在 render 方法中，向浏览器输出一个简单的字符串 "hello,this is my view" 作为自定义页面内容。

在 springmvc.xml 文件中配置视图解析器，使用视图名称实现视图解析。代码如下：

```xml
<bean class="org.springframework.web.servlet.view.BeanNameViewResolver">
    <property name="order" value="50" />
</bean>
```

在 HelloController 类中添加方法 beanNameViewResolver，来使用自定义视图。代码如下：

```java
@RequestMapping("/beanNameViewResolver")
public String beanNameViewResolver(){
    return "myView";
}
```

由于 BeanNameViewResolver 类是根据 Bean 的名称来解析视图的，自定义的视图类 MyView 在 Spring IoC 中的 Bean 实例名为 myView，因此 beanNameViewResolver 方法返回值应该使用 Bean 实例名 myView。

在 index.jsp 页面中添加一个 BeanNameViewResolver 超链接，代码如下：

```html
<a href="springmvc/beanNameViewResolver">BeanNameViewResolver</a><br>
```

重启 Tomcat，浏览页面 index.jsp，单击 BeanNameViewResolver 链接，显示自定义的页面内容：

```
hello,this is my view
```

3. 页面重定向

在前面的示例中，控制器类的方法返回的字符串默认通过转发的方式跳转到目标页面，如果返回的字符串中带 redirect 前缀，则会采用重定向的方式跳转到目标页面。

在 HelloController 类中添加方法 redirect，在返回字符串中使用重定向。代码如下：

```java
@RequestMapping("/redirect")
public String redirect(){
    return "redirect:/index.jsp";
}
```

在 index.jsp 页面中添加一个 Redirect 超链接，代码如下：

```html
<a href="springmvc/redirect">Redirect</a><br><br>
```

重启 Tomcat，浏览页面 index.jsp，单击 Redirect 链接，页面重定向到 index.jsp。

21.4 控制器的类型转换、格式化、数据校验

1. 使用@InitBinder 注解进行类型转换

Spring MVC 默认不支持将表单中的日期字符串和实体类中的日期类型的属性自动转换，必须要手动配置，自定义数据类型的绑定才能实现这个功能。可以使用 Spring MVC 的注解 @initbinder 和 Spring 自带的 WebDataBinder 类和操作来实现这一类型转换。

在实体类 Users.java 中添加一个日期型属性 regDate，并添加该属性的 get 和 set 方法。代码如下：

```java
private Date regDate;
// 省略 regDate 属性的 get 和 set 方法
```

在 HelloController 类中添加方法 initBinder 方法，并用@InitBinder 注解标注，将从表单获取的字符串类型的日期转换成 Date 类型。代码如下：

```java
@InitBinder
public void initBinder(WebDataBinder binder) {
    SimpleDateFormat dateFormat = new SimpleDateFormat("yyyy-MM-dd");
    binder.registerCustomEditor(Date.class, new CustomDateEditor(dateFormat, true));
}
```

@InitBinder 注解标识的方法可以对 WebDataBinder 对象进行初始化，用于完成从表单文本域到实体类属性的绑定。@InitBinder 标识的方法不能有返回值，必须声明为 void。@InitBinder 标识的方法的参数为 WebDataBinder，是 DataBinder 的子类。DataBinder 是数据绑定的核心部件，可用来进行数据类型转换、格式化以及数据校验。

在 HelloController 类中添加方法 testInitBinder，用于测试日期类型转换。代码如下：

```java
@RequestMapping("/testInitBinder")
public String testInitBinder(Users u) {
    System.out.println(u.getRegDate());
```

```
    return "success";
}
```

在 index.jsp 页面中创建一个表单，代码如下：

```
<form action="springmvc/testInitBinder" method="post">
    regDate:<input type="text" name="regDate"><br>
    <input type="submit" value="提交" />
</form>
```

重启 Tomcat，浏览页面 index.jsp，在表单中输入字符串类型日期 2017-6-14，单击"提交"按钮，控制台输出实体对象 u 中的 regDate 属性值，具体如下：

```
Wed Jun 14 00:00:00 CST 2017
```

每次请求都会先调用@initBinder 注解标识的方法，然后再调用控制器类中处理请求的方法。

2. 数据格式化

除了可以使用@initBinder 注解实现数据类型的转换外，还可以通过在实体类的属性上添加相应的注解来实现数据的格式化。在实体类 Users 的 regDate 属性上标识@DateTimeFormat 注解，代码如下：

```
@DateTimeFormat(pattern="yyyy-MM-dd")
private Date regDate;
```

@DateTimeFormat 可将表单中输入的形如 yyyy-MM-dd 的日期字符串格式化为 Date 类型的数据。

将 HelloController 类中使用@InitBinder 注解标识的 testInitBinder 方法注释掉，重启 Tomcat，浏览页面 index.jsp，在表单中输入字符串类型日期 2017-6-14，单击"提交"按钮，控制台依然成功输出实体对象 u 中的 regDate 属性值。

如果在 Float 类型属性上使用@NumberFormat(pattern="#,###,###.#")注解，则将表单中输入的形如"1,234,567.8"的字符串格式化为 Float 类型的数据。

3. 控制器的数据校验

Spring 3 开始支持 JSR-303 验证框架。JSR-303 支持 XML 风格的和注解风格的验证。使用 JSR-303 规范校验只需要在 POJO 字段上加上相应的注解就可以实现校验了。JSR-303 规范校验的步骤如下。

(1) 添加 jar 包。

在项目 springmvc_1 的 WebRoot\WEB-INF\lib 目录下添加 hibernate-validator-5.2.3.Final.jar、validation-api-1.1.0.Final.jar、jboss-logging-3.3.0.Final.jar 和 classmate-1.3.1.jar 这四个 jar 包。

(2) 在 Spring MVC 配置文件中添加对 JSR-303 验证框架的支持。

由于在 Spring MVC 配置文件 springmvc.xml 中已经使用了<mvc:annotation-driven>，因此会自动注册 JSR-303 验证框架。

(3) 使用 JSR-303 验证框架注解为模型对象指定验证信息。

在实体类 Users.java 中，添加 age 和 email 这两个属性及其 get 和 set 方法，再使用 JSR-

303 验证框架注解为 loginName、email 和 age 这三个属性指定验证信息。代码如下：

```
@NotEmpty
@Size(min=6,max=20)
private String loginName;
@Email
@NotEmpty
private String email;
@Range(min = 18, max = 45)
@NotNull
private Integer age;
```

对于 loginName 属性，要求不为空，其长度不小于 6，且不大于 20；对于 email 属性，要求不为空，且格式为 email；对应 age 属性，要求不为空，且输入的年龄范围为 18～45 岁。

(4) 在 HelloController 类中添加方法 testValidate，测试表单数据校验。代码如下：

```
@RequestMapping("/testValidate")
public String testValidate(@Valid Users u, BindingResult result) {
    if (result.getErrorCount() > 0) {
        for (FieldError error : result.getFieldErrors()) {
            System.out.println(error.getField() + ":"
                + error.getDefaultMessage());
        }
    }
    return "success";
}
```

在 testValidate 方法中，通过@Valid 注解来告诉 Spring MVC，Users 类的对象 u 在绑定表单数据后需要进行 JSR-303 验证，绑定的结果保存到 BindingResult 类型的对象 result 中。通过判断 result 就可以知道绑定过程是否发生错误，如果出现错误则输出。

(5) 在 index.jsp 页面中创建一个表单，代码如下：

```
<form action="springmvc/testValidate" method="post">
    loginName:<input type="text" name="loginName"><br>
    email:<input type="text" name="email"><br>
    age:<input type="text" name="age"><br>
    <input type="submit" value="提交" />
</form>
```

重启 Tomcat，浏览页面 index.jsp，如果在表单中未输入任何信息就单击"提交"按钮，控制台输出如下校验错误信息：

```
loginName:不能为空
age:不能为null
loginName:个数必须在 6 和 20 之间
email:不能为空
```

只有表单输入信息满足所有校验要求，控制台才不会输出错误信息。

21.5 Spring MVC 文件上传

Spring MVC 支持 Web 应用程序的上传功能，是通过内置的即插即用的 CommonsMultipartResolver 解析器来实现的。它定义在 org.springframework.web.multipart 包中，Spring 通过使用 Commons FileUpload 插件来完成 MultipartResolver。

在默认情况下，Spring 不会处理 multipart 的 form 信息，因为默认用户会自己去处理这部分信息，当然可以随时打开这个支持。这样对于每一个请求，都会查看它是否包含 multipart 的信息，如果没有则按流程继续执行。如果有，就会交给已经被声明的 MultipartResolver 进行处理，然后就能像处理其他普通属性一样处理文件上传了。

下面使用 CommonsMultipartResolver 实现文件的上传功能，具体流程如下。

(1) 添加 jar 包。

将 commons-fileupload-1.2.jar 和 commons-io-1.3.2.jar 这 2 个 jar 包复制到项目 springmvc_1 的 WebRoot\WEB-INF\lib 目录下。

(2) 在 Spring MVC 配置文件中配置 CommonsMultipartResolver 类，代码如下：

```xml
<bean id="multipartResolver"
    class="org.springframework.web.multipart.commons.CommonsMultipartResolver">
    <!-- 设置上传文件的最大尺寸为 1MB -->
    <property name="maxUploadSize" value="1048576" />
    <!-- 字符编码 -->
    <property name="defaultEncoding" value="UTF-8" />
</bean>
```

(3) 在 index.jsp 页面中创建一个表单，用于上传文件。代码如下：

```html
<form action="springmvc/upload" method="post" enctype="multipart/form-data">
    <input type="file" name="file" />
    <input type="submit" value="上传" />
</form>
```

(4) 在 HelloController 类中添加方法 upload，处理文件上传。代码如下：

```java
@RequestMapping(value = "/upload")
public String upload(@RequestParam(value="file", required=false) MultipartFile file, HttpServletRequest request, ModelMap model) {
    //服务器端 upload 文件夹物理路径
    String path = request.getSession().getServletContext().getRealPath("upload");
    //获取文件名
    String fileName = file.getOriginalFilename();
    //实例化一个 File 对象，表示目标文件(含物理路径)
    File targetFile = new File(path, fileName);
    if(!targetFile.exists()){
        targetFile.mkdirs();
    }
    try {
        //将上传文件写到服务器上指定的文件
```

```
        file.transferTo(targetFile);
    } catch (Exception e) {
        e.printStackTrace();
    }
    model.put("fileUrl",request.getContextPath()+
"/upload/"+fileName);
    return "success";
}
```

在 upload 方法参数中，@RequestParam 注解用于在控制器 HelloController 中绑定请求参数到方法参数。请求参数为 file，将 index.jsp 页面文件上传表单中名为 file 的 value 值赋给 MultipartFile 类型的 file 属性；required=false 表示使用这个注解可以不传这个参数，如果 required=true 时则必须传递该参数，required 默认值是 true。

在 success.jsp 页面中添加用于访问保存到 request 中的 fileUrl 值，具体如下：

上传文件路径：${requestScope.fileUrl }

重启 Tomcat，浏览 index.jsp 页面，在文件上传表单中先通过"浏览"按钮选择一个文件，然后单击"上传"按钮。文件成功上传后，在 Tomcat 的根路径\webapps\springmvc_1\upload 目录下就能看到上传的文件。success.jsp 页面显示的文件路径如下：

上传文件路径：/springmvc_1/upload/01.jpg

21.6 Spring MVC 国际化

在 Web 开发中经常会遇到国际化的问题，除了 Struts 2 框架，Spring MVC 也提供了对国际化的支持。Spring MVC 使用 ResourceBundleMessageSource 实现国际化资源的定义，一个简单的 Spring MVC 国际化示例的实现过程如下。

(1) 在 Spring MVC 配置文件 springmvc.xml 中，首先配置资源文件绑定器 ResourceBundleMessageSource，代码如下：

```
<bean id="messageSource"
class="org.springframework.context.support.ResourceBundleMessageSource">
    <property name="basename" value="mess" />
</bean>
```

配置 messageSource 这个 bean 时，需要注意是 messageSource，而不是 messageResource，也不能是其他，这是 Spring 的规定。ResourceBundleMessageSource 类的 basename 属性指定资源文件的基名，即资源文件以 mess 打头。

然后配置 SessionLocaleResolver，用于将 Locale 对象存储于 Session 中供后续使用。代码如下：

```
<bean id="localeResolver" class="org.springframework.web.servlet.
i18n.SessionLocaleResolver"></bean>
```

最后配置 LocaleChangeInterceptor，用于获取请求中的 locale 信息，将其转为 Locale 对象，获取 LocaleResolver 对象。代码如下：

```xml
<mvc:interceptors>
    <bean class=
"org.springframework.web.servlet.i18n.LocaleChangeInterceptor"></bean>
</mvc:interceptors>
```

（2）在项目的 src 目录下创建国际化资源属性文件 mess_en_US.properties 和 mess_zh_CN.properties。mess_en_US.properties 资源属性文件内容如下：

```
loginName=LoginName
loginPwd=LoginPwd
```

mess_zh_CN.properties 资源属性文件内容如下：

```
loginName=\u7528\u6237\u540D
loginPwd=\u5BC6\u7801
```

（3）在 HelloController 类中添加方法 localeChange 处理国际化，并注入 ResourceBundleMessageSource 的 Bean 实例。代码如下：

```java
@Autowired
private ResourceBundleMessageSource messageSource;
// 国际化
@RequestMapping(value = "/localeChange")
public String localeChange(Locale locale) {
    String u = messageSource.getMessage("loginName", null, locale);
    System.out.println("国际化资源文件 Locale 配置(loginName):" + u);
    return "login";
}
```

（4）创建页面 login.jsp。

在页面头部使用 taglib 指令引入 JSTL 的 fmt 标签，代码如下：

```
<%@ taglib prefix="fmt" uri="http://java.sun.com/jsp/jstl/fmt" %>
```

在<body></body>部分添加用于语言切换的超链接，并通过<fmt:message>元素的 key 属性输出资源属性文件中的 key 所对应的值。代码如下：

```html
<a href="springmvc/localeChange?locale=zh_CH">中文</a> |
<a href="springmvc/localeChange?locale=en_US">英文</a>
<br />
<fmt:message key="loginName"></fmt:message>
<fmt:message key="loginPwd"></fmt:message>
```

重启 Tomcat，浏览页面 login.jsp，依次单击"中文"和"英文"链接，可以看到 <fmt:message>元素显示的文本能够根据所传递的语言来动态展现。

21.7　Spring 整合 Spring MVC 与 Hibernate

Spring MVC 与 Struts 2 一样，也是非常优秀的 MVC 框架，特别是 Spring 3.0 版本之后，越来越多的团队选择了 Spring MVC。在框架整合开发时，也会选择 Spring MVC 替代 Struts 2。下面以登录功能为例，采用 Annotation 注解方式实现 Spring 整合 Spring MVC 与

Hibernate。

21.7.1 环境搭建

在 MyEclipse 中创建一个名为 springmvc_ssh 的 Web 项目，选择 Java version 为 1.8，Target runtime 为 Apache Tomcat v8.0。

将第 20 章中图 20-2 所示的 12 个 Spring 所需的 jar 包，以及 aopalliance-1.0.jar、aspectjweaver-1.8.6.jar、cglib-3.2.0.jar、commons-logging-1.1.3.jar、commons-fileupload-1.3.2.jar 和 commons-io-2.4.jar 这 6 个相关 jar 包。图 20-1 所示的 10 个 Hibernate 所需的 jar 包，以及 MySQL 数据库驱动包 mysql-connector-java-5.1.18-bin.jar、连接池核心包 c3p0-0.9.5.2.jar、c3p0 连接池的依赖包 mchange-commons-java-0.2.11.jar 这 3 个包复制到项目 springmvc_ssh 的 WebRoot\WEB-INF\lib 目录中。

21.7.2 创建实体类

创建包 com.res.entity，在包中创建实体类 Users，采用注解方式实现 Users 类的属性与数据库 restrant 中数据表 users 的字段间的映射关系，代码如下：

```java
package com.res.entity;
import javax.persistence.*;
@Entity
@Table(name = "users", catalog = "restrant")
public class Users {
    private Integer id;
    private String loginName;
    private String loginPwd;
    @Id
    @GeneratedValue(strategy = GenerationType.IDENTITY)
    @Column(name = "Id", unique = true, nullable = false)
    public Integer getId() {
        return id;
    }
    public void setId(Integer id) {
        this.id = id;
    }
    // 此处省略了属性 loginName、loginPwd 的 get 和 set 方法
// 此处省略无参方法和有参构造方法
}
```

21.7.3 Spring 整合 Hibernate

Spring 整合 Hibernate 是在 Spring 配置文件 applicationContext.xml 中通过配置完成的，在 src 目录下创建文件 applicationContext.xml，依次配置数据源；配置 Hibernate 的 sessionFactory 实例；声明 Hibernate 事务管理器；开启注解处理器；开启 Spring 的 Bean 自动扫描机制来检查与管理 Bean 实例；配置基于 @Transactional 注解方式的事务管理。代码如下：

```xml
<?xml version="1.0" encoding="UTF-8"?>
```

```xml
<beans xmlns="http://www.springframework.org/schema/beans"
    xmlns:xsi="http://www.w3.org/2001/XMLSchema-instance"
    xmlns:p="http://www.springframework.org/schema/p"
    xmlns:tx="http://www.springframework.org/schema/tx"
    xmlns:aop="http://www.springframework.org/schema/aop"
    xmlns:context="http://www.springframework.org/schema/context"
    xsi:schemaLocation="http://www.springframework.org/schema/aop
    http://www.springframework.org/schema/aop/spring-aop-4.3.xsd
        http://www.springframework.org/schema/beans
    http://www.springframework.org/schema/beans/spring-beans-4.3.xsd
        http://www.springframework.org/schema/context
    http://www.springframework.org/schema/context/spring-context-4.3.xsd
        http://www.springframework.org/schema/tx
    http://www.springframework.org/schema/tx/spring-tx-4.3.xsd">
    <!-- 配置数据源 -->
    <bean id="dataSource" class="com.mchange.v2.c3p0.ComboPooledDataSource">
        <property name="driverClass" value="com.mysql.jdbc.Driver" />
        <property name="jdbcUrl" value="jdbc:mysql:///restrant" />
        <property name="user" value="root" />
        <property name="password" value="123456" />
        <property name="minPoolSize" value="5" />
        <property name="maxPoolSize" value="10" />
    </bean>
    <!-- 配置Hibernate的sessionFactory实例 -->
    <bean id="sessionFactory"
        class="org.springframework.orm.hibernate5.LocalSessionFactoryBean">
        <!-- 配置数据源属性 -->
        <property name="dataSource">
            <ref bean="dataSource" />
        </property>
        <!-- 配置Hibernate的基本属性 -->
        <property name="hibernateProperties">
            <props>
                <prop key="hibernate.dialect">
                    org.hibernate.dialect.MySQLDialect
                </prop>
            </props>
        </property>
        <!-- 配置Hibernate基于注解的实体类的位置及名称 -->
        <property name="annotatedClasses">
            <list>
                <value>com.res.entity.Users</value>
            </list>
        </property>
    </bean>
    <!-- 声明Hibernate事务管理器 -->
    <bean id="transactionManager"
        class="org.springframework.orm.hibernate5.HibernateTransactionManager">
        <property name="sessionFactory" ref="sessionFactory" />
    </bean>
    <!-- 开启注解处理器 -->
    <context:annotation-config />
    <!-- 开启Spring的Bean自动扫描机制来检查与管理Bean实例 -->
```

```xml
    <context:component-scan base-package="com.res">
      <context:exclude-filter type="annotation"
        expression="org.springframework.stereotype.Controller" />
      <context:exclude-filter type="annotation"  expression=
"org.springframework.web.bind.annotation.ControllerAdvice" />
    </context:component-scan>
    <!-- 配置基于@Transactional 注解方式的事务管理 -->
    <tx:annotation-driven transaction-manager="transactionManager" />
</beans>
```

如果 Spring 的 IoC 容器和 Spring MVC 的 IoC 容器扫描的包有重合的部分，就会导致有的 Bean 会被创建两次。解决的方法是使 Spring 的 IoC 容器扫描的包和 Spring MVC 的 IoC 容器扫描的包没有重合的部分，可使用 exclude-filter 和 include-filter 子节点来规定只能扫描的注解。

21.7.4 DAO 层开发

在项目中创建包 com.res.dao，在包中创建接口 UsersDAO，在 UsersDAO 接口中声明方法 search，用于登录验证，代码如下：

```java
package com.res.dao;
import java.util.List;
import com.res.entity.Users;
public interface UsersDAO {
    public List<Users> search(Users cond);
}
```

创建 UsersDAO 接口的实现类 UsersDAOImpl，存放在 com.res.dao.impl 包中，实现 search 方法。使用@Repository 和@Autowired 注解的 UsersDAOImpl 实现类如下：

```java
package com.res.dao.impl;
import java.util.List;
……
// 在 Spring 容器中注册实例名为 usersDAO 的 UsersDAOImpl 实例
@Repository("usersDAO")
public class UsersDAOImpl implements UsersDAO {
    // 注入 Spring 容器中的 SessionFactory 实例
    @Autowired
    SessionFactory sessionFactory;
    @Override
    @SuppressWarnings("unchecked")
    public List<Users> search(Users cond) {
        List<Users> uList = null;
        // 获得 session
        Session session = sessionFactory.getCurrentSession();
        // 创建 HQL 语句
        String hql = "from Users u where u.loginName = ? and u.loginPwd = ?";
        // 执行查询
        Query<Users> query = session.createQuery(hql);
        query.setParameter(0, cond.getLoginName());
        query.setParameter(1, cond.getLoginPwd());
        uList = query.getResultList();
```

```
            // 返回结果
            return uList;
    }
}
```

21.7.5 Service 层开发

在项目中创建包 com.res.service，在包中创建接口 UsersService，在 UsersService 接口中声明方法 login，用于登录验证，代码如下：

```
package com.res.service;
import java.util.List;
import com.res.entity.Users;
public interface UsersService {
    public List<Users> login(Users cond);
}
```

创建 UsersService 接口的实现类 UsersServiceImpl，存放在 com.res.service.impl 包中，实现 login 方法。使用@Service、@Autowired 和@Transactional 注解的实现类 UsersServiceImpl 如下：

```
package com.res.service.impl;
import com.res.service.UsersService;
……
// 在Spring容器中注册名为usersService的UsersServiceImpl实例
@Service("usersService")
// 使用@Transactional注解实现事务管理
@Transactional
public class UsersServiceImpl implements UsersService {
    // 使用@Autowired注解注入UsersDAOImpl实例
    @Autowired
    UsersDAO usersDAO;
    @Override
    public List<Users> login(Users cond) {
        return usersDAO.search(cond);
    }
}
```

21.7.6 控制器开发

在项目 src 目录下创建包 com.res.controller，在包中创建类 UsersController，代码如下：

```
package com.res.controller;
……
import com.res.service.UsersService;
@RequestMapping("/user")
@Controller
public class UsersController {
    // 注入UsersServiceImpl实例
    @Autowired
    UsersService usersService;
    @RequestMapping("/login")
    public String login(Users u) {
        List<Users> uList = usersService.login(u);
```

```
        if (uList.size() > 0) {
            // 登录成功,转发到 index.jsp
            return "index";
        } else {
            // 登录失败,重定向到 login.jsp
            return "redirect:/login.jsp";
        }
    }
}
```

21.7.7 Spring 整合 Spring MVC

(1) 配置 web.xml。

在 web.xml 配置文件中,依次配置 ContextLoaderListener,加载 Spring 配置文件;配置编码过滤器;配置 Spring MVC 的 DispatcherServlet。代码如下:

```xml
<!-- 配置 ContextLoaderListener,加载 Spring 配置文件 -->
<context-param>
    <param-name>contextConfigLocation</param-name>
    <param-value>classpath:applicationContext.xml</param-value>
</context-param>
<!-- Spring 监听器 -->
<listener>
    <listener-class>org.springframework.web.context.ContextLoaderListener
</listener-class>
</listener>
<!-- 编码过滤器 -->
<filter>
    <filter-name>encodingFilter</filter-name>
    <filter-class>org.springframework.web.filter.CharacterEncodingFilter
</filter-class>
    <async-supported>true</async-supported>
    <init-param>
        <param-name>encoding</param-name>
        <param-value>UTF-8</param-value>
    </init-param>
</filter>
<filter-mapping>
    <filter-name>encodingFilter</filter-name>
    <url-pattern>/*</url-pattern>
</filter-mapping>
<!-- 配置 Spring MVC 的 DispatcherServlet -->
<servlet>
    <servlet-name>dispatcherServlet</servlet-name>
    <servlet-class>org.springframework.web.servlet.DispatcherServlet
</servlet-class>
    <!-- 配置 Spring MVC 配置文件的位置及名称 -->
    <init-param>
        <param-name>contextConfigLocation</param-name>
        <param-value>classpath:springmvc.xml</param-value>
    </init-param>
    <load-on-startup>1</load-on-startup>
</servlet>
```

```xml
<servlet-mapping>
    <servlet-name>dispatcherServlet</servlet-name>
    <url-pattern>/</url-pattern>
</servlet-mapping>
```

(2) 在项目 src 目录下创建 Spring MVC 的配置文件 springmvc.xml，依次配置自动扫描的包；配置视图解析器；启用 MVC 注解驱动；配置 CommonsMultipartResolver 实现文件上传。代码如下：

```xml
<?xml version="1.0" encoding="UTF-8"?>
<beans xmlns="http://www.springframework.org/schema/beans"
    xmlns:xsi="http://www.w3.org/2001/XMLSchema-instance"
    xmlns:aop="http://www.springframework.org/schema/aop"
    xmlns:context="http://www.springframework.org/schema/context"
    xmlns:mvc="http://www.springframework.org/schema/mvc"
    xsi:schemaLocation="http://www.springframework.org/schema/beans
    http://www.springframework.org/schema/beans/spring-beans-4.3.xsd
        http://www.springframework.org/schema/context
    http://www.springframework.org/schema/context/spring-context-4.3.xsd
        http://www.springframework.org/schema/mvc
    http://www.springframework.org/schema/mvc/spring-mvc-4.3.xsd">
    <!-- 配置自动扫描的包,SpringMVC 的 IOC 容器中的 bean 可以来引用 Spring IOC 容器中的 bean.但 Spring IOC 容器中的 bean 却不能来引用 SpringMVC IOC 容器中的 bean! -->
    <context:component-scan base-package="com.res" use-default-filters="false">
        <context:include-filter type="annotation"
            expression="org.springframework.stereotype.Controller" />
        <context:include-filter type="annotation" expression="org.springframework.web.bind.annotation.ControllerAdvice" />
    </context:component-scan>
    <!-- 配置视图解析器,将 handler 方法返回的逻辑视图解析为物理视图 -->
    <bean class="org.springframework.web.servlet.view.InternalResourceViewResolver">
        <property name="prefix" value="/"></property>
        <property name="suffix" value=".jsp"></property>
    </bean>
    <!-- 启用 MVC 注解驱动 -->
    <mvc:annotation-driven></mvc:annotation-driven>
    <mvc:default-servlet-handler />
    <!-- 配置 CommonsMultipartResolver,实现文件上传 -->
    <bean id="multipartResolver"    class="org.springframework.web.multipart.commons.CommonsMultipartResolver">
        <!-- 设置上传文件的最大尺寸为 1MB -->
        <property name="maxUploadSize" value="1048576" />
        <!-- 字符编码 -->
        <property name="defaultEncoding" value="UTF-8" />
    </bean>
</beans>
```

21.7.8 创建登录页

创建登录页 login.jsp，表单部分代码如下：

```
<form action="user/login" method="post">
    <table>
        <tr>
            <td>用户名：</td>
            <td><input type="text" name="loginName" /></td>
        </tr>
        <tr>
            <td>密 码：</td>
            <td><input type="text" name="loginPwd" /></td>
        </tr>
        <tr>
            <td><input type="submit" value="登录" /></td>
            <td></td>
        </tr>
    </table>
</form>
编辑登录成功页 index.jsp
<body>
    欢迎您，登录成功！
</body>
```

部署项目，启动 Tomcat，在浏览器中输入 http://localhost:8080/springmvc_ssh/login.jsp，打开如图 21-3 所示的登录页面。输入数据表 users 中正确的用户名和密码，单击"确定"按钮，页面转到 index.jsp，如图 21-4 所示。

图 21-3 登录页面

图 21-4 登录成功页面

21.8 Spring 整合 Spring MVC 与 MyBatis

MyBatis 与 Hibernate 一样，也是非常优秀的持久层框架。在框架整合开发时，也会选择 MyBatis 替代 Hibernate。以登录功能为例，采用 Annotation 注解方式实现 Spring、Spring MVC 和 MyBatis 的整合。

21.8.1 环境搭建

在 MyEclipse 中创建一个名为 springmvc_ssm 的 Web 项目，选择 Java version 为 1.8，Target runtime 为 Apache Tomcat v8.0。将第 20 章中图 20-2 所示的 12 个 Spring 所需的 jar 包，以及 aopalliance-1.0.jar、aspectjweaver-1.8.6.jar、cglib-3.2.0.jar、commons-logging-1.1.3.jar 这四个相关 jar 包。MySQL 数据库驱动包 mysql-connector-java-5.1.18-bin.jar、连接池核心包 c3p0-0.9.5.2.jar、c3p0 连接池的依赖包 mchange-commons-java-0.2.11.jar 这三个包，打印日志所需要的 jar 包 log4j-1.2.17.jar、mybatis 的必备 jar 包 mybatis-3.3.0.jar 以及整合 Spring 所需的 jar 包 mybatis-spring-1.2.3.jar 复制到项目 springmvc_ssm 的 WebRoot\WEB-INF\lib 目录中。

21.8.2 创建实体类

创建包 com.news.pojo，在包中创建实体类 Users，其属性对应于数据库 news 中数据表 users 的字段，代码如下：

```java
package com.news.pojo;
public class Users {
    private int id;
    private String loginName;
    private String loginPwd;
    // 此处省略上述属性的 get 和 set 方法
}
```

21.8.3 Spring 整合 MyBatis

Spring 整合 MyBatis 是在 Spring 配置文件 applicationContext.xml 中完成的，在 src 目录下创建 applicationContext.xml，依次配置扫描 com.news.dao 包中的所有接口，开启 Spring 的 Bean 自动扫描机制，配置数据源，配置 MapperScannerConfigurer，配置 DataSourceTransactionManager(事务管理)，启用基于注解的声明式事务管理配置。代码如下：

```xml
<?xml version="1.0" encoding="UTF-8"?>
<beans xmlns="http://www.springframework.org/schema/beans"
    xmlns:xsi="http://www.w3.org/2001/XMLSchema-instance"
    xmlns:p="http://www.springframework.org/schema/p"
    xmlns:context="http://www.springframework.org/schema/context"
    xmlns:mvc="http://www.springframework.org/schema/mvc"
    xmlns:mybatis-spring="http://mybatis.org/schema/mybatis-spring"
    xmlns:tx="http://www.springframework.org/schema/tx"
    xsi:schemaLocation="http://www.springframework.org/schema/mvc
    http://www.springframework.org/schema/mvc/spring-mvc-4.3.xsd
        http://www.springframework.org/schema/beans
    http://www.springframework.org/schema/beans/spring-beans-4.3.xsd
        http://mybatis.org/schema/mybatis-spring
    http://mybatis.org/schema/mybatis-spring-1.2.xsd
        http://www.springframework.org/schema/tx
    http://www.springframework.org/schema/tx/spring-tx-4.3.xsd
        http://www.springframework.org/schema/context
```

```xml
    http://www.springframework.org/schema/context/spring-context-4.3.xsd">
<!-- 扫描 com.news.dao 包中的所有接口,当作 Spring 的 Bean 配置,之后可以进行依赖注入 -->
    <mybatis-spring:scan base-package="com.news.dao" />
<!--开启 Spring 的 Bean 自动扫描机制,把 com.news 包中的类注册为 Spring 的 Bean-->
    <context:component-scan base-package="com.news" />
    <!-- 配置数据源 -->
    <bean id="dataSource" class="com.mchange.v2.c3p0.ComboPooledDataSource">
        <property name="driverClass" value="com.mysql.jdbc.Driver" />
        <property name="jdbcUrl" value="jdbc:mysql:///news" />
        <property name="user" value="root" />
        <property name="password" value="123456" />
        <property name="minPoolSize" value="5" />
        <property name="maxPoolSize" value="10" />
    </bean>
    <!-- 配置 SqlSessionFactoryBean -->
    <bean id="sqlSessionFactory" class="org.mybatis.spring.SqlSessionFactoryBean">
        <property name="dataSource" ref="dataSource" />
    </bean>
<!-- 配置 MapperScannerConfigurer,DAO 接口所在包名,Spring 会自动查找其下的类 -->
    <bean class="org.mybatis.spring.mapper.MapperScannerConfigurer">
        <property name="basePackage" value="com.news.dao" />
        <property name="sqlSessionFactoryBeanName" value="sqlSessionFactory"></property>
    </bean>
    <!-- 配置 DataSourceTransactionManager(事务管理) -->
    <bean id="transactionManager" class="org.springframework.jdbc.datasource.DataSourceTransactionManager">
        <property name="dataSource" ref="dataSource" />
    </bean>
    <!-- 启用基于注解的声明式事务管理配置 -->
    <tx:annotation-driven transaction-manager="transactionManager" />
</beans>
```

21.8.4 DAO 层开发

创建包 com.news.dao,在包中创建接口 UsersDAO,在 UsersDAO 接口中声明方法 getUsersByCond,用于登录验证,代码如下:

```java
package com.news.dao;
public interface UsersDAO {
    // 根据登录名和密码查询
    @Select("select * from users where LoginName = #{loginName} and LoginPwd = #{loginPwd}")
    public Users getUsersByCond(@Param("loginName") String loginName,
        @Param("loginPwd") String loginPwd);
}
```

21.8.5 Service 层开发

创建包 com.news.service,在包中创建接口 UsersService,在 UsersService 接口中声明方法

login，用于登录验证，代码如下：

```java
package com.news.service;
public interface UsersService {
    public Users login(String loginName, String loginPwd);
}
```

创建 UsersService 接口的实现类 UsersServiceImpl，存放在 com.news.service.impl 包中，实现 login 方法。使用@Service、@Autowired 注解的实现类 UsersServiceImpl 如下：

```java
package com.news.service.impl;
@Service("usersService")
public class UsersServiceImpl implements UsersService {
    @Autowired
    private UsersDAO usersDAO;
    @Override
    public Users login(String loginName, String loginPwd) {
        return usersDAO.getUsersByCond(loginName, loginPwd);
    }
}
```

21.8.6 控制器开发

创建包 com.news.controller，在包中创建类 UsersController，代码如下：

```java
package com.news.controller;
@Controller
@RequestMapping("/user")
public class UsersController {
    @Autowired
    private UsersService usersService;
    @RequestMapping("/login")
    public String login(Users u) {
        Users user = usersService.login(u.getLoginName(), u.getLoginPwd());
        if (user != null && user.getLoginName() != null) {
            return "index";
        } else {
            return "redirect:/login.jsp";
        }
    }
}
```

21.8.7 Spring 整合 Spring MVC

在项目 src 目录下创建 Spring MVC 的配置文件 springmvc.xml，依次配置自动扫描的包；配置视图解析器；启用 MVC 注解驱动；配置允许对静态资源文件的访问。代码如下：

```xml
<?xml version="1.0" encoding="UTF-8"?>
<beans xmlns="http://www.springframework.org/schema/beans"
    xmlns:xsi="http://www.w3.org/2001/XMLSchema-instance"
xmlns:aop="http://www.springframework.org/schema/aop"
    xmlns:context="http://www.springframework.org/schema/context"
    xmlns:mvc="http://www.springframework.org/schema/mvc"
```

```xml
    xsi:schemaLocation="http://www.springframework.org/schema/beans
http://www.springframework.org/schema/beans/spring-beans-4.3.xsd
       http://www.springframework.org/schema/context
http://www.springframework.org/schema/context/spring-context-4.3.xsd
       http://www.springframework.org/schema/mvc
http://www.springframework.org/schema/mvc/spring-mvc-4.3.xsd">
    <!-- 配置自动扫描的包 -->
    <context:component-scan base-package="com.news.controller" />
    <!-- 配置视图解析器,将handler方法返回的逻辑视图解析为物理视图 -->
    <bean class=
"org.springframework.web.servlet.view.InternalResourceViewResolver">
        <property name="prefix" value="/"></property>
        <property name="suffix" value=".jsp"></property>
    </bean>
    <!-- 启用MVC注解驱动 -->
    <mvc:annotation-driven />
    <!-- 配置允许对静态资源文件的访问 -->
    <mvc:default-servlet-handler />
</beans>
```

在 web.xml 文件中，依次配置 ContextLoaderListener 加载 Spring 配置文件；编码过滤器；配置 Spring MVC 的 DispatcherServlet。代码如下：

```xml
<!-- 配置ContextLoaderListener,加载Spring配置文件 -->
<context-param>
    <param-name>contextConfigLocation</param-name>
    <param-value>classpath:applicationContext.xml</param-value>
</context-param>
<!-- Spring 监听器 -->
<listener>
    <listener-class>org.springframework.web.context.ContextLoaderListener
</listener-class>
</listener>
<!-- 编码过滤器 -->
<filter>
    <filter-name>encodingFilter</filter-name>
    <filter-class>org.springframework.web.filter.CharacterEncodingFilter
</filter-class>
    <async-supported>true</async-supported>
    <init-param>
        <param-name>encoding</param-name>
        <param-value>UTF-8</param-value>
    </init-param>
</filter>
<filter-mapping>
    <filter-name>encodingFilter</filter-name>
    <url-pattern>/*</url-pattern>
</filter-mapping>
<!-- 配置 Spring MVC 的 DispatcherServlet -->
<servlet>
    <servlet-name>dispatcherServlet</servlet-name>
    <servlet-class>org.springframework.web.servlet.DispatcherServlet
</servlet-class>
    <!-- 配置 Spring MVC 配置文件的位置及名称 -->
```

```xml
    <init-param>
        <param-name>contextConfigLocation</param-name>
        <param-value>classpath:springmvc.xml</param-value>
    </init-param>
    <load-on-startup>1</load-on-startup>
</servlet>
<servlet-mapping>
    <servlet-name>dispatcherServlet</servlet-name>
    <url-pattern>/</url-pattern>
</servlet-mapping>
```

21.8.8 创建页面

创建登录页 login.jsp，表单部分代码如下：

```
<form action="user/login" method="post">
    <table>
        <tr><td>用户名: </td>
            <td><input type="text" name="loginName" /></td>    </tr>
        <tr><td>密 码: </td>
            <td><input type="text" name="loginPwd" /></td></tr>
        <tr><td><input type="submit" value="登录" /></td><td></td>
        </tr>
    </table>
</form>
编辑登录成功页 index.jsp
<body>
    欢迎您，登录成功!
</body>
```

部署项目，启动 Tomcat，在浏览器中输入 http://localhost:8080/springmvc_ssm/ login.jsp，打开如图 21-5 所示的登录页面。输入数据表 users 中正确的用户名和密码，单击"确定"按钮，页面转到 index.jsp，如图 21-6 所示。

图 21-5 登录页面

图 21-6 登录成功页面

21.9 小 结

本章主要介绍了 Spring MVC 的基本概念，Spring MVC 的常用注解，类型转换、格式化、数据校验，文件上传，国际化等方面的知识。最后以登录功能为例，详细介绍了基于注解实现 Spring 整合 Spring MVC 与 Hibernate、Spring 整合 Spring MVC 与 MyBatis 的过程。

第 22 章
Spring 整合 Struts 2 与 Hibernate 实现网上订餐系统前台

　　网上订餐系统是电子商务的一个典型案例，由前台和后台这两部分组成，本章将介绍前台功能的实现过程。系统在开发过程中整合了 Spring 4、Hibernate 5 和 Struts 2 框架。其中，Struts 2 框架用来处理页面逻辑，Hibernate 5 框架用来进行持久化操作，Spring 4 对 Struts 2 和 Hibernate 5 进行了整合。Spring 4 与 Hibernate 5 的集成提供了很好的配置方式，同时也简化了 Hibernate 5 的编码。Spring 4 与 Struts 2 的集成实现了系统层与层之间的脱耦，从而使得系统运行效率更高，维护也更方便。

22.1 需求与系统分析

前台即客户端，前台用户进入首页后，可以查看一些形色艳丽的餐品图片。可以通过点击餐品图片来查看餐品的详细信息。在用户看中某一餐品时，可以事先登录，或者注册，然后可以随心订购自己所需要的餐品。可以使用购物车暂存喜爱的餐品，也可以对购物车中的餐品进行管理。最后可以提交订单。系统前台用户的用例图如图 22-1 所示。

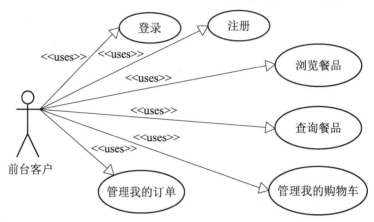

图 22-1 系统前台用户的用例图

根据需求分析，订餐系统前台功能如下。
(1) 前台用户注册为会员。
(2) 登录网上订餐系统浏览餐品。
(3) 用户根据菜系和餐品名称查询餐品。
(4) 用户对自己的个人信息进行更改，如送餐地址、联系电话、账户密码等。
(5) 对暂存入购物车中的餐品进行更改，如选择的数量或者取消选择。
(6) 当用户确定订餐完毕后，将其提交到服务器，生成订单。

根据上述分析，可以得到订餐系统前台的模块结构，如图 22-2 所示。

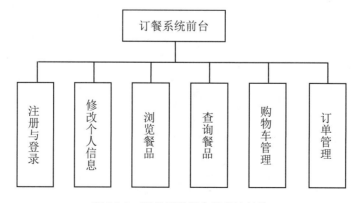

图 22-2 订餐系统前台的模块结构

22.2 数据库设计

数据库设计是系统设计中非常重要的一个环节，数据是设计的基础，直接决定系统的成败。如果数据库设计不合理、不完善，将在系统开发中，甚至到后期的维护时，引起严重的问题。根据系统需求，创建了 8 张表，具体介绍如下。

(1) 用户信息表(users)：用于记录前台用户基本信息。
(2) 管理员信息表(admin)：用于记录管理员基本信息。
(3) 菜系表(mealseries)：用于记录各种菜系。
(4) 餐品信息表(meal)：用于记录餐品信息。
(5) 订单主表(orders)：用于记录订单主要信息。
(6) 订单子表(orderdts)：用于记录订单详细信息。
(7) 系统功能表(functions)：用于记录系统提供的功能信息。
(8) 权限表(powers)：用于记录管理员的权限信息。

其中，用户信息表(users)的字段说明如表 22-1 所示。

表 22-1 用户信息表(users)的字段说明

字 段 名	类 型	说 明
Id	int(4)	用户编号，主键，自增
LoginName	vachar(20)	用户登录名
LoginPwd	varchar(20)	用户登录密码
TrueName	varchar(20)	用户真实姓名
Email	varchar(20)	用户电子邮箱
Phone	varchar(20)	用户联系电话
Address	varchar(50)	用户送货地址
Status	int(4)	用户状态，1 表示启用，0 表示禁用

管理员信息表(admin)的字段说明如表 22-2 所示。

表 22-2 管理员信息表(admin)的字段说明

字 段 名	类 型	说 明
Id	int(4)	管理员编号，主键，自增
LoginName	varchar(20)	管理员登录名
LoginPwd	varchar(20)	管理员登录密码

菜系表(mealseries)的字段说明如表 22-3 所示。

表 22-3 菜系表(mealseries)的字段说明

字 段 名	类 型	说 明
SeriesId	int(4)	菜系编号，主键，自增
SeriesName	varchar(10)	菜系名称

餐品信息表(meal)的字段说明如表 22-4 所示。

表 22-4　餐品信息表(meal)的字段说明

字 段 名	类 型	说 明
MealId	int(4)	餐品编号，主键，自增
MealSeriesId	int(4)	所属菜系，外键，与 mealseries 表 SeriesId 字段关联
MealName	varchar(20)	餐品名称
MealSummarize	varchar(250)	餐品摘要
MealDescription	varchar(250)	餐品详细描述信息
MealPrice	decimal(8,2)	餐品价格
MealImage	varchar(20)	餐品图片文件名
MealStatus	int(4)	餐品状态，1 表示在售，0 表示下架

订单主表(orders)的字段说明如表 22-5 所示。

表 22-5　订单主表(orders)的字段说明

字 段 名	类 型	说 明
OID	int(4)	订单编号，主键，自增
UserId	int(4)	订单用户编号，与 users 表 Id 字段关联
OrderTime	datetime	订单日期
OrderStatus	varchar(20)	订单状态，分为"未付款""已付款""待发货""已发货""已完成"等
OrderPrice	decimal(8,2)	订单合计

订单子表(orderdts)的字段说明如表 22-6 所示。

表 22-6　订单子表(orderdts)的字段说明

字 段 名	类 型	说 明
ODID	int(4)	订单明细编号，主键，自增
OID	int(4)	所属订单编号，与 orders 表 OID 字段关联
MealId	int(4)	餐品编号，与 meal 表的 MealId 字段关联
MealPrice	decimal(8,2)	餐品单价
MealCount	int(4)	餐品购买数量

系统功能表(functions)的字段说明如表 22-7 所示。

表 22-7　系统功能表(functions)的字段说明

字 段 名	类 型	说 明
id	int(4)	系统功能 id，主键，自增
name	varchar(20)	功能菜单名称

续表

字 段 名	类 型	说 明
parentid	int(4)	父结点 id
url	varchar(50)	功能页面
isleaf	bit(1)	是否为叶结点
nodeorder	int(4)	结点顺序

权限表(powers)的字段说明如表 22-8 所示。

表 22-8　权限表(powers)的字段说明

字 段 名	类 型	说 明
aid	int(4)	管理员 id，主键，与 admin 表的 Id 字段关联
fid	int(4)	系统功能 id，主键，与 functions 表 id 字段关联

创建数据表后，设计数据表之间的关系，如图 22-3 所示。

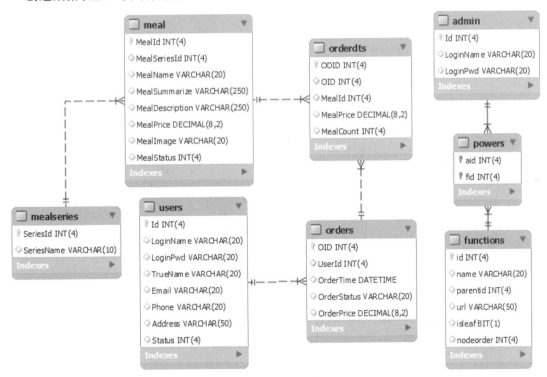

图 22-3　系统数据表之间的关系

22.3　项目环境搭建

在第 20 章的 20.1 节中，以用户登录为例详细介绍了基于 XML 配置文件使用 Spring 整合 Struts 2 与 Hibernate，读者可参照完成网上订餐系统前台的框架搭建。当然，读者也可以直接将 20.1 节创建的项目 s2sh 复制一份并重新命名为 restaurant，再导入 MyEclipse 中。为了避免部署重复，需要修改项目的部署名称。修改过程如下：在 MyEclipse 中右击复制后的项目 restaurant，依次选择 Properties→MyEclipse→Deployment Assembly，将 Web Context Root 修改为 restaurant 即可。

订餐系统前台的目录结构如图 22-4 所示，其中 com.restaurant.action 包用于存放 Action 类，com.restaurant.dao 包用于存放数据访问层接口，com.restaurant.dao.impl 包用于存放数据访问层接口的实现类，com.restaurant.service 包用于存放业务逻辑层接口，com.restaurant.service.impl 包用于存放业务逻辑层接口的实现类，com.restaurant.entity 包用于存放实体类，com.restaurant.filter 包用于存放过滤器类，com.restaurant.interceptor 包用于存放拦截器类。

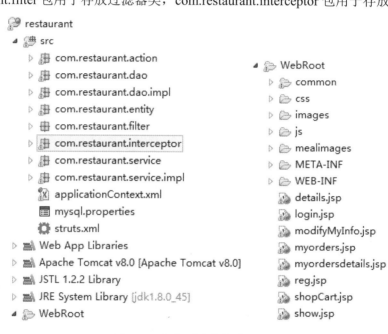

图 22-4 订餐系统前台的目录结构

22.4 Spring 及 Struts 2 配置文件

Spring 框架使用的配置文件为 applicationContext.xml，Struts 2 使用的配置文件为 struts.xml，这些配置文件的含义在 20.1 节中已经具体介绍过，由于篇幅，在此不再赘述。

22.5 创建实体类和映射文件

在 com.restaurant.entity 包中，依次创建实体类 Users.java、Mealseries.java、Meal.java、Orders.java 和 Orderdts.java。

其中，实体类 Users.java 代码如下：

```
package com.restaurant.entity;
```

```java
import java.util.HashSet;
import java.util.Set;
public class Users implements java.io.Serializable {
    private Integer id;
    private String loginName;
    private String loginPwd;
    private String trueName;
    private String email;
    private String phone;
    private String address;
    private Integer status;
    private Set orderses = new HashSet(0);
    // 此处省略了构造方法和上述属性的 get 和 set 方法
}
```

实体类 Mealseries.java 代码如下：

```java
package com.restaurant.entity;
import java.util.HashSet;
import java.util.Set;
public class Mealseries implements java.io.Serializable {
    private Integer seriesId;
    private String seriesName;
    private Set meals = new HashSet(0);
    // 此处省略了构造方法和上述属性的 get 和 set 方法
}
```

实体类 Meal.java 代码如下：

```java
package com.restaurant.entity;
import java.util.HashSet;
import java.util.Set;
public class Meal implements java.io.Serializable {
    private Integer mealId;
    private Mealseries mealseries;
    private String mealName;
    private String mealSummarize;
    private String mealDescription;
    private Double mealPrice;
    private String mealImage;
    private int mealStatus;
    private Set orderdtses = new HashSet(0);
    // 此处省略了构造方法和上述属性的 get 和 set 方法
}
```

实体类 Orders.java 代码如下：

```java
package com.restaurant.entity;
import java.util.Date;
import java.util.HashSet;
import java.util.Set;
public class Orders implements java.io.Serializable {
    private Integer oid;
    private Users users;
    private Date orderTime;
    private String orderStatus;
```

```
    private Double orderPrice;
    private Set orderdtses = new HashSet(0);
    // 此处省略了构造方法和上述属性的get和set方法
}
```

实体类Orderdts.java代码如下：

```
package com.restaurant.entity;
public class Orderdts implements java.io.Serializable {
    private Integer odid;
    private Meal meal;
    private Orders orders;
    private Double mealPrice;
    private Integer mealCount;
    // 此处省略了构造方法和上述属性的get和set方法
}
```

在com.restaurant.entity包中，依次创建上述实体类对应的映射文件Users.hbm.xml、Mealseries.hbm.xml、Meal.hbm.xml、Orders.hbm.xml和Orderdts.hbm.xml。

其中，映射文件Users.hbm.xml如下：

```xml
<?xml version="1.0" encoding="utf-8"?>
<!DOCTYPE hibernate-mapping PUBLIC "-//Hibernate/Hibernate Mapping DTD 3.0//EN" "http://www.hibernate.org/dtd/hibernate-mapping-3.0.dtd">
<hibernate-mapping package="com.restaurant.entity">
    <class name="Users" table="users" catalog="restrant">
        <id name="id" type="java.lang.Integer">
            <column name="Id" />
            <generator class="native"></generator>
        </id>
        <property name="loginName" type="java.lang.String">
            <column name="LoginName" length="20" />
        </property>
        <property name="loginPwd" type="java.lang.String">
            <column name="LoginPwd" length="20" />
        </property>
        <property name="trueName" type="java.lang.String">
            <column name="TrueName" length="20" />
        </property>
        <property name="email" type="java.lang.String">
            <column name="Email" length="20" />
        </property>
        <property name="phone" type="java.lang.String">
            <column name="Phone" length="20" />
        </property>
        <property name="address" type="java.lang.String">
            <column name="Address" length="50" />
        </property>
        <property name="status" type="java.lang.Integer">
            <column name="Status" />
        </property>
        <!-- 配置一对多关联映射 -->
        <set name="orderses" inverse="true" lazy="false">
            <key>
                <column name="UserId" />
```

```xml
            </key>
            <one-to-many class="Orders" />
        </set>
    </class>
</hibernate-mapping>
```

映射文件 Mealseries.hbm.xml 如下：

```xml
<?xml version="1.0" encoding="utf-8"?>
<!DOCTYPE hibernate-mapping PUBLIC "-//Hibernate/Hibernate Mapping DTD
3.0//EN" "http://www.hibernate.org/dtd/hibernate-mapping-3.0.dtd">
<hibernate-mapping package="com.restaurant.entity">
    <class name="Mealseries" table="mealseries" catalog="restrant">
        <id name="seriesId" type="java.lang.Integer">
            <column name="SeriesId" />
            <generator class="native"></generator>
        </id>
        <property name="seriesName" type="java.lang.String">
            <column name="SeriesName" length="10" />
        </property>
        <!-- 配置一对多关联映射 -->
        <set name="meals" inverse="true" lazy="false">
            <key>
                <column name="MealSeriesId" />
            </key>
            <one-to-many class="Meal" />
        </set>
    </class>
</hibernate-mapping>
```

映射文件 Meal.hbm.xml 如下：

```xml
<?xml version="1.0" encoding="utf-8"?>
<!DOCTYPE hibernate-mapping PUBLIC "-//Hibernate/Hibernate Mapping DTD
3.0//EN" "http://www.hibernate.org/dtd/hibernate-mapping-3.0.dtd">
<hibernate-mapping package="com.restaurant.entity">
    <class name="Meal" table="meal" catalog="restrant">
        <id name="mealId" type="java.lang.Integer">
            <column name="MealId" />
            <generator class="native"></generator>
        </id>
        <!-- 配置多对一关联映射 -->
        <many-to-one name="mealseries" class="Mealseries" fetch="select"
lazy="false">
            <column name="MealSeriesId" />
        </many-to-one>
        <property name="mealName" type="java.lang.String">
            <column name="MealName" length="20" />
        </property>
        <property name="mealSummarize" type="java.lang.String">
            <column name="MealSummarize" length="250" />
        </property>
        <property name="mealDescription" type="java.lang.String">
            <column name="MealDescription" length="250" />
        </property>
```

```xml
            <property name="mealPrice" type="java.lang.Double">
                <column name="MealPrice" precision="8" />
            </property>
            <property name="mealImage" type="java.lang.String">
                <column name="MealImage" length="20" />
            </property>
            <property name="mealStatus" type="java.lang.Integer">
                <column name="MealStatus" length="4" />
            </property>
            <!-- 配置一对多关联映射 -->
            <set name="orderdtses" inverse="true" lazy="false" cascade="delete">
                <key>
                    <column name="MealId" />
                </key>
                <one-to-many class="Orderdts" />
            </set>
    </class>
</hibernate-mapping>
```

映射文件 Orders.hbm.xml 如下：

```xml
<?xml version="1.0" encoding="utf-8"?>
<!DOCTYPE hibernate-mapping PUBLIC "-//Hibernate/Hibernate Mapping DTD 3.0//EN" "http://www.hibernate.org/dtd/hibernate-mapping-3.0.dtd">
<hibernate-mapping package="com.restaurant.entity">
    <class name="Orders" table="orders" catalog="restrant">
        <id name="oid" type="java.lang.Integer">
            <column name="OID" />
            <generator class="native"></generator>
        </id>
        <!-- 配置多对一关联映射 -->
        <many-to-one name="users" class="Users" fetch="select" lazy="false">
            <column name="UserId" />
        </many-to-one>
        <property name="orderTime" type="java.sql.Timestamp">
            <column name="OrderTime" length="19" />
        </property>
        <property name="orderStatus" type="java.lang.String">
            <column name="OrderStatus" length="20" />
        </property>
        <property name="orderPrice" type="java.lang.Double">
            <column name="OrderPrice" precision="8" />
        </property>
        <!-- 配置一对多关联映射 -->
        <set name="orderdtses" cascade="all" inverse="true" lazy="false">
            <key>
                <column name="OID" />
            </key>
            <one-to-many class="Orderdts" />
        </set>
    </class>
</hibernate-mapping>
```

映射文件 Orderdts.hbm.xml 如下：

```xml
<?xml version="1.0" encoding="utf-8"?>
<!DOCTYPE hibernate-mapping PUBLIC "-//Hibernate/Hibernate Mapping DTD
3.0//EN" "http://www.hibernate.org/dtd/hibernate-mapping-3.0.dtd">
<hibernate-mapping package="com.restaurant.entity">
    <class name="Orderdts" table="orderdts" catalog="restrant">
        <id name="odid" type="java.lang.Integer">
            <column name="ODID" />
            <generator class="native"></generator>
        </id>
        <!-- 配置多对一关联映射 -->
        <many-to-one name="meal" class="Meal" fetch="select" lazy="false">
            <column name="MealId" />
        </many-to-one>
        <!-- 配置多对一关联映射 -->
        <many-to-one cascade="all" name="orders" class="Orders"
            fetch="select" lazy="false">
            <column name="OID" />
        </many-to-one>
        <property name="mealPrice" type="java.lang.Double">
            <column name="MealPrice" precision="8" />
        </property>
        <property name="mealCount" type="java.lang.Integer">
            <column name="MealCount" />
        </property>
    </class>
</hibernate-mapping>
```

最后，还需要在 Spring 配置文件 applicationContext.xml 中添加对映射文件的引用：

```xml
<property name="mappingResources">
    <list>
        <value>com/restaurant/entity/Meal.hbm.xml</value>
        <value>com/restaurant/entity/Orders.hbm.xml</value>
        <value>com/restaurant/entity/Users.hbm.xml</value>
        <value>com/restaurant/entity/Orderdts.hbm.xml</value>
        <value>com/restaurant/entity/Mealseries.hbm.xml</value>
    </list>
</property>
```

22.6　创建 DAO 接口及实现类

在 com.restaurant.dao 包中，依次创建数据访问层接口 BaseDao.java、UserDAO.java、MealDAO.java、MealSeriesDAO.java、OrdersDAO.java 和 OrderDtsDAO.java。

在接口 BaseDao.java 中声明如下方法：

```java
package com.restaurant.dao;
import java.io.Serializable;
import java.util.List;
public interface BaseDao<T>{
    public Serializable save(T o);
    public void delete(T o);
    public void update(T o);
```

```java
    public void saveOrUpdate(T o);
    public List<T> find(String hql);
    public List<T> find(String hql, Object[] param);
    public List<T> find(String hql, Object[] param, Integer page, Integer rows);
    public T get(Class<T> c, Serializable id);
    public Object findUnique(String hql);
    public Integer executeHql(String hql);
    // 保存或修改 sql
    void saveOrUpdate(String sql);
    // 执行原生 SQL
    public Integer executeSql(String sql, Object[] param);
    // 本地 sql 查询
    List queryBySql(String sql);
}
```

在接口 UserDAO.java 中声明如下方法：

```java
package com.restaurant.dao;
import java.util.List;
import com.restaurant.entity.Users;
public interface UserDAO {
    // 根据用户名获取用户对象
    public List getUserByLoginName(String loginName);
    // 添加新用户
    public void addUsers(Users u);
    // 更新用户信息
    public void modifyUsers(Users u);
}
```

在接口 MealDAO.java 中声明如下方法：

```java
package com.restaurant.dao;
import java.util.List;
import com.restaurant.entity.Meal;
public interface MealDAO {
    // 根据查询条件(封装在 condition 中)和当前页码,获取餐品
    public List getMealByConditionForPager(Meal condition, int page);
    // 根据查询条件(封装在 condition 中),计算餐品总数
    public Integer getCountOfMeal(Meal condition);
    // 根据餐品 id 号,获取餐品对象
    public Meal getMealByMealId(int mealId);
    // 根据多个餐品 id 号,获取餐品
    public List<Meal> getByIds(String ids);
    // 获取餐品销售排行
    public List<Object[]> getSalePaihang();
}
```

在接口 MealSeriesDAO.java 中声明如下方法：

```java
package com.restaurant.dao;
import java.util.List;
public interface MealSeriesDAO {
    // 获取所有菜系
    public List getMealSeries();
}
```

在接口 OrdersDAO.java 中声明如下方法：

```java
package com.restaurant.dao;
import java.util.List;
import com.restaurant.entity.Orders;
public interface OrdersDAO {
    // 根据用户id和当前页码,获取订单
    public List getOrdersByUserIdForPager(int userId, int page);
    // 根据用户id,获取订单总数
    public Integer getCountOfMyOrders(int userId);
    // 根据订单号,获取订单
    public Orders getOrdersByOid(int oid);
    // 删除订单
    public void deleteOrders(Orders orders);
    // 更新订单
    public void updateOrders(Orders orders);
}
```

在接口 OrderDtsDAO.java 中声明如下方法：

```java
package com.restaurant.dao;
import java.util.List;
import com.restaurant.entity.Orderdts;
public interface OrderDtsDAO {
    // 生成订单子表(订单明细)
    public void addOrderDts(Orderdts dts);
    // 根据订单主表编号获取订单明细列表
    public List getOrderDtsByOid(int oid);
}
```

在 com.restaurant.dao 包中，依次创建上述接口的实现类 BaseDaoImpl.java、UserDAOImpl.java、MealDAOImpl.java、MealSeriesDAOImpl.java、OrdersDAOImpl.java、OrderDtsDAOImpl.java。

其中，BaseDaoImpl.java 代码如下：

```java
package com.restaurant.dao.impl;
……
import com.restaurant.dao.BaseDao;
public class BaseDaoImpl<T> implements BaseDao<T> {
    private SessionFactory sessionFactory;
    public SessionFactory getCurrentSessionFactory() {
        return sessionFactory;
    }
    public void setSessionFactory(SessionFactory sessionFactory) {
        this.sessionFactory = sessionFactory;
    }
    public Session getCurrentSession() {
        return this.sessionFactory.getCurrentSession();
    }
    public Serializable save(T o) {
        return this.getCurrentSession().save(o);
    }
    public void delete(T o) {
        this.getCurrentSession().delete(o);
    }
```

```java
public void update(T o) {
    this.getCurrentSession().update(o);
}
public void saveOrUpdate(T o) {
    this.getCurrentSession().saveOrUpdate(o);
}
@SuppressWarnings("unchecked")
public List<T> find(String hql) {
    return this.getCurrentSession().createQuery(hql).getResultList();
}
@SuppressWarnings("unchecked")
public List<T> find(String hql, Object[] param) {
    Query<T> q = this.getCurrentSession().createQuery(hql);
    if (param != null && param.length > 0) {
        for (int i = 0; i < param.length; i++) {
            q.setParameter(i, param[i]);
        }
    }
    return q.getResultList();
}
@SuppressWarnings("unchecked")
public List<T> find(String hql, Object[] param, Integer page, Integer rows) {
    if (page == null || page < 0) {
        page = 0;
    }
    if (rows == null || rows < 1) {
        rows = 9;
    }
    Query<T> q = this.getCurrentSession().createQuery(hql);
    if (param != null && param.length > 0) {
        for (int i = 0; i < param.length; i++) {
            q.setParameter(i, param[i]);
        }
    }
    return q.setFirstResult(page * rows).setMaxResults(rows)
            .getResultList();
}
public T get(Class<T> c, Serializable id) {
    return (T) this.getCurrentSession().get(c, id);
}
public Object findUnique(String hql) {
    return this.getCurrentSession().createQuery(hql).getSingleResult();
}
public Integer executeHql(String hql) {
    return this.getCurrentSession().createQuery(hql).executeUpdate();
}
@Override
public void saveOrUpdate(String sql) {
    Query<?> q = this.getCurrentSession().createNativeQuery(sql);
    q.executeUpdate();
}
@Override
public Integer executeSql(String sql, Object[] param) {
    Query<?> q = this.getCurrentSession().createNativeQuery(sql);
```

```java
            if (param != null && param.length > 0) {
                for (int i = 0; i < param.length; i++) {
                    q.setParameter(i + 1, param[i]);
                }
            }
            return q.executeUpdate();
        }
        @Override
        public List queryBySql(String sql) {
            Query q = this.getCurrentSession().createNativeQuery(sql);
            return q.list();
        }
    }
```

UserDAOImpl.java 代码如下：

```java
package com.restaurant.dao.impl;
……
public class UserDAOImpl extends BaseDaoImpl<Users> implements UserDAO {
    // 根据用户名获取用户对象
    @Override
    public List getUserByLoginName(String loginName) {
        String hql = "from Users u where u.loginName = ?";
        Object[] param = new Object[] { loginName };
        return super.find(hql, param);
    }
    // 添加新用户
    @Override
    public void addUsers(Users u) {
        super.save(u);
    }
    // 更新用户信息
    @Override
    public void modifyUsers(Users u) {
        super.update(u);
    }
}
```

MealDAOImpl.java 代码如下：

```java
package com.restaurant.dao.impl;
……
import com.restaurant.dao.MealDAO;
public class MealDAOImpl extends BaseDaoImpl<Meal> implements MealDAO {
    // 根据查询条件(封装在condition中)和当前页面,获取餐品
    @Override
    public List getMealByConditionForPager(Meal condition, int page) {
        String hql = "from Meal m where m.mealStatus=1";
        Object[] param = null;
        if (condition != null) {
            ArrayList list = new ArrayList();
            if (condition.getMealName() != null
                    && !condition.getMealName().equals("")) {
                hql += " and m.mealName like ?";
                list.add("%" + condition.getMealName() + "%");
```

```java
            }
            if ((condition.getMealseries() != null)
                    && (condition.getMealseries().getSeriesId() != null)) {
                hql += " and m.mealseries.seriesId = ?";
                list.add(condition.getMealseries().getSeriesId());
            }
            if (list.size() > 0)
                param = list.toArray();
        }
        return super.find(hql, param, page - 1, 9);
    }
    // 根据查询条件(封装在condition中),计算餐品总数
    @Override
    public Integer getCountOfMeal(Meal condition) {
        Integer count = null;
        try {
            String hql = "select count(m) from Meal m where m.mealStatus=1";
            if (condition != null) {
                if (condition.getMealName() != null
                        && !condition.getMealName().equals("")) {
                    hql += " and m.mealName like '%" + condition.getMealName() + "%'";
                }
                if ((condition.getMealseries() != null)
                        && (condition.getMealseries().getSeriesId() != null)) {
                    hql += " and m.mealseries.seriesId = "
                            + condition.getMealseries().getSeriesId();
                }
            }
            count = Integer.parseInt(super.findUnique(hql).toString());
        } catch (Exception e) {
            e.printStackTrace();
        }
        return count;
    }
    // 根据餐品id号,获取餐品对象
    @Override
    public Meal getMealByMealId(int mealId) {
        return super.get(Meal.class, mealId);
    }
    // 根据多个餐品id号,获取餐品
    @Override
    public List<Meal> getByIds(String ids) {
        String hql = "from Meal m where m.mealId in " + ids;
        return super.find(hql);
    }
    // 获取餐品销售排行
    @Override
    public List<Object[]> getSalePaihang() {
        String sql = "SELECT DISTINCT mealid, SUM(MealCount) AS mc FROM orderdts od GROUP BY MealId ORDER BY mc DESC";
        return super.queryBySql(sql);
    }
}
```

MealSeriesDAOImpl.java 代码如下：

```java
package com.restaurant.dao.impl;
import java.util.List;
import com.restaurant.dao.MealSeriesDAO;
import com.restaurant.entity.Mealseries;
public class MealSeriesDAOImpl extends BaseDaoImpl<Mealseries> implements MealSeriesDAO {
    // 获取所有菜系
    @Override
    public List getMealSeries() {
        String hql = "from Mealseries ms";
        return super.find(hql);
    }
}
```

OrdersDAOImpl.java 代码如下:

```java
package com.restaurant.dao.impl;
……
import com.restaurant.dao.OrdersDAO;
public class OrdersDAOImpl extends BaseDaoImpl<Orders> implements OrdersDAO {
    // 根据用户id和当前页码,获取订单
    @Override
    public List getOrdersByUserIdForPager(int userId, int page) {
        String hql = "from Orders o where o.users.id=?";
        Object[] param = new Object[] { userId };
        return super.find(hql, param, page - 1, 10);
    }
    // 根据用户id,获取订单总数
    @Override
    public Integer getCountOfMyOrders(int userId) {
        Integer count = null;
        try {
            String hql = "select count(o) from Orders o where o.users.id=" + userId;
            count = Integer.parseInt(super.findUnique(hql).toString());
        } catch (Exception e) {
            e.printStackTrace();
        }
        return count;
    }
    // 根据订单号, 获取订单
    @Override
    public Orders getOrdersByOid(int oid) {
        return (Orders) super.get(Orders.class, oid);
    }
    // 删除订单
    @Override
    public void deleteOrders(Orders orders) {
        super.delete(orders);
    }
    // 更新订单
    @Override
    public void updateOrders(Orders orders) {
```

```java
        super.update(orders);
    }
}
```

OrderDtsDAOImpl.java 代码如下:

```java
package com.restaurant.dao.impl;
……
import com.restaurant.dao.OrderDtsDAO;
public class OrderDtsDAOImpl extends BaseDaoImpl<Orderdts> implements
        OrderDtsDAO {
    // 生成订单子表(订单明细)
    @Override
    public void addOrderDts(Orderdts dts) {
        super.saveOrUpdate(dts);
    }
    // 根据订单主表编号获取订单明细列表
    @Override
    public List getOrderDtsByOid(int oid) {
        String hql = "from Orderdts od where od.orders.oid=" + oid;
        return super.find(hql);
    }
}
```

DAO 接口的实现类还需要在 Spring 配置文件中加以定义,代码如下:

```xml
<!-- 将 BaseDaoImpl 声明成一个 bean 并将 sessionFactory 注入 -->
<bean id="baseDao" class="com.restaurant.dao.impl.BaseDaoImpl">
    <property name="sessionFactory" ref="sessionFactory"></property>
</bean>
<!-- 定义 MealDAOImpl 类 -->
<bean id="mealDAO" class="com.restaurant.dao.impl.MealDAOImpl">
    <property name="sessionFactory" ref="sessionFactory" />
</bean>
<!-- 定义 MealSeriesDAOImpl 类 -->
<bean id="mealSeriesDAO" class="com.restaurant.dao.impl.MealSeriesDAOImpl">
    <property name="sessionFactory" ref="sessionFactory" />
</bean>
<!-- 定义 OrdersDAOImpl 类 -->
<bean id="ordersDAO" class="com.restaurant.dao.impl.OrdersDAOImpl">
    <property name="sessionFactory" ref="sessionFactory" />
</bean>
<!-- 定义 OrderDtsDAOImpl 类 -->
<bean id="orderDtsDAO" class="com.restaurant.dao.impl.OrderDtsDAOImpl">
    <property name="sessionFactory" ref="sessionFactory" />
</bean>
<!-- 定义 UserDAOImpl 类 -->
<bean id="userDAO" class="com.restaurant.dao.impl.UserDAOImpl">
    <property name="sessionFactory" ref="sessionFactory" />
</bean>
```

22.7 创建 Service 接口及实现类

在 com.restaurant.service 包中,依次创建业务逻辑层接口 UserService.java、MealService.java、MealSeriesService.java、OrdersService.java 和 OrderDtsService.java。

在接口 UserService.java 中声明如下方法:

```java
package com.restaurant.service;
import java.util.List;
import com.restaurant.entity.Users;
public interface UserService {
    // 根据用户名获取用户对象
    public List getUserByLoginName(String loginName);
    // 添加新用户
    public void addUsers(Users u);
    // 更新用户信息
    public void modifyUsers(Users u);
}
```

在接口 MealService.java 中声明如下方法:

```java
package com.restaurant.service;
import java.util.List;
……
public interface MealService {
    // 根据查询条件(封装在 condition 中)和当前页码,获取餐品
    public List getMealByConditionForPager(Meal condition, int page);
    // 根据查询条件(封装在 condition 中),初始化分页类 Pager 对象,
    // 设置 perPageRows 和 rowCount 属性
    public Pager getPagerOfMeal(Meal condition);
    // 根据餐品 id 号,获取餐品对象
    public Meal getMealByMealId(int mealId);
    // 获取餐品浏览历史记录
    public List<Meal> getBrowsingMeal(String ids);
    // 获取餐品销售排行
    public List<Meal> getSalePaihang();
// 根据查询条件(封装在 condition 中),计算餐品总数
    public Integer getCountOfMeal(Meal condition);
}
```

为了支持前台餐品列表分页显示,创建了分页实体类 Pager.java,代码如下:

```java
package com.restaurant.entity;
public class Pager {
    private int curPage;          // 当前页码
    private int perPageRows ;     // 每页记录数
    private int rowCount;         //总记录数
    private int pageCount;        // 总页码
    // 此处省略了属性 curPage、perPageRows、rowCount 的 get 和 set 方法
    public int getPageCount() {
        return (rowCount+perPageRows-1)/perPageRows;
    }
}
```

在接口 MealSeriesService.java 中声明如下方法：

```java
package com.restaurant.service;
import java.util.List;
public interface MealSeriesService {
    // 获取所有菜系
    public List getMealSeries();
}
```

在接口 OrdersService.java 中声明如下方法：

```java
package com.restaurant.service;
……
public interface OrdersService {
    // 根据用户id和当前页码,获取订单
    public List getOrdersByUserIdForPager(int userId, int page);
    // 根据用户id,初始化分页类Pager对象,设置perPageRows和rowCount属性
    public Pager getPagerOfMyOrders(int userId);
    // 根据订单号,获取订单
    public Orders getOrdersByOid(int oid);
    // 删除订单
    public void deleteOrdersByOid(int oid);
    // 处理订单
    public void handleOrders(Orders orders);
}
```

在接口 OrderDtsService.java 中声明如下方法：

```java
package com.restaurant.service;
import java.util.List;
import com.restaurant.entity.Orderdts;
public interface OrderDtsService {
    // 生成订单子表(订单明细)
    public void addOrderDts(Orderdts dts);
    // 根据订单主表编号获取订单明细列表
    public List getOrderDtsByOid(int oid);
}
```

在 com.restaurant.service.impl 包中，依次创建业务接口的实现类 UserServiceImpl.java、MealServiceImpl.java、MealSeriesServiceImpl.java、OrdersServiceImpl.java 和 OrderDtsServiceImpl.java，并实现接口中的方法。

其中，UserServiceImpl.java 代码如下：

```java
package com.restaurant.service.impl;
import java.util.List;
……
public class UserServiceImpl implements UserService {
    UserDAO userDAO;
    public void setUserDAO(UserDAO userDAO) {
        this.userDAO = userDAO;
    }
    // 添加新用户
    @Override
    public void addUsers(Users u) {
```

```java
        userDAO.addUsers(u);
    }
// 更新用户信息
    @Override
    public void modifyUsers(Users u) {
        userDAO.modifyUsers(u);
    }
    // 根据用户名获取用户对象
    @Override
    public List getUserByLoginName(String loginName) {
        return userDAO.getUserByLoginName(loginName);
    }
}
```

MealServiceImpl.java 代码如下:

```java
package com.restaurant.service.impl;
……
import com.restaurant.service.MealService;
public class MealServiceImpl implements MealService {
    MealDAO mealDAO;
    public void setMealDAO(MealDAO mealDAO) {
        this.mealDAO = mealDAO;
    }
    // 根据查询条件(封装在 condition 中),初始化分页类 Pager 对象,
    // 设置 perPageRows 和 rowCount 属性
    @Override
    public Pager getPagerOfMeal(Meal condition) {
        int count = mealDAO.getCountOfMeal(condition);
        Pager pager = new Pager();
        pager.setPerPageRows(9);
        pager.setRowCount(count);
        return pager;
    }
    // 根据餐品 id 号,获取餐品对象
    @Override
    public Meal getMealByMealId(int mealId) {
        return mealDAO.getMealByMealId(mealId);
    }
    // 根据查询条件(封装在 condition 中)和当前页码,获取餐品
    @Override
    public List getMealByConditionForPager(Meal condition, int page) {
        return mealDAO.getMealByConditionForPager(condition, page);
    }
    // 获取餐品浏览历史记录
    @Override
    public List<Meal> getBrowsingMeal(String ids) {
        return mealDAO.getByIds(ids);
    }
    // 获取餐品销售排行
    @Override
    public List<Meal> getSalePaihang() {
        List<Meal> mealList = new ArrayList<Meal>();
        List<Object[]> sp = mealDAO.getSalePaihang();
        for (Object[] objects : sp) {
```

```java
            mealList.add(mealDAO.
getMealByMealId(Integer.parseInt(objects[0].toString())));
        }
        return mealList;
    }
// 根据查询条件(封装在condition中),计算餐品总数
    @Override
    public Integer getCountOfMeal(Meal condition) {
        return mealDAO.getCountOfMeal(condition);
    }
}
```

MealSeriesServiceImpl.java 代码如下:

```java
package com.restaurant.service.impl;
import java.util.List;
……
public class MealSeriesServiceImpl implements MealSeriesService {
    MealSeriesDAO mealSeriesDAO;
    public void setMealSeriesDAO(MealSeriesDAO mealSeriesDAO) {
        this.mealSeriesDAO = mealSeriesDAO;
    }
// 获取所有菜系
    @Override
    public List getMealSeries() {
        return mealSeriesDAO.getMealSeries();
    }
}
```

OrdersServiceImpl.java 代码如下:

```java
package com.restaurant.service.impl;
……
import com.restaurant.service.OrdersService;
public class OrdersServiceImpl implements OrdersService {
    OrdersDAO ordersDAO;
    public void setOrdersDAO(OrdersDAO ordersDAO) {
        this.ordersDAO = ordersDAO;
    }
    // 删除订单
    @Override
    public void deleteOrdersByOid(int oid) {
        Orders orders = ordersDAO.getOrdersByOid(oid);
        ordersDAO.deleteOrders(orders);
    }
    // 根据订单号,获取订单
    @Override
    public Orders getOrdersByOid(int oid) {
        return ordersDAO.getOrdersByOid(oid);
    }
    // 处理订单
    @Override
    public void handleOrders(Orders orders) {
        ordersDAO.updateOrders(orders);
```

```
    }
    // 根据用户id,初始化分页类Pager对象,
    // 设置perPageRows和rowCount属性
    @Override
    public Pager getPagerOfMyOrders(int userId) {
        int count = ordersDAO.getCountOfMyOrders(userId);
        Pager pager = new Pager();
        pager.setPerPageRows(10);
        pager.setRowCount(count);
        return pager;
    }
    // 根据用户id和当前页码,获取订单
    @Override
    public List getOrdersByUserIdForPager(int userId, int page) {
        return ordersDAO.getOrdersByUserIdForPager(userId, page);
    }
}
```

OrderDtsServiceImpl.java 代码如下:

```
package com.restaurant.service.impl;
……
import com.restaurant.service.OrderDtsService;
public class OrderDtsServiceImpl implements OrderDtsService {
    OrderDtsDAO orderDtsDAO;
    public void setOrderDtsDAO(OrderDtsDAO orderDtsDAO) {
        this.orderDtsDAO = orderDtsDAO;
    }
    // 生成订单子表(订单明细)
    @Override
    public void addOrderDts(Orderdts dts) {
        orderDtsDAO.addOrderDts(dts);
    }
    // 根据订单主表编号获取订单明细列表
    @Override
    public List getOrderDtsByOid(int oid) {
        return orderDtsDAO.getOrderDtsByOid(oid);
    }
}
```

Service 接口的实现类还需在 Spring 配置文件中加以定义,代码如下:

```xml
<!-- 定义 MealServiceImpl 类,并为其 mealDAO 属性注入值 -->
<bean id="mealService" class="com.restaurant.service.impl.MealServiceImpl">
    <property name="mealDAO" ref="mealDAO" />
</bean>
<!-- 定义 MealSeriesServiceImpl 类,并为其 mealSeriesDAO 属性注入值 -->
<bean id="mealSeriesService"
class="com.restaurant.service.impl.MealSeriesServiceImpl">
    <property name="mealSeriesDAO" ref="mealSeriesDAO" />
</bean>
<!-- 定义 OrdersServiceImpl 类,并为其 ordersDAO 属性注入值 -->
<bean id="ordersService"
```

```
class="com.restaurant.service.impl.OrdersServiceImpl">
    <property name="ordersDAO" ref="ordersDAO" />
</bean>
<!-- 定义 OrderDtsServiceImpl 类,并为其 orderDtsDAO 属性注入值 -->
<bean id="orderDtsService"
class="com.restaurant.service.impl.OrderDtsServiceImpl">
    <property name="orderDtsDAO" ref="orderDtsDAO" />
</bean>
<!-- 定义 UserServiceImpl 类,并为其 userDAO 属性注入值 -->
<bean id="userService" class="com.restaurant.service.impl.UserServiceImpl">
    <property name="userDAO" ref="userDAO" />
</bean>
```

22.8 餐品与菜系展示

网上订餐系统前台餐品与菜系展示页为 show.jsp，如图 22-5 所示。在 show.jsp 页面中，最顶端是网站 logo、登录和注册链接、餐品搜索栏，下面是餐品分类(菜系)，再下面是餐品列表展示，其右侧是餐品浏览历史和销售排行，底端是网站的版权信息。

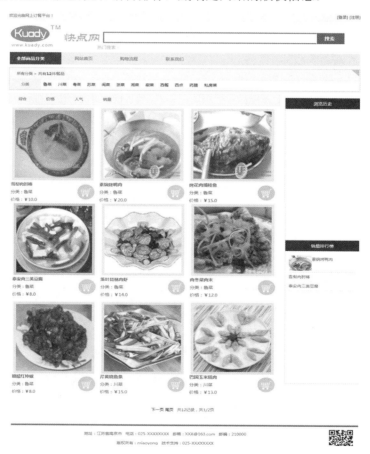

图 22-5 前台餐品与分类展示页

在浏览餐品时，浏览器地址栏中的地址为 http://localhost:8080/restaurant/toShowMeal，实现餐品与菜系展示页的步骤如下。

(1) Action 开发。

在项目的 com.restaurant.action 包中，创建 MealAction 类，让其继承 ActionSupport，并实现 RequestAware、ServletRequestAware、SessionAware 和 ServletResponseAware 接口，在该 Action 中，添加 toShowMeal 方法，获取指定页码、符合查询条件的餐品列表，再转到餐品显示页 show.jsp，代码如下：

```java
package com.restaurant.action;
import java.util.List;
……
public class MealAction extends ActionSupport implements RequestAware,
        ServletRequestAware, SessionAware, ServletResponseAware {
    // 定义 Meal 类型属性，用于封装表单参数
    private Meal meal;
    public Meal getMeal() {
        return meal;
    }
    public void setMeal(Meal meal) {
        this.meal = meal;
    }
    MealService mealService;
    MealSeriesService mealSeriesService;
    public void setMealService(MealService mealService) {
        this.mealService = mealService;
    }
    public void setMealSeriesService(MealSeriesService mealSeriesService) {
        this.mealSeriesService = mealSeriesService;
    }
    // 分页实体类
    private Pager pager;
    public Pager getPager() {
        return pager;
    }
    public void setPager(Pager pager) {
        this.pager = pager;
    }
    Map<String, Object> request;
    @Override
    public void setRequest(Map<String, Object> request) {
        this.request = request;
    }
    HttpServletRequest req;
    @Override
    public void setServletRequest(HttpServletRequest req) {
        this.req = req;
    }
    Map<String, Object> session;
    @Override
    public void setSession(Map<String, Object> session) {
        this.session = session;
    }
```

```java
    HttpServletResponse resp;
    @Override
    public void setServletResponse(HttpServletResponse resp) {
        this.resp = resp;
    }
    // 获取指定页码、符合查询条件的餐品列表，再转到餐品显示页 show.jsp
    public String toShowMeal() throws Exception {
        int curPage = 1;
        if (pager != null)
            curPage = pager.getCurPage();
        List mealList = null;
        if (meal != null) {
            if ((meal.getMealseries() != null)
                    && (meal.getMealseries().getSeriesId() != null))
                request.put("seriesId", new Integer(meal.getMealseries().getSeriesId()));
            if ((meal.getMealName() != null)
                    && !meal.getMealName().equals(""))
                request.put("mealName", meal.getMealName());
        }
        // 获取指定页码、符合查询条件的餐品列表
        mealList = mealService.getMealByConditionForPager(meal, curPage);
        // 将查询获得的列表存入 request 范围
        request.put("mealList", mealList);
        // 获取菜系列表，存入 request 范围
        List mealSeriesList = mealSeriesService.getMealSeries();
        request.put("mealSeriesList", mealSeriesList);
        // 根据查询条件(封装在 meal 中)，初始化分页类 Pager 对象
        pager = mealService.getPagerOfMeal(meal);
        // 设置 Pager 对象中的当前页页码
        pager.setCurPage(curPage);
        // 将餐品总数存入 request 范围
        request.put("totalMealCount", mealService.getCountOfMeal(null));
        // 读取浏览历史
        readingBrowsemeal();
        // 获取销售排行
        List<Meal> spList = mealService.getSalePaihang();
        session.put("spList", spList);
        return "toShowMeal";
    }
}
```

在 MealAction 类中，使用实体类 Meal 声明了属性 meal，用于封装表单参数；使用分页实体类 Pager 声明了属性 pager，用于记录与分页有关的信息；使用 MealService 接口声明了属性 mealService；使用 MealSeriesService 接口声明了属性 mealSeriesService，并为该属性添加了 setter 方法，用于接收 Spring 配置文件中的 MealServiceImpl 类和 MealSeriesServiceImpl 类的相应 Bean 实例注入。在 MealAction 类的 toShowMeal 方法中，执行流程为：通过 pager 对象获取当前页码；将封装在 meal 对象中的查询条件依次取出，并存入 request 范围；调用业务接口 MealService 中的 getMealByConditionForPager 方法，根据查询条件(封装在 meal 对象中)和当前页码，获取餐品列表，并存入 request 范围；调用业务接口 MealSeriesService 中的

getMealSeries 方法，获取菜系列表，并存入 request 范围；调用业务接口 MealService 中的 getPagerOfMeal 方法，根据查询条件(封装在 meal 中)，初始化分页类 Pager 对象，设置 Pager 对象中的当前页页码；将餐品总数存入 request 范围；调用 MealAction 类中的方法 readingBrowsemeal，读取餐品浏览历史；调用业务接口 MealService 中的 getSalePaihang 方法，获取销售排行，并存入 session 范围；将请求转发到 toShowMeal，即页面 show.jsp。

(2) Spring 整合 Struts 2。

Spring 整合 Struts 2 在第 20 章 20.1 节已作介绍，目的在于使用 Spring IoC 容器来管理 Struts 2 的 Action，在 web.xml 配置文件中指定以 Listener 方式启动 Spring，并配置 Struts 2 的 StrutsPrepareAndExecuteFilter。

在 Spring 配置文件中配置 MealAction 类，配置部分如下：

```xml
<!-- 定义 MealAction，并为其 mealService 和 mealSeriesService 属性注入值 -->
<bean name="mealAction" class="com.restaurant.action.MealAction"
  scope="prototype">
  <property name="mealService" ref="mealService" />
  <property name="mealSeriesService" ref="mealSeriesService" />
</bean>
```

为了保证对每个用户的请求都会创建一个新的 Bean 实例，在配置 MealAction 的实例时，需要将<bean>元素的 scope 属性设置为 prototype (原型模式，即非单例)。

修改项目 src 目录下 struts.xml 配置文件，添加 toShowMeal 的 action 请求，代码如下：

```xml
<!-- 为 MealAction 类中的 toShowMeal 方法配置映射 -->
<action name="toShowMeal" class="mealAction" method="toShowMeal">
    <result name="toShowMeal">/show.jsp</result>
</action>
```

Spring 整合 Struts 2 后，<action>元素中 class 属性不再使用 MealAction 的全类名，而是引用 Spring 配置文件中 MealAction 类的 Bean 实例名 mealAction。

(3) 编写页面。

在 show.jsp 页面中，与餐品展示相关的代码如下：

```jsp
<s:if test="#request.mealList==null or #request.mealList.size()==0">
    没有要查询的餐品
</s:if>
<s:else>
<!-- 产品循环开始 -->
<s:iterator var="mealItem" value="#request.mealList">
    <div class="mpro fl">
        <div class="mpro_tp"
            onmouseover="javascript:this.style.background='#fbc837'"
            onmouseout="javascript:this.style.background=''">
            <a href='/restaurant/toShowDetails?meal.mealId=
${mealItem.mealId}' target="_blank"><img width="220px" height="220px"
                src="mealimages/${mealItem.mealImage }" /></a>
        </div>
        <div class="mpro_con">
            <table width="242" border="0" cellpadding="0" cellspacing="0">
                <tr>
                    <td width="180" height="25" align="left" valign="middle"
```

```
class="jiacu">${mealItem.mealName }
</td>
                <td width="62" rowspan="3" align="right"
valign="middle"><a href="addtoshopcart?mealId=
${mealItem.mealId } "> <img src="images/006.jpg" /></a></td>
            </tr>
            <tr align="left" valign="middle">
                <td width="180" height="25">分类：
${mealItem.mealseries.seriesName }</td>
            </tr>
            <tr align="left" valign="middle">
                <td width="180" height="25">价格： <b>
¥</b>${mealItem.mealPrice }</td>
            </tr>
        </table>
    </div>
  </div>
</s:iterator>
<!-- 产品循环结束 -->
</s:else>
```

与餐品分类(菜系)显示相关的代码如下：

```
<!--条件筛选开始-->
<div class="sx mt10">
    <div class="sx_a">
        <span>所有分类 > 共有<span
class="red">${requestScope.totalMealCount }</span>件餐品
        </span>    </div>
    <div class="sx_b">    <p class="tit fl">分类</p><p class="con fl">
        <s:iterator var="mealSeries"
value="#request.mealSeriesList"> 
<a href="/restaurant/toShowMeal?meal.mealseries.seriesId=
${mealSeries.seriesId}">${mealSeries.seriesName }
</a>  
            </s:iterator></p>
    </div>
</div>
<div class="clearall"></div>
<!--条件筛选结束-->
```

部署项目，启动 Tomcat，在浏览器中输入 http://localhost:8080/restaurant/toShowMeal，打开的页面中餐品展示部分，如图 22-6 所示；菜系显示部分，如图 22-7 所示。

图 22-6　餐品展示

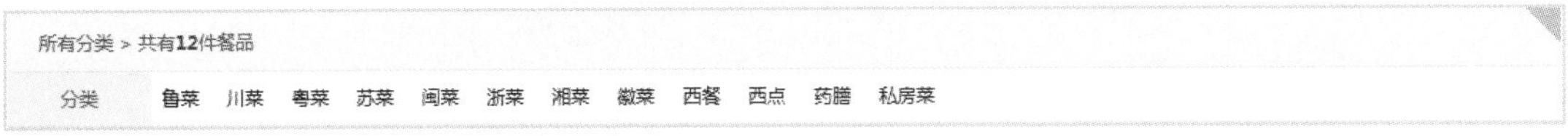

图 22-7 菜系显示

对于餐品展示部分，餐品图片加上相应的超链接，跳转餐品详情页；每件餐品右下角的购物车图片，该图片已添加超级链接，添加至购物车。

此外，餐品展示采用了分页技术，与分页相关的超链接如下：

```
<s:if test="#request.mealList!=null && #request.mealList.size()>0">
<!-- 分页条开始 -->
<s:if test="pager.curPage>1">
    <a    href="/restaurant/toShowMeal?pager.curPage=
1&meal.mealseries.seriesId=${requestScope.seriesId}&meal.mealName=
${requestScope.mealName}">首页</a>
    <a    href="/restaurant/toShowMeal?pager.curPage
=${pager.curPage-1}&meal.mealseries.seriesId=
${requestScope.seriesId}&meal.mealName=${requestScope.mealName}">上一页</a>
</s:if>
<s:if test="pager.curPage < pager.pageCount">
    <a    href="/restaurant/toShowMeal?pager.curPage=
${pager.curPage+1}&meal.mealseries.seriesId=
${requestScope.seriesId}&meal.mealName=${requestScope.mealName}">下一页</a>
    <a    href="/restaurant/toShowMeal?pager.curPage=
${pager.pageCount }&meal.mealseries.seriesId=
${requestScope.seriesId}&meal.mealName=${requestScope.mealName}">尾页</a>
</s:if>  
共${pager.rowCount}记录，共${pager.curPage}/${pager.pageCount}页
<!-- 分页条结束 -->
</s:if>
```

单击分页条上的超链接，会将请求重新发送到 toShowMeal，即再次执行 MealAction 类中的 toShowMeal 方法，并通过 pager 对象传递"首页""上一页""下一页""尾页"的页码，通过 meal 对象传递查询条件。

22.9 查询餐品

在餐品与菜系展示页 show.jsp 中，使用 include 指令包含了餐品查询页 search.jsp。餐品查询页中的搜索表单如下：

```
<div class="search_from">
    <s:form method="post" action="toShowMeal">
        <div>
            <input name="meal.mealName" type="text" value="${requestScope.mealName }" class="s_input fl" />
<!-- 通过隐藏表单域保存用户选择过的菜系，可根据餐品名称和菜系组合查询 -->
            <s:hidden name="meal.mealseries.seriesId"
                value="%{#request.seriesId}" />
        </div>
        <div class="s_botton fl">
```

```
            <input type="image" src="images/002.jpg" />
        </div>
    </s:form>
</div>
```

在餐品查询页中，输入餐品名称，单击"搜索"按钮，会将请求发送到 toShowMeal，即执行 MealAction 类中的 toShowMeal 方法，将餐品名称作为查询条件传递过去，并通过隐藏表单域保存用户选择过的菜系，可根据餐品名称和菜系组合查询。

在餐品分类(菜系)显示中，每个菜系都设置了超链接，代码如下：

```
<a href="/restaurant/toShowMeal?meal.mealseries.seriesId=
${mealSeries.seriesId}">${mealSeries.seriesName }
</a>
```

单击该超链接，也将请求发送到 toShowMeal，并将菜系编号作为查询条件传递到 MealAction 类中的 toShowMeal 方法。

22.10　查看餐品详情

在 show.jsp 页的餐品列表显示部分，餐品图片设置了超链接，代码如下：

```
<a href='/restaurant/toShowDetails?meal.mealId=${mealItem.mealId}'
    target="_blank"><img width="220px" height="220px"
    src="mealimages/${mealItem.mealImage }" /></a>
```

单击餐品图片，即可跳转至该餐品的详情页，如图 22-8 所示。

图 22-8　餐品详情页

实现餐品详情显示的步骤如下。

(1) Action 开发。

在 com.restaurant.action 包中的 MealAction 类中，添加 toShowDetails()方法，再转到餐品详情页 details.jsp，代码如下：

```java
public String toShowDetails() throws Exception {
    // 根据餐品id号,获取餐品对象
    Meal aMeal = mealService.getMealByMealId(meal.getMealId());
    // 将餐品对象存入request
    request.put("aMeal", aMeal);
    // 保存浏览历史信息
    writingbrowsemeal();
    // 读取浏览历史信息
    readingBrowsemeal();
    // 转发到餐品详情页details.jsp
    return "toShowDetails";
}
```

(2) 修改 Struts 2 配置文件。

在 struts.xml 配置文件中，添加 toShowDetails 的 action 请求，代码如下：

```xml
<!-- 为MealAction类中的toShowDetails方法配置映射 -->
<action name="toShowDetails" class="mealAction" method="toShowDetails">
    <result name="toShowDetails">/details.jsp</result>
</action>
```

(3) 编写页面。

在餐品详情页中，显示餐品信息的代码如下：

```html
<!--主体开始-->
<div class="main mt10">
    <div class="mleft fl ah">
        <!--商品详细信息开始-->
        <div class="show_a fl">
            <div class="img fl">
                <img width="353" height="348"
                    src="mealimages/${requestScope.aMeal.mealImage}" />
            </div>
            <div class="canshu fl">
                <p></p>
                <p>编号：${requestScope.aMeal.mealId}</p>
                <p>餐名：${requestScope.aMeal.mealName }</p>
                <p>菜系：${requestScope.aMeal.mealseries.seriesName }</p>
                <p>    <font color="red">价格：
${requestScope.aMeal.mealPrice }</font>     </p>
            </div>
        </div>
        <!--商品详细信息结束-->
        <div class="show_b fl">
            <a href="addtoshopcart?mealId=
${requestScope.aMeal.mealId } "><img
                src="images/d010.jpg" /></a>
        </div>
```

```
            <div class="show_c fl">餐品详情</div>
            <div class="show_d fl ah">
                餐品名称：<br /> ${requestScope.aMeal.mealName }<br /> 餐品简介：<br />
                ${requestScope.aMeal.mealSummarize }<br /> 餐品描述：<br />
                ${requestScope.aMeal.mealDescription }
            </div>
    </div>
    <!-- 排行榜开始 -->
    <%@ include file="common/rankinglist.jsp"%>
    <!-- 排行榜结束 -->
</div>
<!--主体结束-->
```

22.11　用户登录与注册

登录与注册是网上订餐系统不可或缺的功能。当用户注册为系统的用户，并成功登录到系统，确定为本网站合法用户后，可进行网站的浏览，对购物车、订单进行操作；如果没有登录，用户只能浏览餐品，不能进行餐品的交易。

22.11.1　用户登录

网上订餐系统前台登录页为 login.jsp，如图 22-9 所示。

图 22-9　前台登录页 login.jsp

在登录页 login.jsp 中，登录表单的代码如下：

```
<p class="title fl">用户登录</p>
<div class="login_d fl">
```

```html
        <s:form name="loginForm" method="post" action="">
            <table width="350" border="0" cellpadding="0" cellspacing="0">
                <tr>
                    <td width="100" height="40" align="right" valign="middle">用户名: </td>
                    <td width="250" align="left" valign="middle"><input
                        name="u.loginName" type="text" class="login_input" /></td>
                </tr>
                <tr>
                    <td width="100" height="40" align="right" valign="middle">密码: </td>
                    <td width="250" align="left" valign="middle"><input
                        name="u.loginPwd" type="text" class="login_input" /></td>
                </tr>
                <tr>
                    <td height="50" colspan="2" align="center"
valign="middle"><input type="image" src="images/d017.jpg"
                        onclick="submitLogin()" />
                    </td>
                </tr>
                <tr>
                    <td height="50" colspan="2" align="center"
valign="middle"><font
                        color="red" size="3px"> <s:fielderror />
                    </font></td>
                </tr>
            </table>
        </s:form>
</div>
```

在登录表单中,输入登录名和密码,单击"登录"图片按钮,触发该按钮的 onclick 事件,处理该事件的函数为 submitLogin,代码如下:

```html
<script type="text/javascript">
    function submitLogin() {
        document.getElementsByName("loginForm")[0].action = "doLogin";
        document.getElementsByName("loginForm")[0].submit();
    }
</script>
```

在 submitLogin 函数中,将请求提交到 doLogin,即执行 UserAction 类中的 doLogin 方法。实现用户登录的过程如下:

(1) Action 开发。

在 com.restaurant.action 包中,创建 UserAction 类,继承自 ActionSupport 类,并实现 RequestAware、SessionAware 和 ServletResponseAware 接口。在 UserAction 类中添加一个 doLogin()方法,用于处理登录验证请求,代码如下:

```java
package com.restaurant.action;
……
import com.restaurant.entity.Users;
import com.restaurant.service.UserService;
public class UserAction extends ActionSupport implements RequestAware,
        SessionAware, ServletResponseAware {
```

```java
    private Users u;
    private String repassword;
    // 此处省略属性 u 和 repassword 的 get 和 set 方法
    UserService userService;
    public void setUserService(UserService userService) {
        this.userService = userService;
    }
    // 登录验证
    public String doLogin() throws Exception {
        List list = null;
        // 根据用户名获取用户对象
        list = userService.getUserByLoginName(u.getLoginName());
        if (list.size() > 0) {    // 判断用户名是否存在
            Users validUser = (Users) list.get(0);
            if (validUser.getStatus() == 0) {
                this.addFieldError("loginName", "该用户已被禁用,请联系管理员!");
                return "login";
            }
// 判断密码是否正确
            if (u.getLoginPwd().equals(validUser.getLoginPwd())) {
                // 验证通过,将用户信息存入 session
                session.put("user", list.get(0));
            } else {
                this.addFieldError("loginName", "密码不正确!");
                return "login";
            }
        } else {    // 用户名不存在
            this.addFieldError("loginPwd", "用户名不正确!");
            return "login";
        }
        return "show";
    }
    Map<String, Object> session;
    @Override
    public void setSession(Map<String, Object> session) {
        this.session = session;
    }
    Map<String, Object> request;
    @Override
    public void setRequest(Map<String, Object> request) {
        this.request = request;
    }
    HttpServletResponse response;
    @Override
    public void setServletResponse(HttpServletResponse response) {
        this.response = response;
    }
}
```

在 doLogin 方法中,调用 UserService 接口中的 getUserByLoginName 方法根据用户名获取用户对象,判断用户名是否存在。如果存在,则继续判断该用户名是否被禁用,密码是否正确。如果验证没有通过,返回逻辑名为 login,否则返回 show。

(2) 在 Spring 配置文件中定义 UserAction,并为其中属性 userService 注入值,配置如下:

```xml
<!-- 定义UserAction，并为其中属性userService注入值 -->
<bean name="userAction" class="com.restaurant.action.UserAction"
    scope="prototype">
    <property name="userService" ref="userService" />
</bean>
```

(3) 在 Struts 2 配置文件中，为 UserAction 类中的 doLogin()方法配置映射，代码如下：

```xml
<action name="doLogin" class="userAction" method="doLogin">
    <result name="show" type="redirectAction">toShowMeal</result>
    <result name="login" type="dispatcher">login.jsp</result>
</action>
```

在上述配置中，给逻辑名 login 设置了对应的物理名 login.jsp，登录验证失败后回到登录页。给逻辑名 show 设置了对应的物理名 toShowMeal，验证通过后会执行 MealAction 类中的 toShowMeal 方法，进入餐品与菜系展示页。

22.11.2 用户注册

在登录页中，单击"立即注册"图片按钮，或者单击页面顶部的"[注册]"超链接，打开用户注册页 reg.jsp，如图 22-10 所示。

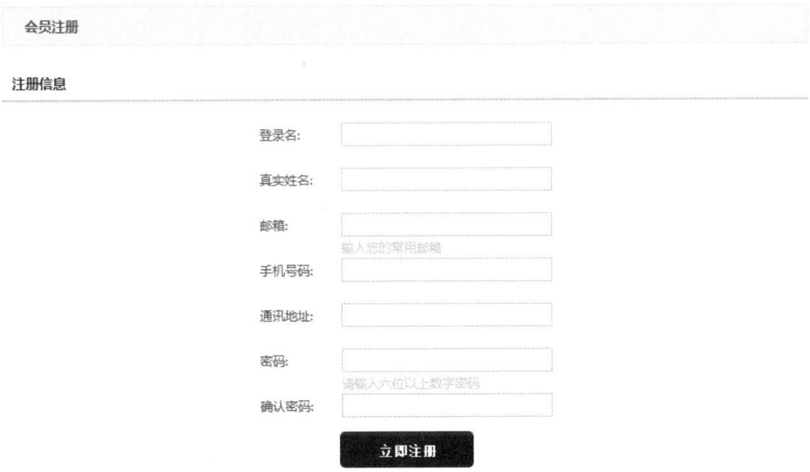

图 22-10 用户注册页 reg.jsp

在页面 reg.jsp 中，注册表单代码如下：

```html
<!--注册开始-->
<div class="reg_a jiacu">会员注册</div>
<div class="reg_b fl">注册信息</div>
<div class="reg_c fl ah">
    <s:form action="" method="post" name="regForm">
        <table width="280" border="0" align="center" cellpadding="0"
            cellspacing="0">
            <tr>
                <s:textfield name="u.loginName" class="login_input" label="登录名" onBlur="validate(this);" />
```

```
            </tr>
            <tr>
                <td width="80" height="20" align="right" valign="middle"> </td>
                <td width="200" align="left" valign="middle" class="hui"></td>
            </tr>
            <tr>
                <s:textfield name="u.trueName" class="login_input" label="真实姓名" />
            </tr>
            <tr>
                <td width="80" height="20" align="right" valign="middle"> </td>
                <td width="200" align="left" valign="middle" class="hui"></td>
            </tr>
            <tr>
                <s:textfield name="u.email" class="login_input" label="邮箱" />
            </tr>
            <tr>
                <td width="80" height="20" align="right" valign="middle"> </td>
                <td width="200" align="left" valign="middle" class="hui">输入您的常用邮箱</td>
            </tr>
            <tr>
                <s:textfield name="u.phone" class="login_input" label="手机号码" />
            </tr>
            <tr>
                <td width="80" height="20" align="right" valign="middle"> </td>
                <td width="200" align="left" valign="middle" class="hui"></td>
            </tr>
            <tr>
                <s:textfield name="u.address" class="login_input" label="通讯地址" />
            </tr>
            <tr>
                <td width="80" height="20" align="right" valign="middle"> </td>
                <td width="200" align="left" valign="middle" class="hui"></td>
            </tr>
            <tr>
                <s:textfield name="u.loginPwd" class="login_input" label="密码" />
            </tr>
            <tr>
                <td width="80" height="20" align="right" valign="middle"> </td>
                <td width="200" align="left" valign="middle" class="hui">请输入六位以上数字密码</td>
            </tr>
```

```html
            <tr>
                <s:textfield name="repassword" class="login_input" label="确认密码" />
            </tr>
            <tr>
                <td height="60" colspan="2" align="center" valign="middle"><input
                    type="image" src="images/d018.jpg" onclick="submitReg()" /></td>
            </tr>
        </table>
    </s:form>
</div>
<!--注册开始-->
```

注册表单的校验使用了 Struts 2 的验证框架，为此在 com.restaurant.action 包中，创建了文件 UserAction-register-validation.xml，对登录名、真实姓名、邮箱、手机号码、通讯地址、密码和确认密码进行了校验。

在注册页面中，填写注册信息后，单击"立即注册"图片按钮，会触发该按钮的 onclick 事件，处理该事件的函数为 submitReg，代码如下：

```html
<script type="text/javascript">
    function submitReg() {
    document.getElementsByName("regForm")[0].action = "register";
    document.getElementsByName("regForm")[0].submit();
    }
</script>
```

在 submitReg 函数中，将请求提交到 register，即执行 UserAction 类中的 register 方法。实现用户注册的过程如下。

(1) Action 开发。

在 UserAction 类中添加一个 register()方法，用于处理登录注册请求，代码如下：

```java
// 用户注册
public String register() throws Exception {
    u.setStatus(1);
    userService.addUsers(u);
    return "show";
}
```

在 register()方法中，调用业务接口 UserService 中的 addUsers 方法，该方法使用了参数 u。之前在 UserAction 类中已经声明了 Users 类型的对象 u，Struts 2 框架会自动将注册表单参数封装到该对象中。

(2) 在 Struts 2 配置文件中，为 UserAction 类中的 register()方法配置映射，代码如下：

```xml
<action name="register" class="userAction" method="register">
<result name="show" type="redirectAction">toShowMeal</result>
    <result name="input">reg.jsp</result>
</action>
```

用户注册并成功登录系统后，就可以使用购物车和订单功能了。

22.12 购物车功能

购物车相当于现实中超市的购物车，不同的是：一个是实体车，一个是虚拟车而已。用户在订餐系统网站中，点击添加至购物车时，该餐品就保存到购物车中，可以多次选购，最后将放在购物车中的所有餐品统一提交订单，这也是尽量让客户体验到现实生活中购物的感觉，购物车的功能包括以下几项。

(1) 把餐品添加到购物车，即订购。
(2) 删除购物车中已选购的餐品。
(3) 修改购物车中某个餐品的订购数量。
(4) 清空购物车。
(5) 显示购物车中餐品清单及数量、价格。

购物车的实现思路如下。

(1) 选中餐品并放进购物车时进入购物车页面。
(2) 进入购物车页面时，判断购物车是否已经存在。如果购物车不存在，添加第一件餐品时，初始化购物车，并把餐品数据放进 HashMap，然后保存在 session 中。如果购物车已经存在，则把购物车数据从存在的购物车数据取出并放在 HashMap 中，并将新的餐品数据插入 HashMap 中，然后存入 session。
(3) 继续购物，选中新的餐品放进购物车，进入第(2)步。

在餐品与分类展示页的餐品列表中，每件餐品显示区域的右下角，都有一个"购物车"图片超链接，超链接的设置如下：

```
<a    href="addtoshopcart?mealId=${mealItem.mealId } ">
<img src="images/006.jpg" /></a>
```

在餐品详情页中，也有一个"加入购物车"图片超链接，超链接的设置如下：

```
<a href="addtoshopcart?mealId=${requestScope.aMeal.mealId } ">
<img src="images/d010.jpg" /></a>
```

用户登录后，单击"购物车"图片超链接，可将该餐品放入购物车暂存。用户购物车显示页效果如图 22-11 所示。

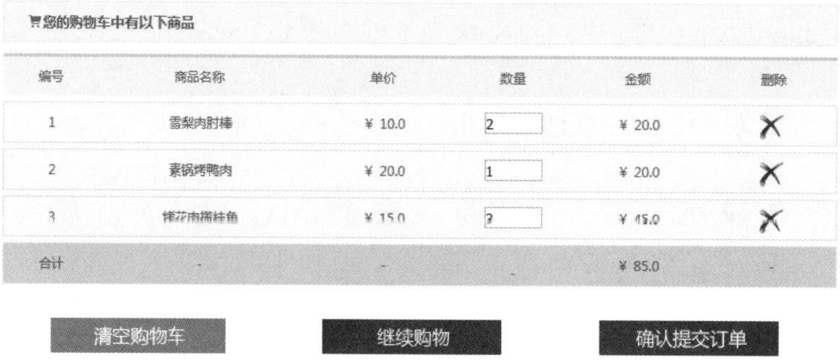

图 22-11 购物车显示页

1. 实现餐品放入购物车功能

购物车实现过程中使用了 Map 来保存顾客购买的商品，Meal、CartItemBean、HashMap 和 session 之间的关系如图 22-12 所示。

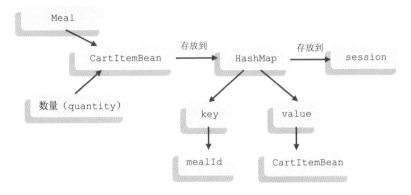

图 22-12　使用 Map 实现购物车原理

由餐品信息类(Meal.java)和购买的数量构成了购物车内商品信息的描述类 (CartItemBean.java)，再将 CartItemBean 类的对象存放到 HashMap 中，其中 HashMap 对象的键是餐品的编号，值是 CartItemBean 类对象。最后将包含了购买商品信息的 HashMap 对象保存到 session 中。这样，就可以操作 session 对象中的数据，来实现不同顾客的购买功能。

在项目 com.restaurant.entity 包中，创建购物车内餐品信息的描述类 CartItemBean.java，代码如下：

```java
package com.restaurant.entity;
import java.io.Serializable;
public class CartItemBean implements Serializable {
    private Meal meal;    // 餐品对象
    private int quantity;  // 购买数量
    // 此处省略了属性 meal、quantity 的 get 和 set 方法
    public CartItemBean(Meal meal, int quantity) {
        this.meal = meal;
        this.quantity = quantity;
    }
}
```

由于购物车使用 Map 来暂存数据，不使用数据库，因此无须编写 DAO 和 Service 层代码，只需要创建 Action 即可。在 com.restaurant.action 包中，创建名为 CartAction 的 Action，让其继承 ActionSupport 类，并实现 SessionAware 接口。在 CartAction 类中添加 addtoshopcart() 方法，实现添加商品到购物车，代码如下：

```java
package com.restaurant.action;
import java.util.HashMap;
import java.util.Map;
……
public class CartAction extends ActionSupport implements SessionAware {
    // 封装表单传递来的餐品编号 mealId
    private Integer mealId;
// 封装从表单传递来的餐品数量 quantity
```

```java
    int quantity;
    // 此处省略了属性 mealId 和 quantity 的 get 和 set 方法
    MealService mealService;
    public void setMealService(MealService mealService) {
        this.mealService = mealService;
    }
    Map<String, Object> session;
    @Override
    public void setSession(Map<String, Object> session) {
        this.session=session;
    }
    // 将餐品添加到购物车
    public String addtoshopcart() throws Exception {
        // 从 Session 中取出购物车，放入 Map 对象的 cart 中
        Map cart = (Map) session.get("cart");
        // 获取当前要添加到购物车的餐品
        Meal meal = mealService.getMealByMealId(mealId);
        // 如果购物车不存在，则创建购物车(实例化 HashMap 类)，
        // 并存入 session 中
        if (cart == null) {
            cart = new HashMap();
            session.put("cart", cart);
        }
        // 如果存在购物车，则判断餐品是否在购物车中
        CartItemBean cartItem = (CartItemBean) cart.get(meal.getMealId());
        if (cartItem != null) {
            // 如果餐品在购物车中，更新其数量
            cartItem.setQuantity(cartItem.getQuantity() + 1);
        } else {
            // 否则，创建一个条目到 Map 中
            cart.put(meal.getMealId(), new CartItemBean(meal, 1));
        }
        return "shopCart";   //页面转到 shopCart.jsp,显示购物车
    }
}
```

在 CartAction 类中，定义了两个属性 mealId 和 quantity，分别用来封装表单传递来的餐品编号 mealId 参数值和餐品数量 quantity 参数值。

在 Spring 配置文件中定义 CartAction，并为其中属性 mealService 注入值，代码如下：

```xml
<!-- 定义 CartAction，并为其中属性 mealService 注入值 -->
<bean name="cartAction" class="com.restaurant.action.CartAction"
      scope="prototype">
   <property name="mealService" ref="mealService" />
</bean>
```

为了保证对每个用户的请求都会创建一个新的 Bean 实例，在配置 CartAction 的实例时，需要将<bean>元素的 scope 属性设置为 prototype(原型模式，即非单例)。

在 Struts 2 配置文件中为 CartAction 类的 addtoshopcart()方法配置映射，代码如下：

```xml
<action name="addtoshopcart" class="cartAction" method="addtoshopcart">
    <result name="shopCart">/shopCart.jsp</result>
</action>
```

无论用户是在餐品与菜系展示页的餐品显示列表中，单击"购物车"图片超链接，还是在餐品详情页中，单击"加入购物车"图片超链接，都将请求提交到 addtoshopcart，即执行 CartAction 类中的 addtoshopcart()方法，并传递一个参数 mealId，addtoshopcart()方法会根据 mealId 参数值，调用业务方法获取餐品对象。再判断 Map 中该对象是否已存在，以决定是将其添加到购物车，还是只增加数量。

addtoshopcart()方法最后将请求转到购物车显示页 shopCart.jsp，在页面上循环显示 Map 中存放的商品信息。新建 shopCart.jsp 页面，页面静态代码可复制提供的静态页，购物车显示页的主体结构与其他页面类似，购物车部分的主要代码如下：

```jsp
<!--主体开始-->
<div class="main mt10">
   <div class="mleft fl ah">
      <s:if test="#session.cart==null or #session.cart.size()==0">
         您的购物车中还没有商品
      </s:if>
      <s:else>
      <!--购物车开始-->
      <div class="car_a jiacu">
    <span><img src="images/d002.jpg" /></span>您的购物车中有以下商品
       </div>
       <div class="car_b fl mt10">
          <p class="bh fl">编号</p>
          <p class="spmc fl">商品名称</p>
          <p class="dj fl">单价</p>
          <p class="sl fl">数量</p>
          <p class="je fl">金额</p>
          <p class="del fl">删除</p>
       </div>
       <s:set var="sumPrice" value="0" />
       <s:iterator var="cartItem" value="#session.cart">
          <div class="car_c fl">
             <p class="bh fl">
                <s:property value="value.meal.mealId" />
             </p>
             <p class="spmc fl">
                <s:property value="value.meal.mealName" />
             </p>
             <p class="dj fl">
                ￥
                <s:property value="value.meal.mealPrice" />
             </p>
             <p class="sl fl">
                <input type="text" value="${value.quantity }" size="5" style="vertical-align: middle;" onchange="window.location='updateSelectedQuantity?mealId=${value.meal.mealId}&quantity='+this.value;" />
             </p>
             <p class="je fl">
                ￥
                <s:property value="value.quantity * value.meal.mealPrice" />
             </p>
             <p class="del fl">
```

```html
            <a href="deleteSelectedOrders?mealId=${value.meal.mealId}"> <img
               src="images/d013.jpg" />
            </a>
         </p>
      </div>
      <s:set var="sumPrice"
         value="#sumPrice+value.quantity*value.meal.mealPrice" />
   </s:iterator>
   <div class="car_c fl" style="background-color:#cccccee;">
      <p class="bh fl">合计</p>
      <p class="spmc fl">-</p>
      <p class="dj fl">-</p>
      <p class="sl fl">-</p>
      <p class="je fl">
         ¥
         <s:property value="#sumPrice" />
         <s:set var="sumPrice" value="#sumPrice" scope="session" />
      </p>
      <p class="del fl">-</p>
   </div>
   <div class="car_d fl">
      <p class="cz fl">
         <a href="clearCart"><img src="images/d014.jpg" /></a>
      </p>
      <p class="cz fl">
         <a href="/restaurant/toShowMeal"><img src="images/d015.jpg" /></a>
      </p>
      <p class="cz fl">
         <a href="/restaurant/addOrders"><img src="images/d016.jpg" /></a>
      </p>
   </div>
</s:else>
            <!-- 购物车结束-->
         </div>
   <!-- 排行榜开始 -->
   <%@ include file="common/rankinglist.jsp"%>
   <!-- 排行榜结束 -->
</div>
<!--主体结束-->
```

部署项目，运行程序，添加商品至购物车，用户没有登录，也能够添加，为阻止未登录用户直接通过 addtoshopcart()方法访问购物车，可以使用登录验证拦截器 loginCheck，这样在需要添加验证的地方，添加该验证拦截器即可，过程如下。

首先，在 src 目录下新建 com.restaurant.interceptor 的包，在该包中创建名为 AuthorityInterceptor 的类，该类继承自 AbstractInterceptor，实现 intercept(ActionInvocation invocation)方法，代码如下：

```java
package com.restaurant.interceptor;
import java.util.Map;
import com.opensymphony.xwork2.ActionInvocation;
import com.opensymphony.xwork2.interceptor.AbstractInterceptor;
import com.restaurant.entity.Users;
public class AuthorityInterceptor extends AbstractInterceptor {
```

```java
@Override
public String intercept(ActionInvocation invocation) throws Exception {
    // 取得用户会话，获取用户会话信息
    Map session = invocation.getInvocationContext().getSession();
    if (session == null) {      // 如果Session为空，则让用户登录
        return "login";
    } else {
        Users user = (Users) session.get("user");
        if (user == null) {
            // 返回login字符串，终止执行，返回登录页面
            return "login";
        } else {
            // 用户登录，放行，继续执行剩余的拦截器和Action
            return invocation.invoke();
        }
    }
}
```

然后，在 struts.xml 文件中定义拦截器，定义全局变量，并在相应需要实现登录验证的 Action 中，引用该拦截器，配置如下：

```xml
<package name="restaurant" namespace="/" extends="struts-default">
    <!-- 配置拦截器 AuthorityInterceptor -->
    <interceptors>
        <interceptor name="loginCheck"
class="com.restaurant.interceptor.AuthorityInterceptor" />
    </interceptors>
    <!-- 设置全局的返回值,返回首页 -->
    <global-results>
        <result name="login" type="redirectAction">toShowMeal</result>
    </global-results>
    <!-- 为类中的方法配置映射，省略其他已经配置的Action -->
    <action name="addtoshopcart" class="cartAction" method="addtoshopcart">
        <result name="shopCart">/shopCart.jsp</result>
        <interceptor-ref name="loginCheck" />
        <interceptor-ref name="defaultStack" />
    </action>
</package>
```

在给 CartAction 类的 addtoshopcart 方法配置映射时，引用了自定义的拦截器 loginCheck，系统就不再引用默认的拦截器，因此需要显式地引用默认拦截器 defaultStack。重新部署项目，运行程序，当用户未登录，添加商品至购物车时，再次跳转回首页。

2. 修改购物车中商品数量

在 shopCart.jsp 页面中，在餐品数量文本框中输入新的数量，文本框失去焦点后，可直接修改购物车中餐品数量，实现修改购物车中餐品数量的流程如下。

首先，在 CartAction 类中添加 updateSelectedQuantity()方法，用以更改购物车中餐品数量，代码如下：

```
// 修改购物车餐品数量
```

```java
public String updateSelectedQuantity() throws Exception {
    //从session中取出购物车,放入Map对象cart中
    Map cart = (Map) session.get("cart");
    CartItemBean cartItem = (CartItemBean) cart.get(mealId);
    cartItem.setQuantity(quantity);
    return "shopCart";
}
```

然后,在 Struts 2 配置文件中为 CartAction 类的 updateSelectedQuantity()方法配置映射,代码如下:

```xml
<!-- 修改购物车中的某个餐品数量 -->
<action name="updateSelectedQuantity" class="cartAction"
    method="updateSelectedQuantity">
    <result name="shopCart">/shopCart.jsp</result>
    <interceptor-ref name="loginCheck" />
    <interceptor-ref name="defaultStack" />
</action>
```

在 shopCart.jsp 页面中,餐品数量文本框的设置如下:

```html
<input type="text" value="${value.quantity }" size="5"
    style="vertical-align: middle;"
onchange="window.location='updateSelectedQuantity?mealId=
${value.meal.mealId}&quantity='+this.value;" />
```

当餐品数量文本框内容发生变化时,会触发 onchange 事件,将请求提交到 updateSelectedQuantity,即执行 CartAction 类的 updateSelectedQuantity()方法,并传递两个参数,mealId 参数为要修改数量的餐品编号,quantity 参数为要修改的数量。updateSelectedQuantity()方法根据这两个参数值,更新购物车中相应餐品的数量。

3. 删除购物车中商品

在 shopCart.jsp 页面的购物车中,每条餐品记录后面都有一个"删除"的叉号图片超链接,其设置如下:

```html
<a href="deleteSelectedMeal?mealId=${value.meal.mealId}"> <img
    src="images/d013.jpg" />    </a>
```

实现删除购物车中餐品的流程如下。
首先,在 CartAction 类中添加方法 deleteSelectedMeal(),代码如下:

```java
// 从购物车中移除指定编号餐品
public String deleteSelectedMeal() throws Exception {
    //从session中取出购物车,放入Map对象cart中
    Map cart = (Map) session.get("cart");
    // 从Map中移除指定键的值
    cart.remove(mealId);
    return "shopCart";
}
```

然后,在 Struts 2 配置文件中为 CartAction 类的 deleteSelectedMeal()方法配置映射,代码如下:

```xml
<!-- 删除购物车中的某个餐品 -->
<action name="deleteSelectedMeal" class="cartAction"
method="deleteSelectedMeal">
    <result name="shopCart">/shopCart.jsp</result>
    <interceptor-ref name="loginCheck" />
    <interceptor-ref name="defaultStack" />
</action>
```

单击"删除"的叉号图片超链接后，将请求提交到 deleteSelectedMeal，即执行 CartAction 类中的 deleteSelectedMeal()方法，并传递一个参数 mealId。deleteSelectedMeal()方法中根据参数 mealId 值，从 Map 中移除相应的商品对象。

4. 清空购物车

在 shopCart.jsp 页面下方，有一个"清空购物车"图片超链接，其设置如下：

```html
<p class="cz fl">
    <a href="clearCart"><img src="images/d014.jpg" /></a></p>
```

实现清空购物车流程如下。

首先，在 CartAction 类中添加方法 clearCart()，清除购物车中全部餐品，代码如下：

```java
// 清空购物车
public String clearCart() throws Exception {
    Map cart = (Map) session.get("cart");
    cart.clear();
    return "shopCart";
}
```

然后，在 Struts 2 配置文件中为 CartAction 的 clearCart()方法配置映射，代码如下：

```xml
<!-- 清空购物车 -->
<action name="clearCart" class="cartAction" method="clearCart">
    <result name="shopCart">/shopCart.jsp</result>
    <interceptor-ref name="loginCheck" />
    <interceptor-ref name="defaultStack" />
</action>
```

单击"清空购物车"图片超链接后，将请求提交到 clearCart，即执行 CartAction 类中的 clearCart()方法，从 Map 中移除所有餐品对象。

22.13 订单功能

订餐系统客户对订单的处理功能包括生成订单、查看我的订单及订单明细、删除订单。

22.13.1 生成订单

购物车只是暂时用来存储客户的购买信息，为了长久保存，需要将购物车中的内容存入订单信息表(订单主表)和订单明细表(订单子表)中。

在 shopCart.jsp 页面购物车下方，有"确认提交订单"图片超链接，其设置如下：

```
<p class="cz fl">
    <a href="/restaurant/addOrders"><img src="images/d016.jpg" /></a>
</p>
```

单击该按钮后，将请求提交到 addOrders，即执行 OrderAction 类中的 addOrders()方法。实现将购物车中的餐品提交生成订单的步骤如下。

(1) Action 开发。

在 com.restaurant.action 包中，创建 OrdersAction 类，继承自 ActionSupport 类，并实现 RequestAware 和 SessionAware 接口。在 OrdersAction 类中添加一个 addOrders()方法，用于处理生成订单请求，代码如下：

```java
package com.restaurant.action;
……
public class OrdersAction extends ActionSupport implements RequestAware,
        SessionAware {
    OrdersService ordersService;
    OrderDtsService orderDtsService;
    public void setOrdersService(OrdersService ordersService) {
        this.ordersService = ordersService;
    }
    public void setOrderDtsService(OrderDtsService orderDtsService) {
        this.orderDtsService = orderDtsService;
    }
    int oid;
    private Orders orders;
    private Pager pager;
    // 此处省略的 oid、orders、pager 属性的 get 和 set 方法
    // 处理生成订单请求
    public String addOrders() throws Exception {
// 封装 Orders 实体对象
        Orders orders = new Orders();
        orders.setOrderStatus("未付款");
        orders.setOrderTime(new Date());
        Users user = (Users) session.get("user");
// 取得当前登录的用户
        orders.setUsers(user);
        orders.setOrderPrice((Double) session.get("sumPrice"));
// 取得购物车对象
        Map cart = (HashMap) session.get("cart");
        Iterator iter = cart.keySet().iterator();
        while (iter.hasNext()) {
            Object key = iter.next();
            CartItemBean cartItem = (CartItemBean) cart.get(key);
// 封装订单明细
            Orderdts orderDts = new Orderdts();
            orderDts.setMeal(cartItem.getMeal());
            orderDts.setMealCount(cartItem.getQuantity());
            orderDts.setMealPrice(cartItem.getMeal().getMealPrice());
            orderDts.setOrders(orders);
            orderDtsService.addOrderDts(orderDts);
```

```
        }
        session.remove("cart");
        return "show";
    }
    Map<String, Object> session;
    @Override
    public void setSession(Map<String, Object> session) {
        this.session = session;
    }
    Map<String, Object> request;
    @Override
    public void setRequest(Map<String, Object> request) {
        this.request = request;
    }
}
```

在 OrdersAction 类的 addOrders()方法调用了 orderDtsService 接口中的 addOrderDts()方法，用来添加订单明细表(订单子表)记录。但由于在映射文件 Orderdts.hbm.xml 中设置了级联属性(cascade="all")，因此订单主表也会执行插入操作。

(2) 在 Spring 配置文件中定义 OrdersAction，并为其中属性 ordersService 和 orderDtsService 注入值，代码如下：

```
<!-- 定义 OrdersAction 类，并为属性 ordersService 和 orderDtsService 注入值 -->
<bean name="ordersAction" class="com.restaurant.action.OrdersAction"
    scope="prototype">
    <property name="ordersService" ref="ordersService" />
    <property name="orderDtsService" ref="orderDtsService" />
</bean>
```

(3) 在 Struts 2 配置文件中为 OrdersAction 类中的 addOrders()方法配置映射，代码如下：

```
<!-- 生成订单 -->
<action name="addOrders" class="ordersAction" method="addOrders">
    <result name="show" type="redirectAction">toShowMeal</result>
<interceptor-ref name="loginCheck" />
    <interceptor-ref name="defaultStack" />
</action>
```

在图 22-11 所示的购物车显示页 shopCart.jsp 中，单击"确认提交订单"图片超链接，在数据中查看订单信息表 orders 和订单明细表 orderdts 中的记录，分别如图 22-13 和图 22-14 所示。

OID	UserId	OrderTime	OrderStatus	OrderPrice
13	1	2017-04-20 10:30:31	未付款	85.00

图 22-13 订单信息表 orders 中的记录

ODID	OID	MealId	MealPrice	MealCount
26	13	1	10.00	2
27	13	2	20.00	1
28	13	3	15.00	3

图 22-14 订单明细表 orderdts 中的记录

22.13.2 查看"我的订单"

用户登录成功后，在页面右上角，有一个"我的订单"超链接，其设置如下：

```
<a href="/restaurant/toMyOrders"><span>我的订单</span></a>
```

单击"我的订单"超链接，可查看登录用户提交的订单列表，如图 22-15 所示。

订单编号	订单时间	订单状态	总额	明细	删除
9	14-5-4 16:07:44.000	未付款	30.0	查看	删除
10	14-5-4 18:54:17.000	未付款	80.0	查看	删除
11	14-5-5 17:30:16.000	未付款	48.0	查看	删除
13	17-4-20 10:30:31.000	未付款	85.0	查看	删除
合计	-	-	¥ 243.0	-	-

图 22-15 我的订单页

单击"我的订单"超链接后，将请求发送到 toMyOrders，即执行 OrdersAction 类中的 toMyOrders 方法。实现查看"我的订单"功能的步骤如下。

(1) Action 开发。

在 OrdersAction 类中添加 toMyOrders ()方法，获取指定用户的订单列表，再转到我的订单页 myorders.jsp，代码如下：

```java
// 获取指定用户的订单列表，再转到我的订单页 myorders.jsp
public String toMyOrders() throws Exception {
    // 获取从分页超链接传递来的页码
    int curPage = 1;
    if (pager != null)
        curPage = pager.getCurPage();
    // 获取登录用户对象
    Users user = (Users) session.get("user");
    // 将用户编号作为查询条件
    Orders condition = new Orders();
    condition.setUsers(user);
    // 获取指定用户和当前页码的订单列表
    List myOrdersList = ordersService.getOrdersByUserIdForPager(
            user.getId(), curPage);
    // 将当前页显示的订单列表存入 request 范围
    request.put("myOrdersList", myOrdersList);
    // 初始化 Pager 对象
    pager = ordersService.getPagerOfMyOrders(user.getId());
    // 设置 Pager 对象中的待显示页页码
    pager.setCurPage(curPage);
    // 转到我的订单页 myorders.jsp
    return "myorders";
}
```

在 toMyOrders()方法中，首先从 Session 范围获取当前登录用户的编号，然后调用业务接口 OrdersService 中的 getOrdersByUserIdForPager(int userId, int page)获取该用户的订单列表，并存入 request 范围，最后转到我的订单页 myorders.jsp，在页面上显示 request 范围中存放的用户订单列表。

（2）配置 Struts 2 映射文件。

在 Struts 2 配置文件中，为 OrdersAction 类中的 toMyOrders()方法配置映射，代码如下：

```xml
<!-- 我的订单信息 -->
<action name="toMyOrders" class="ordersAction" method="toMyOrders">
    <result name="myorders">/myorders.jsp</result>
    <interceptor-ref name="loginCheck" />
    <interceptor-ref name="defaultStack" />
</action>
```

（3）我的订单页 myorders.jsp。

我的订单页 myorders.jsp 的主体结构与购物车页 shopCart.jsp 类似，在 myorders.jsp 页面中显示我的订单列表，代码如下：

```html
<!--订单开始-->
<div class="order_a jiacu">您的订单有以下内容</div>
<div class="order_b fl mt10">
    <p class="bh fl">订单编号</p>
    <p class="ddsj fl">订单时间</p>
    <p class="ddzt fl">订单状态</p>
    <p class="ddze fl">总额</p>
    <p class="mx fl">明细</p>
    <p class="del fl">删除</p>
</div>
<s:set var="total" value="0" />
<s:iterator var="myOrder" value="#request.myOrdersList">
    <div class="order_c fl">
        <p class="bh fl">
            <s:property value="oid" />
        </p>
        <p class="ddsj fl">
            <s:property value="orderTime" />
        </p>
        <p class="ddzt fl">
            <s:property value="orderStatus" />
        </p>
        <p class="ddze fl">
            <s:property value="orderPrice" />
        </p>
        <p class="mx fl">
            <a href="toOrdersDetails?oid=${oid }">查看</a>
        </p>
        <p class="del fl">
            <s:if test="#myOrder.orderStatus=='未付款'">
                <a href="deleteOrders?oid=${oid }">删除</a>
            </s:if>
        </p>
    </div>
```

```
    <s:set var="total" value="#total+orderPrice"></s:set>
</s:iterator>
<div class="order_c fl" style="background-color:#ccccee;">
    <p class="bh fl">-</p>
    <p class="ddsj fl">合计</p>
    <p class="ddzt fl">-</p>
    <p class="ddze fl">
        ￥
        <s:property value="#total" />
    </p>
    <p class="mx fl">-</p>
    <p class="del fl">-</p>
</div>
<!-- 订单列表结束-->
```

在 myorders.jsp 页面中，用于分页的超链接设置如下：

```
<!-- 分页条开始 -->
<s:if test="pager.curPage>1">
    <a
        href="/restaurant/toMyOrders?pager.curPage=1">首页</a>
    <a       href="/restaurant/toMyOrders?pager.curPage=${pager.curPage-1}">
上一页</a>
</s:if>
<s:if test="pager.curPage < pager.pageCount">
    <a       href="/restaurant/toMyOrders?pager.curPage=${pager.curPage+1}">
下一页</a>
    <a
href="/restaurant/toMyOrders?pager.curPage=${pager.pageCount }">尾页</a>
</s:if>  
共${pager.rowCount}记录，共${pager.curPage}/${pager.pageCount}页
<!-- 分页条结束 -->
```

22.13.3　查看订单明细

在图 22-15 所示的我的订单页中，单击“明细”一栏中的“查看”超链接，可查看该订单的明细信息，如图 22-16 所示。

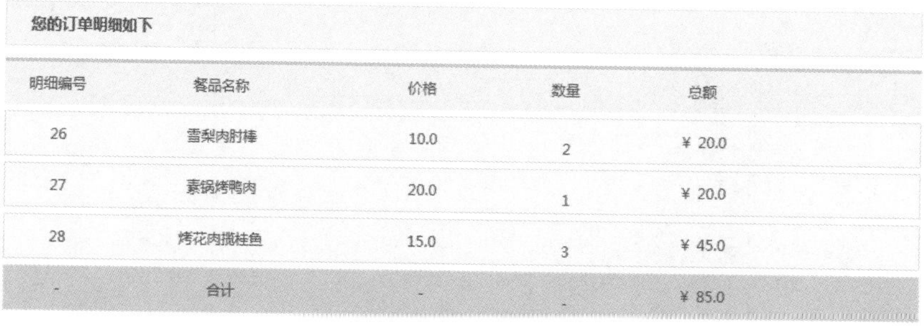

明细编号	餐品名称	价格	数量	总额
26	雪梨肉肘棒	10.0	2	￥ 20.0
27	素锅烤鸭肉	20.0	1	￥ 20.0
28	烤花肉揽桂鱼	15.0	3	￥ 45.0
-	合计	-	-	￥ 85.0

图 22-16　订单明细页

“查看”超链接设置如下：

```
<p class="mx fl">
    <a href="toOrdersDetails?oid=${oid }">查看</a></p>
```

单击"查看"超链接，将请求提交到 toOrdersDetails，即执行 OrdersAction 类中的 toOrdersDetails 方法，并传递参数 oid。实现查看订单明细功能的过程如下。

(1) Action 开发。

在 OrdersAction 类中添加 toOrdersDetails ()方法，根据订单信息表编号获取订单明细列表，再转到我的订单明细页 myordersdetails.jsp，代码如下：

```java
// 根据订单信息表编号获取订单明细列表，再转到订单明细页面
public String toOrdersDetails() throws Exception {
    List ordersDtsList = orderDtsService.getOrderDtsByOid(oid);
    request.put("ordersDtsList", ordersDtsList);
    return "toOrdersDetails";
}
```

在 toOrdersDetails()方法中，使用了变量 oid，该变量用于封装"查看"超链接传递来的参数 oid 的值。之前，在 OrdersAction 类中已经声明该变量，并为其添加 getter 和 setter 方法。通过调用业务接口 OrderDtsService 中的 getOrderDtsByOid 方法，根据传递来的订单信息表编号获取订单明细表的记录，并存入 request 范围，再跳转到我的订单明细页 myordersdetails.jsp，显示 request 范围中存放的订单明细列表。

(2) 在 Struts 2 配置文件中，为 OrdersAction 类中的 toOrdersDetails()方法配置映射，代码如下：

```xml
<action name="toOrdersDetails" class="ordersAction" method="toOrdersDetails">
    <result name="toOrdersDetails">/myordersdetails.jsp</result>
    <interceptor-ref name="loginCheck" />
    <interceptor-ref name="defaultStack" />
</action>
```

(3) 我的订单明细页 myordersdetails.jsp。

在 myordersdetails.jsp 页面中，页面的主体结构与购物车页面、订单页面类似，循环显示订单明细列表的代码如下：

```html
<!--订单明细开始-->
<div class="car_a jiacu">您的订单明细如下</div>
<div class="car_b fl mt10">
    <p class="bh fl">明细编号</p>
    <p class="spmc fl">餐品名称</p>
    <p class="dj fl">价格</p>
    <p class="sl fl">数量</p>
    <p class="je fl">总额</p>
</div>
<s:set var="count" value="0"></s:set>
<s:iterator var="orderDetailItem" value="#request.ordersDtsList">
    <div class="car_c fl">
        <p class="bh fl">
            <s:property value="odid" />
        </p>
        <p class="spmc fl">
            <s:property value="meal.mealName" />
        </p>
        <p class="dj fl">
```

```
            <s:property value="meal.mealPrice" />
        </p>
        <p class="sl fl">
            <s:property value="mealCount" />
        </p>
        <p class="je fl">
            ¥<s:property value="meal.mealPrice*mealCount" />
        </p>
    </div>
    <s:set var="count" value="#count+meal.mealPrice*mealCount" />
</s:iterator>
<div class="car_c fl" style="background-color:#ccccee;">
    <p class="bh fl">-</p>
    <p class="spmc fl">合计</p>
    <p class="dj fl">-</p>
    <p class="sl fl">-</p>
    <p class="je fl">
        ¥<s:property value="#count" />
    </p>
</div>
<!-- 订单明细列表结束-->
```

22.13.4 删除订单

在图 22-15 所示的我的订单页中，单击删除一栏中的"删除"超链接，可将该订单信息表及订单明细表信息删除。"删除"超链接的设置如下：

```
<p class="del fl">
    <s:if test="#myOrder.orderStatus=='未付款'">
        <a href="deleteOrders?oid=${oid }">删除</a>
    </s:if>
</p>
```

单击"删除"超链接，将请求提交到 deleteOrders，即执行 OrdersAction 类中的 deleteOrders 方法，并传递参数 oid。实现删除订单功能的过程如下。

(1) Action 开发。

在 OrdersAction 类中添加 deleteOrders()方法，删除指定编号的订单，再转到 toMyOrders，即再执行 OrdersAction 类中的 toMyOrders 方法，显示我的订单信息，代码如下：

```
public String deleteOrders() throws Exception {
    ordersService.deleteOrdersByOid(oid);
    return "toMyOrders";
}
```

在 deleteOrders()方法中，只删除了订单信息表记录。但由于在映射文件 Orders.hbm.xml 中配置了级联属性(cascade="all")，因此在删除订单信息表时，订单明细表也级联执行删除操作。

(2) 在 Struts 2 配置文件中，为 OrderAction 类中的 deleteOrderInfo()方法配置映射，代码如下：

```xml
<!-- 删除订单 -->
<action name="deleteOrders" class="ordersAction" method="deleteOrders">
    <result name="toMyOrders" type="redirectAction">toMyOrders</result>
    <interceptor-ref name="loginCheck" />
    <interceptor-ref name="defaultStack" />
</action>
```

22.14 小　　结

本章在第 20 章 20.1 节的 Spring、Struts 2 和 Hibernate 这三个框架整合后的项目 s2sh 基础上，实现了网上订餐系统的前台。它的主要功能包括餐品与菜系展示、查询餐品、查看餐品详情、用户登录与注册、购物车功能和订单功能。

通过本章的讲解，希望读者能够掌握使用 Spring、Struts 2 和 Hibernate(S2SH)整合应用开发的基本步骤、方法和技巧。

第 23 章
Spring 整合 Spring MVC 与 Hibernate 实现网上订餐系统后台

在第 22 章中,使用 Spring 整合 Struts 2 与 Hibernate 实现了网上订餐系统前台功能,本章将使用 Spring 整合 Spring MVC 与 Hibernate 实现订餐系统后台功能。

23.1 需求与系统分析

管理员登录系统后，就可以对餐品信息、订单信息、客户信息、权限进行管理。管理员用例图如图 23-1 所示。

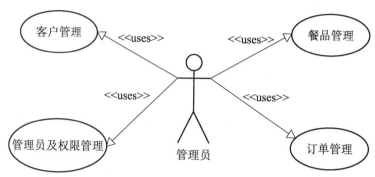

图 23-1 管理员用例图

根据需求分析，管理员后台管理功能如下。
(1) 管理员可以添加餐品、餐品下架、修改餐品、查询餐品。
(2) 管理员可以创建订单、查询订单、修改订单。
(3) 管理员可以添加客户、查询客户、禁用客户。
(4) 超级管理员可以创建普通管理员、设置管理员权限。

根据上述分析，可以得到订餐系统后台的模块结构，如图 23-2 所示。

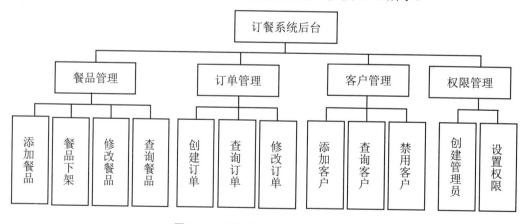

图 23-2 订餐系统后台的模块结构

23.2 数据库设计

在第 22 章中，已经详细介绍过网上订餐系统的数据库，后台数据库与前台一样，读者可以查阅。

23.3 项目环境搭建

在第 21 章 21.7 节中以用户登录为例详细介绍了如何使用 Spring 整合 Spring MVC 与 Hibernate，读者可参照完成网上订餐系统后台的框架搭建。当然，读者也可以直接将 21.7 节创建的项目 springmvc_ssh 复制一份并重新命名为 restaurant-back，再导入 MyEclipse 中。为了避免部署重复，需要修改项目的部署名称。修改过程如下：在 MyEclipse 中右击复制后的项目 restaurant-back，依次选择 Properties→MyEclipse→Deployment Assembly，将 Web Context Root 修改为 restaurant-back 即可。然后将 jackson-annotations-2.6.0.jar、jackson-core-2.6.0.jar 和 jackson-databind-2.6.0.jar 这三个 jar 包复制到项目的 WebRoot\WEB-INF\lib 目录中，用于支持 Spring MVC 实现自动 Json 格式数据转换。

订餐系统后台的目录结构如图 23-3 所示，com.res.controller 包用于存放控制器类，com.res.service 包用于存放业务逻辑层接口，com.res.service.impl 包用于存放业务逻辑层接口的实现类，com.res.dao 包用于存放数据访问层接口，com.res.dao.impl 包用于存放数据访问层接口的实现类，com.res.entity 包用于存放实体类。applicationContext.xml 为 Spring 框架使用的配置文件，springmvc.xml 为 Spring MVC 框架使用的配置文件，admin_login.jsp 为管理员登录页，index.jsp 为后台管理首页面，meallist.jsp 为餐品列表页，createorder.jsp 为创建订单页，searchorder.jsp 为查询订单页，saler.jsp 为订单统计页，userlist.jsp 为用户列表页，adminlist.jsp 为管理员列表页，Easyui 目录下的文件或子目录下的文件为使用 EasyUI 控件所需的 js、css 等文件。echarts 和 echarts-master 目录下的文件或子目录下的文件为使用百度图表控件所需的文件。

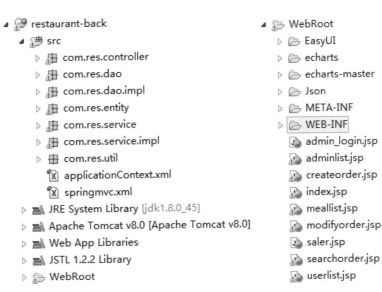

图 23-3 订餐系统后台目录结构

23.4 Spring 及 Spring MVC 配置文件

Spring 框架使用的配置文件为 applicationContext.xml，Spring MVC 使用的配置文件为 springmvc.xml，这些配置文件的含义在 21.7 节中已具体介绍过，由于篇幅在此不再赘述。

23.5 创建实体类

在 com.res.entity 包中，依次创建实体类 Users.java、Admin.java、Functions.java、Powers.java、Meal.java、Mealseries.java、Orders.java 和 Orderdts.java。

其中，实体类 Users.java 代码如下：

```java
package com.res.entity;
......
@Entity
@Table(name = "users", catalog = "restrant")
public class Users {
    private Integer id;
    private String loginName;
    private String loginPwd;
    private String trueName;
    private String email;
    private String phone;
    private String address;
    private int status;
    @Id
    @GeneratedValue(strategy = GenerationType.IDENTITY)
    @Column(name = "Id", unique = true, nullable = false)
    public Integer getId() {
        return id;
    }
    public void setId(Integer id) {
        this.id = id;
    }
    // 此处省略了部分属性的 get 和 set 方法、无参和有参构造方法
}
```

实体类 Admin.java 代码如下：

```java
package com.res.entity;
......
@Entity
@Table(name = "admin", catalog = "restrant")
public class Admin {
    // 基本属性
    private int id;
    private String loginName;
    private String loginPwd;
    // 关联属性
    private Set<Functions> fs = new HashSet<Functions>();
```

```java
    // 配置 Admin 到 Functions 的多对多关联
    @ManyToMany(fetch=FetchType.EAGER)
    @JoinTable(name = "powers", joinColumns = { @JoinColumn(name = "aid") },
inverseJoinColumns = { @JoinColumn(name = "fid") })
    public Set<Functions> getFs() {
        return fs;
    }
    public void setFs(Set<Functions> fs) {
        this.fs = fs;
    }
    @Id
    @GeneratedValue(strategy = GenerationType.IDENTITY)
    @Column(name = "id", unique = true, nullable = false)
    public int getId() {
        return id;
    }
    public void setId(int id) {
        this.id = id;
    }
    // 此处省略了部分属性的 get 和 set 方法、无参和有参构造方法
}
```

实体类 Functions.java 代码如下：

```java
package com.res.entity;
import javax.persistence.*;
@Entity
@Table(name = "functions", catalog = "restrant")
public class Functions implements Comparable<Functions> {
    private int id;
    private String name;
    private int parentid;
    private boolean isleaf;
    @Id
    @GeneratedValue(strategy = GenerationType.IDENTITY)
    @Column(name = "id", unique = true, nullable = false)
    public int getId() {
        return id;
    }
    public void setId(int id) {
        this.id = id;
    }
    // 此处省略了部分属性的 get 和 set 方法、无参和有参构造方法
    @Override
    public int compareTo(Functions o) {
        return this.id - o.getId();
    }
}
```

在实体类 Functions 中，添加了 compareTo 方法，在排序时将两个 Functions 对象的 id 进行比较，根据比较的结果是小于、等于或者是大于而返回一个负数、零或者正数。

实体类 Powers.java 代码如下：

```java
package com.res.entity;
public class Powers {
```

```java
    private Admin admin;
    private Functions f;
    // 此处省略了属性admin、f 的get 和set 方法
}
```

实体类Meal.java 的代码如下：

```java
package com.res.entity;
import javax.persistence.*;
@Entity
@Table(name = "meal", catalog = "restrant")
public class Meal implements java.io.Serializable {
    private Integer mealId;
    private Mealseries mealseries;
    private String mealName;
    private String mealSummarize;
    private String mealDescription;
    private Double mealPrice;
    private String mealImage;
    private Integer mealStatus;
    @Id
    @GeneratedValue(strategy = GenerationType.IDENTITY)
    @Column(name = "MealId", unique = true, nullable = false)
    public Integer getMealId() {
        return this.mealId;
    }
    public void setMealId(Integer mealId) {
        this.mealId = mealId;
    }
    // 使用@ManyToOne 和@JoinColumn 注解实现Meal 到Mealseries 的多对一关联
    @ManyToOne(fetch=FetchType.EAGER)
    @JoinColumn(name = "MealSeriesId")
    public Mealseries getMealseries() {
        return this.mealseries;
    }
    public void setMealseries(Mealseries mealseries) {
        this.mealseries = mealseries;
    }
    // 此处省略了属性mealName、mealSummarize、mealDescription、
    // mealPrice、mealImage、mealStatus 的get 和set 方法
    public Meal() {
    }
}
```

实体类Mealseries.java 的代码如下：

```java
package com.res.entity;
import javax.persistence.*;
@Entity
@Table(name = "mealseries", catalog = "restrant")
public class Mealseries implements java.io.Serializable {
    private Integer seriesId;
    private String seriesName;
    @Id
    @GeneratedValue(strategy = GenerationType.IDENTITY)
```

```java
    @Column(name = "SeriesId", unique = true, nullable = false)
    public Integer getSeriesId() {
        return this.seriesId;
    }
    public void setSeriesId(Integer seriesId) {
        this.seriesId = seriesId;
    }
    // 此处省略了属性 seriesName 的 get 和 set 方法
    public Mealseries() {
    }
}
```

实体类 Orders.java 的代码如下：

```java
package com.res.entity;
……
@Entity
@Table(name = "orders", catalog = "restrant")
public class Orders {
    // 订单基本属性
    private int oid;
    private int userId;
    private String orderTime;
    private String orderStatus;
    private double orderPrice;
    // 附加属性,用于订单查询
    private String orderTimeFrom;
    private String orderTimeTo;
    // 关联属性
    private Users u;
    @ManyToOne(fetch = FetchType.EAGER)
    @JoinColumn(name = "UserId")
    public Users getU() {
        return u;
    }
    public void setU(Users u) {
        this.u = u;
    }
    // 关联属性
    @JsonIgnoreProperties(value = { "o", "m" })
    private Set<Orderdts> ods = new HashSet<Orderdts>();
    @OneToMany(mappedBy = "o", fetch = FetchType.EAGER, cascade = { CascadeType.ALL })
    public Set<Orderdts> getOds() {
        return ods;
    }
    public void setOds(Set<Orderdts> ods) {
        this.ods = ods;
    }
    @Id
    @GeneratedValue(strategy = GenerationType.IDENTITY)
    @Column(name = "OID", unique = true, nullable = false)
    public int getOid() {
        return oid;
    }
```

```java
    public void setOid(int oid) {
        this.oid = oid;
    }
    // 此处省略了属性 orderTime、orderStatus、orderPrice、
    // userId、orderTimeFrom、orderTimeTo 的 get 和 set 方法
}
```

在数据表 orders 中，没有与实体类 Orders 中 userId、orderTimeFrom、orderTimeTo 属性对应的字段。因此，这些属性的 get 方法上需要使用@Transient 注解修饰，以表示这些属性不需要映射到数据表中的字段。此外，在 Set<Orderdts>类型的属性 ods 的 get 方法上，使用了 @JsonIgnoreProperties 注解修改，以表示在进行 JSON 转化时，忽略 Orderdts 对象中包含的 Orders 类型的属性 o 和 Meal 类型的属性 m，以避免无限递归转化。

实体类 Orderdts.java 的代码如下：

```java
package com.res.entity;
import javax.persistence.*;
@Entity
@Table(name = "orderdts", catalog = "restrant")
public class Orderdts {
    // 基本属性
    private int odid;
    private int mealCount;
    private double mealPrice;
    private double totalprice;
    // 关联属性
    private Orders o;
    private Meal m;
    private int mealId;
    @Id
    @GeneratedValue(strategy = GenerationType.IDENTITY)
    @Column(name = "ODID", unique = true, nullable = false)
    public int getOdid() {
        return odid;
    }
    public void setOdid(int odid) {
        this.odid = odid;
    }
    // 此处省略了属性 mealId、mealCount、mealPrice、
    // totalprice 的 get 和 set 方法
    @JsonIgnoreProperties(value={"u","ods"})
    @ManyToOne(fetch = FetchType.EAGER)
    @JoinColumn(name = "OID")
    public Orders getO() {
        return o;
    }
    public void setO(Orders o) {
        this.o = o;
    }
    @ManyToOne(fetch = FetchType.EAGER)
    @JoinColumn(name = "MealId", unique = true)
    public Meal getM() {
        return m;
    }
```

```java
    public void setM(Meal m) {
        this.m = m;
    }
}
```

在实体类 Orderdts 中，属性 mealId、totalprice 的 get 方法上使用了@Transient 注解修饰，以表示这些属性不需要映射到数据表中的字段。

最后，在 Spring 配置文件 applicationContext.xml 中添加对基于注解的实体类的引用：

```xml
<!-- 配置 Hibernate 基于注解的实体类的位置及名称 -->
<property name="annotatedClasses">
    <list>
        <value>com.res.entity.Admin</value>
        <value>com.res.entity.Functions</value>
        <value>com.res.entity.Meal</value>
        <value>com.res.entity.Mealseries</value>
        <value>com.res.entity.Users</value>
        <value>com.res.entity.Orders</value>
        <value>com.res.entity.Orderdts</value>
    </list>
</property>
```

23.6　创建 DAO 接口及实现类

在 com.res.dao 包中，依次创建数据访问层接口 BaseDao.java、UserDAO.java、AdminDAO.java、FunctionsDAO.java、PowersDAO.java、MealDAO.java、MealSeriesDAO.java、OrderDAO.java 和 OrderdtsDAO.java。其中，接口 BaseDao.java 中声明的方法与第 22 章中相同。在接口 UserDAO.java 中声明如下方法：

```java
package com.res.dao;
import java.util.List;
import com.res.entity.Users;
public interface UserDAO {
    // 获取所有合法用户(即未禁用)
    public List<Users> getValidUser();
    // 根据 id 获取用户对象
    public Users getUserById(int id);
//根据查询条件(封装在对象 u 中)、当前页码和每页记录数，分页获取用户列表
    public List<Users> getUsersByConditionForPager(Users u, int pageIndex, int pageSize);
    // 根据查询条件(封装在对象 u 中)计算用户总记录数
    public int getTotalCount(Users u);
    // 更新用户状态
    public void updateUserStatus(String uids, String flag);
}
```

在接口 AdminDAO.java 中声明如下方法：

```java
package com.res.dao;
……
public interface AdminDAO {
    // 管理员登录验证
```

```java
    public List<Admin> adminLogin(Admin admin);
    // 根据id获取管理员对象及功能权限
    public Admin getAdminFunctions(int id);
    // 获取所有管理员
    public List<Admin> getAllAdmin();
    // 新增管理员
    public void addAdmin(Admin admin);
}
```

在接口 FunctionsDAO.java 中声明如下方法：

```java
package com.res.dao;
……
public interface FunctionsDAO {
    // 获取所有功能对象
    public List<Functions> getAllFunctions();
}
```

在接口 PowersDAO.java 中声明如下方法：

```java
package com.res.dao;
public interface PowersDAO {
    // 删除指定管理员的权限
    public void delPowersByAdminid(int adminid);
    // 添加权限
    public void addPowers(int aid, int fid);
}
```

在接口 MealDAO.java 中声明如下方法：

```java
package com.res.dao;
……
public interface MealDAO {
    // 根据查询条件和每页记录数，获取指定页显示的餐品列表
    public List<Meal> getMealByConditionForPager(Meal meal, int pageIndex, int pageSize);
    //根据查询条件和每页记录数,计算总页数
    public int getTotalPages(int pageSize,Meal meal);
    //获取指定条件的餐品总数
    public int getTotalCount(Meal meal);
    // 根据id号获取餐品
    public Meal getMealByMealId(int mealId);
    // 添加餐品
    public int addMeal(Meal meal);
    // 修改餐品对象
    public void updateMeal(Meal meal);
    // 修改餐品状态
    public int updateMealStatus(String ids);
    // 获取在售餐品列表
    public List<Meal> getOnSaleMeal();
}
```

在接口 MealSeriesDAO.java 中声明如下方法：

```java
package com.res.dao;
import java.util.List;
```

```java
public interface MealSeriesDAO {
    // 获取菜系列表
    public List getMealSeries();
}
```

在接口 OrderDAO.java 中声明如下方法：

```java
package com.res.dao;
……
public interface OrderDAO {
// 根据查询条件(封装在对象o中)、指定页码、每页记录数，获取当前页的订单列表
    public List<Orders> getOrderByConditionForPager(Orders o, int pageIndex, int pageSize);
    // 根据查询条件(封装在对象o中)，获取订单总记录数
    public int getTotalCount(Orders o);
    // 根据订单主表编号获取订单对象
    public Orders getOrdersByOid(int oid);
    // 新增订单
    public int addOrder(Orders o);
    // 删除订单
    public void deleteOrder(Orders o);
    // 修改订单
    public int modifyOrder(Orders o);
    // 餐品销量统计
    public List findSalerStandby();
}
```

在接口 OrderdtsDAO.java 中声明如下方法：

```java
package com.res.dao;
……
public interface OrderdtsDAO {
    // 根据订单主表编号，获取订单明细列表
    public List<Orderdts> getOrderdtsByOid(int oid);
    // 删除订单明细
    public int deleteOrderdts(Orderdts od);
}
```

在 com.res.dao 包中，依次创建上述接口的实现类 BaseDaoImpl.java、UserDAOImpl.java、AdminDAOImpl.java、FunctionsDAOImpl.java、PowersDAOImpl.java、MealDAOImpl.java、MealSeriesDAOImpl.java、OrderDAOImpl.java 和 OrderdtsDAOImpl.java。其中，实现类 BaseDaoImpl.java 与第 22 章中基本相同，只是使用了@Repository 注解，在 Spring 容器中注册实例名为 baseDao 的 BaseDaoImpl 实例，并通过@Autowired 注解注入 Spring 容器中的 SessionFactory 实例。代码如下：

```java
package com.res.dao.impl;
……
// 在Spring容器中注册实例名为baseDao的BaseDaoImpl实例
@Repository("baseDao")
public class BaseDaoImpl<T> implements BaseDao<T> {
    // 通过@Autowired注解注入Spring容器中的SessionFactory实例
    @Autowired
    SessionFactory sessionFactory;
    public Session getCurrentSession() {
```

```java
        return this.sessionFactory.getCurrentSession();
    }
    ……    // 此处省略了接口 BaseDao 中方法的实现
}
```

UserDAOImpl.java 代码如下：

```java
package com.res.dao.impl;
……
@Repository("userDAO")
public class UserDAOImpl extends BaseDaoImpl<Users> implements UserDAO {
    @Override
    public List<Users> getValidUser() {
        String hql = "from Users u where u.status=1";
        return super.find(hql);
    }
    @Override
    public Users getUserById(int id) {
        return super.get(Users.class, id);
    }
    @Override
    public List<Users> getUsersByConditionForPager(Users u, int pageIndex, int pageSize) {
        String hql = "from Users u where 1=1";
        Object[] param = null;
        if (u != null) {
            List list = new ArrayList();
            if (u.getLoginName() != null && !"".equals(u.getLoginName())) {
                hql += " and u.loginName like ?";
                list.add("%" + u.getLoginName() + "%");
            }
            if (list.size() > 0)
                param = list.toArray();
        }
        return super.find(hql, param, pageIndex - 1, pageSize);
    }
    @Override
    public int getTotalCount(Users u) {
        Integer count = null;
        try {
            String hql = "select count(u) from Users u where 1=1";
            if (u != null) {
                if (u.getLoginName() != null && !"".equals(u.getLoginName())) {
                    hql += " and u.loginName like '%" + u.getLoginName() + "%'";
                }
            }
            count = Integer.parseInt(super.findUnique(hql).toString());
        } catch (Exception e) {
            e.printStackTrace();
        }
        return count;
    }
    @Override
    public void updateUserStatus(String uids, String flag) {
```

```java
        String sql = "update users u set u.status=" + Integer.parseInt(flag);
        sql += " where u.id in " + uids;
        super.saveOrUpdate(sql);
    }
}
```

AdminDAOImpl.java 代码如下：

```java
package com.res.dao.impl;
……
// 在 Spring 容器中注册实例名为 adminDAO 的 AdminDAOImpl 实例
@Repository("adminDAO")
public class AdminDAOImpl extends BaseDaoImpl<Admin> implements AdminDAO {
    // 管理员登录验证
    @Override
    public List<Admin> adminLogin(Admin admin) {
        String hql = "from Admin ad where ad.loginName = ? and ad.loginPwd = ?";
        Object[] param = new Object[] { admin.getLoginName(),
                admin.getLoginPwd() };
        return super.find(hql, param);
    }
    // 根据 id 获取管理员对象及功能权限
    @Override
    public Admin getAdminFunctions(int id) {
        return super.get(Admin.class, id);
    }
    // 获取所有管理员
    @Override
    public List<Admin> getAllAdmin() {
        String hql = "from Admin";
        return super.find(hql);
    }
    // 新增管理员
    @Override
    public void addAdmin(Admin admin) {
        super.save(admin);
    }
}
```

FunctionsDAOImpl.java 代码如下：

```java
package com.res.dao.impl;
……
//在 Spring 容器中注册实例名为 functionsDAO 的 FunctionsDAOImpl 实例
@Repository("functionsDAO")
public class FunctionsDAOImpl extends BaseDaoImpl<Functions> implements
FunctionsDAO {
    // 获取所有功能对象
    @Override
    public List<Functions> getAllFunctions() {
        String hql = "from Functions";
        return super.find(hql);
    }
}
```

PowersDAOImpl.java 代码如下：

```java
package com.res.dao.impl;
……
@Repository("powersDAO")
public class PowersDAOImpl extends BaseDaoImpl<Powers> implements PowersDAO
{
    // 删除指定管理员的权限
    @Override
    public void delPowersByAdminid(int adminid) {
        String sql = "delete from powers where aid=" + adminid;
        super.executeSql(sql, null);
    }
    // 添加权限
    @Override
    public void addPowers(int aid, int fid) {
        String sql = " insert into powers(aid,fid) values(?,?)";
        Object[] params = new Object[] { aid, fid };
        super.executeSql(sql, params);
    }
}
```

MealDAOImpl.java 的代码如下：

```java
package com.res.dao.impl;
……
// 在 Spring 容器中注册实例名为 mealDAO 的 MealDAOImpl 实例
@Repository("mealDAO")
public class MealDAOImpl extends BaseDaoImpl<Meal> implements MealDAO {
    // 根据查询条件和每页记录数，获取指定页显示的餐品列表
    @Override
    public List<Meal> getMealByConditionForPager(Meal meal, int pageIndex,
int pageSize) {
        String hql = "from Meal m where 1=1";
        Object[] param = null;
        if (meal != null) {
            List list = new ArrayList();
            if (meal.getMealName() != null && !meal.getMealName().equals(""))
{
                hql += " and m.mealName like ?";
                list.add("%" + meal.getMealName() + "%");
            }
            if ((meal.getMealseries() != null)
                    && (meal.getMealseries().getSeriesId() != null)) {
                hql += " and m.mealseries.seriesId = ?";
                list.add(meal.getMealseries().getSeriesId());
            }
            if (list.size() > 0)
                param = list.toArray();
        }
        return super.find(hql, param, pageIndex - 1, pageSize);
    }
    //根据查询条件和每页记录数,计算总页数
    @Override
    public int getTotalPages(int pageSize, Meal meal) {
```

```java
        int count = 0;
        int totalPages = 0;
        count = getTotalCount(meal);
        totalPages = (count % pageSize == 0) ? (count / pageSize) : (count
                / pageSize + 1);
        return totalPages;
    }
    //获取指定条件的餐品总数
    @Override
    public int getTotalCount(Meal meal) {
        Integer count = null;
        try {
            String hql = "select count(m) from Meal m where 1=1";
            if (meal != null) {
                if (meal.getMealName() != null
                        && !meal.getMealName().equals("")) {
                    hql += " and m.mealName like '%" + meal.getMealName()
                            + "%'";
                }
                if ((meal.getMealseries() != null)
                        && (meal.getMealseries().getSeriesId() != null)) {
                    hql += " and m.mealseries.seriesId = "
                            + meal.getMealseries().getSeriesId();
                }
            }
            count = Integer.parseInt(super.findUnique(hql).toString());
        } catch (Exception e) {
            e.printStackTrace();
        }
        return count;
    }
    //根据id号获取餐品
    @Override
    public Meal getMealByMealId(int mealId) {
        return (Meal) super.get(Meal.class, mealId);
    }
    //添加餐品
    @Override
    public int addMeal(Meal meal) {
        return (Integer) super.save(meal);
    }
    //修改餐品
    @Override
    public void updateMeal(Meal meal) {
        super.update(meal);
    }
    //修改餐品状态
    @Override
    public int updateMealStatus(String ids) {
        String hql = "update Meal m set m.mealStatus=0 where m.mealId in " + ids;
        return super.executeHql(hql);
    }
    //获取在售餐品列表
    @Override
```

```java
    public List<Meal> getOnSaleMeal() {
        String hql = "from Meal m where m.mealStatus=1";
        return super.find(hql);
    }
}
```

MealSeriesDAOImpl.java 代码如下:

```java
package com.res.dao.impl;
......
//在 Spring 容器中注册实例名为 mealSeriesDAO 的 MealSeriesDAOImpl 实例
@Repository("mealSeriesDAO")
public class MealSeriesDAOImpl extends BaseDaoImpl<Mealseries> implements MealSeriesDAO {
    // 获取菜系列表
    @Override
    public List getMealSeries() {
        String hql = "from Mealseries";
        return super.find(hql);
    }
}
```

OrderDAOImpl.java 的代码如下:

```java
package com.res.dao.impl;
......
//在 Spring 容器中注册实例名为 orderDAO 的 OrderDAOImpl 实例
@Repository("orderDAO")
public class OrderDAOImpl extends BaseDaoImpl<Orders> implements OrderDAO {
// 根据查询条件(封装在对象 o 中)、指定页码、每页记录数,获取当前页的订单列表
    @Override
    public List<Orders> getOrderByConditionForPager(Orders o, int pageIndex, int pageSize) {
        String hql = "from Orders o where 1=1";
        Object[] param = null;
        if (o != null) {
            if (o.getOid() > 0) {
                hql += " and o.oid=" + o.getOid();
            } else {
                List list = new ArrayList();
                if (o.getOrderStatus() != null
                        && !"请选择".equals(o.getOrderStatus())) {
                    hql += " and o.orderStatus = ?";
                    list.add(o.getOrderStatus());
                }
                if (o.getOrderTimeFrom() != null
                        && !"".equals(o.getOrderTimeFrom())) {
                    hql += " and o.orderTime >= ?";
                    list.add(o.getOrderTimeFrom());
                }
                if (o.getOrderTimeTo() != null
                        && !"".equals(o.getOrderTimeTo())) {
                    hql += " and o.orderTime <? ";
                    list.add(o.getOrderTimeTo());
                }
```

```java
                if (o.getUserId() > 0) {
                    hql += " and o.u.id= ? ";
                    list.add(o.getUserId());
                }
                if (list.size() > 0)
                    param = list.toArray();
            }
        }
        return super.find(hql, param, pageIndex - 1, pageSize);
    }
    // 根据查询条件(封装在对象o中),获取订单总记录数
    @Override
    public int getTotalCount(Orders o) {
        Integer count = null;
        try {
            String hql = "select count(o) from Orders o where 1=1";
            if (o != null) {
                if (o.getOid() > 0) {
                    hql += " and o.oid=" + o.getOid();
                } else {
                    if (o.getOrderStatus() != null
                            && !"请选择".equals(o.getOrderStatus())) {
                        hql += " and o.orderStatus = '" + o.getOrderStatus() + "'";
                    }
                    if (o.getOrderTimeFrom() != null
                            && !"".equals(o.getOrderTimeFrom())) {
                        hql += " and o.orderTime >= '" + o.getOrderTimeFrom() + "'";
                    }
                    if (o.getOrderTimeTo() != null
                            && !"".equals(o.getOrderTimeTo())) {
                        hql += " and o.orderTime < '" + o.getOrderTimeTo()
                                + "'";
                    }
                    if (o.getUserId() > 0) {
                        hql += " and o.u.id= " + o.getUserId();
                    }
                }
            }
            count = Integer.parseInt(super.findUnique(hql).toString());
        } catch (Exception e) {
            e.printStackTrace();
        }
        return count;
    }
    // 根据订单主表编号获取订单对象
    @Override
    public Orders getOrdersByOid(int oid) {
        return super.get(Orders.class, oid);
    }
    // 新增订单
    @Override
    public int addOrder(Orders o) {
        return (Integer) super.save(o);
    }
```

```java
    // 删除订单
    @Override
    public void deleteOrder(Orders o) {
        super.delete(o);
    }
    // 修改订单
    @Override
    public int modifyOrder(Orders o) {
        try {
            super.saveOrUpdate(o);
        } catch (Exception e) {
            return 0;
        }
        return 1;
    }
    // 餐品销量统计
    @Override
    public List findSalerStandby() {
        String sql = "SELECT DISTINCT MealName, SUM(MealCount * m.MealPrice) AS mc FROM orderdts od, meal m WHERE od.MealId=m.MealId GROUP BY od.MealId";
        return super.queryBySql(sql);
    }
}
```

OrderdtsDAOImpl.java 的代码如下：

```java
package com.res.dao.impl;
......
//在 Spring 容器中注册实例名为 orderdtsDAO 的 OrderdtsDAOImpl 实例
@Repository("orderdtsDAO")
public class OrderdtsDAOImpl extends BaseDaoImpl<Orderdts> implements OrderdtsDAO {
    // 根据订单主表编号，获取订单明细列表
    @Override
    public List<Orderdts> getOrderdtsByOid(int oid) {
        String hql = "from Orderdts od where od.o.oid=" + oid;
        return super.find(hql);
    }
    // 删除订单明细
    @Override
    public int deleteOrderdts(Orderdts od) {
        try {
            super.delete(od);
        } catch (Exception e) {
            return 0;
        }
        return 1;
    }
}
```

23.7 创建 Service 接口及实现类

在 com.res.service 包中，依次创建业务逻辑层接口 UserService.java、AdminService.java、

FunctionsService.java、PowersService.java、MealService.java、MealseriesService.java 和 OrderService.java。

在接口 UserService.java 中声明如下方法：

```java
package com.res.service;
import java.util.List;
import com.res.entity.Users;
public interface UserService {
    public List<Users> getValidUser();
    public Users getUserById(int id);
    public List<Users> getUsersByConditionForPager(Users u, int pageIndex, int pageSize);
    public int getTotalCount(Users u);
    public void updateUserStatus(String uids, String flag);
}
```

在接口 AdminService.java 中声明如下方法：

```java
package com.res.service;
import java.util.List;
import com.res.entity.Admin;
public interface AdminService {
    // 管理员登录验证
    public List<Admin> adminLogin(Admin admin);
    // 根据id获取管理员对象及功能权限
    public Admin getAdminFunctions(int id);
    // 获取所有管理员
    public List<Admin> getAllAdmin();
    // 新增管理员
    public void addAdmin(Admin admin);
}
```

在接口 FunctionsService.java 中声明如下方法：

```java
package com.res.service;
import java.util.List;
import com.res.entity.Functions;
public interface FunctionsService {
    // 获取所有功能对象
    public List<Functions> getAllFunctions();
}
```

在接口 PowersService.java 中声明如下方法：

```java
package com.res.service;
public interface PowersService {
    // 删除指定管理员的权限
    public void delPowersByAdminid(int adminid);
    // 添加权限
    public void addPowers(int adminId, String[] fids);
}
```

在接口 MealService.java 中声明如下方法：

```java
package com.res.service;
import java.util.List;
```

```java
import com.res.entity.Meal;
public interface MealService {
    // 根据id号获取餐品
    public Meal getMealByMealId(int mealId);
    // 根据查询条件和每页记录数,获取指定页显示的餐品列表
    public List<Meal> getMealByConditionForPager(Meal meal, int pageIndex, int pageSize);
    // 根据查询条件和每页记录数,计算总页数
    public int getTotalPages(int pageSize,Meal meal);
    // 获取指定条件的餐品总数
    public int getTotalCount(Meal meal);
    // 添加餐品
    public int addMeal(Meal meal);
    // 修改餐品状态(下架)
    public int updateMealStatus(String ids);
    // 修改餐品对象
    public void updateMeal(Meal meal);
    // 获取在售餐品列表
    public List<Meal> getOnSaleMeal();
}
```

在接口 MealseriesService.java 中声明如下方法:

```java
package com.res.service;
import java.util.List;
public interface MealseriesService {
    // 获取菜系列表
    public List getMealSeries();
}
```

在接口 OrderService.java 中声明如下方法:

```java
package com.res.service;
import java.util.List;
import com.res.entity.*;
public interface OrderService {
// 根据查询条件(封装在对象o中)、指定页码、每页记录数,获取当前页的订单列表
    public List<Orders> getOrderByConditionForPager(Orders o, int pageIndex, int pageSize);
    // 根据订单主表编号获取订单对象
    public Orders getOrdersByOid(int oid);
    // 根据查询条件(封装在对象o中),获取订单总记录数
    public int getTotalCount(Orders o);
    // 新增订单
    public int addOrder(Orders o);
    // 删除订单
    public void deleteOrder(Orders o);
    // 根据订单号获取订单明细
    public List<Orderdts> getOrderdtsByOid(int oid);
    // 删除订单明细
    public int deleteOrderdts(Orderdts od);
    // 修改订单
    public int modifyOrder(Orders o);
    // 餐品销量统计
    public List findSalerStandby();
}
```

在 com.res.service.impl 包中，依次创建上述接口的实现类 UserServiceImpl.java、AdminServiceImpl.java、FunctionsServiceImpl.java、PowersServiceImpl.java、MealServiceImpl.java、MealseriesServiceImpl.java 和 OrderServiceImpl.java，并实现接口中的方法。

其中，UserServiceImpl.java 的代码如下：

```java
package com.res.service.impl;
……
@Service("userService")
@Transactional           // 使用@Transactional注解实现事务管理
public class UserServiceImpl implements UserService {
    @Autowired
    UserDAO userDAO;
    @Override
    public List<Users> getValidUser() {
        return userDAO.getValidUser();
    }
    @Override
    public Users getUserById(int id) {
        return userDAO.getUserById(id);
    }
    @Override
    public List<Users> getUsersByConditionForPager(Users u, int pageIndex, int pageSize) {
        return userDAO.getUsersByConditionForPager(u, pageIndex, pageSize);
    }
    @Override
    public int getTotalCount(Users u) {
        return userDAO.getTotalCount(u);
    }
    @Override
    public void updateUserStatus(String uids, String flag) {
        userDAO.updateUserStatus(uids, flag);
    }
}
```

AdminServiceImpl.java 的代码如下：

```java
package com.res.service.impl;
……
// 在Spring容器中注册名为adminService的AdminServiceImpl实例
@Service("adminService")
// 使用@Transactional注解实现事务管理
@Transactional
public class AdminServiceImpl implements AdminService {
    // 使用@Autowired注解注入UserInfoDAOImpl实例
    @Autowired
    AdminDAO adminDAO;
    // 管理员登录验证
    @Override
    public List<Admin> adminLogin(Admin admin) {
        return adminDAO.adminLogin(admin);
    }
    // 根据id获取管理员对象及功能权限
    @Override
```

```java
public Admin getAdminFunctions(int id) {
    return adminDAO.getAdminFunctions(id);
}
// 获取所有管理员
@Override
public List<Admin> getAllAdmin() {
    return adminDAO.getAllAdmin();
}
// 新增管理员
@Override
public void addAdmin(Admin admin) {
    adminDAO.addAdmin(admin);
}
}
```

FunctionsServiceImpl.java 的代码如下：

```java
package com.res.service.impl;
……
@Service("functionsService")
@Transactional
public class FunctionsServiceImpl implements FunctionsService {
    @Autowired
    private FunctionsDAO functionsDAO;
    // 获取所有功能对象
    @Override
    public List<Functions> getAllFunctions() {
        return functionsDAO.getAllFunctions();
    }
}
```

PowersServiceImpl.java 的代码如下：

```java
package com.res.service.impl;
……
@Service("powersService")
@Transactional
public class PowersServiceImpl implements PowersService {
    @Autowired
    private PowersDAO powersDAO;
    @Autowired
    private AdminDAO adminDAO;
    // 删除指定管理员的权限
    @Override
    public void delPowersByAdminid(int adminid) {
        powersDAO.delPowersByAdminid(adminid);
    }
    // 添加权限
    @Override
    public void addPowers(int adminId, String[] fids) {
        for (String fid : fids) {
            powersDAO.addPowers(adminId, Integer.parseInt(fid));
        }
    }
}
```

MealServiceImpl.java 的代码如下：

```java
package com.res.service.impl;
……
// 在 Spring 容器中注册名为 mealService 的 MealServiceImpl 实例
@Service("mealService")
// 使用@Transactional 注解实现事务管理
@Transactional
public class MealServiceImpl implements MealService {
    // 使用@Autowired 注解注入 MealDAOImpl 实例
    @Autowired
    MealDAO mealDAO;
    // 根据查询条件和每页记录数，获取指定页显示的餐品列表
    @Override
    public List<Meal> getMealByConditionForPager(Meal meal, int pageIndex,
int pageSize) {
        return mealDAO.getMealByConditionForPager(meal, pageIndex, pageSize);
    }
    // 根据查询条件和每页记录数,计算总页数
    @Override
    public int getTotalPages(int pageSize, Meal meal) {
        return mealDAO.getTotalPages(pageSize, meal);
    }
    // 获取指定条件的餐品总数
    @Override
    public int getTotalCount(Meal meal) {
        return mealDAO.getTotalCount(meal);
    }
    // 添加餐品
    @Override
    public int addMeal(Meal meal) {
        return mealDAO.addMeal(meal);
    }
    // 修改餐品状态
    @Override
    public int updateMealStatus(String ids) {
        return mealDAO.updateMealStatus(ids);
    }
    // 根据 id 号获取餐品
    @Override
    public Meal getMealByMealId(int mealId) {
        return mealDAO.getMealByMealId(mealId);
    }
    // 修改餐品对象
    @Override
    public void updateMeal(Meal meal) {
        mealDAO.updateMeal(meal);
    }
    // 获取在售餐品列表
    @Override
    public List<Meal> getOnSaleMeal() {
        return mealDAO.getOnSaleMeal();
    }
}
```

MealseriesServiceImpl.java 的代码如下:

```java
package com.res.service.impl;
……
//在 Spring 容器中注册名为 mealseriesService 的 MealseriesServiceImpl 实例
@Service("mealseriesService")
// 使用@Transactional 注解实现事务管理
@Transactional
public class MealseriesServiceImpl implements MealseriesService {
    // 使用@Autowired 注解注入 mealSeriesDAO 实例
    @Autowired
    MealSeriesDAO mealSeriesDAO;
    // 获取菜系列表
    @Override
    public List getMealSeries() {
        return mealSeriesDAO.getMealSeries();
    }
}
```

OrderServiceImpl.java 的代码如下:

```java
package com.res.service.impl;
……
@Service("orderService")
@Transactional
public class OrderServiceImpl implements OrderService {
    @Autowired
    OrderDAO orderDAO;
    @Autowired
    OrderdtsDAO orderdtsDAO;
// 根据查询条件(封装在对象 o 中)、指定页码、每页记录数,获取当前页的订单列表
    @Override
    public List<Orders> getOrderByConditionForPager(Orders o, int pageIndex,
int pageSize) {
        return orderDAO.getOrderByConditionForPager(o, pageIndex, pageSize);
    }
    // 根据订单主表编号获取订单对象
    @Override
    public Orders getOrdersByOid(int oid) {
        return orderDAO.getOrdersByOid(oid);
    }
    // 根据查询条件(封装在对象 o 中),获取订单总记录数
    @Override
    public int getTotalCount(Orders o) {
        return orderDAO.getTotalCount(o);
    }
    // 新增订单
    @Override
    public int addOrder(Orders o) {
        return orderDAO.addOrder(o);
    }
    // 删除订单
    @Override
    public void deleteOrder(Orders o) {
        orderDAO.deleteOrder(o);
```

```
    }
    // 根据订单号获取订单明细
    @Override
    public List<Orderdts> getOrderdtsByOid(int oid) {
        return orderdtsDAO.getOrderdtsByOid(oid);
    }
    // 删除订单明细
    @Override
    public int deleteOrderdts(Orderdts od) {
        return orderdtsDAO.deleteOrderdts(od);
    }
    // 修改订单
    @Override
    public int modifyOrder(Orders o) {
        return orderDAO.modifyOrder(o);
    }
    // 餐品销量统计
    @Override
    public List findSalerStandby() {
        return orderDAO.findSalerStandby();
    }
}
```

23.8　后台登录与管理首页面

网上订餐系统后台登录页 admin_login.jsp 如图 23-4 所示。

图 23-4　网上订餐系统后台登录页

页面 admin_login.jsp 使用 Easy UI 框架进行布局，代码如下：

```
<%@ page language="java" import="java.util.*" pageEncoding="UTF-8"%>
<html>
<head>
<title>订餐系统——后台登录页</title>
```

```html
<link href="EasyUI/themes/default/easyui.css" rel="stylesheet"
    type="text/css" />
<link href="EasyUI/themes/icon.css" rel="stylesheet" type="text/css" />
<link href="EasyUI/demo.css" rel="stylesheet" type="text/css" />
<script src="EasyUI/jquery.min.js" type="text/javascript"></script>
<script src="EasyUI/jquery.easyui.min.js" type="text/javascript"></script>
<script src="EasyUI/easyui-lang-zh_CN.js" type="text/javascript"></script>
</head>
<body>
    <script type="text/javascript">
        function clearForm() {
            $('#adminLoginForm').form('clear');
        }
        function checkAdminLogin() {
            $("#adminLoginForm").form("submit", {
                url : 'admin/login',
                success : function(result) {
                    var result = eval('(' + result + ')');
                    if (result.success == 'true') {
                        window.location.href = 'index.jsp';
                        $("#adminLoginDlg").dialog("close");
                    } else {
                        $.messager.show({
                            title : "提示信息",
                            msg : result.message
                        });
                    }
                }
            });
        }
    </script>
    <div id="adminLoginDlg" class="easyui-dialog"
        style="top: 150;left: 550;width: 250;height: 200"
        data-options="title:'后台登录',buttons:'#bb',modal:true">
        <form id="adminLoginForm" method="post">
            <table>
                <tr>
                    <td>用户名</td>
                    <td><input class="easyui-textbox" type="text"
id="loginName" name="loginName" value="admin" data-
options="required:true"></input></td>
                </tr>
                <tr>
                    <td>密码</td>
                    <td><input class="easyui-textbox" type="text"
id="loginPwd"   name="loginPwd" value="123456" data-
options="required:true"></input></td>
                </tr>
            </table>
        </form>
    </div>
    <div id="bb">
        <a href="javascript:void(0)" class="easyui-linkbutton"
            onclick="checkAdminLogin()">登录</a> <a href="javascript:void(0)"
```

```
        class="easyui-linkbutton" onclick="clearForm();">重置</a>
    </div>
</body>
</html>
```

为了使用 Easy UI 框架，在页面开始部分的<head></head>标签中，需要引入 Easy UI 的相关 css 和 js 文件。id 为 adminLoginDlg 的<div>标签使用 Easy UI 对话框控件 Dialog 定义了一个对话框，对话框中包含一个 id 为 adminLoginForm 的登录表单，表单中使用 Easy UI 文本框控件 TextBox 定义了用户名和密码两个文本域；id 为 bb 的<div>标签中定义了登录和重置两个按钮。

在后台登录表单中输入用户名和密码，单击"登录"按钮，执行 JavaScript 函数 checkAdminLogin()，函数中使用 jQuery 将请求提交到 admin/login，即执行 com.res.controller 包中的控制类 AdminController 中的 login 方法，代码如下：

```
package com.res.controller;
……
@SessionAttributes(value = { "admininfo" })
@Controller
@RequestMapping("/admin")
public class AdminController {
    // 使用@Autowired 注解注入 AdminServiceImpl 实例
    @Autowired
    AdminService adminService;
    // 后台登录验证
    @RequestMapping(value = "/login", produces = "text/html;charset=UTF-8")
    @ResponseBody
    public String login(Admin admin, ModelMap model) {
        List<Admin> adminList = adminService.adminLogin(admin);
        if (adminList.size() > 0) {
            // 验证通过后，再判断是否已为该管理员分配功能权限
            if (adminList.get(0).getFs().size() > 0) {
                // 验证通过且已分配功能权限，则将 admininfo 对象存入 model 中
                model.put("admininfo", adminList.get(0));
                // 以 JSON 格式向页面发送成功信息
                return "{\"success\":\"true\",\"message\":\"登录成功\"}";
            } else {
                return "{\"success\":\"false\",\"message\":\"您没有权限，请联系超级管理员设置权限！\"}";
            }
        } else {
            // 登录失败，重定向到 login.jsp
            return "{\"success\":\"false\",\"message\":\"登录失败\"}";
        }
    }
}
```

login 方法包含两个参数，Admin 类型的参数 admin 用于接收后台登录表单传递来的用户名和密码，ModelMap 类型的参数 model 存放登录成功后的管理员对象信息。

在 login 方法中，首先调用业务接口 adminService 中的 adminLogin 方法进行登录验证；在验证通过后，再判断是否已为该管理员分配功能权限，如果没有，则以 JSON 格式返回

"您没有权限，请联系超级管理员设置权限！"的提示信息。只有验证通过且分配了权限的管理员登录，才返回"登录成功"的提示，并将管理员对象以 admininfo 为名称存入 ModelMap 类型的参数 model 中，并通过在 AdminController 类名上修饰的@SessionAttributes 注解将其存入 Session 范围。

由于 login 方法使用了@ResponseBody 注解修饰，通过 return 语句返回的 JSON 格式的字符串将发送回前端页面中的 JavaScript 函数 checkAdminLogin()，函数再判断返回的 JSON 格式字符串中 success 所对应的值是否等于 true，等于 true 表示登录成功，则打开后台管理首页面 index.jsp，并关闭后台登录表单对话框；否则通过消息框给出错误提示。后台管理首页面 index.jsp 如图 23-5 所示。

图 23-5　后台管理首页面 index.jsp

index.jsp 页面主要代码如下：

```jsp
<%@ page language="java" import="java.util.*" pageEncoding="UTF-8"%>
<%
    if (session.getAttribute("admininfo") == null)
        response.sendRedirect("/restaurant-back/admin_login.jsp");
%>
<html>
<head>
<title>后台管理首页面</title>
<link href="EasyUI/themes/default/easyui.css" rel="stylesheet"
    type="text/css" />
<link href="EasyUI/themes/icon.css" rel="stylesheet" type="text/css" />
<link href="EasyUI/demo.css" rel="stylesheet" type="text/css" />
<script src="EasyUI/jquery.min.js" type="text/javascript"></script>
<script src="EasyUI/jquery.easyui.min.js"
 type="text/javascript"></script>
<script src="EasyUI/easyui-lang-zh_CN.js"
 type="text/javascript"></script>
</head>
```

```jsp
<body class="easyui-layout">
    <div data-options="region:'north',border:false"
        style="height:60px;background:#B3DFDA;padding:10px">
        <div align="left">
            <font style="font-family: cursive;;font-size: 20">网上订餐系统后台管理</font>
        </div>
        <div align="right">欢迎您,${sessionScope.admininfo.loginName}</div>
    </div>
    <div data-options="region:'west',split:true,title:'功能菜单'"
        style="width:200px;padding:10px;">
        <!-- 定义tree -->
        <ul id="tt"></ul>
    </div>
    <div data-options="region:'south',border:false"
        style="height:50px;background:#A9FACD;padding:10px;"
align="center">Powered By MiaoYong</div>
    <div data-options="region:'center',title:'主界面'">
        <div id="tabs" data-options="fit:true" class="easyui-tabs"
            style="width:500px;height:250px"></div>
    </div>
    <script type="text/javascript">
        // 为tree指定数据
        $('#tt').tree({
            url : 'admin/getTree?adminid=${sessionScope.admininfo.id}'
        });
        $('#tt').tree({
            onClick : function(node) {
                if ("餐品列表" == node.text) {
                    if ($('#tabs').tabs('exists', '餐品列表')) {
                        $('#tabs').tabs('select', '餐品列表');
                    } else {
                        $('#tabs').tabs('add', {
                            title : node.text,
                            href : 'meallist.jsp',
                            closable : true
                        });
                    }
                } else if ("查询订单" == node.text) {
                    if ($('#tabs').tabs('exists', '查询订单')) {
                        $('#tabs').tabs('select', '查询订单');
                    } else {
                        $('#tabs').tabs('add', {
                            title : node.text,
                            href : 'searchorder.jsp',
                            closable : true
                        });
                    }
                } else if ("创建订单" == node.text) {
                    if ($('#tabs').tabs('exists', '创建订单')) {
                        $('#tabs').tabs('select', '创建订单');
                    } else {
                        $('#tabs').tabs('add', {
```

```javascript
                    title : node.text,
                    href : 'createorder.jsp',
                    closable : true
                });
            }
        } else if ("订单统计" == node.text) {
            if ($('#tabs').tabs('exists', '订单统计')) {
                $('#tabs').tabs('select', '订单统计');
            } else {
                $('#tabs').tabs('add', {
                    title : node.text,
                    href : 'saler.jsp',
                    closable : true
                });
            }
        } else if ("用户列表" == node.text) {
            if ($('#tabs').tabs('exists', '用户列表')) {
                $('#tabs').tabs('select', '用户列表');
            } else {
                $('#tabs').tabs('add', {
                    title : node.text,
                    href : 'userlist.jsp',
                    closable : true
                });
            }
        } else if ("管理员列表" == node.text) {
            if ($('#tabs').tabs('exists', '管理员列表')) {
                $('#tabs').tabs('select', '管理员列表');
            } else {
                $('#tabs').tabs('add', {
                    title : node.text,
                    href : 'adminlist.jsp',
                    closable : true
                });
            }
        } else if ("退出系统" == node.text) {
            $.ajax({
                url : 'admin/loginout',
                success : function(data) {
                    window.location.href = "admin_login.jsp";
                }
            })
        }
    }
    });
    </script>
</body>
</html>
```

为了使用 Easy UI 框架，需要在页面开始部分的<head></head>标签中引入 Easy UI 的相关 css 和 js 文件。<body></body>部分通过 class 属性来使用 Easy UI 的 Layout 控件，用于生成页面的布局。Layout 控件的左侧定义了 id 为 tt 的<ul>标签，再使用 JavaScript 将其包装成 Easy UI 的 Tree 控件，并给 Tree 控件指定数据源，用来显示系统功能菜单。Tree 控件的数据加载通过其 url 属性值 "admin/getTree?adminid=${sessionScope.admininfo.id}" 完成，即调用 AdminController 类的 getTree 方法，代码如下：

```java
@RequestMapping("/getTree")
@ResponseBody
public List<TreeNode> getTree(
        @RequestParam(value = "adminid") String adminid) {
    // 根据管理员 id 号获取 Admin 对象及关联的 Functions 对象集合
    Admin admin = adminService.getAdminFunctions(Integer.parseInt(adminid));
    List<TreeNode> nodes = new ArrayList<TreeNode>();
    List<Functions> functionsList = new ArrayList<Functions>();
    // 获取 Admin 对象关联的 Functions 对象集合
    Set<Functions> functionsSet = admin.getFs();
    //将 Functions 对象集合类型从 Set<Functions>转换为 List<Functions>类型
    for (Functions functions : functionsSet) {
        functionsList.add(functions);
    }
    // 对 List<Functions>类型的 Functions 对象集合排序
    Collections.sort(functionsList);
    // 将排序后的 Functions 对象集合转换到 List<TreeNode>类型的列表 nodes 中
    for (Functions f : functionsList) {
        TreeNode treeNode = new TreeNode();
        treeNode.setId(f.getId());
        treeNode.setFid(f.getParentid());
        treeNode.setText(f.getName());
        nodes.add(treeNode);
    }
    // 调用自定义工具类 JsonFactory 的 buildtree 方法，为 nodes 列表中各
    // TreeNode 元素中的 children 赋值(即该节点包含的子节点)
    List<TreeNode> treeNodes = JsonFactory.buildtree(nodes, 0);
    // 以 JSON 格式向页面返回绑定 tree 所需的数据
    return treeNodes;
}
```

getTree 方法的参数为管理员的 id 号，首先调用业务接口 AdminService 中的 getAdminFunctions 方法，根据管理员 id 号获取 Admin 对象及关联的 Functions 对象集合；然后获取 Admin 对象关联的 Functions 对象集合，由于该集合类型为 HashSet，其元素没有根据 Functions 对象的 id 进行排序，因此将其转换为 List<Functions>类型，再使用 Collections 类的 sort 方法对其排序，从而保证绑定 Easy UI 的 Tree 控件时顺序一致；接着将排序后的 Functions 对象集合转换到 List<TreeNode>类型的列表 nodes 中；最后调用自定义工具类 JsonFactory 的 buildtree 方法，为 nodes 列表中各 TreeNode 元素中的 children 赋值，即该节点包含的子节点。TreeNode 是用于描述菜单树的每个节点的实体类，代码如下：

```java
package com.res.entity;
import java.util.List;
// 描述菜单树的每个节点的实体类
public class TreeNode {
    private int id;
    private String text;
    private int fid; // 节点的父 id
    private List<TreeNode> children; // 节点的孩子节点
    // 此处省略上述属性的 get 和 set 方法
}
```

JsonFactory 类位于 com.res.util 包中，buildtree 方法代码如下：

```java
package com.res.util;
import java.util.ArrayList;
import java.util.List;
import com.res.entity.TreeNode;
public class JsonFactory {
    public static List<TreeNode> buildtree(List<TreeNode> nodes, int id) {
        List<TreeNode> treeNodes = new ArrayList<TreeNode>();
        for (TreeNode treeNode : nodes) {
            TreeNode node = new TreeNode();
            node.setId(treeNode.getId());
            node.setText(treeNode.getText());
            if (id == treeNode.getFid()) {
                //递归调用 buildtree 方法给 TreeNode 中的 children 属性赋值
                node.setChildren(buildtree(nodes, node.getId()));
                treeNodes.add(node);
            }
        }
        return treeNodes;
    }
}
```

为了给 nodes 列表的 TreeNode 元素中的 children 属性赋值，递归调用了 buildtree 方法。

执行完 AdminController 类的 getTree 方法后，List<TreeNode>类型的 treeNodes 列表被 getTree 方法上修饰的@ResponseBody 注解自动转换成 JSON 格式数据返回前端页面，从而实现对 Tree 控件的数据绑定。

在 index.jsp 页面中，还为 Tree 控制添加了 onClick 事件。单击"餐品列表"节点时，如果"餐品列表"标签页不存在，则添加一个标签页，用于显示餐品列表页 meallist.jsp，否则选中该标签页。单击"查询订单""创建订单""订单统计""用户列表""管理员列表"节点时，效果与"餐品列表"节点类似。

单击"退出系统"节点时，通过 Ajax 方式发送请求，请求地址为 admin/loginout，即执行 AdminController 类的 loginout 方法，代码如下：

```java
@RequestMapping(value = "/loginout", method = RequestMethod.GET)
@ResponseBody
public String loginout(SessionStatus status) {
    // @SessionAttributes 清除
    status.setComplete();
    return "{\"success\":\"true\",\"message\":\"注销成功\"}";
}
```

退出系统后，页面跳转到 admin_login.jsp。

23.9　餐品管理

餐品管理主要功能包括餐品列表显示、查询餐品、添加餐品、餐品下架、修改餐品。

23.9.1 餐品列表显示

在 index.jsp 页面中,单击 Tree 控件上的"餐品列表"节点,打开餐品列表页 meallist.jsp,如图 23-6 所示。

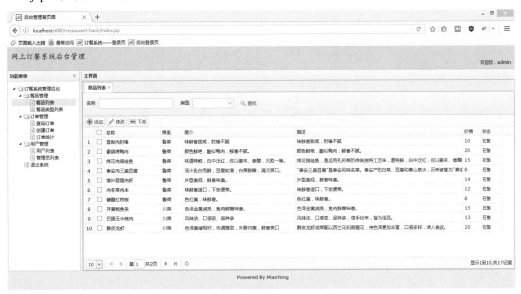

图 23-6 餐品列表页 meallist.jsp

在页面 meallist.jsp 的<body></body>部分,首先定义了一个 id 为 meal_dg 的 table,代码如下:

```
<table id="meal_dg" class="easyui-datagrid"></table>
```

该 table 标签用于创建 Easy UI 的 DataGrid 控件,用来显示餐品记录。

然后,创建 DataGrid 控件的工具栏,代码如下:

```
<div id="meal_tb" style="padding: 2px 5px;">
<a href="javascript:void(0)" iconCls="icon-add"
    class="easyui-linkbutton" onclick="openAddMealDlg()">添加</a>
<a  href="javascript:void(0)" iconCls="icon-edit"
    class="easyui-linkbutton" onclick="openUpdateMealDlg()">修改</a>
<a  href="javascript:void(0)" iconCls="icon-remove"
    class="easyui-linkbutton" onclick="soldOut()">下架</a>
</div>
```

接着,创建搜索栏,代码如下:

```
<div id="meal_searchtb">
    <form id="meal_searchForm">
        <div style="padding: 10px  5px  1px    5px;">
        名称  <input class="easyui-textbox" name="mealName"
id="mealName" style="width: 200px" />    类型
  <input class="easyui-combobox" name="mealseries.seriesId"
id="mealseries.seriesId"    style="width: 110px"         data-
```

```
options="valueField:'seriesId',textField:'seriesName',
url:'mealSeries/list'" />     <a href="javascript:void(0)"
iconCls="icon-search" plain="true" class="easyui-linkbutton"
onclick="searchMeal();">查找</a>
        </div>
    </form>
</div>
```

使用 JavaScript 将 id 为 meal_dg 的 table 创建为 DataGrid 控件的代码如下:

```
<script type="text/javascript">
    $(function() {
        $('#meal_dg').datagrid({
            singleSelect : false,
            url : 'meal/list', //为datagrid控件设置数据源
            queryParams : {}, //查询条件
            pagination : true, // 启用分页
            pageSize : 10, // 初始每页记录数
            pageList : [ 10, 15, 20 ], // 设置可供选择的页大小
            rownumbers : true, //显示行号
            fit : true, //设置自适应
            toolbar : '#meal_tb', //给datagrid添加工具栏
            header : '#meal_searchtb', //给datagrid添加搜索工具栏
            columns : [ [ {
                title : '序号',
                field : 'mealId',
                align : 'center',
                checkbox : true
            }, {
                title : '名称',
                field : 'mealName',
                width : 150
            }, {
                title : '菜系',
                field : 'mealseries',
                width : 60,
                formatter : function(value, row, index) {
                    if (row.mealseries) {
                        return row.mealseries.seriesName;
                    } else {
                        return value;
                    }
                }
            }, {
                title : '简介',
                field : 'mealSummarize',
                width : 300
            }, {
                title : '描述',
                field : 'mealDescription',
                width : 450
            }, {
                title : '价格',
                field : 'mealPrice',
                width : 50
```

```
            }, {
                title : '状态',
                field : 'mealStatus',
                width : 60,
                formatter : function(value, row, index) {
                    if (row.mealStatus == 1) {
                        return "在售";
                    } else {
                        return "下架";
                    }
                }
            } ] ]
        });
    });
    ……
</script>
```

将 DataGrid 控件的 singleSelect 属性设置为 false 以允许多选，pagination 属性设置为 true 以允许分页，pageSize 属性设置初始每页记录数(即页大小)，pageList 设置可供选择的页大小，rownumbers 属性设置为 true 以显示行号，fit 属性设置为 true 以自适应显示数据，toolbar 属性设置为#meal_tb 为 DataGrid 控件添加工具栏，header 属性设置为#meal_searchtb 为 DataGrid 标头添加搜索栏，设置 columns 属性以指定 DataGrid 显示的列，DataGrid 控件数据源通过 url 属性来指定，此处设置为 meal/list，即将请求发送到 MealController 类的 list 方法，代码如下：

```
package com.res.controller;
……
import com.res.service.MealService;
@Controller
@RequestMapping("/meal")
public class MealController {
    // 使用@Autowired注解注入 MealServiceImpl 实例
    @Autowired
    MealService mealService;
    // 显示餐品列表
    @ResponseBody
    @RequestMapping("/list")
    public Map<String, Object> list(Meal meal, int page, int rows) {
        Map<String, Object> result = new HashMap<String, Object>(2);
        // 根据查询条件获取餐品记录总数
        int totalCount = mealService.getTotalCount(meal);
        // 根据当前页码、每页记录数、查询条件获取当前页的餐品列表
        List<Meal> mealList = mealService.getMealByConditionForPager(meal,
            page, rows);
        result.put("total", totalCount);
        result.put("rows", mealList);
        return result;
    }
    ……
}
```

list 方法的参数 meal 用于封装表单参数，参数 page 和 rows 用于接收从 DataGrid 控件传递来的页码和每页显示记录数(可选)。在 list 方法中，首先调用业务接口 MealService 的

getTotalCount 方法，根据查询条件获取餐品记录总数。然后调用 getMealByConditionForPager 方法根据当前页码、每页记录数、查询条件获取当前页的餐品列表。再将餐品列表转为 JSON 格式，最后将 JSON 格式字符串发送到前端页面 meallist.jsp，作为 DataGrid 控件的数据源。

23.9.2 查询餐品

在餐品列表页 meallist.jsp 的搜索栏中，输入餐品名称、餐品类型(菜系)，单击"查找"按钮，将执行 JavaScript 函数 searchMeal()，代码如下：

```javascript
function searchMeal() {
    $('#meal_dg').datagrid('load',
        convertArray($('#meal_searchForm').serializeArray()));
}
```

在函数 searchMeal()中，通过执行 DataGrid 控件的 load 方法，再次将请求发送到 meal/list，即再次执行 MealController 类的 list 方法，并将搜索栏中输入的查询条件传递过去。搜索栏表单中的参数通过 serializeArray 方法编译成拥有 name 和 value 对象组成的数组，再通过调用自定义的 convertArray 方法将序列化后的值转为 "name:value" 的形式进行传递。serializeArray 方法通过序列化表单值来创建对象数组，即将表单元素的值编译成由 name 和 value 对象组成的数组，形如 "[{name: 'firstname', value: 'Hello'}, {name: 'lastname', value: 'World'}…]"。再通过调用自定义的函数 convertArray，进一步将序列化后的值转化为 "name:value" 的形式。自定义方法 convertArray 代码如下：

```javascript
function convertArray(o) {
    var v = {};
    for ( var i in o) {
        if (typeof (v[o[i].name]) == 'undefined')
            v[o[i].name] = o[i].value;
        else
            v[o[i].name] += "," + o[i].value;
    }
    return v;
}
```

MealController 类的 list 方法根据传递来的查询参数重新获取餐品列表以更新 DataGrid 控件的数据源。

23.9.3 添加餐品

在餐品列表页 meallist.jsp 的工具栏中，单击"添加"按钮，执行 JavaScript 函数 openAddMealDlg()，打开新增餐品对话框，代码如下：

```javascript
var urls;
// 打开新增餐品对话框
function openAddMealDlg() {
    $('#meal_dlg').dialog('open').dialog('setTitle', '新增餐品');
    $('#meal_ff').form('clear');
    urls = 'meal/addMeal';  //将请求提交到Action中的方法addMeal
}
```

"新增餐品"对话框通过 id 为 meal_dlg 的 Easy UI Dialog 控件创建，对话框中包含 id 为 meal_ff 的 form 表单，效果如图 23-7 所示。

图 23-7 "新增餐品"对话框

输入餐品信息，单击"保存"按钮，执行 JavaScript 函数 saveMeal()，代码如下：

```
// 保存餐品
function saveMeal() {
    $('#meal_ff').form("submit", {
        url : urls,
        success : function(result) {
            // eval 方法将 JSON 字符串转成 JSON 对象
            var result = eval('(' + result + ')');
            if (result.success == 'true') {
                $('#meal_dg').datagrid("reload");
                $('#meal_dlg').dialog("close");
            }
            $.messager.show({
                title : "提示信息",
                msg : result.message
            });
        }
    });
}
```

在函数 saveMeal()中，通过"$("#meal_ff ").form("submit", {});"将表单提交到 url 属性指定的 urls(meal/addMeal)，即执行 MealController 类的 addMeal 方法，代码如下：

```
// 添加餐品
@RequestMapping(value = "/addMeal", produces = "text/html;charset=UTF-8")
@ResponseBody
public String addMeal(Meal meal,  @RequestParam(value = "pic",
required = false) MultipartFile pic,  HttpServletRequest request) {
    // 服务器端 upload 文件夹物理路径
    String path = request.getSession().getServletContext()
            .getRealPath("mealimages");
    // 获取文件名
    String fileName = pic.getOriginalFilename();
    // 实例化一个 File 对象，表示目标文件(含物理路径)
    File targetFile = new File(path, fileName);
```

```
    if (!targetFile.exists()) {
        targetFile.mkdirs();
    }
    try {
        // 将上传文件写到服务器上指定的文件
        pic.transferTo(targetFile);
    } catch (Exception e) {
        e.printStackTrace();
    }
    meal.setMealImage(fileName);
    int result = mealService.addMeal(meal);
    String str = "";
    if (result > 0) {
        str = "{\"success\":\"true\",\"message\":\"添加成功!\"}";
    } else {
        str = "{\"success\":\"false\",\"message\":\"添加失败!\"}";
    }
    return str;
}
```

addMeal 方法 Meal 类型的参数 meal 用于封装"新增餐品"对话框中输入的餐品信息。在 addProduct 方法中，MultipartFile 类型的参数 pic 用于接收"新增餐品"对话框中上传的图片文件。在 addMeal 方法中，除了上传餐品的图片外，还在对象 meal 中设置了图片的路径信息。最后调用业务接口 MealService 的 addMeal 方法，将新增餐品添加到数据表 meal 中。如果执行成功，则向前端页面发送"添加成功"信息，否则发送"添加失败"信息。

23.9.4 餐品下架

在餐品列表页 meallist.jsp 的 DataGrid 控件中，选中某条记录前的复选框，然后单击工具栏中的"下架"按钮，将执行 JavaScript 函数 soldOut ()，代码如下：

```javascript
function soldOut() {
    // 获取 datagrid 控件中选中的行
    var rows = $('#meal_dg').datagrid('getSelections');
    if (rows.length > 0) {
        $.messager.confirm('Confirm', '确认要下架么?', function(r) {
            if (r) {
                var ids = "";
                for (var i = 0; i < rows.length; i++) {
                    ids += rows[i].mealId + ",";
                }
                $.post('meal/modifyMealStatus', {
                    id : ids
                }, function(result) {
                    if (result.success == 'true') {
                        // 重新加载 datagrid
                        $('#meal_dg').datagrid('reload');
                        $.messager.show({
                            title : "提示信息",
                            msg : result.message
                        });
                    } else {
```

```
                    $.messager.show({
                         title : "提示信息",
                         msg : result.message
                    });
                }
            }, 'json');
        }
    });
} else {
    $.messager.alert('提示', '请选择要下架的餐品', 'info');
}
}
```

在函数 soldOut 中，首先获取 DataGrid 控件上选中的餐品记录，然后将选中的餐品编号保存到变量 ids 中，然后使用$.post 向 MealController 类的 modifyMealStatus 方法发送请求，并将参数 id 传递过去。modifyMealStatus 方法代码如下：

```
@RequestMapping(value = "/modifyMealStatus", produces =
"text/html;charset=UTF-8")
@ResponseBody
public String modifyMealStatus(String id) {
   String ids = "(" + id.substring(0, id.length() - 1) + ")";
   int result = mealService.updateMealStatus(ids);
   String str = "";
   if (result > 0) {
      str = "{\"success\":\"true\",\"message\":\"下架成功!\"}";
   } else {
      str = "{\"success\":\"false\",\"message\":\"下架失败!\"}";
   }
   return str;
}
```

在 modifyMealStatus 方法中，调用业务接口 MealService 的 updateMealStatus 方法，根据选中餐品的 id 号，将其状态设置为 0。如果执行成功，向前端页面发送"下架成功"信息，否则发送"下架失败"信息。前端页面判断返回的结果，如果成功，则通过调用 DataGrid 的 reload 方法，重新向 MealController 类的 list 方法发送请求，以更新数据源。

23.9.5 修改餐品

在餐品列表页 meallist.jsp 的 DataGrid 控件中，选中某条记录前的复选框，然后单击工具栏中的"修改"按钮，将执行 JavaScript 函数 openUpdateMealDlg()，代码如下：

```
// 打开餐品修改对话框
function openUpdateMealDlg() {
   var row = $('#meal_dg').datagrid('getSelected');
   if (row != null) {
      // 打开修改餐品对话框(即添加餐品对话框)
      $('#meal_dlg').dialog('open').dialog('setTitle', '修改餐品');
      // 给对话框中的 Form 表单中的文本域绑定值
      $('#meal_ff').form("load", {
         "mealStatus" : row.mealStatus,
         "mealseries.seriesId" : row.mealseries.seriesId,
```

```
            "mealName" : row.mealName,
            "mealSummarize" : row.mealSummarize,
            "mealDescription" : row.mealDescription,
            "mealPrice" : row.mealPrice
        });
        urls = "meal/updateMeal?mealId=" + row.mealId;
    } else {
        $.messager.alert('提示', '请选择要修改的餐品', 'info');
    }
}
```

在函数 openUpdateMealDlg 中，首先获取 DataGrid 控件中选中的行，然后打开"修改餐品"对话框(与添加餐品使用同一个对话框和表单)，并将要修改的餐品信息绑定到对话框中的表单文本域中，修改完餐品信息后，单击对话框中的"保存"按钮时，会向 MealController 类 updateMeal 方法发送请求，并将参数 mealId(要修改餐品编号)传递过去。updateMeal 方法代码如下：

```
@RequestMapping(value = "/updateMeal", produces = "text/html;charset=UTF-8")
@ResponseBody
public String updateMeal(Meal meal, int mealId,
        @RequestParam(value = "pic", required = false) MultipartFile pic,
        HttpServletRequest request) {
    // 获取修改前的餐品对象
    Meal editMeal = mealService.getMealByMealId(mealId);
    // 修改餐品对象信息
    editMeal.setMealStatus(meal.getMealStatus());
    editMeal.setMealseries(meal.getMealseries());
    editMeal.setMealName(meal.getMealName());
    editMeal.setMealSummarize(meal.getMealSummarize());
    editMeal.setMealDescription(meal.getMealDescription());
    editMeal.setMealPrice(meal.getMealPrice());
    String str = "";
    String fileName = "";
    if (pic.getSize() > 0) { // 重新上传餐品图片
        // 服务器端 upload 文件夹物理路径
        String path = request.getSession().getServletContext()
                .getRealPath("mealimages");
        // 获取文件名
        fileName = pic.getOriginalFilename();
        // 实例化一个 File 对象，表示目标文件(含物理路径)
        File targetFile = new File(path, fileName);
        if (!targetFile.exists()) {
            targetFile.mkdirs();
        }
        try {
            // 将上传文件写到服务器上指定的文件
            pic.transferTo(targetFile);
            editMeal.setMealImage(fileName);
        } catch (Exception e) {
            e.printStackTrace();
            return "{\"success\":\"false\",\"message\":\"图片上传失败!\"}";
        }
    }
}
```

```
    try {
        mealService.updateMeal(editMeal);
    } catch (Exception e) {
        e.printStackTrace();
        return "{\"success\":\"true\",\"message\":\"修改失败!\"}";
    }
    return "{\"success\":\"true\",\"message\":\"修改成功!\"}";
}
```

updateMeal 方法中的 Meal 类型的参数 meal 用于封装"修改餐品"对话框表单中的餐品信息。在 updateMeal 方法中,首先调用业务接口 MealService 的 getMealByMealId 方法获取要修改餐品对象 editMeal,然后使用对象 meal 修改 editMeal 对象中的属性值,最后调用接口 MealService 的 updateMeal 方法将对象 editMeal 中的属性值更新到数据表中。

餐品类型(菜系)管理与餐品管理实现过程类似,由于篇幅有限,在此不再赘述。

23.10 订单管理

订单管理主要功能包括创建订单、查询订单、删除订单、修改订单与查看订单明细。

23.10.1 创建订单

在 index.jsp 页面中,单击 Tree 控件上的"创建订单"节点,打开创建订单页 createorder.jsp,如图 23-8 所示。

图 23-8 创建订单页 createorder.jsp

页面 createorder.jsp 的<body></body>部分,首先定义 id 为 odbox 的 table,代码如下:

```
<table id="odbox" class="easyui-datagrid"></table>
```

该 table 标签用于创建 Easy UI 的 DataGrid 控件，用来输入订单明细信息。

然后，创建 DataGrid 控件的工具栏，代码如下：

```
<div id="ordertb" style="padding: 2px 5px;">
    <a href="javascript:void(0)" iconCls="icon-add"
class="easyui-linkbutton" onclick="addOrderDts()">添加订单明细</a>
    <a href="javascript:void(0)" iconCls="icon-edit"
class="easyui-linkbutton" onclick="saveorder()">保存订单</a>
    <a   href="javascript:void(0)" iconCls="icon-remove"
class="easyui-linkbutton" onclick="removeOrderDts()">删除订单明细</a>
</div>
```

接着，创建订单主表输入布局，代码如下：

```
<div id="divOrder">
    <div style="padding:3px">
        客户名称 <input style="width:115px;" id="create_userid"
            class="easyui-combobox" name="create_userid" value="0"
data-options="valueField:'id',textField:'loginName',
url:'user/getValidUser'">   
        订单金额 <input type="text" name="create_orderprice"
            id="create_orderprice" class="easyui-textbox" readonly="readonly"
style="width:115px" />   
    </div>
    <div style="padding:3px">
        订单日期 <input type="text" name="create_ordertime"
            id="create_ordertime" value="<%=new Date() %>" class="easyui-
datebox" style="width:115px" />
           订单状态 <select id="create_orderstatus"
            class="easyui-combobox" name="create_orderstatus"
style="width:115px;">
            <option value="未付款" selected>未付款</option>
            <option value="已付款">已付款</option>
            <option value="待发货">待发货</option>
            <option value="已发货">已发货</option>
            <option value="已完成">已完成</option>
        </select>
    </div>
</div>
```

其中，客户名称下拉列表绑定的数据源来自控制器类 UserController 中的 getValidUser 方法执行后返回的 JSON 格式的结果。

再使用"$(function(){....});"，将 id 为 odbox 的 table 创建为 Easy UI 的 DataGrid 控件，代码如下：

```
<script type="text/javascript">
var $odbox = $('#odbox');
$(function() {
    $odbox.datagrid({
        rownumbers : true,
        singleSelect : false,
        fit : true,
        toolbar : '#ordertb',
        header : '#divOrder',
```

```
        columns : [ [ {
            title : '序号',
            field : '',
            align : 'center',
            checkbox : true
        }, {
            field : 'mealId',
            title : '餐品名',
            width : 300,
            editor : {
                type : 'combobox',
                options : {
                    valueField : 'mealId',
                    textField : 'mealName',
                    url : 'meal/getOnSaleMeal',
                    onChange : function(newValue, oldValue) {
                    // 当选择了不同的餐品，更新当前行的餐品价格、小计、订单金额
                        var rows = $odbox.datagrid('getRows');
                        var orderprice = 0;
                        for (var i = 0; i < rows.length; i++) {
                //获取餐品下拉列表框编辑器,getEditor方法用于获取指定编辑器
                            var mealIdEd=$odbox.datagrid('getEditor', {
                                index : i,
                                field : 'mealId'
                            });
                            // 获取单价文本框编辑器
                            var priceEd=$odbox.datagrid('getEditor', {
                                index : i,
                                field : 'mealPrice'
                            });
                            // 获取数量文本框编辑器
                            var mealCountEd=
$odbox.datagrid('getEditor', {
                                index : i,
                                field : 'mealCount'
                            });
                            // 获取小计文本框编辑器
                            var totalpriceEd=
$odbox.datagrid('getEditor', {
                                index : i,
                                field : 'totalprice'
                            });
                            if(mealIdEd!=null){
                                // 获取选择餐品的id
                                var mealId=
$(mealIdEd.target).combobox('getValue');
// 采用Ajax向MealController类的getMealPriceByMealId方法发送请求,
// 根据餐品id获取餐品价格
                                $.ajax({
                                    type: 'POST',
                                    url:'meal/getMealPriceByMealId',
                                    data:{mealId:mealId},
                                    success: function(result){
                            // 根据获取的价格，设置单价数字框编辑器的值
```

```
                    $(priceEd.target).numberbox('setValue',result);
                                 // 设置小计数字框编辑器的值
$(totalpriceEd.target).numberbox('setValue',
result*$(mealCountEd.target).numberbox('getValue'));
                                   // 设置 orderprice 的值
                                   orderprice=Number(orderprice)+
Number($(totalpriceEd.target).numberbox('getValue'));
                                },
                                dataType: 'json',
                                async: false
                            });
                        }
// 循环结束后，给订单主表部分的名称为 create_orderprice 的文本域设置值
$('#create_orderprice').textbox('setValue',orderprice);
                    } // end of onChange
                 }
             }
        },{
            field: 'mealPrice',
            title: '单价',
            width: 80,
            editor: {
                type:"numberbox",
                options: {
                    editable: false
                }
            }
        },{
            field: 'mealCount',
            title: '数量',
            width: 50,
            editor: {
                type:"numberbox",
                options: {
                    onChange: function(newValue, oldValue) {
                        var rows = $odbox.datagrid('getRows');
                        var orderprice = 0;
                        for (var i = 0; i < rows.length; i++) {
                            // 获取单价文本框编辑器
                            var priceEd=$odbox.datagrid('getEditor', {
                                index : i,
                                field : 'mealPrice'
                            });
                            // 获取数量文本框编辑器
                            var mealCountEd=$odbox.datagrid('getEditor', {
                                index : i,
                                field : 'mealCount'
                            });
                            // 获取小计文本框编辑器
                            var totalpriceEd=$odbox.datagrid('getEditor', {
                                index : i,
                                field : 'totalprice'
                            });
```

```
                        // 设置小计数字框编辑器的值
$(totalpriceEd.target).numberbox('setValue',$(priceEd.target).
numberbox('getValue')*$(mealCountEd.target).numberbox('getValue'));
                        // 设置 orderprice 的值
orderprice=Number(orderprice)+Number($(totalpriceEd.target).
numberbox('getValue'));
                    }
// 循环结束后，给订单主表部分的名称为 create_orderprice 的文本域设置值
$('#create_orderprice').textbox('setValue',orderprice);
                } // end of onChange
            }
        }
    },{
        field: 'totalprice',
        title: '小计',
        width: 100,
        editor: {
            type:"numberbox",
            options: {
                editable: false
            }
        }
    } ] ]
});
});
……
</script>
```

在$odbox.datagrid 中，设置 rownumbers 为 true 以显示行号，设置 singleSelect 属性为 false 以允许多选，fit 属性设置为 true 以自适应显示数据，toolbar 属性设置为#ordertb 为 DataGrid 添加工具栏，header 属性设置为#divOrder 为 DataGrid 标头添加输入订单主表数据部分，设置 columns 属性以指定 DataGrid 控件显示的列。其中，在"餐品名"列中使用了 Easy UI 的 ComboBox 控件，通过 url 属性指定其绑定的数据源为 meal/getOnSaleMeal，即执行控制器类 MealController 的 getOnSaleMeal 方法，代码如下：

```
@ResponseBody
@RequestMapping("/getOnSaleMeal")
public List<Meal> getOnSaleProduct() {
    List<Meal> onSaleMealList = mealService.getOnSaleMeal();
    return onSaleMealList;
}
```

在 getOnSaleMeal 方法中，调用业务接口 MealService 的 getOnSaleMeal 方法获取在售餐品列表。并通过@ResponseBody 注解，将 List<Meal>类型的返回值 onSaleMealList 自动转换为 JSON 格式，再发送到前端页面。

在显示"餐品名"的 ComboBox 控件中，添加了 onChange 事件处理代码，实现根据所选择的餐品，显示餐品价格列数据，并更新"小计"列数据和"订单金额"文本域值。在显示"数量"的 Easy UI NumberBox 控件中，也添加了 onChange 事件代码，实现根据所填写的数量更新"小计"列数据和"订单金额"文本域值。

(1) 添加订单明细。

在创建订单页 createorder.jsp 中，单击"添加订单明细"按钮，Easy UI 的 DataGrid 控件上会增加一个记录行。处理"添加订单明细"按钮 onclick 事件的 JavaScript 函数 addOrderDts()如下：

```javascript
function addOrderDts(){
    $odbox.datagrid('appendRow',{
        mealCount:'1',
        totalprice:'0'
    });
    var rows=$odbox.datagrid('getRows');
//让添加的行处于可编辑状态
    $odbox.datagrid('beginEdit',rows.length-1);
}
```

在函数 addOrderDts()中，使用了 DataGrid 控件的 appendRow 方法，从 DataGrid 控件的尾部添加一个记录行，同时对 mealCount、totalprice 列进行初始化。然后使用 DataGrid 控件的 beginEdit 方法将该行设置为编辑状态。

(2) 删除订单明细。

在创建订单页 createorder.jsp 的 DataGrid 控件中，选中要删除的记录(支持多选)，单击"删除订单明细"按钮，可以删除所选择的行。处理"删除订单明细"按钮 onclick 事件的 JavaScript 函数 removeOrderDts()如下：

```javascript
function removeOrderDts(){
    var rows=$odbox.datagrid('getSelections');
    if(rows.length>0){
        // 获取订单主表部分文本框中的订单金额
        var create_orderprice=
$('#create_orderprice').textbox("getValue");
        for(var i=0;i<rows.length;i++){
            var index=$odbox.datagrid('getRowIndex',rows[i]);
            // 获取该行中的小计数字框编辑器
            var totalpriceEd = $odbox.datagrid('getEditor', {
                index : index,
                field : 'totalprice'
            });          create_orderprice=create_orderprice-
Number($(totalpriceEd.target).
numberbox('getValue'));
            $odbox.datagrid('deleteRow',index);
        }
        // 重新给订单主表部分文本框中的订单金额设置值
        $('#create_orderprice').textbox('setValue',create_orderprice);
    } else {
        $.messager.alert('提示', '请选择要删除的行', 'info');
    }
}
```

在函数 removeOrderDts()中，首先使用 DataGrid 控件的 getSelections 方法获取所选择的行记录，然后获取"订单金额"文本域的值，再遍历选中的行记录，以更新订单金额。

(3) 保存订单。

在创建订单页 createorder.jsp 中，填写订单主表数据，添加并填写订单明细记录后，单击"保存订单"按钮，执行 JavaScript 函数 saveorder()保存订单。函数 saveorder()如下：

```javascript
function saveorder(){
    // 获取订单用户
    var userid=$('#create_userid').combobox('getValue');
    if(userid==0){
        $.messager.alert('提示','请选择客户名称','info');
    }else{
        // 调用自定义方法取消 datagrid 控件的行编辑状态
        create_endEdit();
        // 定义 orderinfo 数组存放订单主表数据
        var order=[];
        // 获取订单时间
        var ordertime=$('#create_ordertime').datebox('getValue');
        // 获取订单状态
        var status=$('#create_orderstatus').combobox('getValue');
        // 向数组的末尾添加元素(订单主表数据)
        order.push({
            orderTime:ordertime,
            userId:userid,
            orderStatus:status
        });
        // 获取订单明细(datagrid 控件中的记录)
        if($odbox.datagrid('getChanges').length){
            // 获取 datagrid 控件中插入的记录行
            var inserted=$odbox.datagrid('getChanges',"inserted");
            // 获取 datagrid 控件中删除的记录行
            var deleted=$odbox.datagrid('getChanges',"deleted");
            // 获取 datagrid 控件中更新的记录行
            var updated=$odbox.datagrid('getChanges',"updated");
            // 定义 effectRow,保存 inserted 和 order
            var effectRow=new Object();
            if(inserted.length){
                // JSON.stringify 将对象数组转换成 JSON 字符串
                effectRow["inserted"]=JSON.stringify(inserted);
            }
            effectRow["order"]=JSON.stringify(order);
            // 提交请求
            $.post(
                "order/commitOrder",
                effectRow,
                function(data){
                    if(data=='success'){
                        $.messager.alert('提示','创建成功！','info');
                        // 提交 datagrid 控件的当前行
                        $odbox.datagrid('acceptChanges');
                        if($('#tabs').tabs('exists','创建订单')){
                            $('#tabs').tabs('close','创建订单');
                        }
                        // 重新加载"查询订单页"中用于显示订单记录的 datagrid 控件的数据
                        $('#orderDg').datagrid('reload');
                    }else{
```

```
                     $.messager.alert('提示','创建失败！','info');
                }
            }
        );
    }
}
```

在函数 saveorder 中，首先通过调用自定义的 JavaScript 函数 create_endEdit()取消 DataGrid 控件的行编辑状态，函数 create_endEdit 代码如下：

```javascript
function create_endEdit() {
    var rows = $odbox.datagrid('getRows');
    for (var i = 0; i < rows.length; i++) {
        $odbox.datagrid('endEdit', i);
    }
}
```

然后定义了变量 order，用于存放订单用户、订单时间、订单状态等订单主表数据，并通过 DataGrid 控件的 getChanges 方法，获取控件上发生的改变情况，包括 inserted、updated、deleted 三种类型。由于 DataGrid 控件初始时没有任何记录行，是通过单击"添加订单明细"按钮创建的，此时即使删除或者修改某个订单明细，DataGrid 控件上发生的改变都属于 inserted 类型。接下来定义 effectRow 保存 inserted 和 orderinfo，最后通过$.post 将请求提交到 order/commitOrder，即执行控制器类 OrderController 的 commitOrder 方法，代码如下：

```java
package com.res.controller;
......
import com.res.service.*;
@Controller
@RequestMapping("/order")
public class OrderController {
    @Autowired
    OrderService orderService;
    @Autowired
    MealService mealService;
    @Autowired
    UserService userService;
    // 提交订单
    @ResponseBody
    @RequestMapping(value = "/commitOrder")
    public String commitOrder(String inserted, String order)
        throws JsonParseException, JsonMappingException, IOException {
        try {
            // 创建 ObjectMapper 对象，实现 JavaBean 和 JSON 的转换
            ObjectMapper mapper = new ObjectMapper();
            // 设置输入时忽略在 JSON 字符串中存在但 JavaBean 对象实际没有的属性
            mapper.disable(DeserializationFeature.FAIL_ON_UNKNOWN_PROPERTIES);
            mapper.configure(SerializationFeature.FAIL_ON_EMPTY_BEANS, false);
            // 将 json 字符串 order 转换为 Orders 对象 o (订单主表)
            Orders o = mapper.readValue(order, Orders[].class)[0];
            // 给对象 o 设置关联的 ui 属性值
```

```java
            o.setU(userService.getUserById(o.getUserId()));
            // 将json字符串inserted转换为List<Orderdts>(订单子表或明细)
            List<Orderdts> odList = mapper.readValue(inserted,
                    new TypeReference<ArrayList<Orderdts>>() {
                    });
            Meal meal = null;
            double orderprice = 0;
            // 给订单子表对象的其他属性赋值
            for (Orderdts od : odList) {
                meal = mealService.getMealByMealId(od.getMealId());
                orderprice += meal.getMealPrice() * od.getMealCount();
                // 设置订单明细对象关联属性
                od.setO(o);
                od.setM(meal);
            // 设置订单主表与订单明细的关联(将订单子表对象添加到订单主表对象中)
                o.getOds().add(od);
            }
            // 将计算出的订单价格设置到订单主表对象o中
            o.setOrderPrice(orderprice);
            // 保存订单主表，级联保存订单子表记录
            if (orderService.addOrder(o) > 0) {
                return "success";
            } else {
                return "failure";
            }
        } catch (Exception e) {
            return "failure";
        }
    }
    ……
}
```

在commitOrder方法中，首先创建ObjectMapper对象，用于实现JavaBean和JSON的转换，并设置输入时忽略在JSON字符串中存在但Java对象实际没有的属性。然后将JSON字符串order转换为Orders对象o(对应订单主表)，将JSON字符串inserted转换成List<Orderdts>集合(对应订单子表)。再给订单子表对象的其他属性赋值，并将订单子表对象集合添加到订单主表对象中。最后保存订单主表，级联保存订单子表记录。如果成功执行则向前端页面发送success，否则发送failure。前端页面中的函数saveorder()对返回值进行判断，如果成功，则关闭"创建订单"标签页，并调用DataGrid控件的reload方法重新获取数据源，如果失败，则提示错误信息。

23.10.2 查询订单

在index.jsp页面中，单击Tree控件上的"查询订单"节点，打开订单查询页searchorder.jsp，如图23-9所示。

图 23-9 订单查询页 searchorder.jsp

在订单查询页中，可根据订单编号、客户名称、订单状态和订单日期查询订单。查询窗体代码如下：

```html
<div id="searchOrderTb" style="padding:2px 5px;">
    <form id="searchOrderForm">
        <div style="padding:3px">
            订单编号 <input class="easyui-textbox" name="search_oid" id="search_oid" style="width:110px" /></div>
        <div style="padding:3px">    客户名称 <input style="width:115px;" id="search_userid"    class="easyui-combobox" value="0" name="search_userid"    data-options="valueField:'id', textField:'loginName', url:'user/getValidUser'">   
            订单状态 <select id="search_orderstatus" class="easyui-combobox" name="search_orderstatus" style="width:115px;">
            <option value="请选择" selected>--请选择--</option>
            <option value="未付款">未付款</option>
            <option value="已付款">已付款</option>
            <option value="待发货">待发货</option>
            <option value="已发货">已发货</option>
            <option value="已完成">已完成</option>
        </select>    订单时间  <input class="easyui-datebox" name="orderTimeFrom" id="orderTimeFrom" style="width:115px;" /> ~
<input class="easyui-datebox" name="orderTimeTo" id="orderTimeTo" style="width:115px;" /> <a href="javascript:void(0)" class="easyui-linkbutton" iconCls="icon-search" plain="true" onclick="searchOrder();">查找</a>
        </div></form></div>
```

在查询窗体中，客户名称组合框的数据来源于控制器类 UserController 的 getValidUser 方法执行后返回的 JSON 格式的结果。

为了创建 Easy UI DataGrid 控件，需要先创建<table>标签，代码如下：

```
<table id="orderDg" class="easyui-datagrid"></table>
```

DataGrid 控件上的工具栏代码如下:

```
<div id="orderTb" style="padding:2px 5px;">
  <a href="javascript:void(0)" class="easyui-linkbutton"
    iconCls="icon-edit" plain="true" onclick="editOrder();">修改订单/查看明细</a>
  <a href="javascript:void(0)" class="easyui-linkbutton"
iconCls="icon-remove" onclick="removeOrder();" plain="true">删除订单
</a></div>
```

通过 table 标签使用 JavaScript 来创建 DataGrid 控件的代码如下:

```
<script type="text/javascript">
$(function() {
    $('#orderDg').datagrid({
        singleSelect : false,
        url : 'order/getAllOrder', //为datagrid 设置数据源
        queryParams : {}, //查询条件
        pagination : true, //启用分页
        pageSize : 5, //设置初始每页记录数(页大小)
        pageList : [ 5, 10, 15 ], //设置可供选择的页大小
        rownumbers : true, //显示行号
        fit : true, //设置自适应
        toolbar : '#orderTb', //为datagrid 添加工具栏
        header : '#searchOrderTb', //为datagrid 标头添加搜索栏
        columns : [ [ { //编辑 datagrid 的列
            title : '序号',
            field : 'oid',
            align : 'center',
            checkbox : true
        }, {
            field : 'u',
            title : '订单客户',
            formatter : function(value, row, index) {
                if (row.u) {
                    return row.u.loginName;
                } else {
                    return value;
                }
            },
            width : 100
        }, {
            field : 'orderStatus',
            title : '订单状态',
            width : 80
        }, {
            field : 'orderTime',
            title : '订单时间',
            width : 150
        }, {
            field : 'orderPrice',
            title : '订单金额',
            width : 100
```

```
        } ] ]
    });
});
</script>
```

将 DataGrid 控件的 singleSelect 属性设置为 false 以允许多选，pagination 属性设置为 true 以允许分页，pageSize 属性设置初始每页记录数(即页大小)，pageList 设置可供选择的页大小，rownumbers 属性设置为 true 以显示行号，fit 属性设置为 true 以自适应显示数据，toolbar 属性设置为#orderTb 为 datagrid 添加工具栏，header 属性设置为#searchOrderTb 为 datagrid 标头添加搜索栏，设置 columns 属性以指定 DataGrid 显示的列，设置 queryParams 属性以指定查询条件(默认值为{})，DataGrid 控件数据源通过 url 属性来指定，此处设置为 order/getAllOrder，即将请求发送到控制器类 OrderController 的 getAllOrder 方法，获取所有订单列表。代码如下：

```
@ResponseBody
@RequestMapping("/getAllOrder")
public Map<String, Object> getAllOrder(Orders o, String search_oid,
        int page, int rows) {
    Map<String, Object> result = new HashMap<String, Object>(2);
    if (search_oid != null && !"".equals(search_oid)) {
        o.setOid(Integer.parseInt(search_oid));
    }
    // 根据查询条件获取订单记录总数
    int totalCount = orderService.getTotalCount(o);
    // 根据当前页码、每页记录数、查询条件获取指定页显示的订单列表
    List<Orders> oList = orderService.getOrderByConditionForPager(o, page,
            rows);
    result.put("total", totalCount);
    result.put("rows", oList);
    return result;
}
```

getAllOrder 方法有四个参数，Orders 类参数 o 封装查询窗体中的查询条件，参数 search_oid 保存查询窗体中的订单编号，参数 page 代表当前页码，参数 rows 代表每页显示记录数，参数 page 和 rows 值来自 DataGrid 控件分页条。在 getAllOrder 方法中，首先创建 Map 类型对象 result，用于向前端页面发送数据。然后调用业务接口 OrderService 的 getTotalCount 方法，根据查询条件获取订单记录总数，接着调用 getOrderByConditionForPager 方法，根据当前页码、每页显示记录数和查询条件获取当前页显示的订单列表。再向对象 result 中放入键值对，键为 total，值为 totalCount，向对象 result 中放入键值对，键为 rows，值为 oList。最后通过@ResponseBody 注解自动将 Map<String, Object>类型 result 转换为 JSON 格式，并向前端页面发送。前端页面的 DataGrid 控件使用返回结果，给 DataGrid 控件的列绑定数据。

初始时 DataGrid 控件显示所有订单记录，如果在查询窗体中输入查询条件，单击"查找"按钮，将执行 JavaScript 函数 searchOrder()，代码如下：

```
function searchOrder(){
    var search_oid=$('#search_oid').val();
    var orderStatus=$('#search_orderstatus').combobox("getValue");
    var userId=$('#search_userid').combobox("getValue");
```

```
    var orderTimeFrom=$('#orderTimeFrom').datebox("getValue");
    var orderTimeTo=$('#orderTimeTo').datebox("getValue");
    // 重新加载datagrid控件的数据源
    $('#orderDg').datagrid('load',{
        search_oid:search_oid,
        orderStatus:orderStatus,
        userId:userId,
        orderTimeFrom:orderTimeFrom,
        orderTimeTo:orderTimeTo
    });
}
```

在函数 searchOrder 中，首先获取输入的查询条件，然后调用 DataGrid 控件的 load 方法，此时会向控制器类 OrderController 的 getAllOrder 方法再次发送请求，并将参数传递过去。

23.10.3 删除订单

在订单查询页 searchorder.jsp 中，选中 DataGrid 控件中的某条记录，单击工具栏中的"删除订单"按钮，可将订单主表和明细记录同时删除。单击"删除订单"按钮时，将执行 JavaScript 函数 removeOrder()，代码如下：

```
function removeOrder(){
    // 获取选中的订单记录行
    var rows=$('#orderDg').datagrid('getSelections');
    if(rows.length>0){
        $.messager.confirm('Confirm','确认要删除么?',function(r){
            if(r){
                var ids="";
                // 获取选中订单记录的订单id，保存到ids中
                for(var i=0;i<rows.length;i++){
                    ids+=rows[i].oid+",";
                }
                // 发送请求到服务器，执行删除操作
                $.post('order/deleteOrder',{
                    oids:ids
                },function(result){
                    if(result.success='true'){
                        // 删除成功，重新加载datagrid控件
                        $('#orderDg').datagrid('reload');
                        $.messager.show({
                            title:'提示信息',
                            msg: result.message
                        });
                    }else{
                        $.messager.show({
                            title:'提示信息',
                            msg: result.message
                        });
                    }
                },'json');
            }
```

```
            });
        }else{
            $.messager.alert('提示','请选择要删除的行！','info');
        }
    }
```

在函数 removeOrder 中，首先获取 DataGrid 控件中选中的订单记录，然后将选中订单记录的订单编号保存到 ids 中，再使用$.post 发送请求 order/deleteOrder，即向控制器类 OrderController 的 deleteOrder 方法发送请求，并将参数 oids 传递过去。deleteOrder 方法代码如下：

```
@ResponseBody
@RequestMapping(value = "/deleteOrder", produces = "text/html;charset=UTF-8")
public String deleteOrder(String oids) {
    String str = "";
    try {
        oids = oids.substring(0, oids.length() - 1);
        String[] ids = oids.split(",");
        for (String id : ids) {
            // 循环删除订单记录
            Orders o = orderService.getOrdersByOid(Integer.parseInt(id));
            orderService.deleteOrder(o);
        }
        str = "{\"success\":\"true\",\"message\":\"删除成功！\"}";
    } catch (Exception e) {
        str = "{\"success\":\"false\",\"message\":\"删除失败！\"}";
    }
    return str;
}
```

在 deleteOrder 方法中，调用业务接口 OrderService 的 getOrdersByOid 方法，根据订单编号循环删除订单记录，再根据执行情况向前端页面发送信息。在前端页面中，根据结果进行判断，如果执行成功，则通过调用 DataGrid 的 reload 方法，重新向控制器类 OrderController 的 getAllOrder 方法发送请求，以更新数据源。

23.10.4 修改订单/查看明细

在订单查询页 searchorder.jsp 中，选中 DataGrid 控件中的某条记录，单击工具栏中的"修改订单/查看明细"按钮，将执行 JavaScript 函数 editOrder()，代码如下：

```
function editOrder(){
    var rows=$('#orderDg').datagrid('getSelections');
    if(rows.length>0){
        var row=$('#orderDg').datagrid('getSelected');
        if($('#tabs').tabs('exists','修改订单')){
            $('#tabs').tabs('close','修改订单');
        }
        $('#tabs').tabs('add',{
            title:"修改订单",
            href: 'order/getOrder?oid='+row.oid,
```

```
            closable: true
        });
    }else{
        $.messager.alert('提示','请选择要修改的订单！','info');
    }
}
```

在函数 editOrder 中，首先获取 DataGrid 控件中选中的行，然后向控制器类 OrderController 的 getOrder 方法发送请求，并将参数 oid 传递过去，参数值为选中的订单编号。getOrder 方法根据订单编号获取要修改的订单对象，最后再返回订单修改页，代码如下：

```
@RequestMapping("/getOrder")
public ModelAndView getOrder(String oid) {
    String viewName = "modifyorder";
    ModelAndView mv = new ModelAndView(viewName);
    Orders o = orderService.getOrdersByOid(Integer.parseInt(oid));
    // 对象o是存放request范围，在订单修改页modifyorder.jsp中
    // 可通过requestScope来获得对象o
    mv.addObject("o", o);
    return mv;
}
```

在 getOrder 方法中，首先定义了 ModelAndView 对象，然后调用业务接口 OrderService 的 getOrdersByOid 方法获取指定编号的订单对象，再将得到的订单对象和视图存入 ModelAndView 对象中。getOrder 方法执行结束后，跳转到订单修改页 modifyorder.jsp。订单修改页布局与创建订单页基本相同，此处不再列出相关布局代码。

在订单修改页 modifyorder.jsp 中，从 ModelAndView 对象中取出保存的订单对象，用来显示订单信息。订单主表部分的数据绑定是在 id 为 edit_divOrder 的<div>标签中完成的，订单明细数据是在 "$(function() { });" 中完成的。绑定订单明细数据的代码如下：

```
var $editodbox = $('#edit_odbox');
$(function() {
    $editodbox.datagrid({
        // 获取订单明细数据
        url : 'order/getOrderDts?oid=${requestScope.o.oid }',
        // 当数据加载成功后，使得datagrid控件上所有行都处于可编辑状态
        onLoadSuccess : function(data) {
            var rows = $editodbox.datagrid('getRows');
            for (var i = 0; i < rows.length; i++) {
                var index = $editodbox.datagrid('getRowIndex',
                    rows[i]);
                $editodbox.datagrid('beginEdit', index);
            }
        },
        rownumbers : true,
        singleSelect : false,
        fit : true,
        toolbar : '#edit_ordertb',
        header : '#edit_divOrder',
        columns : [ [ {
            // 此处DataGrid控件列与创建订单页中基本相同,故省略相关代码
        } ] ]
```

		});
	});
```

使用 DataGrid 控件显示订单明细数据时，数据源通过 url 属性指定为 order/getOrderDts?oid=${requestScope.o.oid }，即执行控制器类 OrderController 的 getOrderDts 方法，并将参数 oid 传递过去。代码如下：

```java
@ResponseBody
@RequestMapping("/getOrderDts")
public List<Orderdts> getOrderDts(String oid) {
    List<Orderdts> ods = orderService.getOrderdtsByOid(Integer
            .parseInt(oid));
    for (Orderdts od : ods) {
        od.setMealId(od.getM().getMealId());
        od.setMealPrice(od.getM().getMealPrice());
        od.setTotalprice(od.getM().getMealPrice() * od.getMealCount());
    }
    return ods;
}
```

在 getOrderDts 方法中，调用了业务接口 OrderService 的 getOrderdtsByOid 方法，根据订单编号获取订单明细列表。然后遍历订单明细列表，将关联的餐品编号、价格、小计等信息保存到每一个订单明细对象中。再通过@ResponseBody 注解自动将 List<Orderdts>类型的 ods 转变为 JSON 格式，发送到前端页面，作为前端页面中 DataGrid 控件的数据源。通过 DataGrid 控件的 onLoadSuccess 事件，在数据加载成功后，将 DataGrid 控件中的数据行设置为编辑状态。

与创建订单页类似，订单修改页面中也可以使用"添加订单明细"在 DataGrid 控件最后添加一个记录行，使用"删除订单明细"按钮删除选中的记录行。

完成订单主表数据和明细数据的修改后，单击工具栏中的"保存订单"按钮，执行 JavaScript 函数 edit_saveorder()，代码如下：

```javascript
function edit_saveorder() {
    // 获取订单客户
    var userid = $('#edit_userid').combobox('getValue');
    if (userid == 0) {
        $.messager.alert('提示', '请选择客户名称', 'info');
    } else {
        // 调用自定义方法取消 datagrid 控件的行编辑状态
        edit_endEdit();
        // 定义 order 数组存放订单主表数据
        var order = [];
        // 获取订单时间
        var ordertime = $('#edit_ordertime').datebox('getValue');
        // 获取订单状态
        var orderstatus = $('#edit_orderstatus').datebox('getValue');
        // 获取订单的 id 号
        var oid = $('#oid').val();
        // 获取订单总金额
        var orderprice = $('#edit_orderprice').textbox('getValue');
        // 向数组的末尾添加元素(订单主表数据)
        order.push({
```
```

```javascript
            orderTime : ordertime,
            userId : userid,
            orderStatus : orderstatus,
            oid : oid,
            orderPrice : orderprice
        });
        // 获取订单明细(datagrid控件中的记录)
        if ($editodbox.datagrid('getChanges').length) {
            // 获取datagrid控件中插入的记录行
            var inserted = $editodbox
                    .datagrid('getChanges', "inserted");
            // 获取datagrid控件中删除的记录行
            var deleted = $editodbox.datagrid('getChanges', "deleted");
            // 获取datagrid控件中更新的记录行
            var updated = $editodbox.datagrid('getChanges', "updated");
            // 定义effectRow,保存inserted和order
            var effectRow = new Object();
            if (inserted.length) {
                // JSON.stringify将对象数组转换成JSON字符串
                effectRow["inserted"] = JSON.stringify(inserted);
            }
            if (updated.length) {
                // JSON.stringify将对象数组转换成JSON字符串
                effectRow["updated"] = JSON.stringify(updated);
            }
            if (deleted.length) {
                // JSON.stringify将对象数组转换成JSON字符串
                effectRow["deleted"] = JSON.stringify(deleted);
            }
            effectRow["order"] = JSON.stringify(order);
            // 提交请求
            $.post("order/commitModifyOrder", effectRow,
                function(data) {
                    if (data == 'success') {
                        $.messager.alert('提示', '修改成功!', 'info');
                        // 提交datagrid控件的当前行
                        $editodbox.datagrid('acceptChanges');
                        if ($('#tabs').tabs('exists', '修改订单')) {
                            $('#tabs').tabs('close', '修改订单');
                        }
            // 重新加载"查询订单页"中用于显示订单记录的datagrid控件的数据
                        $('#orderDg').datagrid('reload');
                    } else {
                        $.messager.alert('提示', '修改失败!', 'info');
                    }
            });
        }
    }
}
```

在函数 edit_saveorder 中，首先通过调用自定义的 JavaScript 函数 edit_endEdit()取消 DataGrid 控件的行编辑状态。然后定义了 order 存放订单客户编号、订单时间和订单状态、订单编号和订单总金额等订单主表数据，并通过 DataGrid 控件的 getChanges 方法，获取控件上发生的改变情况，包括 inserted、updated、deleted 三种类型。由于 DataGrid 控件初始时绑定了订单明细数据，如果是单击"添加订单明细"按钮添加的行，则 DataGrid 控件上发生的改变属于 inserted 类型；如果是删除原有的行，则改变属于 deleted；如果是修改原有行中的数据，则改变属于 updated。接下来定义 effectRow 保存 inserted、updated、deleted 和 order，最后通过 $.post 将请求提交到 order/commitModifyOrder，即执行控制器类 OrderController 的 commitModifyOrder 方法，代码如下：

```java
@ResponseBody
@RequestMapping(value = "/commitModifyOrder")
public String commitModifyOrder(String inserted, String updated,
        String deleted, String order) throws JsonParseException,
        JsonMappingException, IOException {
    try {
        // 定义要插入的、要更新的、要删除的订单明细集合
        List<Orderdts> insertedOdList = null;
        List<Orderdts> updatedOdList = null;
        List<Orderdts> deletedOdList = null;
        // 创建 ObjectMapper 对象，实现 JavaBean 和 JSON 的转换
        ObjectMapper mapper = new ObjectMapper();
        // 设置输入时忽略在 JSON 字符串中存在但 JavaBean 对象实际没有的属性
mapper.disable(DeserializationFeature.FAIL_ON_UNKNOWN_PROPERTIES);
        mapper.configure(SerializationFeature.FAIL_ON_EMPTY_BEANS, false);
        // 将 json 字符串 order 转换为 JavaBean 对象(订单主表)
        Orders tempoi = mapper.readValue(order, Orders[].class)[0];
        // 订单主表对象(原始的、修改前)
        Orders o = orderService.getOrdersByOid(tempoi.getOid());
        o.setU(userService.getUserById(tempoi.getUserId()));
        o.setOrderStatus(tempoi.getOrderStatus());
        o.setOrderTime(tempoi.getOrderTime());
        o.setOrderPrice(tempoi.getOrderPrice());
        // 此时的 o 即为更新后的订单主表对象,下面处理订单明细部分
        // 处理删除的订单明细
        if (deleted != null) {
            deletedOdList = mapper.readValue(deleted,
                new TypeReference<ArrayList<Orderdts>>() {    });
            for (Orderdts dod : deletedOdList) {
// dod 就是要删除的订单明细对象
                Set<Orderdts> odset = o.getOds();
                Iterator<Orderdts> itor = odset.iterator();
                // 定义 delods,用于临时保存要从订单对象 oi 中移除的关联的订单明细
                List<Orderdts> delods = new ArrayList<Orderdts>();
                while (itor.hasNext()) {
                    Orderdts odd = itor.next();
                    if (dod.getOdid() == odd.getOdid()) {
                        orderService.deleteOrderdts(odd);
                        delods.add(odd);
                    }
                }
                for (Orderdts delod : delods) {
                    o.getOds().remove(delod);
                }
            }
```

```java
        }
        // 处理要修改的订单明细
        if (updated != null) {
            updatedOdList = mapper.readValue(updated,
                new TypeReference<ArrayList<Orderdts>>() { });
            for (Orderdts uod : updatedOdList) {
                Set<Orderdts> odset = o.getOds();
                Iterator<Orderdts> itor = odset.iterator();
        // 定义 removeods,用于临时保存要从订单对象 o 中移除的关联的订单明细
                List<Orderdts> removeods = new ArrayList<Orderdts>();
                // 定义 addods,用于临时保存要添加到订单对象 o 中关联的订单明细
                List<Orderdts> addods = new ArrayList<Orderdts>();
                while (itor.hasNext()) {
                    Orderdts odd = itor.next();
                    if (uod.getOdid() == odd.getOdid()) {
                        // 将要移除的修改前的订单明细对象添加到 removeods
                        removeods.add(odd);
                        uod.setM(mealService.getMealByMealId(uod
                            .getMealId()));
                        // 将修改后的订单明细对象添加到 addods 中
                        addods.add(uod);
                    }
                }
                // 从订单对象 o 关联的订单明细集合中移除 removeods 中的对象
                for (Orderdts removeod : removeods) {
                    o.getOds().remove(removeod);
                }
                // 向订单对象 o 关联的订单明细集合中添加 addods 中保存的对象
                for (Orderdts addod : addods) {
                    o.getOds().add(addod);
                }
            }
        }
        // 处理新增的订单明细
        if (inserted != null) {
            insertedOdList = mapper.readValue(inserted,
                new TypeReference<ArrayList<Orderdts>>() { });
            Meal m = null;
            for (Orderdts iod : insertedOdList) {
                m = mealService.getMealByMealId(iod.getMealId());
                iod.setM(m);
                iod.setO(o);
                // 向订单对象 oi 关联的订单明细集合中添加新增的订单明细对象
                o.getOds().add(iod);
            }
        }
        // 最后判断订单对象 o 关联的订单明细集合中是否还有记录
        // 修改后,没有订单明细数据,此时需要将订单主表一起删除
        if (o.getOds().size() == 0) {
            orderService.deleteOrder(o);
        } else {
            orderService.modifyOrder(o);
        }
    } catch (Exception e) {
```

```
        return "failure";
    }
    return "success";
}
```

在 commitModifyOrder 方法中，首先生成订单主表的对象 o，然后处理要删除的订单明细(先调用业务接口 orderService 的 deleteOrderdts 方法删除订单明细，然后从订单主表对象 o 中移除关联的订单明细对象)，接着处理要修改的订单明细(先定义 removeods，将要移除的修改前的订单明细对象添加到 removeods 中，定义 addods，将修改后的订单明细对象添加到 addods 中，再依次从订单对象 o 关联的订单明细集合中移除 removeods 集合中的对象，添加 addods 集合中的订单明细对象)，接着处理新增的订单明细(向订单对象 o 关联订单明细集合中添加新增的订单明细对象)，接着判断订单对象 o 关联的订单明细集合中是否有记录，如果没有则调用业务接口 orderService 的 deleteOrder 方法将订单主表和明细同时删除，否则调用业务接口 orderService 的 modifyOrder 方法修改订单。最后，根据执行的情况，向前端页面发送信息。前端页面根据返回结果进行判断，如果执行成功，则关闭"修改订单"标签页，并通过订单查询页 searchorder.jsp 中的 id 为 orderDg 的 DataGrid 的 reload 方法向 order/getAllOrder 发送请求，以更新数据源。

23.10.5 使用 Echarts 显示销售统计

在 index.jsp 页面中，单击 Tree 控件上的"订单统计"节点，打开订单统计页，如图 23-10 所示。

图 23-10 订单统计页 saler.jsp

下面结合 Echarts 图表控件，介绍如何在页面 saler.jsp 中以柱状图显示销售统计结果。

首先将存放使用 Echarts 所需 js 文件的 echarts 和 echarts-master 两个文件夹复制到项目的

WebRoot 目录下。

然后在页面 saler.jsp 中引入 js 文件，代码如下：

```
<script src="${pageContext.request.contextPath}/echarts-master/
build/dist/echarts.js"></script>
```

在页面 saler.jsp 中为 ECharts 准备一个具备高宽的 DOM 容器，代码如下：

```
<div id="main" style="width:600px;height:400px;margin: 20px;"></div>
```

通过 ec.init 方法初始化一个 Echarts 实例并通过 setOption 方法生成一个简单的柱状图，代码如下：

```
<script type="text/javascript">
    setTimeout(DayNumOfMonth, 1000);
    function DayNumOfMonth() {
        $.ajax({
            url : "order/salerStatistics",
            type : "post",
            async : false,
            datatype : "text",
            success : function(data) {
                // alert(data);
                mealNameArr = data[0];
                accountArr = data[1];
// 使用柱状图就加载 bar 模块，按需加载
                require([ 'echarts', 'echarts/chart/bar',
                ], function(ec) {
                    // 基于准备好的 dom，初始化 echarts 图表
                    var myChart = ec.init(document
                        .getElementById('main'));
                    //设置数据
                    var option = {
                        title : {
                            text : '销售统计'
                        },
                        tooltip : {},
                        legend : {
                            data : [ '销售总额' ]
                        },
                        //设置坐标轴
                        xAxis : [ {
                            type : 'category',
                            data : mealNameArr
                        } ],
                        yAxis : [ {
                            type : 'value'
                        } ],
                        //设置数据
                        series : [ {
                            "name" : "销售总额",
```

```
                        "type" : "bar",
                        data : accountArr
                    } ]
                };
                // 为 echarts 对象加载数据
                myChart.setOption(option);
            })
        }
    });
}
</script>
```

这样一个图表就诞生了，该柱状图的数据源为 order/salerStatistics，即执行控制器类 OrderController 中的 salerStatistics 方法，代码如下：

```
// 销售统计，给订单统计页 saler.jsp 中的 Echarts 柱状图提供数据源
@ResponseBody
@RequestMapping(value="/salerStatistics", produces = 
"application/json;charset=UTF-8")
public List<List> salerStatistics() {
    List list = orderService.findSalerStandby();
    List<List> result=new ArrayList<List>();
    Iterator itor = list.iterator();
    List<String> list1 =new ArrayList<String>();
    List<Double> list2 =new ArrayList<Double>();
    while (itor.hasNext()) {
        Object[] obj = (Object[]) itor.next();
        list1.add(obj[0].toString());
        list2.add(Double.parseDouble(obj[1].toString()) );
    }
    result.add(list1);
    result.add(list2);
    return result;
}
```

salerStatistics 方法返回的字符串作为 Echarts 柱状图数据源，格式如下：

```
[["雪梨肉肘棒","素锅烤鸭肉","泰安肉三美豆腐"],[30.0,180.0,16.0]]
```

23.11 客户管理

客户管理主要功能包括客户列表显示、查询客户、启用和禁用客户。

23.11.1 客户列表显示

在 index.jsp 页面中，单击 Tree 控件上的"用户管理"下的"用户列表"节点，打开客户列表显示页 userlist.jsp，如图 23-11 所示。

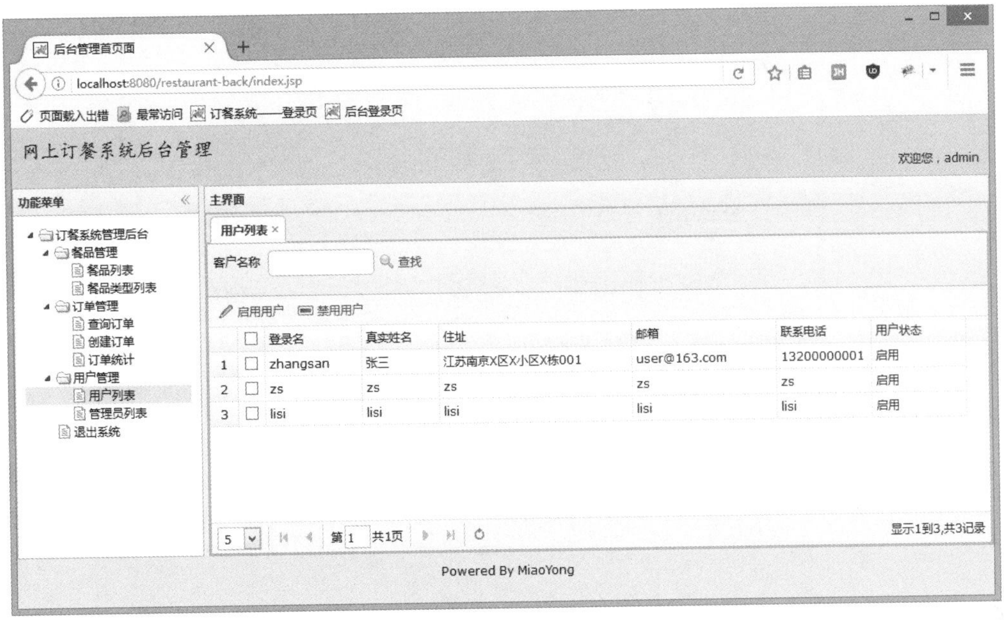

图 23-11 客户列表显示页 userlist.jsp

在 userlist.jsp 页面中使用了 Easy UI 的 DataGrid 控件来显示客户列表，该控件是通过 table 标签来创建的，其定义如下：

```
<table id="userListDg" class="easyui-datagrid"></table>
```

页面上的工具栏和搜索栏代码如下：

```
<!-- 创建工具栏 -->
<div id="userListTb" style="padding:2px 5px;">
    <a href="javascript:void(0)" class="easyui-linkbutton"
       iconCls="icon-edit" plain="true" onclick="SetIsEnableUser(1);">启用用户</a>
    <a href="javascript:void(0)" class="easyui-linkbutton"
       iconCls="icon-remove" onclick="SetIsEnableUser(0);" plain="true">禁用用户</a>
</div>
<!-- 创建搜索栏 -->
<div id="searchUserListTb" style="padding:4px 3px;">
    <form id="searchUserListForm">
        <div style="padding:3px ">
            客户名称  <input class="easyui-textbox" name="search_loginName" id="search_loginName" style="width:110px" /><a href="javascript:void(0)" class="easyui-linkbutton" iconCls="icon-search" plain="true" onclick="searchUsers();">查找</a>
        </div>
    </form>
</div>
```

通过 table 标签使用 JavaScript 来创建 DataGrid 控件的代码如下：

```
<script type="text/javascript">
$(function() {
```

```javascript
    $('#userListDg').datagrid({
        singleSelect : false,
        url : 'user/list',
        queryParams : {}, //查询条件
        pagination : true, //启用分页
        pageSize : 5, //设置初始每页记录数(页大小)
        pageList : [ 5, 10, 15 ], //设置可供选择的页大小
        rownumbers : true, //显示行号
        fit : true, //设置自适应
        toolbar : '#userListTb', //为datagrid添加工具栏
        header : '#searchUserListTb', //为datagrid标头添加搜索栏
        columns : [ [ { //编辑datagrid的列
            title : '序号',
            field : 'id',
            align : 'center',
            checkbox : true
        }, {
            field : 'loginName',
            title : '登录名',
            width : 100
        }, {
            field : 'trueName',
            title : '真实姓名',
            width : 80
        }, {
            field : 'address',
            title : '住址',
            width : 200
        }, {
            field : 'email',
            title : '邮箱',
            width : 150
        }, {
            field : 'phone',
            title : '联系电话',
            width : 100
        }, {
            field : 'status',
            title : '用户状态',
            width : 100,
            formatter : function(value, row, index) {
                if (row.status == 1) {
                    return "启用";
                } else {
                    return "禁用";
                }
            }
        } ] ]
    });
});
......
</script>
```

将 DataGrid 控件的 singleSelect 属性设置为 false 以允许多选，pagination 属性设置为 true

以允许分页，pageSize 属性设置初始每页记录数(即页大小)，pageList 设置可供选择的页大小，rownumbers 属性设置为 true 以显示行号，fit 属性设置为 true 以自适应显示数据，toolbar 属性设置为#userListTb 为 datagrid 添加工具栏，header 属性设置为#searchUserListTb 为 DataGrid 标头添加搜索栏，设置 columns 属性以指定 DataGrid 显示的列，设置 queryParams 属性以指定查询条件(默认值为{})，DataGrid 控件数据源通过 url 属性来指定，此处设置为 user/list，即将请求发送到 UserController 类的 list 方法，代码如下：

```java
package com.res.controller;
……
@RequestMapping("/user")
@Controller
public class UserController {
    // 使用@Autowired注解注入UserServiceImpl实例
    @Autowired
    UserService userService;
    ……
    @RequestMapping("/list")
    @ResponseBody
    public Map<String, Object> list(int page, int rows, Users u) {
        // 创建 Map 类型对象 result,用于向前端页面发送数据
        Map<String, Object> result = new HashMap<String, Object>(2);
        // 根据登录名参数模糊查询所有客户记录数
        int totalCount = userService.getTotalCount(u);
        // 查询指定页显示的客户列表
        List<Users> uList = userService.getUsersByConditionForPager(u, page, rows);
        // 向 Map 类型的对象 result 中放入键值对，键为total,值为totalCount
        result.put("total", totalCount);
        // 向对象 result 中放入键值对，键为rows,值为uList
        result.put("rows", uList);
        // 通过@ResponseBody,发送到前端页面的 result 自动转成 JSON 格式
        return result;
    }
    ……
}
```

在 list 方法中，首先创建 Map 类型对象 result，用于向前端页面发送数据；然后调用业务接口 UserService 的 getTotalCount 方法，根据登录名参数模糊查询所有客户记录数；接着调用 getUsersByConditionForPager 方法，查询指定页显示的客户列表；接着依次向 Map 类型的对象 result 中放入键值对，键为 total，值为 totalCount，键为 rows，值为 uList(即当前页显示的客户列表)；最后通过 list 方法上修饰的@ResponseBody 注解，自动将 Map<String, Object>类型的数据 result 转成 JSON 格式，再发送到前端页面作为 DataGrid 控件的数据源。

23.11.2　查询客户

在 userlist.jsp 页面的搜索栏中，输入客户名称，单击"查找"按钮，会根据客户名称进行模糊查询，并将查询结果显示在 DataGrid 控件中。例如，在客户名称栏中输入 z，单击"查找"按钮，查询结果如图 23-12 所示。

图 23-12 客户查询

单击"查找"按钮时，会执行 JavaScript 函数 searchUsers()，代码如下：

```
function searchUsers() {
    var loginName = $('#search_loginName').textbox("getValue");
    $('#userListDg').datagrid('load', {
        loginName : loginName
    });
}
```

在函数 searchUsers 中，首先获取输入的客户名称，然后使用 DataGrid 控件的 load 方法，因为指定了参数 loginName，将取代 queryParams 属性。当传递参数执行查询时，load 方法将从服务器加载新的数据，此时将重新执行 UserController 类的 list 方法，并将参数 loginName 传递到方法中，再使用查询获得的新数据重新绑定 DateGrid 控件，以显示查询结果。

23.11.3 启用和禁用客户

在 userlist.jsp 页面显示用户列表的 DateGrid 控件中，选中若干条记录，单击"启用"或"禁用"按钮，可以修改用户的状态。单击"启用"或"禁用"按钮后，将执行 JavaScript 函数 SetIsEnableUser，代码如下：

```
function SetIsEnableUser(flag) {
    var rows = $("#userListDg").datagrid('getSelections');
    if (rows.length > 0) {
        $.messager.confirm('Confirm', '确认要设置么?', function(r) {
            if (r) {
                var uids = "";
                for (var i = 0; i < rows.length; i++) {
                    uids += rows[i].id + ",";
```

```javascript
            }
            $.post('user/setIsEnableUser', {
                uids : uids,
                flag : flag
            }, function(result) {
                if (result.success == 'true') {
                    $("#userListDg").datagrid('reload');
                    $.messager.show({
                        title : '提示信息',
                        msg : result.message
                    });
                } else {
                    $.messager.show({
                        title : '提示信息',
                        msg : result.message
                    });
                }
            }, 'json');
        }
    });
} else {
    $.messager.alert('提示', '请选择要启用或禁用的客户', 'info');
}
}
```

SetIsEnableUser 函数的参数 flag 表示执行启用还是禁用操作，等于 1 时启用用户，等于 0 时禁用用户。在 SetIsEnableUser 函数中，首先获取选中行的客户编号，并将其保存到变量 uids 中，然后通过 $.post 方法提交请求 user/setIsEnableUser，即执行 UserController 类的 setIsEnableUser 方法，并将参数 uids 和 flag 传递过去。setIsEnableUser 方法的代码如下：

```java
@RequestMapping(value = "/setIsEnableUser", produces = "text/html;charset=UTF-8")
@ResponseBody
public String setIsEnableUser(@RequestParam(value = "uids") String uids,
@RequestParam(value = "flag") String flag) {
    uids = "(" + uids.substring(0, uids.length() - 1) + ")";
    String str = "";
    try {
        userService.updateUserStatus(uids, flag);
        str = "{\"success\":\"true\",\"message\":\"设置成功！\"}";
    } catch (Exception e) {
        str = "{\"success\":\"false\",\"message\":\"设置失败！\"}";
    }
    return str;
}
```

在 setIsEnableUser 方法中，调用业务接口 UserService 的 updateUserStatus 方法，将数据表 users 的 Status 字段设置为 1(启用)或 0(禁用)。最后根据执行结果，向前端页面发送信息。

前端页面 userlist.jsp 中的 JavaScript 函数 SetIsEnableUser 根据服务器的返回信息进行判断，如果执行成功，则调用 DataGrid 控件的 reload 方法，重新执行 UserController 类的 list 方法，重新获取数据以更新用户列表。

23.12 管理员及其权限管理

管理员及其权限管理主要功能包括管理员列表显示、新增管理员、设置/修改管理员权限。

23.12.1 管理员列表显示

在 index.jsp 页面中，单击 Tree 控件上的"用户管理"下的"管理员列表"节点，打开管理员列表显示页 adminlist.jsp，如图 23-13 所示。

图 23-13　管理员列表显示页 adminlist.jsp

在 adminlist.jsp 页面中使用了 Easy UI 的 DataGrid 控件来显示管理员列表，该控件是通过 table 标签来创建的，其定义如下：

```
<table id="adminListDg" class="easyui-datagrid"></table>
```

页面上的工具栏代码如下：

```
<div id="adminListTb" style="padding:2px 5px;">
    <a href="javascript:void(0)" class="easyui-linkbutton"
        iconCls="icon-add" onclick="openAddAdminDlg();" plain="true">添加管理员</a> <a href="javascript:void(0)" class="easyui-linkbutton"
        iconCls="icon-edit" plain="true" onclick="editPowers();">设置/修改管理员权限</a> </div>
```

通过 table 标签使用 JavaScript 来创建 DataGrid 控件的代码如下：

```
<script type="text/javascript">
$(function() {
    $('#adminListDg').datagrid({
        singleSelect : false,
        url : 'admin/list',
```

```
        rownumbers : true, //显示行号
        fit : true, //设置自适应
        toolbar : '#adminListTb', //为datagrid添加工具栏
        columns : [ [ { //编辑datagrid的列
            title : '序号',
            field : 'id',
            align : 'center',
            checkbox : true
        }, {
            field : 'loginName',
            title : '登录名',
            width : 100
        }, {
            field : 'loginPwd',
            title : '密码',
            width : 80
        } ] ]
    });
});
</script>
```

将 DataGrid 控件的属性含义与客户列表时的 DataGrid 控件相同，在此不再赘述。显示管理员列表显示的 DataGrid 控件数据源通过 url 属性来指定，此处设置为 admin/list，即将请求发送到 AdminController 类的 list 方法，代码如下：

```
@RequestMapping("/list")
@ResponseBody
public List<Admin> list() {
    List<Admin> aList = adminService.getAllAdmin();
    return aList;
}
```

在 list 方法中，首先调用业务接口 AdminService 的 getAllAdmin 方法获取所有管理员列表，再通过 list 方法修饰的@ResponseBody 注解，自动将 List<Admin>类型的数据转换为 JSON 格式，再发送到前端页面 adminlist.jsp 作为 DataGrid 控件的数据源。

23.12.2 新增管理员

在 adminlist.jsp 页面中，单击工具栏中的"添加管理员"按钮，打开"新增管理员"对话框，如图 23-14 所示。

图 23-14 "新增管理员"对话框

在 adminlist.jsp 页面中，生成"新增管理员"对话框的代码如下：

```html
<div id="addAdminDlg" class="easyui-dialog" title="添加管理员" closed="true"
 style="width:500px;">
    <div style="padding:10px 60px 20px 60px">
        <form id="addAdminForm" method="POST" action="">
            <table cellpadding="5">
                <tr>
                    <td>登录名</td>
                    <td><input class="easyui-textbox" type="text"
id="loginName" name="loginName" data-options="required:true">
</input></td></tr>
                <tr>
                    <td>密码:</td>
                    <td><input class="easyui-textbox" type="text"
id="loginPwd" name="loginPwd" data-options="required:true">
</input></td></tr>
            </table>
        </form>
        <div style="text-align:center;padding:5px">
            <a href="javascript:void(0)" class="easyui-linkbutton"
                onclick="saveAdmin();">保存</a> <a href="javascript:void(0)"
class="easyui-linkbutton" onclick="clearAddAdminForm();">清空</a>
        </div>
    </div>
</div>
```

单击"添加管理员"按钮，将执行 JavaScript 函数 openAddAdminDlg()，代码如下：

```javascript
function openAddAdminDlg() {
    $('#addAdminDlg').dialog('open').dialog('setTitle', '新增管理员');
    $('#addAdminDlg').form('clear');
    urls = 'admin/addAdmin';
}
```

该函数用来打开"新增管理员"对话框，并给变量 urls 设置了值 admin/addAdmin。在"新增管理员"对话框中，输入"登录名"和"密码"，单击"保存"按钮，将执行 JavaScript 函数 saveAdmin()，saveAdmin 代码如下：

```javascript
function saveAdmin() {
    $("#addAdminForm").form("submit", {
        url : urls, //使用参数
        success : function(result) {
            var result = eval('(' + result + ')');
            if (result.success == 'true') {
                $("#adminListDg").datagrid("reload");
                $("#addAdminDlg").dialog("close");
            }
            $.messager.show({
                title : "提示信息",
                msg : result.message
            });
        }
    });
}
```

在函数 saveAdmin()中，使用$("#addAdminForm").form 提交表单，将请求发送到 urls 变量保存的 admin/addAdmin，即执行 AdminController 类的 addAdmin 方法。addAdmin 方法代

码如下：

```
@RequestMapping(value = "/addAdmin", produces = "text/html;charset=UTF-8")
@ResponseBody
public String addAdmin(Admin ai) {
    try {
        adminService.addAdmin(ai);
        return "{\"success\":\"true\",\"message\":\"新增成功\"}";
    } catch (Exception e) {
        return "{\"success\":\"false\",\"message\":\"新增失败\"}";
    }
}
```

在 addAdmin 方法中，参数 ai 用来封装对话框中填写的"登录名"和"密码"文本域值，通过调用业务接口 AdminService 的 addAdmin 方法，将对象 ai 中的数据添加到数据表 admin 中，再根据执行情况，向前端页面中的 JavaScript 函数 saveAdmin()发送信息。函数 saveAdmin()根据收到的信息进行判断，如果执行成功，则调用 DateGrid 控件的 reload 方法，重新加载管理员列表数据，并关闭"新增管理员"对话框。

23.12.3 设置/修改管理员权限

在 adminlist.jsp 页面中，在 DateGrid 控件上选择一条管理员记录，然后单击"设置/修改管理员权限"按钮，打开"设置/修改管理员权限"对话框，如图 23-15 所示。

图 23-15 "设置/修改管理员权限"对话框

单击"设置/修改管理员权限"按钮，执行 JavaScript 函数 editPowers()，代码如下：

```
function editPowers() {
    var rows = $("#adminListDg").datagrid('getSelections');
    if (rows.length > 0) {
        var row = $("#adminListDg").datagrid("getSelected");
        // 打开"设置/修改权限"对话框
        $("#powerDlg").dialog("open").dialog('setTitle', '设置/修改管理员权限');
        // 通过 id 为 editAdminId 的隐藏文本域保存选中管理员编号
        $('#editAdminId').val(row.id);
// 使用 id 为 powerTree 的 Tree 控件显示系统功能，并绑定该管理员已分配的权限
        $('#powerTree').tree( {
```

```
            checkbox : true,
            url : 'functions/getTree',
            onLoadSuccess : function() {
                // 绑定权限
                $.ajax({
                    url : 'admin/getFidByAdminId?',
                    // cache 必须设置为 false,意思为不缓存当前页，
// 否则更改权限后绑定的权限还是上一次的操作结果
                    cache : false,
                    // dataType 表示获取服务器发送的数据,text 表示纯文本
                    dataType : 'text',
                    data : {
                        adminid : row.id
                    },
                    success : function(data) {
                        if (data != "") {
                            var array = data
                                    .split(',');
                            for (var i = 0; i < array.length; i++) {
                                var node = $('#powerTree')
                                        .tree('find', array[i]);
                                $('#powerTree')
                                        .tree('check', node.target);
                            }
                        }
                    }
                })
            }
        });
    } else {
        $.messager.alert('提示', '请选择要修改的管理员', 'info');
    }
}
```

在 editPowers 函数中，首先判断是否选择了要设置或修改权限的管理员记录，然后打开 id 为 powerDlg 的"设置/修改权限"对话框，生成该对话框的代码如下：

```
<div id="powerDlg" class="easyui-dialog" title="设置/修改管理员权限"
    closed="true" style="width:500px;height: 350px">
    <div style="padding:10px 60px 20px 60px">
        <form id="adminForm" method="POST" action="">
            <table cellpadding="5">
                <tr><td>管理员权限:</td>
                    <td><ul id="powerTree"></ul> <input type="hidden"
                        name="editAdminId" id="editAdminId" style="width:10px"
/></td></tr>
            </table></form>
        <div style="text-align:center;padding:5px">
            <a href="javascript:void(0)" class="easyui-linkbutton"
                onclick="savePowers();">保存</a></div>
    </div>
</div>
```

然后通过 id 为 editAdminId 的隐藏文本域保存选中管理员编号，再使用$('#powerTree').tree，通过 id 为 powerTree 的 Tree 控件显示所有系统功能，并绑定该管理员已分配的权限。将 Tree 的属性 checkbox 设置为 true 以显示复选框，url 属性为 functions/getTree，即执行控制器类

FunctionsController 的 getTree 方法，获取所有的系统功能列表。

getTree 方法代码如下：

```java
package com.res.controller;
import java.util.ArrayList;
……
@Controller
@RequestMapping("/functions")
public class FunctionsController {
    @Autowired
    private FunctionsService functionsService;
    @RequestMapping("/getTree")
    @ResponseBody
    public List<TreeNode> getTree(){
        List<TreeNode> nodes=new ArrayList<TreeNode>();
        List<Functions> fs=functionsService.getAllFunctions();
        for (Functions f : fs) {
            TreeNode treeNode=new TreeNode();
              treeNode.setId(f.getId());
              treeNode.setFid(f.getParentid());
              treeNode.setText(f.getName());
              nodes.add(treeNode);
        }
        List<TreeNode> treeNodes=JsonFactory.buildtree(nodes,0);
        return treeNodes;
    }
}
```

通过 getTree 方法修改的@ResponseBody 注解，自动将 List<TreeNode>类型的功能列表转换为 JSON 格式发送到前端页面，作为 Tree 控件的数据源。再通过 Tree 控件的 onLoadSuccess 事件，当 Tree 数据源加载成功后，使用$.ajax 发送请求，绑定该管理员已分配的功能权限。通过$.ajax 将请求发送到 "admin/getFidByAdminId?"，并使用 data 指定传递的参数 adminid，值为 DateGrid 控件选中行的管理员编号，即执行 AdminController 类的 getFidByAdminId 方法，并将参数 adminid 传递过去。getFidByAdminId 方法的代码如下：

```java
@RequestMapping("/getFidByAdminId")
@ResponseBody
public String getFidByAdminId(
        @RequestParam(value = "adminid") String adminid) {
    // 根据管理员id号获取管理员对象及关联的Functions对象集合
    Admin admin = adminService.getAdminFunctions(Integer.parseInt(adminid));
    // 定义变量sb,用于保存该管理员已分配的功能且属于叶子节点的功能id号
    StringBuffer sb = new StringBuffer();
    // 获取关联的功能集合(即已分配的功能权限)
    Set<Functions> fsSet = admin.getFs();
// 遍历该功能集合，将属于叶子节点的功能id号添加到变量sb中，并以","号分隔
    for (Functions f : fsSet) {
        if (f.isIsleaf())
            sb.append(f.getId()).append(",");
    }
    if (sb.length() > 0)
        return sb.substring(0, sb.length() - 1).toString();
    else
```

```
        return "";
}
```

在 getFidByAdminId 方法中，先调用业务接口 AdminService 的 getAdminFunctions 方法，根据管理员 id 号获取管理员对象及关联的 Functions 对象集合(即已分配的功能权限)。然后获取关联的功能集合(即已分配的功能权限)，再遍历该功能集合，将属于叶子节点的功能 id 号添加到变量 sb 中，并以逗号分隔，最后将 sb 发送到前端页面。前端页面接收到 sb 后，按照逗号进行分隔，分割后的字符串数组为 array，数组中的元素就是该管理员已分配的功能的 id 号。遍历数组 array，通过 Tree 控件的 find 方法，从 Tree 上依次找出相应的节点，然后将其前面的复选框选中，这样就完成了权限的绑定。

在"设置/修改管理员权限"对话框中，完成权限绑定后，还可以重新设置功能权限。只需要重新选中或取消选中复选框，然后单击"保存"按钮，将执行 JavaScript 函数 savePowers()，函数代码如下：

```javascript
function savePowers() {
    var fids = getChecked();
    $.ajax({
        url : 'powers/savePowers?fids=' + fids + '&editadminid='
            + $('#editAdminId').val(),
        cache : false,
        success : function(data) {
            eval('data=' + data);
            if (data.success == "true") {
                $("#powerDlg").dialog("close");
                $.messager.alert('提示', '权限修改成功', 'info');
            } else {
                $.messager.alert('提示', '权限修改失败', 'info');
            }
        }
    })
}
```

在函数 savePowers()中，首先调用 JavaScript 函数 getChecked()获取"设置/修改管理员权限"对话框中 Tree 控件上选中的功能节点的 id 号，函数 getChecked()代码如下：

```javascript
function getChecked() {
    var node = $('#powerTree').tree('getChecked');
    var cnodes = '';
    var pnodes = '';
    var prevNode = ''; //保存上一步所选父节点
    for (var i = 0; i < node.length; i++) {
        if ($('#powerTree').tree('isLeaf', node[i].target)) {
            cnodes += node[i].id + ',';
            var pnode = $('#powerTree').tree('getParent',
                node[i].target);  //获取当前节点的父节点
            if (prevNode != pnode.id)  //保证当前父节点与上一次父节点不同
            {
                pnodes += pnode.id + ',';
                prevNode = pnode.id;  //保存当前节点
            }
        }
    }
```

```
            cnodes = cnodes.substring(0, cnodes.length - 1);
            pnodes = pnodes.substring(0, pnodes.length - 1);
            return pnodes + "," + cnodes;
}
```

在 JavaScript 函数 savePowers()中,接着再使用$.ajax 发送请求到 powers/savePowers,即执行控制器类 PowersController 的 savePowers 方法,并将参数 fids 和 editadminid 传递到 savePowers 方法中。其中,参数 fids 表示选中的功能节点的 id 号的集合,参数 editadminid 表示待设置或修改功能权限的管理员 id 号。savePowers 方法的代码如下:

```
package com.res.controller;
……
@Controller
@RequestMapping("/powers")
public class PowersController {
    @Autowired
    private PowersService powersService;
    @RequestMapping(value = "/savePowers", produces = "text/html;charset=UTF-8")
    @ResponseBody
    public String savePowers(@RequestParam(value = "fids") String fids,
            @RequestParam(value = "editadminid") String editadminid) {
        try {
            // 从数据表powers中将待修改或设置的管理员功能权限全部删除
            powersService.delPowersByAdminid(Integer.parseInt(editadminid));
            if (!",".equals(fids)) {
                if (fids.indexOf("1") < 0)
                    fids = fids + ",1";
                String[] fidArray = fids.split(",");
                powersService.addPowers(
Integer.parseInt(editadminid), fidArray);
            }
        } catch (Exception e) {
            return "{\"success\":\"failure\",\"message\":\"保存失败\"}";
        }
        return "{\"success\":\"true\",\"message\":\"保存成功\"}";
    }
}
```

在 savePowers 方法中,先调用业务接口 PowersService 的 delPowersByAdminid 方法从数据表 powers 中将待修改或设置的管理员功能权限全部删除,再调用业务接口 PowersService 的 addPowers 方法向数据表 powers 中添加功能权限。

23.13 小　　结

本章在第 21 章 21.7 节的项目 springmvc_ssh 的基础上,基于注解使用 Spring 整合 Spring MVC 与 Hibernate,并结合前端 Easy UI 框架实现网上订餐系统后台。主要功能包括餐品管理、订单管理、客户管理和管理员及权限管理。

通过本章的讲解,希望读者能够掌握 Spring 整合 Spring MVC 与 Hibernate 进行应用开发的基本步骤、方法和技巧。

第 24 章 Spring 整合 Spring MVC 与 MyBatis 实现新闻发布系统

　　新闻发布系统是一个信息传播平台，系统主要功能包括新闻浏览功能、新闻发布功能和新闻管理功能。任何用户都可以通过本系统来阅读新闻信息。管理员登录系统后，可以使用新闻管理功能，包括新闻的添加、修改和删除。

　　本系统在开发过程中整合了 Spring 4、Spring MVC 和 Mybatis 三框架。其中，Spring MVC 用来处理页面逻辑，Mybatis 用来进行持久化操作，Spring 则对 Spring MVC 和 Mybatis 进行了整合。

24.1　系统概述及需求分析

本章实现的是一个简易的新闻发布系统，主要分为两个部分：前台与后台。在前台，未登录用户可以通过选择主题，分页查看该主题的所有新闻标题，单击新闻标题可浏览新闻详细内容；登录用户还可以发表评论。在后台，管理员可以对主题和新闻进行管理，具体包括新闻管理、主题管理、评论管理和用户管理。管理员在后台添加的新闻，前台的新闻列表会自动更新。

新闻发布系统中普通用户和管理员的用例图如图 24-1 和图 24-2 所示。

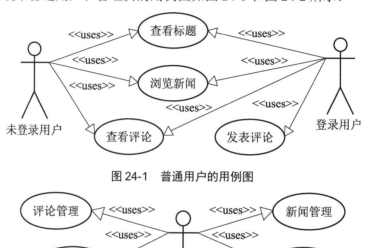

图 24-1　普通用户的用例图

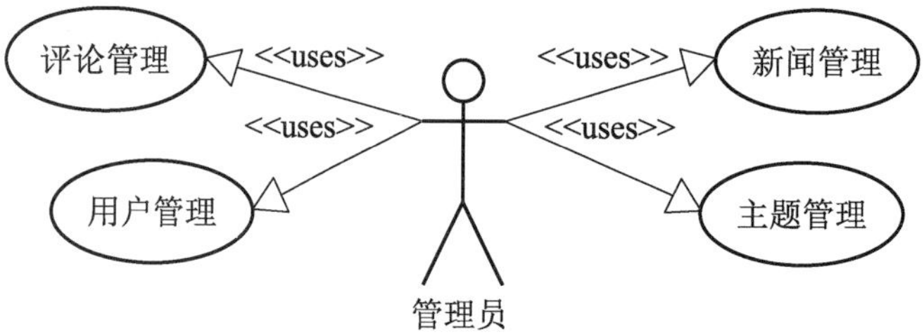

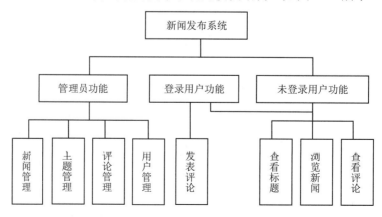

图 24-2　管理员的用例图

根据系统需求分析，可以得到新闻发布系统的模块结构，如图 24-3 所示。

图 24-3　新闻发布系统的模块结构

24.2 数据库设计

数据库设计是系统设计中非常重要的一个环节，数据是设计的基础，直接决定系统的成败。如果数据库设计不合理、不完善，将在系统开发中，甚至到后期的维护时，引起严重的问题。根据系统需求，创建了7张表，具体介绍如下。

(1) 主题表(topic)：用于记录新闻主题。
(2) 新闻信息表(newsinfo)：用于记录新闻相关信息。
(3) 新闻评论表(comment)：用于记录新闻评论信息。
(4) 用户信息表(users)：用于记录新闻前台的用户信息。
(5) 管理员信息表(admin)：用于记录管理员的信息。
(6) 系统功能表(functions)：用于记录系统可供使用的功能菜单。
(7) 权限表(powers)：用于记录各管理员所拥有的系统功能。

其中，主题表(topic)的字段说明如表24-1所示。

表24-1 主题表(topic)的字段说明

字 段 名	类 型	说 明
Id	int(4)	主题编号，主键，自增
Name	varchar(10)	主题名称

新闻信息表(newsinfo)的字段说明如表24-2所示。

表24-2 新闻信息表(newsinfo)的字段说明

字 段 名	类 型	说 明
Id	int(4)	新闻编号，主键，自增
Title	varchar(100)	新闻标题
Author	varchar(10)	新闻发布人
CreateDate	datetime	发布时间
Content	varchar(10000)	新闻内容
Summary	varchar(500)	新闻摘要
Tid	int(4)	所属主题，外键

新闻评论表(comment)的字段说明如表24-3所示。

表24-3 新闻评论表(comment)的字段说明

字 段 名	类 型	说 明
Id	int(4)	编号，主键，自增
Nid	int(4)	所属新闻编号，外键
Content	varchar(200)	评论内容

续表

字 段 名	类 型	说 明
Uid	int(4)	评论人，外键
CreateDate	datetime	评论日期
Status	int(4)	状态，1：未审核，2：已审核

用户信息表(users)的字段说明如表 24-4 所示。

表 24-4　用户信息表(users)的字段说明

字 段 名	类 型	说 明
Id	int(4)	编号，主键，自增
LoginName	varchar(20)	登录名
LoginPwd	varchar(20)	登录密码
Status	int(4)	状态，1：禁用，2：启用

管理员信息表(admin)的字段说明如表 24-5 所示。

表 24-5　管理员信息表(admin)的字段说明

字 段 名	类 型	说 明
Id	int(4)	编号，主键，自增
LoginName	varchar(20)	登录名
LoginPwd	varchar(20)	登录密码

系统功能表(functions)的字段说明如表 24-6 所示。

表 24-6　系统功能表(functions)的字段说明

字 段 名	类 型	说 明
id	int(4)	系统功能 id，主键，自增
name	varchar(20)	功能菜单名称
parentid	int(4)	父结点 id
url	varchar(50)	功能页面
isleaf	bit(1)	是否叶结点
nodeorder	int(4)	结点顺序

权限表(powers)的字段说明如表 24-7 所示。

表 24-7　权限表(powers)的字段说明

字 段 名	类 型	说 明
aid	int(4)	管理员 id，主键，与 admin 表的 Id 字段关联
fid	int(4)	系统功能 id，主键，与 functions 表的 id 字段关联

创建数据表后，设计数据表之间的关系，如图24-4所示。

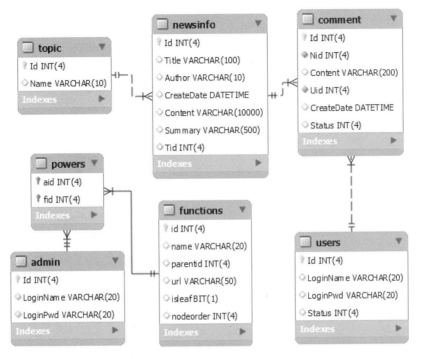

图24-4 系统数据表之间的关系

24.3 系统环境搭建

在第 21 章的 21.8 节中，以用户登录为例详细介绍了 Spring 整合 Spring MVC 与 MyBatis，读者可参照完成新闻发布系统的框架搭建。当然，读者也可以直接将 21.8 节创建的项目 springmvc_ssm 复制一份并重新命名为 news，再导入 MyEclipse 中。为避免部署重复，需要修改项目的部署名称。修改过程如下：在 MyEclipse 中右击项目 news，依次选择 Properties→MyEclipse→Deployment Assembly，将 Web Context Root 修改为 news 即可。然后将 jackson-annotations-2.6.0.jar、jackson-core-2.6.0.jar 和 jackson-databind-2.6.0.jar 这三个 jar 包复制到项目的 WebRoot\WEB-INF\lib 目录中，用于支持 Spring MVC 实现自动 Json 格式数据转换。

新闻发布系统的目录结构如图 24-5 所示，其中 com.news.pojo 包用于存放实体类，com.news.dao 包用于存放数据访问层接口，com.news.dao.provider 包用于存放构建动态 SQL 语句的类，com.news.service 包用于存放业务逻辑层接口，com.news.service.impl 包用于存放业务逻辑层接口的实现类，com.news.controller 包用于存放控制器类，com.news.interceptor 包用于存放登录权限验证的拦截器类。

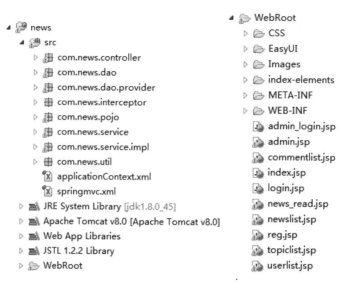

图 24-5 新闻发布系统目录结构

24.4 系统配置文件

Spring 使用的配置文件为 applicationContext.xml，Spring MVC 使用的配置文件为 springmvc.xml，这些配置文件的含义在第 21 章的 21.8 节中已具体介绍过，由于篇幅，在此不再赘述。

24.5 创建实体类

在 com.news.pojo 包中，依次创建实体类 Topic.java、Newsinfo.java、Comment.java、Users.java、Admin.java、Functions.java、Powers.java、Pager.java 和 TreeNode.java。其中，Topic.java 代码如下：

```
package com.news.pojo;
……
public class Topic {
    private int id;
    private String name;
    private Set newsinfos = new HashSet();
    // 此处省略了上述属性的 get 和 set 方法
}
```

Newsinfo.java 代码如下：

```
package com.news.pojo;
public class Newsinfo {
    private int id;
    private String title;
    private String author;
    private Date createDate;
```

```java
    private String content;
    private String summary;
    private Topic topic;
    // 此处省略了其他属性的 get 和 set 方法
    @JsonFormat(pattern="yyyy-MM-dd HH:mm:ss")
    public Date getCreateDate() {
        return createDate;
    }
}
```

Comment.java 代码如下：

```java
package com.news.pojo;
public class Comment {
    private int id;
    private Newsinfo newinfo;
    private String content;
    private Users user;
    private Date createDate;
    private int status;
    private Date commentTimeFrom;
    private Date commentTimeTo;
    // 此处省略了其他属性的 get 和 set 方法
    @JsonFormat(pattern="yyyy-MM-dd HH:mm:ss")
    public Date getCreateDate() {
        return createDate;
    }
    @Transient
    @DateTimeFormat(pattern="yyyy-MM-dd")
    public Date getCommentTimeFrom() {
        return commentTimeFrom;
    }
    @Transient
    @DateTimeFormat(pattern="yyyy-MM-dd")
    public Date getCommentTimeTo() {
        return commentTimeTo;
    }
}
```

Users.java 代码如下：

```java
package com.news.pojo;
public class Users {
    private int id;
    private String loginName;
    private String loginPwd;
    private int status;
    // 此处省略了上述属性的 get 和 set 方法
}
```

Admin.java 代码如下：

```java
package com.news.pojo;
import java.util.List;
public class Admin {
    private int id;
```

```java
    private String loginName;
    private String loginPwd;
    // 管理员和功能是多对多关系
    private List<Functions> fs;
    // 此处省略了上述属性的 get 和 set 方法
}
```

Functions.java 代码如下：

```java
package com.news.pojo;
public class Functions implements Comparable<Functions> {
    private int id;
    private String name;
    private int parentid;
    private boolean isleaf;
    // 此处省略了上述属性的 get 和 set 方法
    @Override
    public int compareTo(Functions arg0) {  //比较两个 Functions 对象 id
        return ((Integer) this.getId()).compareTo((Integer) (arg0.getId()));
    }
}
```

Powers.java 代码如下：

```java
package com.news.pojo;
public class Powers {
    private Admin a;
    private Functions f;
    // 此处省略了上述属性的 get 和 set 方法
}
```

分页实体类 Pager 用于记录与分页相关的属性，代码如下：

```java
package com.news.pojo;
public class Pager {
    private int curPage;// 待显示页页码
    private int perPageRows;// 每页显示的记录数
    private int rowCount;  // 记录总数
    private int pageCount; // 总页数
    // 省略了其他属性的 get 和 set 方法
    // 根据 rowCount 和 perPageRows 计算总页数
    public int getPageCount() {
        return (rowCount + perPageRows - 1) / perPageRows;
    }
    // 分页显示时，获取当前页的第一条记录的索引
    public int getFirstLimitParam() {
        return (this.curPage - 1) * this.perPageRows;
    }
}
```

实体类 TreeNode 用于描述树型功能菜单的节点信息，代码如下：

```java
package com.news.pojo;
import java.util.List;
public class TreeNode {
    private int id;        //功能 id
```

```
    private String text;    //功能名称
    private int fid;    //节点的父节点 id
    private List<TreeNode> children;    //节点的孩子节点
    // 省略了属性的 get 和 set 方法
}
```

24.6 创建 DAO 接口及动态提供类

在 com.news.dao 包中,依次创建数据访问层接口 TopicDAO.java、NewsinfoDAO.java、CommentDAO.java、UserDAO.java、AdminDAO.java、FunctionDAO.java。在这些 DAO 接口中基于 MyBatis 注解完成数据库的操作。其中,TopicDAO 接口代码如下:

```
package com.news.dao;
……
public interface TopicDAO {
    // 根据 id 查询新闻主题
    @Select("select * from topic where Id = #{id}")
    Topic selectById(int id);
    // 查询所有新闻主题
    @Select("select * from topic")
    List<Topic> selectAllTopic();
    // 分页动态查询
    @SelectProvider(type = TopicDynaSqlProvider.class, method = "selectWithParam")
    List<Topic> selectByPage(Map<String, Object> params);
    // 根据条件动态查询主题总记录数
    @SelectProvider(type = TopicDynaSqlProvider.class, method = "count")
    Integer count(Map<String, Object> params);
    // 添加主题
    @Insert("insert into topic(Name) values(#{name})")
    @Options(useGeneratedKeys = true, keyProperty = "id")
    int save(Topic topic);
    // 修改主题
    @Update("update topic set Name=#{name} where Id=#{id}")
    void edit(Topic topic);
}
```

TopicDAO 接口中使用了动态 SQL 提供类 TopicDynaSqlProvider,代码如下:

```
package com.news.dao.provider;
……
public class TopicDynaSqlProvider {
    // 根据条件动态查询主题总记录数
    public String count(Map<String, Object> params) {
        return new SQL() {
            {
                SELECT("count(*)");
                FROM("topic");
                if (params.get("topic") != null) {
                    Topic topic = (Topic) params.get("topic");
                    if (topic.getName() != null && !"".equals(topic.getName()))
                    {
```

```java
                    WHERE(" Name LIKE CONCAT ('%',#{topic.name},'%') ");
                }
            }
        }
    }.toString();
}
// 分页动态查询
public String selectWithParam(Map<String, Object> params) {
    String sql = new SQL() {
        {
            SELECT("*");
            FROM("topic");
            if (params.get("topic") != null) {
                Topic topic = (Topic) params.get("topic");
                if (topic.getName() != null && !"".equals(topic.getName())) {
                    WHERE(" Name LIKE CONCAT ('%',#{topic.name},'%') ");
                }
            }
        }
    }.toString();
    if (params.get("pager") != null) {
        sql += " limit #{pager.firstLimitParam} , #{pager.perPageRows} ";
    }
    return sql;
}
```

NewsinfoDAO 接口代码如下:

```java
package com.news.dao;
……
public interface NewsinfoDAO {
    // 根据条件查询新闻总数
    @SelectProvider(type = NewsinfoDynaSqlProvider.class, method = "count")
    Integer count(Map<String, Object> params);
    // 分页动态查询
    @SelectProvider(type = NewsinfoDynaSqlProvider.class, method = "selectWithParam")
    @Results({
            @Result(id = true, column = "id", property = "id"),
            @Result(column = "Title", property = "title"),
            @Result(column = "Author", property = "author"),
            @Result(column = "CreateDate", property = "createDate", javaType = java.util.Date.class),
            @Result(column = "Content", property = "content"),
            @Result(column = "Summary", property = "summary"),
            @Result(column = "Tid", property = "topic", one = @One(select = "com.news.dao.TopicDAO.selectById", fetchType = FetchType.EAGER)) })
    List<Newsinfo> selectByPage(Map<String, Object> params);
    // 根据主题获取前 5 条新闻
    @Select("select * from newsinfo where Tid = #{tid} limit 5")
    List<Newsinfo> selectTop5ByTid(int tid);
    // 根据新闻编号获取新闻对象
```

```java
    @Results({
            @Result(id = true, column = "id", property = "id"),
            @Result(column = "Title", property = "title"),
            @Result(column = "Author", property = "author"),
            @Result(column = "CreateDate", property = "createDate",
javaType = java.util.Date.class),
            @Result(column = "Content", property = "content"),
            @Result(column = "Summary", property = "summary"),
            @Result(column = "Tid", property = "topic",
one = @One(select = "com.news.dao.TopicDAO.selectById",
fetchType = FetchType.EAGER)) })
    // 根据新闻编号获取新闻对象
    @Select("select * from newsinfo where Id = #{id}")
    Newsinfo selectById(int id);
    // 添加新闻
    @Insert("insert into newsinfo(Title,Author,CreateDate,
Content,Summary,Tid) values(#{title},#{author},#{createDate},
#{content},#{summary},#{topic.id})")
    @Options(useGeneratedKeys = true, keyProperty = "id")
    void save(Newsinfo ni);
    // 修改新闻
    @Update("update newsinfo set
Title=#{title},Author=#{author},CreateDate=#{createDate},
Content=#{content},Summary=#{summary},Tid=#{topic.id} where Id=#{id}")
    void edit(Newsinfo ni);
    // 删除新闻
    @Delete("delete from newsinfo where id=#{id}")
    void deleteById(int id);
}
```

NewsinfoDAO 接口中使用了动态 SQL 提供类 NewsinfoDynaSqlProvider，代码如下：

```java
package com.news.dao.provider;
……
public class NewsinfoDynaSqlProvider {
    // 根据条件动态查询新闻总记录数
    public String count(Map<String, Object> params) {
        return new SQL() {
            {
                SELECT("count(*)");
                FROM("newsinfo");
                if (params.get("newsinfo") != null) {
                    Newsinfo newsinfo = (Newsinfo) params.get("newsinfo");
                    if (newsinfo.getTopic() != null
                            && newsinfo.getTopic().getId() != 0) {
                        WHERE(" Tid = #{newsinfo.topic.id} ");
                    }
                    if (newsinfo.getTitle() != null
                            && !newsinfo.getTitle().equals("")) {
                        WHERE(" Title LIKE CONCAT ('%',#{newsinfo.title},'%')
");
                    }
                }
            }
        }.toString();
```

```java
    }
    // 分页动态查询
    public String selectWithParam(Map<String, Object> params) {
        String sql = new SQL() {
            {
                SELECT("*");
                FROM("newsinfo");
                if (params.get("newsinfo") != null) {
                    Newsinfo newsinfo = (Newsinfo) params.get("newsinfo");
                    if (newsinfo.getTopic() != null
                            && newsinfo.getTopic().getId() != 0) {
                        WHERE(" Tid = #{newsinfo.topic.id} ");
                    }
                    if (newsinfo.getTitle() != null
                            && !newsinfo.getTitle().equals("")) {
                        WHERE(" Title LIKE CONCAT ('%',#{newsinfo.title},'%') ");
                    }
                }
            }
        }.toString();
        if (params.get("pager") != null) {
            sql += " limit #{pager.firstLimitParam} , #{pager.perPageRows} ";
        }
        return sql;
    }
}
```

CommentDAO 接口代码如下：

```java
package com.news.dao;
……
public interface CommentDAO {
    // 添加评论
    @Insert("insert into comment(Nid,Content,Uid,CreateDate,Status) values(#{newinfo.id},#{content},#{user.id},#{createDate},#{status})")
    @Options(useGeneratedKeys = true, keyProperty = "id")
    void save(Comment comment);
    // 根据新闻编号，分页动态查询该新闻的评论
    @SelectProvider(type = CommentDynaSqlProvider.class, method = "selectWithParam")
    @Results({
            @Result(id = true, column = "id", property = "id"),
            @Result(column = "Content", property = "content"),
            @Result(column = "CreateDate", property = "createDate", javaType = java.util.Date.class),
            @Result(column = "Status", property = "status"),
            @Result(column = "Uid", property = "user", one = @One(select = "com.news.dao.UserDAO.selectById", fetchType = FetchType.EAGER)) })
    List<Comment> selectByPage(Map<String, Object> params);
    // 根据条件查询评论总数
    @SelectProvider(type = CommentDynaSqlProvider.class, method = "count")
    Integer count(Map<String, Object> params);
    // 根据新闻 id 删除评论
    @Delete("delete from comment where Nid=#{nid}")
```

```java
    void deleteByNid(int nid);
    // 评论审核
    @Update("update comment set Status=2 where id in (${ids})")
    void updateState(@Param("ids") String ids);
    // 删除评论
    @Delete("delete from comment where id in (${ids})")
    void deleteByIds(@Param("ids") String ids);
}
```

CommentDAO 接口中使用了动态 SQL 提供类 CommentDynaSqlProvider，代码如下：

```java
package com.news.dao.provider;
……
public class CommentDynaSqlProvider {
    // 分页动态查询
    public String selectWithParam(Map<String, Object> params) {
        String sql = new SQL() {
            {
                SELECT("*");
                FROM("comment");
                if (params.get("comment") != null) {
                    Comment comment = (Comment) params.get("comment");
                    if (comment.getNewinfo() != null
                            && comment.getNewinfo().getId() > 0) {
                        WHERE(" Nid = #{comment.newinfo.id} ");
                    }
                    if (comment.getStatus() > 0) {
                        WHERE(" Status = #{comment.status} ");
                    }
                    if (comment.getUser() != null
                            && comment.getUser().getId() > 0) {
                        WHERE(" Uid = #{comment.user.id} ");
                    }
                    if (comment.getCommentTimeFrom() != null
&& !"".equals(comment.getCommentTimeFrom())) {
                        WHERE(" CreateDate >= #{comment.commentTimeFrom} ");
                    }
                    if (comment.getCommentTimeTo() != null
                            && !"".equals(comment.getCommentTimeTo())) {
                        WHERE(" CreateDate < #{comment.commentTimeTo} ");
                    }
                }
            }
        }.toString();
        if (params.get("pager") != null) {
            sql += " limit #{pager.firstLimitParam} , #{pager.perPageRows}  ";
        }
        return sql;
    }
    // 动态查询总记录数
    public String count(Map<String, Object> params) {
        return new SQL() {
            {
                SELECT("count(*)");
                FROM("comment");
```

```java
            if (params.get("comment") != null) {
                Comment comment = (Comment) params.get("comment");
                if (comment.getNewinfo() != null
                        && comment.getNewinfo().getId() > 0) {
                    WHERE(" Nid = #{comment.newinfo.id} ");
                }
                if (comment.getStatus() > 0) {
                    WHERE(" Status = #{comment.status} ");
                }
                if (comment.getUser() != null
                        && comment.getUser().getId() > 0) {
                    WHERE(" Uid = #{comment.user.id} ");
                }
                if (comment.getCommentTimeFrom() != null
&& !"".equals(comment.getCommentTimeFrom())) {
                    WHERE(" CreateDate >= #{comment.commentTimeFrom} ");
                }
                if (comment.getCommentTimeTo() != null
                        && !"".equals(comment.getCommentTimeTo())) {
                    WHERE(" CreateDate < #{comment.commentTimeTo} ");
                }
            }
        }
    }.toString();
    }
}
```

UserDAO 接口代码如下:

```java
package com.news.dao;
……
public interface UserDAO {
    // 添加用户
    @Insert("insert into users(LoginName,LoginPwd,Status) values(#{loginName},#{loginPwd},#{status})")
    @Options(useGeneratedKeys = true, keyProperty = "id")
    int save(Users user);
    // 根据登录名和密码查询合法用户
    @Select("select * from users where LoginName = #{loginName} and LoginPwd = #{loginPwd} and Status=2")
    public Users selectByLoginNameAndPwd(@Param("loginName") String loginName,  @Param("loginPwd") String loginPwd);
    // 根据用户编号获取用户对象
    @Select("select * from users where Id = #{id}")
    Users selectById(int id);
    // 获取所有用户
    @Select("select * from users")
    List<Users> selectAll();
    // 根据登录名，分页动态查询用户
    @SelectProvider(type = UserDynaSqlProvider.class, method = "selectWithParam")
    List<Users> selectByPage(Map<String, Object> params);
    // 根据条件查询用户总数
    @SelectProvider(type = UserDynaSqlProvider.class, method = "count")
    Integer count(Map<String, Object> params);
```

```java
    // 更新用户状态
    @Update("update users set Status=${flag} where id in (${ids})")
    void updateState(@Param("ids") String ids, @Param("flag") int flag);
}
```

UserDAO 接口中使用了动态 SQL 提供类 UserDynaSqlProvider，代码如下：

```java
package com.news.dao.provider;
……
public class UserDynaSqlProvider {
    // 分页动态查询
    public String selectWithParam(Map<String, Object> params) {
        String sql = new SQL() {
            {
                SELECT("*");
                FROM("users");
                if (params.get("user") != null) {
                    Users user = (Users) params.get("user");
                    if (user.getLoginName() != null
                            && !"".equals(user.getLoginName())) {
                        WHERE(" LoginName LIKE CONCAT ('%',#{user.loginName},'%') ");
                    }
                    if (user.getStatus() > 0) {
                        WHERE(" Status = #{user.status} ");
                    }
                }
            }
        }.toString();
        if (params.get("pager") != null) {
            sql += " limit #{pager.firstLimitParam} , #{pager.perPageRows} ";
        }
        return sql;
    }
    // 动态查询总记录数
    public String count(Map<String, Object> params) {
        return new SQL() {
            {
                SELECT("count(*)");
                FROM("users");
                if (params.get("user") != null) {
                    Users user = (Users) params.get("user");
                    if (user.getLoginName() != null
                            && !"".equals(user.getLoginName())) {
                        WHERE(" LoginName LIKE CONCAT ('%',#{user.loginName},'%') ");
                    }
                    if (user.getStatus() > 0) {
                        WHERE(" Status = #{user.status} ");
                    }
                }
            }
        }.toString();
    }
}
```

AdminDAO 接口代码如下：

```java
package com.news.dao;
……
public interface AdminDAO {
    // 根据登录名和密码查询用户
    @Select("select * from admin where LoginName = #{loginName} and LoginPwd = #{loginPwd}")
    public Admin selectByLoginNameAndPwd(@Param("loginName") String loginName,   @Param("loginPwd") String loginPwd);
    // 根据管理员id获取管理员对象及关联的功能集合
    @Select("select * from admin where Id=#{id}")
    @Results({
        @Result(id = true, column = "id", property = "id"),
        @Result(column = "LoginName", property = "loginName"),
        @Result(column = "LoginPwd", property = "loginPwd"),
        @Result(column = "id", property = "fs", many = @Many(select = "com.news.dao.FunctionDAO.selectByAdminId", fetchType = FetchType.EAGER)) })
    Admin selectById(Integer id);
}
```

FunctionDAO 接口代码如下：

```java
package com.news.dao;
……
public interface FunctionDAO {
    @Select("select * from functions where id in (select fid from powers where aid=#{id} ) ")
    public List<Functions> selectByAdminId(Integer aid);
}
```

24.7　创建 Service 接口及实现类

在 com.news.service 包中，创建业务逻辑层接口 TopicService.java、NewsinfoService.java、CommentService.java、AdminService.java 和 UserService.java。

其中，在 TopicService 接口中声明了如下方法：

```java
package com.news.service;
……
public interface TopicService {
    List<Topic> selectAllTopic();
    List<Topic> findTopic(Topic topic,Pager pager);
    Integer count(Map<String, Object> params);
    public int addTopic(Topic topic);
    void modify(Topic topic);
    public List<Topic> getAllTopic();
}
```

在 NewsinfoService 接口中声明了如下方法：

```java
package com.news.service;
……
```

```java
public interface NewsinfoService {
    // 前台分页获得新闻
    List<Newsinfo> findNewsinfo(Newsinfo newsinfo,Pager pager);
    List<Newsinfo> selectTop5ByTid(int tid);
    Newsinfo selectById(int id);
    // 后台新闻列表
    List<Newsinfo> findNewsinfoForBackstage(Newsinfo newsinfo,Pager pager);
    Integer count(Map<String, Object> params);
    public void addNewsinfo(Newsinfo ni);
    void modify(Newsinfo ni);
    void removeNewsinfoById(int id);
}
```

在 CommentService 接口中声明了如下方法：

```java
package com.news.service;
……
public interface CommentService {
    public void addComment(Comment comment);
    List<Comment> findComment(Comment comment, Pager pager);
    Integer count(Map<String, Object> params);
    void removeCommentByNid(int nid);
    public List<Comment> findCommentForBackstage(Comment comment, Pager pager);
    void modifyStatus(String ids);
    void deleteCommentByIds(String ids);
}
```

在 AdminService 接口中声明了如下方法：

```java
package com.news.service;
import com.news.pojo.Admin;
public interface AdminService {
    public Admin login(String loginName,String loginPwd);
    public Admin getAdminAndFunctions(Integer id);
}
```

在 UserService 接口中声明了如下方法：

```java
package com.news.service;
……
public interface UserService {
    public int addUser(Users user);
    public Users login(String loginName, String loginPwd);
    public List<Users> getAllUsers();
    List<Users> findUsers(Users user, Pager pager);
    Integer count(Map<String, Object> params);
    void modifyStatus(String ids, int flag);
}
```

在 com.news.service.impl 包中，创建上述接口的实现类 TopicServiceImpl.java、NewsinfoServiceImpl.java、CommentServiceImpl.java、AdminServiceImpl.java 和 UserServiceImpl.java。其中，TopicServiceImpl 实现类代码如下：

```java
package com.news.service.impl;
……
```

```java
@Service("topicService")
@Transactional(propagation = Propagation.REQUIRED, isolation =
Isolation.DEFAULT)
public class TopicServiceImpl implements TopicService {
    @Autowired
    TopicDAO topicDAO;
    @Transactional(readOnly = true)
    @Override
    public List<Topic> selectAllTopic() {
        return topicDAO.selectAllTopic();
    }
    @Override
    public List<Topic> findTopic(Topic topic, Pager pager) {
        Map<String, Object> params = new HashMap<>();
        params.put("topic", topic);
        int recordCount = topicDAO.count(params);
        pager.setRowCount(recordCount);
        if (recordCount > 0) {
            /** 开始分页查询数据：查询第几页的数据 */
            params.put("pager", pager);
        }
        List<Topic> topics = topicDAO.selectByPage(params);
        return topics;
    }
    @Override
    public Integer count(Map<String, Object> params) {
        return topicDAO.count(params);
    }
    @Override
    public int addTopic(com.news.pojo.Topic topic) {
        return topicDAO.save(topic);
    }
    @Override
    public void modify(Topic topic) {
        topicDAO.edit(topic);
    }
    @Override
    public List<Topic> getAllTopic() {
        return topicDAO.selectAllTopic();
    }
}
```

NewsinfoServiceImpl 实现类代码如下：

```java
package com.news.service.impl;
......
@Service("newsinfoService")
@Transactional(propagation=Propagation.REQUIRED,
isolation=Isolation.DEFAULT)
public class NewsinfoServiceImpl implements NewsinfoService {
    @Autowired
    NewsinfoDAO newsinfoDAO;
    @Override
public List<Newsinfo> findNewsinfo(Newsinfo newsinfo,Pager pager) {
        /** 当前需要分页的总数据条数 */
```

```java
        Map<String,Object> params = new HashMap<>();
        params.put("newsinfo", newsinfo);
        int recordCount = newsinfoDAO.count(params);
        pager.setPerPageRows(10);
        pager.setRowCount(recordCount);
        if(recordCount > 0){
            /** 开始分页查询数据: 查询第几页的数据 */
            params.put("pager", pager);
        }
        List<Newsinfo> newsinfos = newsinfoDAO.selectByPage(params);
        return newsinfos;
    }
    @Override
    public List<Newsinfo> selectTop5ByTid(int tid) {
        return newsinfoDAO.selectTop5ByTid(tid);
    }
    @Override
    public Newsinfo selectById(int id) {
        return newsinfoDAO.selectById(id);
    }
    @Override
    public List<Newsinfo> findNewsinfoForBackstage(Newsinfo newsinfo,
            Pager pager) {
        Map<String,Object> params = new HashMap<>();
        params.put("newsinfo", newsinfo);
        int recordCount = newsinfoDAO.count(params);
        pager.setRowCount(recordCount);
        if(recordCount > 0){
            /** 开始分页查询数据: 查询第几页的数据 */
            params.put("pager", pager);
        }
        List<Newsinfo> newsinfos = newsinfoDAO.selectByPage(params);
        return newsinfos;
    }
    @Override
    public Integer count(Map<String, Object> params) {
        return newsinfoDAO.count(params);
    }
    @Override
    public void addNewsinfo(Newsinfo ni) {
        newsinfoDAO.save(ni);
    }
    @Override
    public void modify(Newsinfo ni) {
        newsinfoDAO.edit(ni);
    }
    @Override
    public void removeNewsinfoById(int id) {
        newsinfoDAO.deleteById(id);
    }
}
```

CommentServiceImpl 实现类代码如下:

```java
package com.news.service.impl;
```

```java
......
@Service("commentService")
@Transactional(propagation = Propagation.REQUIRED, isolation = Isolation.DEFAULT)
public class CommentServiceImpl implements CommentService {
    @Autowired
    CommentDAO commentDAO;
    @Override
    public void addComment(Comment comment) {
        commentDAO.save(comment);
    }
    @Override
    public List<Comment> findComment(Comment comment, Pager pager) {
        /** 当前需要分页的总数据条数 */
        Map<String, Object> params = new HashMap<>();
        params.put("comment", comment);
        int recordCount = commentDAO.count(params);
        pager.setPerPageRows(2);
        pager.setRowCount(recordCount);
        if (recordCount > 0) {
            /** 开始分页查询数据：查询第几页的数据 */
            params.put("pager", pager);
        }
        List<Comment> comments = commentDAO.selectByPage(params);
        return comments;
    }
    @Override
    public void removeCommentByNid(int nid) {
        commentDAO.deleteByNid(nid);
    }
    @Override
    public Integer count(Map<String, Object> params) {
        return commentDAO.count(params);
    }
    @Override
    public List<Comment> findCommentForBackstage(Comment comment, Pager pager) {
        Map<String, Object> params = new HashMap<>();
        params.put("comment", comment);
        int recordCount = commentDAO.count(params);
        pager.setRowCount(recordCount);
        if (recordCount > 0) {
            /** 开始分页查询数据：查询第几页的数据 */
            params.put("pager", pager);
        }
        List<Comment> commens = commentDAO.selectByPage(params);
        return commens;
    }
    @Override
    public void modifyStatus(String ids) {
        commentDAO.updateState(ids);
    }
    @Override
    public void deleteCommentByIds(String ids) {
```

```java
        commentDAO.deleteByIds(ids);
    }
}
```

AdminServiceImpl 实现类代码如下：

```java
package com.news.service.impl;
……
@Service("adminService")
@Transactional(propagation=Propagation.REQUIRED,
isolation=Isolation.DEFAULT)
public class AdminServiceImpl implements AdminService {
    @Autowired
    private AdminDAO adminDAO;
    @Override
    public Admin login(String loginName, String loginPwd) {
        return adminDAO.selectByLoginNameAndPwd(loginName, loginPwd);
    }
    @Override
    public Admin getAdminAndFunctions(Integer id) {
        return adminDAO.selectById(id);
    }
}
```

UserServiceImpl 实现类代码如下：

```java
package com.news.service.impl;
……
@Service("userService")
@Transactional(propagation = Propagation.REQUIRED, isolation = Isolation.DEFAULT)
public class UserServiceImpl implements UserService {
    @Autowired
    UserDAO userDAO;
    @Override
    public int addUser(Users user) {
        return userDAO.save(user);
    }
    @Override
    public Users login(String loginName, String loginPwd) {
        return userDAO.selectByLoginNameAndPwd(loginName, loginPwd);
    }
    @Override
    public List<Users> getAllUsers() {
        return userDAO.selectAll();
    }
    @Override
    public List<Users> findUsers(Users user, Pager pager) {
        Map<String, Object> params = new HashMap<>();
        params.put("user", user);
        int recordCount = userDAO.count(params);
        pager.setRowCount(recordCount);
        if (recordCount > 0) {
            /** 开始分页查询数据：查询第几页的数据 */
            params.put("pager", pager);
```

```
        }
        List<Users> users = userDAO.selectByPage(params);
        return users;
    }
    @Override
    public Integer count(Map<String, Object> params) {
        return userDAO.count(params);
    }
    @Override
    public void modifyStatus(String ids, int flag) {
        userDAO.updateState(ids, flag);
    }
}
```

24.8 新闻浏览

未登录用户可以浏览新闻，通过选择主题，分页查看该主题的新闻标题，单击新闻标题可以浏览新闻详细内容。

24.8.1 新闻首页

在浏览器地址栏中输入 http://localhost:8080/news/newsinfo/list，显示新闻首页，如图 24-6 所示。

图 24-6　新闻首页 index.jsp

新闻首页的实现流程如下。

(1) 实现控制器类。

在 com.news.controller 包中创建一个控制器类 NewsinfoController.java，再添加一个 selectNewsinfo 方法，根据条件和指定页码获取新闻列表，再转到新闻首页 index.jsp。代码如下：

```
package com.news.controller;
……
@Controller
@RequestMapping("/newsinfo")
public class NewsinfoController {
    @Autowired
    private NewsinfoService newsinfoService;
    @Autowired
    private TopicService topicService;
    @Autowired
    private CommentService commentService;
    // 根据条件和指定页码获取新闻列表
    @RequestMapping("/list")
    public String selectNewsinfo(Integer pageIndex, Integer topicId,
            @ModelAttribute Newsinfo newsinfo, Model model) {
        // 创建分页对象
        Pager pager = new Pager();
        pager.setCurPage(1);
        // 如果参数 pageIndex 不为 null，设置 pageIndex，即显示第几页
        if (pageIndex != null) {
            pager.setCurPage(pageIndex);
        }
        if (topicId != null) {
            Topic topic = new Topic();
            topic.setId(topicId);
            newsinfo.setTopic(topic);
        }
        // 获取所有新闻主题
        List<Topic> topics = topicService.selectAllTopic();
        // 查询新闻信息
        List<Newsinfo> newsinfos = newsinfoService
                .findNewsinfo(newsinfo, pager);
        // 获取前 5 条国内新闻
        List<Newsinfo> domesticNewsList = newsinfoService.selectTop5ByTid(1);
        // 获取前 5 条国际新闻
        List<Newsinfo> internationalNewsList = newsinfoService
                .selectTop5ByTid(2);
        // 设置 Model 数据
        model.addAttribute("newsinfos", newsinfos);
        model.addAttribute("domesticNewsList", domesticNewsList);
        model.addAttribute("internationalNewsList", internationalNewsList);
        model.addAttribute("pager", pager);
        model.addAttribute("topics", topics);
        // 返回新闻首页 index.jsp
        return "index";
    }
}
```

(2) 新闻首页设计。

在新闻首页 index.jsp 中，循环显示新闻标题的代码如下：

```
<!-- 循环显示当前页新闻列表 -->
<c:forEach var="newsinfo" items="${requestScope.newsinfos }">
    <li><a href="newsread?id=${newsinfo.id}">
```

```
            ${newsinfo.title} </a> <span><fmt:formatDate
              value="${newsinfo.createDate}" pattern="yyyy-MM-dd HH:mm:ss"
/></span></li>
</c:forEach>
```

分页超链接代码如下:

```
<!-- 分页超链接部分 -->
<c:if test="${requestScope.pager.curPage>1}">
   <p align="center">
      <a    href='list?pageIndex=1&topicId=
${requestScope.newsinfo.topic.id}'>首页</a>
      <a    href='list?pageIndex=${requestScope.pager.curPage-1 }&
topicId=${requestScope.newsinfo.topic.id}'>上一页</a></p>
</c:if>
<c:if test="${requestScope.pager.curPage < requestScope.pager.pageCount}">
<p align="center">
      <a    href='list?pageIndex=${requestScope.pager.curPage+1}&
topicId=${requestScope.newsinfo.topic.id}'>下一页</a>
      <a    ref='list?pageIndex=${requestScope.pager.pageCount }&
topicId=${requestScope.newsinfo.topic.id}'>尾页</a>    </p>
</c:if>
```

单击"首页""上一页""下一页"和"尾页"分页超链接时,会再次将请求提交到 list,即再次执行 NewsinfoController 类中的 selectNewsinfo 方法。但此时会将待显示页页码 pageIndex 和新闻主题编号 topicId 这两个参数传递到 selectNewsinfo 方法参数中,selectNewsinfo 方法用这两个参数的值作为条件,重新获取满足条件的新闻列表,将其显示在 index.jsp 页面中。

在 index.jsp 页面中,循环显示主题的代码如下:

```
<!-- 循环显示主题列表 -->
<c:forEach var="topic" items="${requestScope.topics }">
   <a href="list?topicId=${topic.id }"><b>${topic.name }</b></a>
</c:forEach>
```

在 index.jsp 页面中,使用了<jsp:include>标准动作将页面 index_sidebar.jsp 包含其中,代码如下:

```
<jsp:include page="index-elements/index_sidebar.jsp" />
```

该页面用于显示新闻首页左侧的国内新闻和国际新闻,国内新闻显示代码如下:

```
<div class="side_list">
   <ul>
      <!-- 循环显示 5 条国内新闻   -->
      <c:forEach var="domesticNews"
         items="${requestScope.domesticNewsList }">
         <li><a href='newsread?id=${domesticNews.id }'><b>
                  ${domesticNews.title }</b></a></li>
      </c:forEach>
   </ul>
</div>
```

国际新闻显示代码如下:

```
<div class="side_list">
   <ul><c:forEach var="internationalNews"
         items="${requestScope.internationalNewsList }">
      <li><a href='newsread?id=${internationalNews.id }'><b>
            ${internationalNews.title }</b></a></li>
      </c:forEach></ul></div>
```

24.8.2　浏览新闻

在新闻首页中，单击新闻标题，可以浏览新闻内容，新闻浏览页 news_read.jsp 如图 24-7 所示。

图 24-7　新闻浏览页 news_read.jsp

新闻标题超链接设置如下：

```
<a href="newsread?id=${newsinfo.id}">${newsinfo.title}</a>
```

单击新闻标题链接时，将请求发送到 newsread，即执行控制器类 NewsinfoController 中的 newsread 方法，并将参数 id(新闻编号)传递过去。Newsread 方法接收到请求后，会根据新闻编号获取新闻对象，再转到新闻浏览页，显示新闻详细内容。

实现新闻浏览的流程如下。

(1) 实现控制器类。

在控制器类 NewsinfoController 中添加一个方法 newsread，根据新闻编号获取新闻，然后转到新闻浏览页，显示新闻详细内容。代码如下：

```
@RequestMapping("/newsread")
public String newsread(Integer id, Model model,
      @ModelAttribute Comment comment, Integer pageIndex) {
   Newsinfo newsinfo = newsinfoService.selectById(id);
   // 获取前 5 条国内新闻
   List<Newsinfo> domesticNewsList = newsinfoService.selectTop5ByTid(1);
```

```
    // 获取前 5 条国际新闻
    List<Newsinfo> internationalNewsList = newsinfoService
            .selectTop5ByTid(2);
    // 获取所有新闻主题
    List<Topic> topics = topicService.selectAllTopic();
    model.addAttribute("newsinfo", newsinfo);
    model.addAttribute("domesticNewsList", domesticNewsList);
    model.addAttribute("internationalNewsList", internationalNewsList);
    model.addAttribute("topics", topics);
    // 创建分页对象
    Pager pager = new Pager();
    pager.setCurPage(1);
    // 如果参数 pageIndex 不为 null, 设置 pageIndex, 即显示第几页
    if (pageIndex != null) {
        pager.setCurPage(pageIndex);
    }
    if (id != null) {
        Newsinfo ni = new Newsinfo();
        ni.setId(id);
        comment.setNewinfo(ni);
        comment.setStatus(2); // 查询已审核评论
    }
    List<Comment> comments = commentService.findComment(comment, pager);
    model.addAttribute("comments", comments);
    model.addAttribute("pager", pager);
    return "news_read";
}
```

在 newsread 方法中，依次根据新闻 id 获取了新闻对象、获取了前 5 条国内新闻和前 5 条国际新闻、获取了所有新闻主题、获取了该新闻的所有已审核评论，并将这些结果存入 model 中，以便在新闻浏览页面中访问。

(2) 新闻浏览页设计。

在新闻浏览页 news_read.jsp 中，显示新闻内容的代码如下：

```
<ul class="classlist">
    <table width="98%">
        <tr width="100%">
            <td colspan="2" align="center">${newsinfo.title }</td>
        </tr>
        <tr><td colspan="2"><hr /></td>    </tr>
        <tr>
            <td align="center">作者: ${newsinfo.author }   类型: <a href="list?topicId=${topic.id }">
${newsinfo.topic.name }</a>
                发布时间: <fmt:formatDate value="${newsinfo.createDate}" pattern="yyyy-MM-dd HH:mm:ss" /></td>
        </tr>
        <tr><td align="left"><strong>摘要:
${newsinfo.summary }</strong></td></tr>
            <tr><td colspan="2" align="center"></td>    </tr>
            <tr><td colspan="2">${newsinfo.content }</td></tr>
            <tr><td colspan="2"><hr /></td></tr>
    </table>
</ul>
```

显示该新闻评论的代码如下：

```html
<!-- 新闻评论开始 -->
<div class="content" style="padding-top:50px;">
    <ul class="class_date">新闻评论
    </ul>
    <c:forEach var="comment" items="${requestScope.comments }">
        <ul class="classlist">
            <table width="98%">
                <tr>
                    <td align="center">评论人：
${comment.user.loginName }   评论时间：<fmt:formatDate value="${comment.createDate}" pattern="yyyy-MM-dd HH:mm:ss" />
                    </td>
                </tr>
                <tr><td colspan="2">${comment.content }</td></tr>
                <tr><td colspan="2"><hr /></td></tr>
            </table>
        </ul>
    </c:forEach>
    <!-- 分页超链接部分 -->
    <c:if test="${requestScope.pager.curPage>1}">
        <p align="center">
            <a href='newsread?pageIndex=1&id=${requestScope.comment.newinfo.id}'>首页</a>
            <a href='newsread?pageIndex=${requestScope.pager.curPage-1 }&id=${requestScope.comment.newinfo.id}'>上一页</a>
        </p>
    </c:if>
    <c:if test="${requestScope.pager.curPage < requestScope.pager.pageCount}">
        <p align="center">
            <a href='newsread?pageIndex=${requestScope.pager.curPage+1}&id=${requestScope.comment.newinfo.id}'>下一页</a>
            <a href='newsread?pageIndex=${requestScope.pager.pageCount }&id=${requestScope.comment.newinfo.id}'>尾页</a>
        </p>
    </c:if>
</div>
<!-- 新闻评论结束 -->
```

24.9 发表评论

未登录用户可以浏览新闻及评论，但不能评论新闻，只有成功登录后才能发表新闻评论。

24.9.1 普通用户登录

新闻浏览首页 index.jsp 使用了<jsp:include>将登录页面 index_top.jsp 包含在其头部，登录表单代码如下：

```html
<div id="top_login">
   <c:if test="${sessionScope.u==null }">
      <form action="/news/user/login" method="post"
         onsubmit="return check()">
         <label>用户名</label> <input type="text" id="loginName"
name="loginName" value="" class="login_input" /> <label>  密  码
</label> <input type="password" id="loginPwd"name="loginPwd" value=""
class="login_input" /> <input
type="submit" class="login_sub" value="登录" /> <input
type="button" onclick="openReg()" class="login_sub" value="注册" />
         <label id="error"> </label>
      </form>
   </c:if>
   <c:if test="${sessionScope.u!=null }">
欢迎您：${sessionScope.u.loginName}   <a
         href="${pageContext.request.contextPath}/admin">登录控制台</a>
  <a href="/news/user/loginout">退出</a>
   </c:if>
</div>
```

如果没有登录，则显示如图 24-8 所示的登录表单；如果登录成功，则会显示如图 24-9 所示的欢迎信息。

图 24-8 登录表单

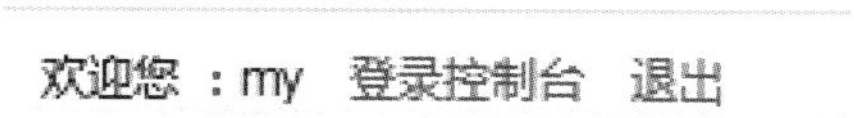

图 24-9 登录成功

实现普通用户登录验证的流程如下。

在 com.news.controller 包中创建一个控制器类 UserController.java，再添加一个 login 方法，用来处理普通用户的登录请求。代码如下：

```java
package com.news.controller;
……
@Controller
@RequestMapping("/user")
public class UserController {
   @Autowired
   private UserService userService;
   // 处理用户登录的请求
   @RequestMapping("/login")
   public ModelAndView login(@RequestParam("loginName") String loginName,
@RequestParam("loginPwd") String loginPwd, HttpSession session,
ModelAndView mv) {
      Users u = userService.login(loginName, loginPwd);
      if (u != null && u.getLoginName() != null) {
         // 将用户保存到HttpSession中
```

```
            session.setAttribute("u", u);
        }
        // 执行 NewsinfoController 类中的 list 方法
        mv.setViewName("redirect:/newsinfo/list");
        return mv;
    }
}
```

在登录表单中输入用户名和密码，单击"登录"按钮，将登录请求提交到 user/login，即执行 UserController 类中的 login 方法。执行结束后，重定向到名为 newsinfo/list，即再执行控制器类 NewsinfoController 中的 selectNewsinfo 方法，重新显示新闻首页。

24.9.2 发表评论

在新闻浏览页 news_read.jsp 中，发表评论的表单如图 24-10 所示。

图 24-10 发表评论的表单

实现发表评论的流程如下。

在 com.news.controller 包中创建一个控制器类 CommentController.java，再添加一个 addComment 方法，用来处理用户提交新闻评论请求。代码如下：

```
package com.news.controller;
……
@Controller
@RequestMapping("/comment")
public class CommentController {
    @Autowired
    private CommentService commentService;
    @RequestMapping("/addComment")
    public ModelAndView addComment(@ModelAttribute Comment comment,
            HttpSession session, ModelAndView mv) {
        Users u = (Users) session.getAttribute("u");
        comment.setUser(u);
        comment.setCreateDate(new Date());
        // 初始时，评论状态为未审核，前端页面不显示该评论
        comment.setStatus(1);
        // 执行添加操作
        commentService.addComment(comment);
        // 执行 NewsinfoController 的 list 方法
```

```
            mv.setViewName("redirect:/newsinfo/list");
            return mv;
        }
    }
```

在发表评论表单中，填写评论内容，单击"提交"按钮，将请求提交到comment/ addComment，即执行 CommentController 类中的 addComment 方法。执行结束后重定向到 newsinfo/list，即再次执行控制器类 NewsinfoController 中的 selectNewsinfo 方法，重新显示新闻首页。

24.10 新闻系统后台

管理员成功登录系统后，可以进行新闻管理、主题管理、评论管理、用户管理。后台功能页面使用了 Easy UI 框架布局，在第 23 章中结合网上订餐系统后台功能实现过程详细介绍了如何使用 Easy UI 框架，本章不再赘述，读者可以参阅第 23 章相关内容。

24.10.1 管理员登录与后台管理首页

管理员后台登录页为 admin_login.jsp，运行效果如图 24-11 所示。

管理员登录页使用了 Easy UI 框架进行布局，与第 23 章中相同，此处不再赘述。

实现管理员登录功能的流程如下。

在 com.news.controller 包中创建一个控制器类 AdminController.java，再添加一个 login 方法，用来处理用户提交新闻评论请求。代码如下：

图 24-11 管理员登录页

```
package com.news.controller;
……
@SessionAttributes(value = { "admininfo" })
@Controller
@RequestMapping("/admin")
public class AdminController {
    @Autowired
    private AdminService adminService;
    // 处理管理员登录请求
    @RequestMapping(value = "/login", produces = "text/html;charset=UTF-8")
    @ResponseBody
    public String login(@RequestParam("loginName") String loginName,
        @RequestParam("loginPwd") String loginPwd, ModelMap model) {
        Admin admininfo = adminService.login(loginName, loginPwd);
        if (admininfo != null && admininfo.getLoginName() != null) {
            // 验证通过后，再判断是否已为该管理员分配功能权限
            if (adminService.getAdminAndFunctions(admininfo.getId()).
getFs()   .size() > 0) {
                // 验证通过且已分配功能权限，则将 admininfo 对象存入 model 中
                model.put("admininfo", admininfo);
                // 以 JSON 格式向页面发送成功信息
```

```
                return "{\"success\":\"true\",\"message\":\"登录成功\"}";
            } else {
                return "{\"success\":\"false\",\"message\":\"您没有权限,请联系超
级管理员设置权限!\"}";
            }
        } else
            return "{\"success\":\"false\",\"message\":\"登录失败\"}";
    }
}
```

在管理员登录表单中填写用户名和密码,单击"登录"按钮,将请求提交到admin/login,即执行控制器类 AdminController 中的 login 方法。验证通过后,会调用业务接口 AdminService 中的 getAdminAndFunctions 方法,获取该管理员拥有的功能权限,并进行判断。只有验证通过且已分配功能权限,才将该管理员对象存入 model 中,同时跳转到后台管理首页 admin.jsp,页面效果如图 24-12 所示。

图 24-12 后台管理首页

后台管理首页使用了 Easy UI 框架的 Layout 控件布局,与第 23 章中相同,此处不再赘述。页面左侧的树型菜单使用了 Easy UI 框架的 Tree 控件,其数据源设置如下:

```
$('#tt').tree({
    url : 'admin/getTree?adminid=${sessionScope.admininfo.id}'
});
```

即通过执行控制器类 AdminController 中的 getTree 方法,方法返回的 JSON 格式的结果作为绑定 Tree 控件的数据源。getTree 方法如下:

```
@RequestMapping("/getTree")
@ResponseBody
public List<TreeNode> getTree(
        @RequestParam(value = "adminid") String adminid) {
    // 根据管理员 id 号获取 AdminInfo 对象及关联的 Functions 对象集合
    Admin admininfo = adminService.getAdminAndFunctions(Integer
            .parseInt(adminid));
    List<TreeNode> nodes = new ArrayList<TreeNode>();
    List<Functions> functionsList = admininfo.getFs();
```

```
        Collections.sort(functionsList);
        // 将排序后的 Functions 对象集合转换到 List<TreeNode>类型的列表 nodes 中
        for (Functions f : functionsList) {
            TreeNode treeNode = new TreeNode();
            treeNode.setId(f.getId());
            treeNode.setFid(f.getParentid());
            treeNode.setText(f.getName());
            nodes.add(treeNode);
        }
        // 调用自定义工具类 JsonFactory 的 buildtree 方法，为 nodes 列表中各 TreeNode 元素
        中的 children 赋值(即该节点包含的子节点)
        List<TreeNode> treeNodes = JsonFactory.buildtree(nodes, 0);
        // 以 JSON 格式向页面返回绑定 tree 所需的数据
        return treeNodes;
}
```

24.10.2 新闻管理

新闻管理功能包括显示新闻列表、查询新闻、添加新闻、编辑新闻、删除新闻。

(1) 显示新闻列表。

新闻列表页为 newslist.jsp，如图 24-13 所示。新闻列表页 newslist.jsp 中使用了 Easy UI 框架的 Datagrid 控件布局，与第 23 章中相同，此处不再赘述。通过 Datagrid 控件的 url 属性指定数据源为 newsinfo/newslist，即执行控制器类 NewsinfoController 中的 newslist 方法，方法返回的 JSON 格式的结果作为绑定 Datagrid 控件的数据源。

图 24-13 新闻列表页 newslist.jsp

newslist 方法如下：

```
// 后台管理新闻列表显示
@RequestMapping("/newslist")
@ResponseBody
```

```java
public Map<String, Object> newslist(Integer page, Integer rows,
        @ModelAttribute Newsinfo newsinfo, Integer topicId) {
    // 创建分页对象
    Pager pager = new Pager();
    pager.setCurPage(page);
    pager.setPerPageRows(rows);
    // 封装查询条件
    Map<String, Object> params = new HashMap<>();
    if (topicId != null) {
        Topic topic = new Topic();
        topic.setId(topicId);
        newsinfo.setTopic(topic);
    }
    params.put("newsinfo", newsinfo);
    // 根据查询条件计算所有新闻记录数
    int totalCount = newsinfoService.count(params);
    // 根据 Map 中的条件查询指定页显示的新闻列表
    List<Newsinfo> newsinfos =
newsinfoService.findNewsinfoForBackstage(newsinfo, pager);
    // 创建 Map 类型对象 result,用于向前端页面发送数据
    Map<String, Object> result = new HashMap<String, Object>(2);
    // 向 Map 类型的对象 result 中放入键值对，键为 "total",值为 totalCount
    result.put("total", totalCount);
    // 向对象 result 中放入键值对，键为 "rows",
//   值为 newsinfos,即当前页显示的新闻列表
    result.put("rows", newsinfos);
    // 通过@ResponseBody,发送到前端页面的 result 自动转成 JSON 格式
    return result;
}
```

(2) 查询新闻。

在 newslist.jsp 页面中，输入新闻标题或选择新闻主题，单击"查找"按钮，将执行 JavaScript 函数 searchNewsinfo，代码如下：

```javascript
function searchNewsinfo() {
    var newsinfo_search_title = $('#newsinfo_search_title').textbox(
            "getValue");
    var newsinfo_search_tid = $('#newsinfo_search_tid').combobox(
            "getValue");
    $('#newsinfoDg').datagrid('load', {
        title : newsinfo_search_title,
        topicId : newsinfo_search_tid
    });
}
```

在函数 searchNewsinfo()中，执行 DataGrid 控件的 load 方法，再次将请求发送到 newsinfo/newslist，即再次执行控制器类 NewsinfoController 中的 newslist 方法，并将搜索栏中输入的查询条件传递过去。

(3) 添加和编辑新闻。

在 newslist.jsp 页面中，新闻添加和编辑使用同一个 Easy UI 的 Dialog 控件，与第 23 章中相同，此处不再赘述。新闻添加和编辑对话框如图 24-14 所示。

图 24-14 新闻添加和编辑对话框

在 newslist.jsp 页面中,单击工具栏上的"添加新闻"按钮,打开添加新闻对话框的 JavaScript 函数为 openAddNewsinfoDlg,代码如下:

```
function openAddNewsinfoDlg() {
    $('#addNewsinfoDlg').dialog('open').dialog('setTitle', '添加新闻');
    $('#addNewsinfoForm').form('clear');
// 保存新闻时,将请求提交到NewsinfoController类中的addNewsinfo方法
    urls = 'newsinfo/addNewsinfo';
}
```

填写新闻相关内容,单击"保存"按钮,执行 JavaScript 函数 saveNewsinfo,代码如下:

```
function saveNewsinfo() {
    $("#addNewsinfoForm").form("submit", {
        url : urls, //使用参数
        success : function(result) {
            var result = eval('(' + result + ')');
            if (result.success == 'true') {
                $("#newsinfoDg").datagrid("reload");
            }
            $("#addNewsinfoDlg").dialog("close");
            $.messager.show({
                title : "提示信息",
                msg : result.message
            });
        }
    });
}
```

此时将请求提交到 newsinfo/addNewsinfo,即执行控制器类 NewsinfoController 中的 addNewsinfo 方法。addNewsinfo 方法代码如下:

```
@RequestMapping(value = "/addNewsinfo", produces = "text/html;charset=UTF-
```

```
8")
@ResponseBody
public String addNewsinfo(Newsinfo ni) {
    try {
        ni.setCreateDate(new Date());
        newsinfoService.addNewsinfo(ni);
        return "{\"success\":\"true\",\"message\":\"新闻添加成功\"}";
    } catch (Exception e) {
        e.printStackTrace();
        return "{\"success\":\"false\",\"message\":\"新闻添加失败\"}";
    }
}
```

单击工具栏上的"编辑新闻"按钮,打开编辑新闻对话框的 JavaScript 函数为 openEditNewsinfoDlg,代码如下:

```
function openEditNewsinfoDlg() {
    var row = $("#newsinfoDg").datagrid("getSelected");
    if (row) {
        $("#addNewsinfoDlg").dialog("open").dialog('setTitle', '编辑新闻');
        $("#addNewsinfoForm").form("load", {
            "id" : row.id,
            "topic.id" : row.topic.id,
            "title" : row.title,
            "author" : row.author,
            "summary" : row.summary,
            "content" : row.content
        });
        urls = "newsinfo/modifyNewsinfo?id=" + row.id;
    }
}
```

在编辑新闻对话框中,首先要绑定新闻内容。修改新闻后,再单击"保存"按钮,此时请求被提交到 newsinfo/modifyNewsinfo, 即执行控制器类 NewsinfoController 中的 modifyNewsinfo 方法。modifyNewsinfo 方法如下:

```
@RequestMapping(value = "/modifyNewsinfo", produces = 
"text/html;charset=UTF-8")
@ResponseBody
public String modifyNewsinfo(Newsinfo ni) {
    try {
        ni.setCreateDate(new Date());
        newsinfoService.modify(ni);
        return "{\"success\":\"true\",\"message\":\"新闻修改成功\"}";
    } catch (Exception e) {
        e.printStackTrace();
        return "{\"success\":\"false\",\"message\":\"新闻修改失败\"}";
    }
}
```

(4) 删除新闻。

在新闻列表页中,选择要删除的新闻,再单击"删除新闻"按钮,执行 JavaScript 函数 removeNews,代码如下:

```javascript
function removeNews() {
    var rows = $("#newsinfoDg").datagrid('getSelections');
    if (rows.length > 0) {
        $.messager.confirm('Confirm', '确认要删除么?', function(r) {
            if (r) {
                var ids = "";
                for (var i = 0; i < rows.length; i++) {
                    ids += rows[i].id + ",";
                }
                $.post('newsinfo/deleteNewsinfo', {
                    id : ids
                }, function(result) {
                    if (result.success) {
                        $("#newsinfoDg").datagrid('reload');
                        $.messager.show({
                            title : '提示信息',
                            msg : result.message
                        });
                    } else {
                        $.messager.show({
                            title : '提示信息',
                            msg : result.message
                        });
                    }
                }, 'json');
            }
        });
    } else {
        $.messager.alert('提示', '请选择要删除的行', 'info');
    }
}
```

在 removeNews 函数中，会发送请求到 newsinfo/deleteNewsinfo，即执行控制器类 NewsinfoController 中的 deleteNewsinfo 方法。deleteNewsinfo 方法如下：

```java
@ResponseBody
@RequestMapping(value = "/deleteNewsinfo", produces = "text/html;charset=UTF-8")
public String deleteNewsinfo(String id) {
    String str = "";
    try {
        id = id.substring(0, id.length() - 1);
        String[] ids = id.split(",");
        for (String nid : ids) {
            commentService.removeCommentByNid(Integer.parseInt(nid));
            newsinfoService.removeNewsinfoById(Integer.parseInt(nid));
        }
        str = "{\"success\":\"true\",\"message\":\"删除成功！\"}";
    } catch (Exception e) {
        e.printStackTrace();
        str = "{\"success\":\"false\",\"message\":\"删除失败！\"}";
    }
    return str;
}
```

24.10.3 评论管理

评论管理功能包括显示评论列表、查询评论、评论审核、删除评论。

(1) 显示评论列表。

评论列表页为 commentlist.jsp，如图 24-15 所示。

图 24-15 评论列表页 commentlist.jsp

评论列表页中使用了 Easy UI 的 Datagrid 控件，与第 23 章中相同，此处不再赘述。通过 Datagrid 控件的 url 属性指定数据源为 comment/list，即执行 CommentController 类中的 list 方法，方法返回的 JSON 格式的结果作为绑定 Datagrid 控件的数据源。list 方法如下：

```java
@RequestMapping("/list")
@ResponseBody
public Map<String, Object> list(Integer page, Integer rows,
        @ModelAttribute Comment comment, Integer userId) {
    // 创建分页对象
    Pager pager = new Pager();
    pager.setCurPage(page);
    pager.setPerPageRows(rows);
    // 封装查询条件
    Map<String, Object> params = new HashMap<>();
    if (userId != null) {
        Users user = new Users();
        user.setId(userId);
        comment.setUser(user);
    }
    params.put("comment", comment);
    // 根据查询条件计算评论记录数
    int totalCount = commentService.count(params);
    // 根据 Map 中的条件查询指定页显示的评论列表
```

```
        List<Comment> comments = commentService.findCommentForBackstage(
                comment, pager);
        // 创建 Map 类型对象 result,用于向前端页面发送数据
        Map<String, Object> result = new HashMap<String, Object>(2);
        // 向 Map 类型的对象 result 中放入键值对, 键为 "total", 值为 totalCount
        result.put("total", totalCount);
        // 向对象 result 中放入键值对, 键为 "rows", 值为 comments
        result.put("rows", comments);
        // 通过@ResponseBody,发送到前端页面的 result 自动转成 JSON 格式
        return result;
}
```

(2) 查询评论。

在评论列表页中，选择评论人或评论状态、或输入评论时间，单击"查找"按钮，将执行 JavaScript 函数 searchComment，代码如下：

```
function searchComment() {
    var comment_search_uid = $('#comment_search_uid').combobox(
            "getValue");
    var comment_search_status = $('#comment_search_status').combobox(
            "getValue");
    var commentTimeFrom = $("#commentTimeFrom").datebox("getValue");
    var commentTimeTo = $("#commentTimeTo").datebox("getValue");
    $('#commentDg').datagrid('load', {
        userId : comment_search_uid,
        status : comment_search_status,
        commentTimeFrom : commentTimeFrom,
        commentTimeTo : commentTimeTo
    });
}
```

在函数 searchComment()中，执行 DataGrid 控件的 load 方法，再次将请求发送到 newsinfo/newslist，即再次执行控制器类 CommentController 中的 list 方法，并将搜索栏中输入的查询条件传递过去。

(3) 评论审核。

在评论列表页中，选择要评论的记录，单击"评论审核"按钮，将执行 JavaScript 函数 searchComment，代码如下：

```
function auditComment() {
    var rows = $("#commentDg").datagrid('getSelections');
    if (rows.length > 0) {
        $.messager.confirm('Confirm', '确认审核么?', function(r) {
            if (r) {
                var cids = "";
                for (var i = 0; i < rows.length; i++) {
                    cids += rows[i].id + ",";
                }
                $.post('comment/commentAudit', {
                    ids : cids
                }, function(result) {
                    if (result.success == 'true') {
                        $("#commentDg").datagrid('reload');
```

```
                    $.messager.show({
                        title : '提示信息',
                        msg : result.message
                    });
                } else {
                    $.messager.show({
                        title : '提示信息',
                        msg : result.message
                    });
                }
            }, 'json');
        }
    });
} else {
    $.messager.alert('提示', '请选择要审核的评论', 'info');
}
}
```

在函数 searchComment()中，首先获取要审核的新闻评论编号，然后将请求提交到 comment/commentAudit，即执行 CommentController 类中的 commentAudit 方法，代码如下：

```
@RequestMapping(value = "/commentAudit", produces = "text/html;charset=UTF-8")
@ResponseBody
public String commentAudit(@RequestParam(value = "ids") String ids) {
    try {
        commentService.modifyStatus(ids.substring(0, ids.length() - 1));
        return "{\"success\":\"true\",\"message\":\"审核成功\"}";
    } catch (Exception e) {
        e.printStackTrace();
        return "{\"success\":\"false\",\"message\":\"审核失败\"}";
    }
}
```

(4) 删除评论。

在评论列表页中，选择要删除的记录，单击"删除评论"按钮，将执行 JavaScript 函数 deleteComment，代码如下：

```
function deleteComment() {
    var rows = $("#commentDg").datagrid('getSelections');
    if (rows.length > 0) {
        $.messager.confirm('Confirm', '确认删除么?', function(r) {
            if (r) {
                var cids = "";
                for (var i = 0; i < rows.length; i++) {
                    cids += rows[i].id + ",";
                }
                $.post('comment/deleteComment', {
                    ids : cids
                }, function(result) {
                    if (result.success == 'true') {
                        $("#commentDg").datagrid('reload');
                        $.messager.show({
                            title : '提示信息',
```

```
                    msg : result.message
                });
            } else {
                $.messager.show({
                    title : '提示信息',
                    msg : result.message
                });
            }
        }, 'json');
    }
    });
    } else {
        $.messager.alert('提示', '请选择要删除的评论', 'info');
    }
}
```

在函数 deleteComment() 中,首先获取要删除的新闻评论编号,然后将请求提交到 comment/deleteComment,即执行控制器类 CommentController 中的 deleteComment 方法,代码如下:

```
@RequestMapping(value = "/deleteComment", produces = 
"text/html;charset=UTF-8")
@ResponseBody
public String deleteComment(@RequestParam(value = "ids") String ids) {
    try {
        commentService.deleteCommentByIds(ids.substring(0, ids.length() - 1));
        return "{\"success\":\"true\",\"message\":\"删除成功\"}";
    } catch (Exception e) {
        e.printStackTrace();
        return "{\"success\":\"false\",\"message\":\"删除失败\"}";
    }
}
```

24.10.4 用户管理

用户管理功能包括显示用户列表、查询用户、启用和禁用用户。

(1) 显示用户列表。

用户列表页为 userlist.jsp,如图 24-16 所示。

用户列表页中使用了 Easy UI 的 Datagrid 控件,与第 23 章中相同,此处不再赘述。通过 Datagrid 控件的 url 属性指定数据源为 user/list,即执行控制器类 UserController 中的 list 方法,方法返回的 JSON 格式的结果作为绑定 Datagrid 控件的数据源。list 方法如下:

```
@RequestMapping("/list")
@ResponseBody
public Map<String, Object> list(Integer page, Integer rows,
        @ModelAttribute Users user) {
    // 创建分页对象
    Pager pager = new Pager();
    pager.setCurPage(page);
    pager.setPerPageRows(rows);
```

```java
// 封装查询条件
Map<String, Object> params = new HashMap<>();
params.put("user", user);
// 根据查询条件计算用户记录数
int totalCount = userService.count(params);
// 根据 Map 中的条件查询指定页显示的用户列表
List<Users> users = userService.findUsers(user, pager);
// 创建 Map 类型对象 result,用于向前端页面发送数据
Map<String, Object> result = new HashMap<String, Object>(2);
// 向 Map 类型的对象 result 中放入键值对，键为"total",值为 totalCount
result.put("total", totalCount);
// 向对象 result 中放入键值对，键为"rows",值为 comments
result.put("rows", users);
// 通过@ResponseBody,发送到前端页面的 result 自动转成 JSON 格式
return result;
}
```

图 24-16　用户列表页 userlist.jsp

(2) 查询用户。

在用户列表页中，输入用户名称或选择客户状态，单击"查找"按钮，将执行 JavaScript 函数 searchUsers，代码如下：

```javascript
function searchUsers() {
  var users_search_loginName =
$('#users_search_loginName').textbox("getValue");
  var users_search_status = $('#users_search_status').combobox(
      "getValue");
  $('#userListDg').datagrid('load', {
     loginName : users_search_loginName,
     status : users_search_status
  });
}
```

在函数 searchUsers()中，执行 DataGrid 控件的 load 方法，再次将请求发送到 user/list，即执行控制器类 UserController 中的 list 方法，并将搜索栏中输入的查询条件传递过去。

(3) 启用和禁用用户。

在用户列表页中，选择要启用或禁用的记录，单击启用用户或禁用用户按钮，将执行 JavaScript 函数 setIsEnableUser(flag)，代码如下：

```javascript
function setIsEnableUser(flag) {
    var rows = $("#userListDg").datagrid('getSelections');
    if (rows.length > 0) {
        $.messager.confirm('Confirm', '确认要设置么?', function(r) {
            if (r) {
                var uids = "";
                for (var i = 0; i < rows.length; i++) {
                    uids += rows[i].id + ",";
                }
                $.post('user/setIsEnableUser', {
                    uids : uids,
                    flag : flag
                }, function(result) {
                    if (result.success == 'true') {
                        $("#userListDg").datagrid('reload');
                        $.messager.show({
                            title : '提示信息',
                            msg : result.message
                        });
                    } else {
                        $.messager.show({
                            title : '提示信息',
                            msg : result.message
                        });
                    }
                }, 'json');
            }
        });
    } else {
        $.messager.alert('提示', '请选择要启用或禁用的客户', 'info');
    }
}
```

在函数 setIsEnableUser(flag)中，首先获取要启用或禁用用户的编号，参数 flag=1 时表示禁用用户，flag=2 时表示启用用户。然后将请求提交到 user/setIsEnableUser，即执行控制器类 UserController 中的 setIsEnableUser 方法，代码如下：

```java
// 启用或禁用用户
@RequestMapping(value = "/setIsEnableUser", produces = "text/html;charset=UTF-8")
@ResponseBody
public String setIsEnableUser(@RequestParam(value = "uids") String uids,
    @RequestParam(value = "flag") String flag) {
    try {
        userService.modifyStatus(uids.substring(0, uids.length() - 1),
            Integer.parseInt(flag));
        return "{\"success\":\"true\",\"message\":\"设置成功! \"}";
    } catch (Exception e) {
        e.printStackTrace();
```

```
        return "{\"success\":\"false\",\"message\":\"设置失败！\"}";
    }
}
```

24.11 小　　结

  本章将 Spring 4、Spring MVC 和 Mybatis 三框架进行整合，实现了新闻发布系统的前后台功能模块。通过本章的讲解，希望读者能够掌握使用 SSM(Spring、Spring MVC 和 MyBatis)整合应用开发的基本步骤、方法和技巧。